Computational Techniques for Modeling Atmospheric Processes

Vitaliy Prusov
University of Kyiv, Ukraine

Anatoliy Doroshenko
National Academy of Sciences, Ukraine

A volume in the Advances in Environmental Engineering and Green Technologies (AEEGT) Book Series

IGI Global
DISSEMINATOR OF KNOWLEDGE

Published in the United States of America by
IGI Global
Information Science Reference (an imprint of IGI Global)
701 E. Chocolate Avenue
Hershey PA, USA 17033
Tel: 717-533-8845
Fax: 717-533-8661
E-mail: cust@igi-global.com
Web site: http://www.igi-global.com

Library of Congress Cataloging-in-Publication Data

Names: Prusov, V. A., 1940- author. | Doroshenko, Anatoliy, 1948- author.
Title: Computational techniques for modeling atmospheric processes / by Vitaliy Prusov and Anatoliy Doroshenko.
Description: Hershey, PA : Engineering Science Reference, [2018] | Includes bibliographical references.
Identifiers: LCCN 2017008479| ISBN 9781522526360 (hardcover) | ISBN 9781522526377 (ebook)
Subjects: LCSH: Atmospheric physics--Mathematical models. | Atmospheric models.
Classification: LCC QC874.5 .P78 2018 | DDC 551.5101/1--dc23 LC record available at https://lccn.loc.gov/2017008479

This book is published in the IGI Global book series Advances in Environmental Engineering and Green Technologies (AEEGT) (ISSN: 2326-9162; eISSN: 2326-9170)

British Cataloguing in Publication Data
A Cataloguing in Publication record for this book is available from the British Library.

Advances in Environmental Engineering and Green Technologies (AEEGT) Book Series

ISSN:2326-9162
EISSN:2326-9170

Mission

Growing awareness and an increased focus on environmental issues such as climate change, energy use, and loss of non-renewable resources have brought about a greater need for research that provides potential solutions to these problems. Research in environmental science and engineering continues to play a vital role in uncovering new opportunities for a "green" future.

The **Advances in Environmental Engineering and Green Technologies (AEEGT)** book series is a mouthpiece for research in all aspects of environmental science, earth science, and green initiatives. This series supports the ongoing research in this field through publishing books that discuss topics within environmental engineering or that deal with the interdisciplinary field of green technologies.

Coverage

- Contaminated Site Remediation
- Air Quality
- Industrial Waste Management and Minimization
- Green Transportation
- Green Technology
- Renewable Energy
- Sustainable Communities
- Cleantech
- Radioactive Waste Treatment
- Pollution Management

IGI Global is currently accepting manuscripts for publication within this series. To submit a proposal for a volume in this series, please contact our Acquisition Editors at Acquisitions@igi-global.com or visit: http://www.igi-global.com/publish/.

The Advances in Environmental Engineering and Green Technologies (AEEGT) Book Series (ISSN 2326-9162) is published by IGI Global, 701 E. Chocolate Avenue, Hershey, PA 17033-1240, USA, www.igi-global.com. This series is composed of titles available for purchase individually; each title is edited to be contextually exclusive from any other title within the series. For pricing and ordering information please visit http://www.igi-global.com/book-series/advances-environmental-engineering-green-technologies/73679. Postmaster: Send all address changes to above address.

Titles in this Series

For a list of additional titles in this series, please visit:
http://www.igi-global.com/book-series/advances-environmental-engineering-green-technologies/73679

Advanced Nanomaterials for Water Engineering, Treatment, and Hydraulics
Tawfik A. Saleh (King Fahd University of Petroleum and Minerals, Saui Arabia)
Engineering Science Reference • ©2017 • 384pp • H/C (ISBN: 9781522521365) • US $200.00

Handbook of Research on Inventive Bioremediation Techniques
Jatindra Nath Bhakta (University of Kalyani, India)
Engineering Science Reference • ©2017 • 624pp • H/C (ISBN: 9781522523253) • US $260.00

Handbook of Research on Entrepreneurial Development and Innovation Within Smart Cities
Luisa Cagica Carvalho (Universidade Aberta, Portugal & CEFAGE - Universidade de Évora, Portugal)
Information Science Reference • ©2017 • 661pp • H/C (ISBN: 9781522519782) • US $235.00

Applied Environmental Materials Science for Sustainability
Takaomi Kobayashi (Nagaoka University of Technology, Japan)
Information Science Reference • ©2017 • 416pp • H/C (ISBN: 9781522519713) • US $205.00

Environmental Issues Surrounding Human Overpopulation
Rajeev Pratap Singh (Banaras Hindu University, India) Anita Singh (University of Allahabad, India) and Vaibhav Srivastava (Banaras Hindu University, India)
Information Science Reference • ©2017 • 325pp • H/C (ISBN: 9781522516835) • US $200.00

Reconsidering the Impact of Climate Change on Global Water Supply, Use, and Management
Prakash Rao (Symbiosis International University, India) and Yogesh Patil (Symbiosis International University, India)
Information Science Reference • ©2017 • 430pp • H/C (ISBN: 9781522510468) • US $215.00

Environmental Sustainability and Climate Change Adaptation Strategies
Wayne Ganpat (The University of the West Indies, Trinidad and Tobago) and Wendy-Ann Isaac (The University of the West Indies, Trinidad and Tobago)
Information Science Reference • ©2017 • 406pp • H/C (ISBN: 9781522516071) • US $200.00

Technologies for the Treatment and Recovery of Nutrients from Industrial Wastewater
Ángeles Val del Río (University of Santiago de Compostela, Spain) José Luis Campos Gómez (University Adolfo Ibáñez, Chile) and Anuska Mosquera Corral (University of Santiago de Compostela, Spain)
Information Science Reference • ©2017 • 391pp • H/C (ISBN: 9781522510376) • US $200.00

For an enitre list of titles in this series, please visit:
http://www.igi-global.com/book-series/advances-environmental-engineering-green-technologies/73679

IGI Global
DISSEMINATOR OF KNOWLEDGE

701 East Chocolate Avenue, Hershey, PA 17033, USA
Tel: 717-533-8845 x100 • Fax: 717-533-8661
E-Mail: cust@igi-global.com • www.igi-global.com

Table of Contents

Foreword

An atmospheric model is a mathematical model constructed around the full set of dynamical equations which govern atmospheric motions. These equations are often supplemented with parameterizations for turbulent diffusion, heat exchange, vegetation, the kinematic effects of terrain, convection, and many others. Typically, we analyze discrete models, which are constructed from the discretized equations of motion.

This is because the models are constructed on the basis of mathematical equations describing the physics and dynamics of the atmosphere. These equations are nonlinear and are impossible to solve exactly. Therefore, numerical methods are used to compute a numerical solution, which should be the approximation of the true solution.

One of the main fields of numerical mathematics is its application to the analysis of environmental phenomena, particular of those occurring in the atmosphere. The basic goal is to provide better understanding of the processes determining the dynamical and chemical state, and to enable the assignment of individual natural and anthropogenic forces, which are relevant for the variability and long-term change of atmospheric quantities.

In this book the authors develop a complete theoretical model of atmospheric dynamics and its adaptation to operational weather forecast. We note that, in general, modeling complex atmospheric systems is very time-consuming. The models must give sufficiently accurate results, and should run efficiently on the available computer architectures. Therefore, the development and parallelizing of efficient numerical methods play an important role.

The book is devoted to present several numerical methods aimed at overcoming this and other challenges. It consists of relevant and important knowledge for understanding the topic, and includes all the basic knowledge of the meteorological discipline. The book provides the Reader the tools they need to pursue study and research using numerical methods. The book contains not only classical results, but also several new scientific findings which were obtained by the authors in the last decades. The authors are well-known and highly recognized experts of the field. Therefore, the book is an excellent and useful tool for a wider range of readership, for students as well as for instructors and professional researchers. It will enable them to write computer programs to solve fairly complex problems, to explore and understand the current literature in which numerical methods are used, and to have the confidence to delve into texts on numerical methods to extend their knowledge and solve new problems.

István Faragó
Eötvös Loránd University, Hungary
November 2016

István Faragó *was born in Budapest, in 1950. He is professor at the Institute of Mathematics, Eötvös Loránd University, at Technical University, Budapest. In the period 2005-2015 he was head of the Department of Applied Analysis and Computational Mathematics at Eötvös Loránd University. He is head of the Research Group "Numerical Analysis and Large Networks" at the Hungarian Academy of Sciences, too. He has got the degree of candidate of sciences in Kyiv, and he is Doctor of the Hungarian Academy of Sciences from 2009. He was guest and visiting professor in more than 25 universities (University of Jyvaskyla (Finland), Jakutsk State University (Russia), Otago State University (New Zealand), Valladolid University (Spain), etc.) He works on different fields of mathematics and its applications, mainly in numerical analysis of differential equations, mathematical modelling, numerical linear algebra, operator splitting, parallel algorithms. He is author of 5 monographs, editor of 11 books and journals, 4 parts of a book, 12 lecture notes. He wrote more than 100 papers in scientific journals, and 50 papers in refereed proceedings published by outstanding publisher houses. He is member of several Editorial boards of 6 scientific journals (International Journal of Computational Science and Engineering, Open Mathematical Journal, Advances in Numerical Analysis, etc.). He has had 9 PhD students with degree. In the last years he was leader of several international projects in applied mathematics.*

Preface

The problem of the instrumental and theoretical study of real physical processes in the atmosphere (and the reliable prediction of their impact) is fundamental for both the improvement of qualitative and quantitative analysis of the physical phenomena and embodiment of derived models in the operational practice of weather forecasts. Physical, mathematical and numerical aspects of this complex problem have been addressed by many researchers and definitely, remain the actual problem at this time.

Students, professionals, and researchers in the field of hydrometeorology are probably the only category of researchers who are unable to carry out laboratory tests and experimental works, which are widely used in physics. This disadvantage is certainly not compensated by the observations in the free atmosphere because meteorologists cannot repeat the study at the same synoptic processes for obtaining quantitative relationships between the different components of the circulation or to define the role of individual factors affecting the change in the circulation. Analysis of long-term observations of the results is not allowed to reveal the presence of similarity in atmospheric processes.

Despite the challenges, meteorology as a science has made great strides successes due to the invention of radio-meteorological instruments, the development of networks of stations for radio sounding of the atmosphere and increase the height of the radiosonde ascent. Almost fifty years ago the radar and rocket sounding of the atmosphere and 10-15 years thereafter – the weather and artificial satellites have been launched. Meteorology also contributed to the development of the use of computers to process observational data. It is the combination of a new large global information on the thermodynamic, optical and radiative processes in the atmosphere, obtained using remote sensing equipment, with high-speed computer technology ensured the rapid development of meteorological science.

This book presents recent advances in the theoretical study of the collection of basic physical and thermodynamic processes in the troposphere, the stratosphere, and the mesosphere. In these three interrelated spheres of atmospheric processes, there are valid hydrodynamic equations for modeling the various units of the general circulation of the atmosphere and weather forecasting. Upper discharged atmosphere (thermosphere and exosphere) are quite different from the lower areas by the nature of the physical and chemical processes, are little associated with them, and therefore are not considered here.

Meteorological studies aim to identify and explain the causality of movements of the atmosphere of different temporal and spatial scales, to solve the fundamental problems of the theory of interaction of the atmosphere with the underlying surface by the radiation, heat, moisture, and others. These studies allow determining the role of different factors affecting the change in the circulation, the properties of the pressure and thermal fields, precipitation regimes.

Currently, the general circulation of the atmosphere means the combination of the main air movements that provide a thermodynamic exchange between the large air masses in the horizontal and vertical directions in the lower three spheres. These movements include air flows caused by the temperature difference between the different latitudes, jet flows and circulation of air masses in the cyclonic and anticyclonic formations. The total circulation of the atmosphere plays an important role in the formation of the Earth's weather and climate.

Regional, smaller-scale movements are of local importance and are characteristic of specific geographic areas. Usually, these small movements are born under the influence of the general circulation of the atmosphere, or, on the contrary, are destroyed under its influence. Obviously, the mesoscale phenomena (squalls, tornadoes, breezes, mountain-valley winds, etc.) may affect the large-scale circulation mode as small perturbations.

In order to study such complex atmospheric processes, along with the instrumental methods of observation, it is widely used synoptic, statistical and hydrodynamic methods. The main objective of these studies is developing theoretical models of atmospheric dynamics and adapting them to the weather forecast. In recent years, studies of the problem of air pollution have had a great public attention. It has become clear that addressing the air pollution problem requires a comprehensive approach, and this problem itself is a part of the more general problem of preservation of the environment and natural resources. The number of professionals solving this problem is growing rapidly, the scope of their mutual interests is expanding because of not only the need to understand but also to use the results obtained in related fields of research. Solving this problem, physicists, chemists, biologists, mathematicians, and, of course, the meteorologists are working together on the base of multilateral evaluation of effects of air pollution. This is one of the main factors that determine the need for inclusion in the book assumptions of the theory of atmospheric diffusion, distribution patterns and principles of calculation of dispersion in the atmosphere of gasses and aerosols from the permanent and periodical emission sources.

In this book, the authors focus on the hydro-mechanical, thermodynamic, environmental and mathematical aspects of dynamic meteorology. The content of the material has been formed gradually, over the years, in the process of research and teaching, and reflects not only the authors' point of view but also modern requirements. Dynamic meteorology is usually studied in higher education institutions at the meteorological departments as a full-fledged discipline. In this book, the authors make an attempt to link dynamic meteorology with other related disciplines to show how dynamic meteorology is used as a working machine in weather forecasts. It is widely used in the book the information from the theory of fluid mechanics, thermodynamics, mathematical analysis and numerical methods; we examine in detail some of the concepts that are the basis of various approximations and abstractions.

The content of the chapters of the book is unusual for several reasons. We are deeply convinced that the study of simplified linear systems, that allow the use of analytical methods, should be carried out hand in hand with the studying the full non-linear systems. Linearization in the neighborhood of some solution or other known methods should be considered in the light of reasonable prospects of research and it should be considered only as a way to shed a light on the solution of practical problems arisen in the decision process.

In each chapter, which deals with one or the other issue of some physical or mathematical problem, a review of existing approaches to the solution of this particular problem is given. Considerable attention is paid to the modeling of regional atmospheric processes, numerical solution of differential equations with partial derivatives, the theory of convective diffusion parameterization of sub-scale processes that reflect the experience of the authors relating to the comprehensive study of weather problems. Naturally, it is impossible in a single book to fully describe the different approaches and, of course, much of the material is devoted to the exposition of the methods proposed and implemented by the authors.

The book consists of nine chapters. The content of the opening chapters may serve as reference material for those who would like to refresh the basic knowledge of the meteorological disciplines. This material allows recalling all sorts of physical quantities, better understanding the nature of the various concepts, and looking for new ideas in static equilibrium, adiabatic process, etc.

The first chapter analyzes the system properties of the earth-atmosphere system, the basic meteorological variables and thermodynamic processes related to atmospheric circulation.

The second chapter deals with the physical basis of the general circulation of the atmosphere, heat and moisture cycle in the atmosphere, the influence of local physical and geographical conditions of atmospheric air movements, patterns of development of man-made and natural processes in the boundary layer.

The third chapter presents the basic equations of mechanics of a viscous stratified fluid used in the simulation of atmospheric circulation. The necessary initial and boundary conditions are analyzed for the correct formulation of the hydrodynamic problems in the modeling of atmospheric processes.

The fourth and fifth chapters analyze the problems related to the modeling of atmospheric processes of various scales (from the planetary to the mesoscale). The basic equations of hydrodynamics are formulated, depending on the adopted scale, in the coordinate systems mostly used in the application of meteorological problems. In the simulation of atmospheric processes of synoptic and macro-synoptic scale, the assumption of hydrostatic equilibrium is justified. It is this assumption that allows selecting as the vertical coordinate a number of variables other than geometric height. A common vertical coordinate is a pressure because in this system the basic equations of the atmospheric circulation take rather a simple form, what makes a much easier comparison of synoptic weather maps for different heights. Selecting the vertical coordinate and accounting orography, especially sub-scale orography (very steep mountains) are interrelated. Due to this relationship, there appear significant errors in the formulation of the boundary conditions on the underlying surface. Some basic points are formulated that are common to many existing models, despite the diversity of approaches to the formulation of initial and boundary conditions at modeling the atmospheric circulation of various scales. The methods of the parameterization of sub-scale processes, methods of parameterization processes of horizontal and vertical turbulent exchange, dry and moist convection, large-scale condensation and radiative heat fluxes are considered. Physical and mathematical foundations of the theory of atmospheric diffusion, laws and principles of the spread dispersion modeling in the atmosphere of gasses, aerosols, and particulate matter are consistently outlined. There are simulated mesoscale and macroscale effects of the spread of contaminants from separate sources with huge emissions and permanent industrial sources under real weather conditions, taking into account the topography of the underlying surface and atmospheric stratification.

In the sixth and seventh chapters, a brief historical overview of the development and use of parameterization methods of sub-scale processes is presented. Particular attention is paid to the development of new methods of parameterization of turbulence, large-scale condensation, clouds and precipitation, convection processes and radiation. Numerical methods used in the hydrodynamics of viscous continuum mechanics and atmospheric models of different scales are presented. We consider finite difference and spectral methods for the numerical solution of differential equations, which describe the mesoscale and macroscale processes in the atmosphere.

The eighth and ninth chapters are devoted to the analysis of numerical methods for reproducing the specific physical processes. The problems associated with the numerical solution of equations predictive models of regional atmospheric circulation are considered. There is a systematic exposition of construction, research, and application of numerical methods developed by the authors for effective regional solutions of boundary value problems with a prehistory on the basis of the "one-way interaction" approach. Within this approach, a method is proposed for filling the data, given on the macro-scale grid nodes, in the mesoscale network based on spline interpolation and precise (the fourth order of accuracy) numerical method for the approximation of the first and second order differential equations. There are proposed a numerical method for solving the non-stationary problem with a prehistory on the basis of interpolation with multiple knots, and absolutely stable implicit difference scheme, that is as effective as explicit scheme, for solving the convective diffusion problem. The results of numerical experiments are exposed on modeling and forecasting of weather conditions in the regions of all sizes and with varying lead times.

The methods are considered for solving evolutionary problems of transfer and diffusion of dust-gas mixture injected into the atmosphere by industrial enterprises under the conditions of the project mode of operation and during accidents. Area contamination spread is divided into "near" and "distant" areas. This hypothetical division allows using both analytical and numerical methods for solving the problem. The method of splitting is used as the basic algorithm of realization of the forecast tasks impurity concentrations in the atmosphere. There are presented computer scenarios of technogenic pollution of the atmosphere in real meteorological conditions, industrial emissions in normal operation, as well as in fires and accidents.

Vitaliy Prusov
University of Kyiv, Ukraine

Anatoliy Doroshenko
National Academy of Sciences, Ukraine
October 2016

Acknowledgment

This book is the result of a decade of the authors' cooperation in the field of development of efficient numerical methods for solving problems in meteorological forecasting. However, it could not be born without the help and support of the people of our Departments of Meteorology and Climatology in the Kyiv National University of Taras Shevchenko, Ukrainian Hydrometeorological Institute of National Academy of Sciences of Ukraine, Department of Automation and Control in the National Technical University of Ukraine "Igor Sikorsky Kyiv Polytechnic Institute" and Institute of Software Systems of National Academy of Sciences of Ukraine. The authors are grateful to all of them.

Special thanks to Prof. István Faragó who has kindly accepted the invitation to write the Foreword to the book. The authors are also grateful to their colleagues Olena Yatsenko and Iurii Tyrchak for the technical assistance in preparing the manuscript of the book.

Chapter 1
Brief Description of the Main Characteristics of the Earth–Atmosphere System

ABSTRACT

The chapter gives introductory concepts of the science of meteorology including astronomical data, basic information about the composition and structure of the atmosphere, determination of the main meteorological variables, solar radiation, longwave radiation of the Earth and atmosphere. There are given fundamentals of atmospheric thermodynamics and polytrophic changes in atmospheric air that include thermodynamic properties of dry and moist air and conditions for vertical air stability.

INTRODUCTORY CONCEPTS

Meteorology (from Greek *Meteora* – atmospheric and celestial phenomena) is the science of the Earth's atmosphere and the processes occurring in it. Meteorology applies close disciplines: actinometry, dynamic and synoptic meteorology, atmospheric optics, atmospheric electricity, aerology et al. that study:

- The composition and structure of the atmosphere.
- Heat exchange and thermal conditions in the atmosphere and at the Earth's surface, including the radiation processes and different mechanisms of non-radiation exchange between the atmosphere and the underlying surface and the atmosphere inside.
- Hydrologic cycle and phase transformation of water into the atmosphere during its interaction with the Earth's surface.
- Atmospheric movement – the general circulation of the atmosphere, part of its mechanism and local circulation.
- The electric field of the atmosphere.
- Optical and acoustic phenomena in the atmosphere.

DOI: 10.4018/978-1-5225-2636-0.ch001

Figure 1. Ecliptic of the Earth

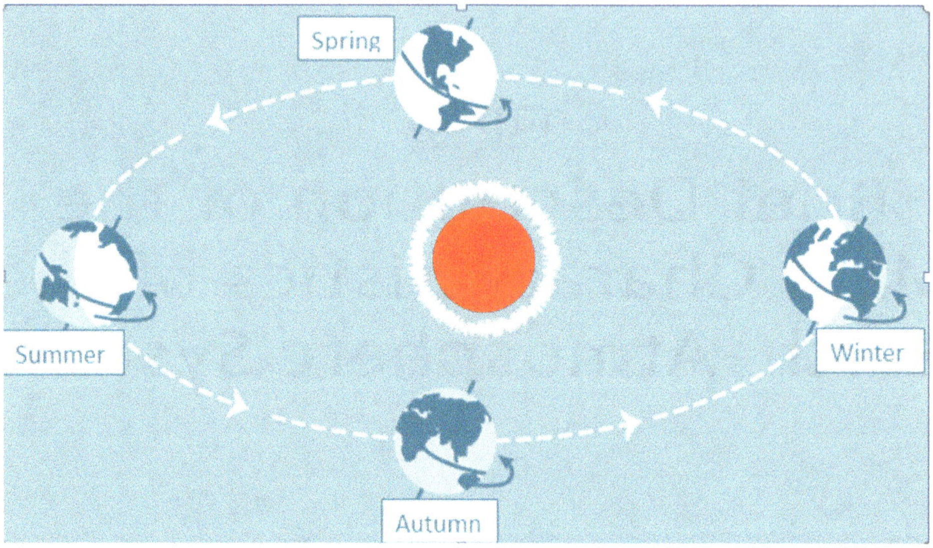

An important task of meteorology is to build a physical-mathematical theory of atmospheric processes, which has the ultimate goal of the forecast atmospheric phenomena.

Astronomical Data

The Earth is the third planet from the Sun in the Solar system. The average distance from the Sun is 149,597,890 km. It is the largest of the terrestrial planets (the equatorial radius is 6 378.14 km, the polar radius is 6 356.78 km, the mean radius is R=6 371.3 km, the equator circumference is 40,075 km, the surface area is 510,065,700 square kilometers) that has been formed about 4.5 billion years ago. Approximately 70.8% (361,132,106 km^2) of the surface of the planet is covered with water, and 29.2% (148,094,106 km^2) is the land. Surface temperature: minimal – 185°K, maximal – 331°K.

The Earth rotates uniformly around its axis passing through the north and south poles, with the speed of 465 m/s (1674 km/h) at the equator. The rotation period in the average (mean solar day) is 24 hours. In addition, the Earth moves around the Sun in an ellipsoid orbit at the average rate of 29,785.9 m/s (10.006,104 km/h). The sidereal orbital period is equal to 365.24 Earth days. Center of the Earth describes an ellipse around the Sun (perihelion of 147.1×10^6 km, the aphelion of 152.1×10^6 km, the eccentricity of 0.01671022), which has the Sun in its focus. This ellipse is situated in the plane (the ecliptic) passing through the Sun (see. Figure 1). The angle between the plane of the Earth's orbit and the ecliptic is 0,00005° and between the plane of the Earth's orbit and the equatorial plane of the ecliptic plane is 23°27´. Since the secular variation of the last angle is very slow and small, and the annual changes are negligible, we can assume that the declination of the Sun varies from –23°27´ (winter solstice) to +23°27´ (summer solstice), passing through zero at the time of the equinox.

The time is associated with the angle formed by two planes, one passes through the center of the solar disk and another passes through Earth's equator (Figure 2). There is a direct connection between the units of these two values: 24 hours (1 day) and 360°; 1h corresponds to 15°; 0.15h (4 min) corresponds to 1°. *The true solar time* (see. Figure 2) is determined by the hour angle τ formed by the plane of the

Figure 2. Solar time and hour angle: φ – latitude, δ – declination of the Sun, τ – hour angle (time)

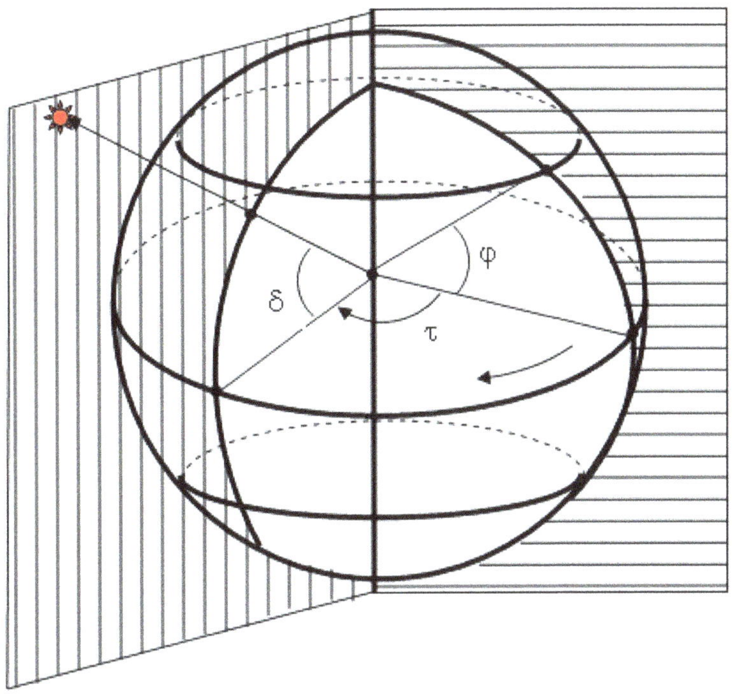

declination circle (circle time) of the center of the solar disk and the plane of the meridian of given place. Consequently, in the noon of the true solar time τ=0°; at 6 AM τ=270°; at 6 PM τ=90° and at 12 PM (0 AM) τ=180°.

The mean solar time (local time) is close to the true solar time. Their origins (0 AM) are at the maximum deviate from each other by an amount of ±16min depending on the date. Local time varies with the longitude of the place, but it is the same for all points of the same meridian. In meteorology, there is used *the universal time*, i.e., the mean solar time at the Greenwich meridian (the longitude λ=0), the initial point of which is midnight. Usually, it is denoted GMT (Greenwich Mean Time). To ensure the comfort living of inhabitants of some country (or some area of the Earth) the local (zonal) time set is associated with the global time by difference, usually expressed in hours. For example, GMT +1 means the local time for the zone of 15° East from the Greenwich meridian.

Based on the above definitions, we can calculate the angular coordinates of the Sun at any point on the Earth and at any given time. Indeed, the formula of spherical trigonometry gives the opportunity to obtain a relatively simple dependence on the angular coordinates of the Sun on the latitude φ and longitude λ of geographical location and the declination of the Sun δ and the value of hour angle τ, associated with the true solar time. Denoting the Sun height by β (see Figure 3) and the azimuth by a we obtain

$$\sin\beta = \sin\varphi\sin\delta + \cos\varphi\cos\delta\cos\tau\,, \tag{1.1.1}$$

$$\sin a = \frac{\cos\delta\cos\tau}{\cos\beta}\,.$$

Figure 3. The path of sunlight through the atmosphere, depending on the position of the Sun

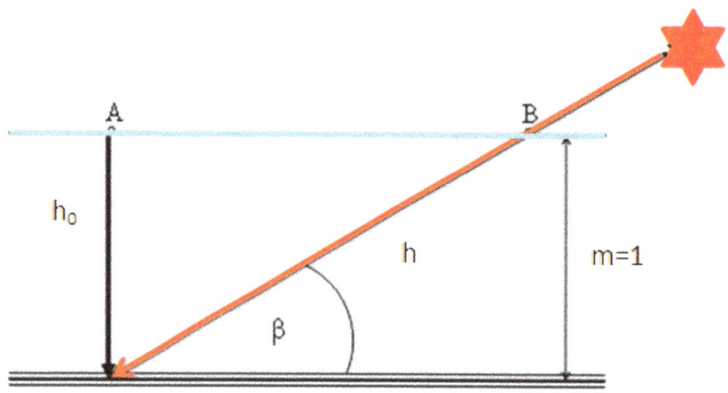

It follows directly from these classical formulas that, given the place φ and the time δ, sinus of the Sun height β reaches maximal value with the maximum of cosτ, i.e. when τ=0. This point corresponds to the true solar noon time, i.e. the Sun zenith. At sunrise and sunset, we have β=0, so sinβ=0 and, consequently, $\cos\tau = -tg\varphi tg\delta$. At the equator ($tg\varphi = 0$) the Sun always rises at 6 and sets at 18 o'clock of the true solar time since cosτ=0. At the time of the equinox ($tg\delta = 0$) corresponding clock values of the true solar time of sunrise (6 AM) and sunset (6 PM) are the same for all points of the globe. And finally, when $tg\varphi tg\delta \geq 1$ the Sun never sets, and at $tg\varphi tg\delta \leq -1$ the Sun never rises.

The Sun emits energy in the surrounding area of approximately $3.71 \bullet 10^{26}$ W. Of this amount, about $3.3 \bullet 10^{8}$ W per 1 km^2 reaches upper boundary of the Earth atmosphere.

Brief Information About the Composition and Structure of the Atmosphere

The Earth's atmosphere is the gas spherical envelope surrounding the Earth and rotating along with it. The height of the atmosphere to some extent is arbitrary. There are no clear atmosphere boundaries since the density of its constituent gasses is gradually approaching the gas density of interplanetary space. As there is no sharp boundary between the atmosphere and interplanetary space, the atmosphere is usually considered to be the area around the celestial body, in which the gaseous medium rotates with it.

The weight of the atmosphere per area unit is about 104 kg/m^2. As the gravitational acceleration is about 10 m/sec^2, the weight per area unit, or atmospheric pressure, at sea level is about 105 Pa ≡105 N/m^2 ≡1 bar. Pressure and density of air decrease with height rapidly. Although the atmosphere extends upwards for hundreds of kilometers, the bulk of it is located in a relatively thin bottom layer. Thus, 50% of the atmosphere mass is concentrated in the layer from sea level to altitudes of 5-6 km, and 90% is concentrated in the layer of the height of 30 km. The mass of 1 m^3 of air at the surface is equal to 1033 g, but at the height of 40 km, it is only 4 g.

The lower layers of the atmosphere consist of a mixture of gasses (see Table 1).

The chemical composition of the atmosphere remains unchanged for approximately 100 km of altitude. This constancy of the gas composition is due to intense air circulation accompanied by turbulent motions that cause the continuous mixing of gasses not only in the lower layers but also at high altitudes. Continuous airflow prevents separation of gasses according to their specific gravity. It becomes virtually impossible also the process of slow diffusion penetration of one gas to another.

Table 1. The gas composition of the atmosphere

Gas	Volume Concentration (%)	Molecular Mass
Nitrogen	78,084	28,0134
Oxygen	20,9476	31,9988
Argon	0,934	39,948
Carbon dioxide	0,0314	44,00995
Neon	0,001818	20,179
Helium	0,000524	4,0026
Methane	0,0002	16,04303
Krypton	0,000114	83,80
Hydrogen	0,00005	2,01594
Nitrous oxide	0,00005	44,0128
Xenon	0,0000087	131,30
Sulfur dioxide	from 0 unto 0,0001	64,0628
Ozone	from 0 unto 0,000007 in summer	47,9982
Nitrogen dioxide	from 0 unto 0,000002 in winter	46,0055
Ammonia	from 0 unto 0,000002	17,03061
Carbon monoxide	the traces	28,01055
Dry air	100	28,9644

As for the chemical composition, the atmosphere is usually divided into two vertical layers: *homosphere* that is the layer up to 100-110 km altitude, and *heterosphere* that is conventional layer extending at an altitude of 100-110 km above. Heterosphere and homosphere are separated by the intermediate layer, called *homopause*, which is located at altitudes of 80-90 km.

The air temperature is subject to significant changes with altitude, and by the nature of these changes, the atmosphere is divided into several areas with some the transition layers between. Names and some quantitative characteristics of the main areas and transition layers are shown in Figure 4. Let us briefly characterize the basic data on the physical properties of the atmosphere areas.

The *troposphere* is the lowest and most dense layer of the atmosphere in which the height of the temperature decreases rapidly. It contains up to 80% of the total mass of the atmosphere and extends into the polar and middle latitudes to the heights of 8–10 km, and in the tropics up to 16–18 km. Here are developing almost all weather-forming processes: heat and moisture exchange between the Earth and its atmosphere; cloud formation; various meteorological phenomena like fog and rainfall. The troposphere is in convective equilibrium and, due to the active mixing, it has a homogeneous composition, primarily of molecular nitrogen (78%) and oxygen (21%). It is the troposphere where the vast amount of concentrated natural and anthropogenic aerosol and gaseous air pollutants are located.

The physical properties of the troposphere and the dynamics of the lower part of the atmosphere up to 2 km are largely determined by the influence of solar radiation and the heterogeneity of the underlying surface, causing the appearance of zones of significant horizontal gradients of temperature, pressure, and wind speed. These horizontal gradients (altitude planetary frontal zones), encircling the Earth, play an important role in the development of atmospheric processes and the formation of the weather. The

Figure 4. The vertical structure of the atmosphere

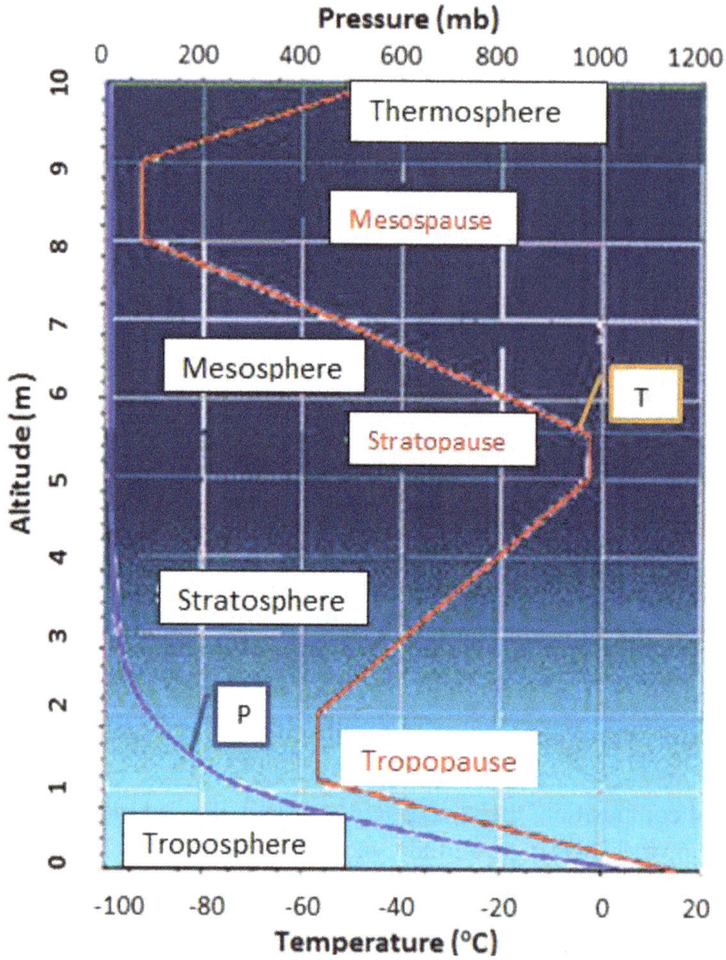

system of high-altitude frontal zones is the place of origin of planetary tropospheric jet streams (air flow of strictly defined direction) that cause intense horizontal circulation of the atmosphere. Vertical air movements and turbulent motion accompany horizontal flows.

The troposphere contains most of the atmospheric water vapor. Due to the raising and lowering of large volumes of air, clouds are formed and scattered. Therefore, the troposphere contains the bulk of the cloud, although some of them, composed of ice crystals, are found in the higher layers.

The temperature in the troposphere decreases with the height. The rate at which the temperature decreases in the layer of dry air is called the dry adiabatic vertical gradient. It is equal to about 10°C/km. If the layer of air saturated with water vapor then, due to the condensation of steam, latent heat is released, so that the rate of temperature decrease with height will be less than for dry air. In this case, the rate of decrease of temperature with height is called the moist-adiabatic vertical gradient that has the value dependent on temperature and pressure. The value of the moist-adiabatic vertical gradient in the troposphere is about 4°C/km at the temperature of 20°C and 5°C/km at the temperature of 10°C. Typically, the rate of decrease of temperature with height in the troposphere is adopted observed the average lapse rate, namely 6.5°C/km.

For the time intervals of the order of one day influence of the earth's surface is limited to a thin layer, called the atmospheric boundary layer. This layer is characterized by well-developed stirring generated by turbulent friction resistance occurring when the air flows along the uneven rough underlying surface, and air particles, pop up on the heated surface of the earth under the influence of convection. The boundary layer absorbs most of the heat and moisture, which are carried in the process of turbulent exchange. In the absence of the large horizontal pressure gradient in the atmosphere, the height of the boundary layer of the terrain is variable in time and depends on the intensity of turbulent exchange. During the day, when the sun heats the earth's surface, due to convective processes due to thermal reasons, the boundary layer can reach a height of 1-2 km. At night, when the ground surface cools faster than air, and the heat flow is directed downward, the boundary layer thickness can be less than 100 m.

Inside the boundary layer, there are two sub-layers, which are especially determined by the structure of the underlying surface. In direct contact with it, there is a laminar sublayer whose thickness does not exceed a few millimeters. A turbulent surface layer is above and characterized by the intense small-scale turbulent exchange of momentum, heat, and moisture. Day thickness of this layer may be up to 50 m, and at night, it is less than 10 m.

The tropopause is a transitional layer of a thickness of a few hundred meters up to 1.5 - 2 km, in which the temperature drop is slowed or stopped.

The stratosphere is the layer of the atmosphere, located above the tropopause, which extends to an altitude of 50-55 km. It is characterized by an increase in temperature with height in an average of 1-2°C/km. The heating atmosphere in this area is due to the presence of ozone molecules. The ozone layer serves as a solar battery, which allows accumulating solar energy. In addition, it is a kind of umbrella, which enables life on the Earth. Seasonal variations in temperature in the stratosphere are caused mainly by radiation conditions (change of the polar night by polar day, and vice versa), and nonperiodic ones are caused by dynamic processes.

More recently, it was believed that the stratosphere is a relatively peaceful environment, where there is no mixing of air, gasses are arranged in layers (stratus means laminate) according to their specific gravity, and the temperature is formed only under the influence of radiative equilibrium. However, new data received from radiosonde and weather rockets showed that in the stratosphere there holds intensive circulation of air which promotes significant vertical displacements, turbulent mixing, and hence a change in temperature.

Water vapor in the stratosphere is contained in minute quantity, so normally clouds in this layer are not formed. Above the stratosphere, which ends stratopause, the temperature drops again.

The mesosphere is a layer of the atmosphere located above the stratosphere and is a shell, in which up to a height of 80-85 km there is a decrease in temperature to the lowest values for the atmosphere as a whole. Record low temperature of –110°C has been recorded by a meteorological rocket launched from the US-Canadian installation in Fort Churchill (Canada).

Typical for the mesosphere temperature decrease with height occurs differently at different latitudes during the year. At low latitudes, the temperature drops more slowly (2.3°C/km) than at the high ones (3.1°C/km). In accordance with the meridional temperature distribution, thermal winds are observed in the lower layer in the mesosphere throughout the year. Thus, western winds in this layer in winter are weakened with height and eastern winds are intensified in the summer.

The upper bound of the mesosphere (*mesopause*) coincides with the lower boundary of the active absorption of X-rays and the most short-wave ultraviolet radiation from the Sun, which is accompanied by heating and ionization of the gas.

The thermosphere is a layer of the atmosphere in which the pressure and thus the gas density is continuously decreasing with height, and the temperature rises rapidly. The height of this layer can reach 600 km. At an altitude of about 100 km, the temperature passes through the 0°C. In the layer of 150-200 km, it rises to 500°C, and at the level of 500-600 km, it is over 1500°C. Such a high temperature in the thermosphere due to the high-speed motion of molecules.

In the thermosphere, the temperature regime is strongly influenced by ultraviolet, X-ray, and corpuscular solar radiation spectra. The observed temperature fluctuations are associated with changes in solar activity. When the maximum and minimum solar activity, the temperature difference can reach hundreds of degrees. About the same size, the temperature difference can reach in the upper thermosphere during daylight and night-time due to daily changes in temperature. These temperature fluctuations are the cause of the instability of the wind in the thermosphere.

The exosphere is the outer layer of the atmosphere located at the hight more than 700-800 km and indicated based on changes in temperature and the properties of the neutral gas. It is assumed that the temperature of this layer increases with height up to 2000°C. Theoretical consideration shows that the gas particles speed of about 12 km/sec, what is possible at such high temperature, gasses can overcome the effect of Earth's gravity and gradually escape into interplanetary space. In this layer, in the absorption of ultraviolet and corpuscular solar radiation, there is occurred photochemical decomposition reaction of the gas molecules into electrically charged atoms. Therefore, this layer is highly ionized and has a very high electrical conductivity. It is observed such phenomena as auroras and permanent air-glow, creating the so-called night-sky light, there are also occurred complex microphysical processes associated with cosmic radiation. Methods of research of these processes are unique; their study is very little related to the study of the atmosphere near the Earth's surface and in the lower layers but it is closely connected with the study of terrestrial magnetism. Therefore, recently it is adopted to consider the subject of physical (and chemical) processes in the upper layers of the atmosphere as a particular scientific discipline, called Aeronomy. This science is no longer a part of meteorology and stands next to it.

The layer located between the thermosphere and exosphere (that is overlying) is called *thermopause*. Thermopause is separated from the surface of the earth at an altitude of about 650 km.

Definition of the Main Meteorological Variables

Quantitative and qualitative characteristics of the atmosphere and weather phenomena obtained by observations and measurements in the network of meteorological stations and posts are called *meteorological variables*. These include temperature, pressure, humidity, wind speed and direction, cloudiness, precipitation, visibility (transparency of the atmosphere) as well as the temperature of the soil and surface water, solar radiation, long-wave radiation from the earth and the atmosphere. For meteorological parameters also include a variety of weather phenomena: droughts, dust storms, blizzards, sleet, etc. Changes in meteorological variables are the result of atmospheric processes and phenomena occurring in the Earth's atmosphere. They are extremely complex and are interdependent on one another. Their generalizations are possible only on the basis of an extensive, perhaps the exact material obtained by observations at meteorological stations. Here are the main ones, since they will have to occur throughout the book.

The mass density distribution, or, for short, *medium density* in the given point mass density distribution, or, for short, medium density in the given point $\mathcal{M}$ is called a scalar quantity equal to the limit of the ratio of the numerical value of the mass Δm of the small volume $\Delta\wp$ medium containing within itself the point $\mathcal{M}$, to the value of at $\Delta\wp$ when $\Delta\wp \to 0$:

$$\rho = \lim_{\Delta\wp \to 0} \frac{\Delta m}{\Delta \wp}. \tag{1.1.2}$$

It is assumed here that as the volume $\Delta\wp$ tends to zero, the point $\mathcal{M}$ always remains inside the volume. The reciprocal $\vartheta = 1 / \rho$ is called *the specific volume.*

The medium density of moving material depends on its composition, temperature, and pressure, as well as the nature of the medium. Recall that under the Boyle's law at a given temperature, the gas density is directly proportional to the pressure, and under the law of Gay-Lussac at a given pressure, gas density increases inversely proportional to its absolute temperature. Under normal conditions, dry air density is 1,293 kg/m^3.

In general, the density of the medium is represented by the function of position and time $\rho = \rho(x, y, z, t)$ that constitutes a scalar field, which may be either stationary or nonstationary. The surface of the density of the scalar field is called *the isosteric surface* and the line of some scalar field density level (in the particular case of a flat distribution) is called *the isostere.*

Air is heterogeneous medium representing a mechanical mixture of a number of chemical values (Table 1). Because of condensation or sublimation of water vapor in the atmosphere, there is a visible aggregate of suspended water droplets or ice crystals, which forms a mist or cloud, depending on the height above the Earth's surface. Experimental studies have shown that natural fogs and clouds are composed of particles of different sizes, that is, are polydisperse. The concentration of particles is very variable, from a few dozen to thousands in 1 cm^3. Furthermore, the air includes a large amount of suspended solid and liquid particles (aerosols) of natural and industrial origin.

In theoretical fluid mechanics, a medium called *homogeneous* if all substances contained in it in solid, liquid or gaseous are of the same phase in its composition, otherwise it is called *heterogeneous.* Regardless of hypotheses concerning air (whether it is homogeneous or heterogeneous medium), it should be taken to follow the basic assumption about the common properties of individual components and mixtures in general. Below, we assume that in the elementary volume of air, in spite of the smallness of this volume, there is contained a sufficiently large number of particles in order to make it possible to apply a statistical averaging of the physical parameters of these particles in their set.

A heterogeneous medium is a mechanical mixture of a variety of chemical composition and phase state of substances which can be both neutral and interact with one another. The state of this medium is completely determined by the coordinates and velocities (momentum) of all the particles of the mixture. At each time point, each of the particles of the mixture has a certain velocity, so every part must be attributed to a specific position in the velocity space which is characterized by the vector $\mathbf{V}_i = (u_i, v_i, w_i)$. Each particle position in physical space is defined by the vector $r = (x,y,z)$ If the mixture is not in equilibrium, the properties of each ingredient of the mixture can be expressed in terms of the distribution function $f_i = (r, \mathbf{V}_i, t)$.

The total number of particles of the i-th ingredient which are contained in the elementary volume unit of physical space in time moment t can be obtained by integrating distribution function f_i over all possible values of $\mathbf{V}_i = (u_i, v_i, w_i)$:

$$n_i(r, t) = \iiint_{\mp\infty} f_i du_i dv_i dw_i. \tag{1.1.3}$$

Then, by definition of the density of the *i*-th ingredient which we consider a continuous medium, we have $\rho_i = n_i m_i$, where m_i is the mass of a separate particle. In accordance with the definition of density as the mass per unit volume, by summing up all the ingredients of the mixture we will have the following expression density of the mixture at a point (r,t):

$$\rho = \sum_i n_i m_i = \sum_i \rho_i \, . \tag{1.1.4}$$

The relation

$$\frac{\rho_i}{\rho} = \frac{n_i m_i}{\sum n_i m_i} = c_i, \sum_i c_i = 1 \, . \tag{1.1.5}$$

is called the mass concentration of particles of the *i*-th ingredient.

In meteorology, excluding radiant radiation problems, clean air is represented as a mixture of two components: dry air that is approximately described by the ideal gas laws, and steam. Consequently, in accordance with (1.1.4), the air density ρ is the sum of the densities of the dry air ρ_d and steam ρ_s.

$$\rho = \rho_d + \rho_s \, . \tag{1.1.6}$$

The pressure p is a physical quantity which is equal to the numerical value of the limit of ratio of the normal force ΔF_n acting on a portion of the surface area ΔS to the value of ΔS when $\Delta S \to 0$:

$$p = \lim_{\Delta S \to 0} \frac{\Delta F_n}{\Delta S} = \frac{dF_n}{dS} \, .$$

This formula implies that *the atmospheric pressure* is the force equal to the weight of the vertical column of air acting on an area unit. In the International System of Units (SI), the unit for measuring pressure is 1 Pa = 1 kg•m^{-1}•s^{-2}. Atmospheric pressure is measured with the help of devices, the most common of which are:

- Liquid barometer, based on a balancing of the atmospheric pressure and the weight of the liquid column.
- Deformation barometers (e.g., aneroid), whose operation is based on the elastic deformation of the membrane box.
- Hypsometer, based on the dependence of the boiling point of some liquids, such as water, on external pressure.

The most common standard devices are mercury barometers.

The normal (standard) air pressure is taken the value p=101325 Pa measured by the weight of mercury column of 760 mm high with a base of 1 cm^2 at the temperature of 0°C at the latitude φ=45°, where the gravitational acceleration is g=9.806m•c^2. From this definition, it follows that in a quiescent atmosphere, air pressure depends on the height of the corresponding column of air, its density, and the acceleration of gravity which varies by latitude and altitude. This dependence of the vertical distribution of atmospheric pressure is described by the equation of statics:

$$\frac{dp}{dz} = -\rho g \,, \tag{1.1.7}$$

where ρ is the density of air.

If we consider air as a mixture of dry air and steam, the atmospheric pressure p is equal to the sum of the pressure of dry air p_d and steam e, i.e.

$$p = p_d + e \,. \tag{1.1.8}$$

The atmospheric pressure at the Earth's surface varies from place to place and time. There are known fluctuations of atmospheric pressure (at sea level) in the range of 684-809 mm Hg. Pressure fluctuations correspond to the arrival of the main phenomena, determining the weather: anticyclones – associated with the origin, development and destruction of the slow-moving high-pressure area, and cyclones – associated with relatively fast-moving large vortices, which is dominated by low pressure.

Despite the great variability, distribution of average monthly values of atmospheric pressure on the surface of the globe every year is about the same. The average annual atmospheric pressure is reduced near the equator and has a minimum at 10° North. Further from the equator, the atmospheric pressure increases and reaches a maximum at 30°-35° North and South. The atmospheric pressure then decreases again, reaching a minimum at 60°-65°, and rises again toward the poles. This latitudinal distribution of atmospheric pressure is significantly affected by time of a year and the nature of the distribution of continent and ocean areas. Over the winter cold continents, there are areas of high atmospheric pressure. Thus, the latitudinal distribution of atmospheric pressure is disturbed, and the pressure field is divided into several areas of high and low pressures, called *centers of atmospheric action*. The horizontal pressure distribution becomes more simple with a height, approaching the latitudinal distribution. Starting from a height of about 5 km, the atmospheric pressure around the globe decreases from the equator to the poles.

One can observe two maxima (at 9-10 and 21-22 o'clock) and two minima (at 3-4 and 15-16 o'clock) in diurnal pressure variation. Especially the right diurnal variation of atmospheric pressure is in the tropics, where the daily fluctuation reaches 2.4 mm Hg and nightly is 1.6 mm Hg. With the increase of latitude, the amplitude of atmospheric pressure change decreases, but at the same time, non-periodic changes become more powerful.

The temperature t is a physical quantity that characterizes the average kinetic energy of the particles of a macroscopic system in a state of thermodynamic equilibrium. In equilibrium, the temperature is the same for all parts of the macroscopic system.

Temperature measurement is possible only indirectly due to the dependence on temperature of some physical properties of substances that are amenable to direct measurement. In general, changing the selected thermodynamic parameter of thermometric substance is uniquely associated with a change in temperature. In meteorology, to establish the beginning of the reference temperature and its unit of measurement (*the degree*), there is taken a transition temperature of some chemically pure substance from one state to another. For Celsius Centigrade, it is taken the temperature of ice melting $t_0 = 0°C$ as a beginning value, and as the upper value, it is taken the boiling point of water $t_b = 100°C$ at a normal atmospheric pressure of 760 mm Hg. The Kelvin temperature T is measured from absolute zero $t_0 = -273.15°C$ called absolute temperature. The relationship between the values of Kelvin and Celsius temperature in is given by

$$T = t + 273.15 \tag{1.1.9}$$

Some countries use the Fahrenheit scale, which has as its origin the temperature $t_0 = 32°F$, and as the high value the temperature $t_t = 212°F$. The relationship between the values of the temperature in Fahrenheit and Celsius is given by

$$t°F = 32 + 1.8\, t°C. \tag{1.1.10}$$

The term *humidity* is broadly used in meteorology to refer to the content of water vapor in the air. The following characteristics are used for depicting atmospheric humidity:

- **Elasticity e of Water Vapor:** Is the partial pressure of the vapor contained in the air;
- **Absolute Humidity of the Wet Air:** Is the mass of water vapor m_s contained in a unit volume;
- **The Mixing Ratio of Wet Air *s*:** Is the ratio $s = m_s / m_d$ of the mass of water vapor to the mass of dry air in which the water vapor is associated;
- **Specific Humidity *q* of Wet Air:** Is the ratio $q = m_s / (m_s + m_d)$ of the mass of water vapor to the mass of wet air, in which the water vapor is contained;
- **Saturation:** Is a condition where the wet air at a pressure p and temperature T has such a composition that can be in equilibrium with the neutral plane surface of the clean condensed water or ice under the same temperature and pressure (saturated vapor pressure will be referred to as E);
- **Relative Humidity *f* of Wet Air at a Temperature *T*:** Is the ratio f=100e/E% of the actual vapor pressure e to the saturation vapor pressure E, corresponding to the temperature T;
- **Dew Point Temperature T_d of Wet Air at a Pressure *p*, Temperature T, and Mixing Ratio *s*:** Is the temperature to which the air must be cooled to become saturated with respect to water at an initial pressure p and mixing ratio s;
- **Humidity Deficiency *d* of Wet Air at a Temperature *T*:** Is the difference between the saturated vapor elasticity E and steam e.

These characteristics of the humidity are linked all together and those, that cannot be measured, can be expressed by those that can be measured.

By definition, the mixing ratio of s we have

$$s = \frac{m_s}{m_d} = \frac{\rho_s}{\rho_d} = \frac{R_d}{R_s}\frac{e}{p-e} = 0.62197\frac{e}{p-e}, \tag{1.1.11}$$

because the ratio R_d / R_s of the gas constant dry air R_d to the gas constant of water vapor R_s is 0.62197.

Using the definition of specific humidity q, in a similar manner, we will have

$$q = \frac{m_s}{m_s + m_d} = \frac{\rho_s}{\rho_s + \rho_d} = \frac{R_d}{R_s} \frac{e}{p - \left(1 - R_d / R_s\right)e} = 0.62197 \frac{e}{p - e}, \tag{1.1.12}$$

Comparing (1.1.11) and (1.1.12), we obtain another useful relationship

$$q = \frac{s}{1 + s}. \tag{1.1.13}$$

Since $e \ll p$, then $s \approx q$. Formulas (1.1.11), (1.1.12) that the mixing ratio s and the specific humidity q are conservative with respect to adiabatic and the non-adiabatic changes of temperature, and are non-conservative with respect to the evaporation and condensation.

The wind in meteorology is the movement of air relative to the earth surface, characterized by speed (the value of its modulus) and direction. Usually, it refers to the horizontal component of the movement that is determined by the station equipment (weather vane, an anemometer, and others) or, as in the free atmosphere, with the help of pilot balloon observations. The main reason for the movement of air is the difference in temperature of adjacent areas and, thus, their uneven air pressure. The air from the high-pressure area tends to lower, but undergoes a deviation to the right in the northern hemisphere, and to the left in the south hemisphere (the law of Buys-Ballot) due to the rotational motion of the earth. Wind velocity depends on the difference between the barometric pressure at two points spaced at a distance, i.e. pressure gradient. Since the pressure changes both horizontally and vertically, the air flows at an angle to the surface. Essentially, they consider only the horizontal component of the movement. This is justified by the fact that the vertical component of the wind is usually very small, and becomes noticeable only when the strong convection.

The light of the country indicates wind direction where it blows. Usually, they distinguish 8 directions or points of the compass: N – North, E – East, S – South, W – West, and NE – North-East, SE – South-East, SW – South-West, NW – North-West. The velocity is indicated by the number of meters per second or kilometers per hour, and for short it is marked from 0 (calm) to 12 (hurricane) in the so-called Beaufort's scale score (see Table 2.), Developed by the British Admiral F. Beaufort in 1806 and 1874 accepted for use in international synoptic practice.

Due to continuous changes in pressure over time and space, wind velocity and direction are constantly changing. With a height, wind velocity changes due to friction decrease.

The wind velocity usually is measured as the *averaged* (for the 2 or 10 minutes, depending on the type of device) and as the *instantaneous* (being averaged for 2–5 seconds). Wind direction is also averaged over an interval of about 2 minutes. Averaging the instantaneous velocity for the interval 2-5 s. is achieved by the automatic sensor wind-measuring devices inertia factor which lie within these limits. The maximum value of the instantaneous velocity for any period is called the gust. Gusty wind is highly dependent on the velocity: the higher the velocity, the more is the wind gustiness.

The diurnal variation of wind is observed at the surface boundary layer of the atmosphere. Minimum wind velocity is observed at night. In the afternoon, the wind velocity and its gustiness reach their peak. The diurnal variation of wind has a place to heights of 100-300 m in the summer, and to a height of 20-30 m in the winter. The cause of the diurnal variation is the daily change of the turbulent mixing intensity.

Table 2. Beaufort's scale

The Mark for Scale of Beaufort	Speed (m/s)	Characteristic	Visible Effect
0	0.0–0.2	Calm	Smoke rises vertically, the leaves still on the trees
1	0.3–1.5	Light air	Smoke slightly deviates, light air movement
2	1.6–3.3	Light breeze	Wind felt face, leaves rustle
3	3.4–5.4	Gentle breeze	Swaying leaves and thin branches
4	5.5–7.9	Moderate breeze	The tops of the trees bend stir small twigs, dust rises
5	8.0–10.7	Fresh breeze	Teeter branches and thin trunks of trees
6	10.8–13.8	Strong wind	Sway branches, humming telephone wires
7	13.9–17.1	Moderate gale	Rock the trunks, large branches bend, hard to go against the wind
8	17.2–20.7	Fresh gale	Rock the big trees, small branches break, very hard to walk
9	20.8–24.4	Storm	Slight damage to buildings breaks thick branches
10	24.5–28.4	Heavy storm	Trees are broken or uprooted, great damage to buildings
11	28.5–32.6	Fierce storm	Great destruction
12	32.7–36.9	Hurricane	The devastating destruction
13–17	37.0 and more		

The wind over large area forms vast air currents that make up the general and local atmospheric circulation (see below).

Air flows are often seen as arising and getting typical properties under the influence of local physiographic and thermal conditions (local winds). According to these features of the set of existing local winds, one can identify four main types of the most common:

- Breeze is the wind with a daily frequency, occurring along the shores of the ocean, sea, and a large lake, as well as some of the larger rivers. Day (sea) breeze is directed from the water surface to land, night (coast) breeze – on the contrary, from the land to the water surface. The sea breeze starts from 10-11 AM and extends inland to 20-40 km. Its vertical thickness reaches an average of 1000 m. A coastal breeze begins after sunset, extends deep over the water surface for 8-10 km, reaching a height of about 250 m.
- Mountain-valley wind is a local air circulation between the ridge and the valley with the diurnal period. During the day, the air flow is directed from the valley up a heated slope, at night – on the contrary, from the slopes of the mountain to the valley. The mountain-valley winds blow in all mountain systems and are particularly well expressed in clear weather in the summer.
- *Bora* is a strong cold wind directed from the coastal low mountains (up to 1000 m) on the water surface. Bora extends deep into the surface of the water for a few kilometers, and along the coast – for a few tens of kilometers. The vertical flow rate is approximately 200 m.
- Foehn is a warm, dry wind directed from the mountains, often strong and gusty. When the foehn, there are difficult weather conditions on the windward side of the ridge (clouds, precipitation, poor visibility), on the leeward side, on the contrary, there is dry, overcast weather.

Cloudiness

Water vapor enters the atmosphere because of the thermal instability of the air mass, and purely dynamic factors effect. The cooling steam reaches dew point and condenses into cloud droplets and ice crystals. Thus, the cloud is the accumulation of water droplets suspended in the atmosphere, or ice crystals, or a mixture of both, resulting from the condensation of water vapor. Clouds and limited visibility are the main meteorological factors, complicating the activities of almost all types of transport. Furthermore, under favorable conditions, the cloud droplets and crystals reach such an extent that they can no longer remain in suspension and fall as rain. Heavy downpours, floods, mudslides, hurricanes and other phenomena associated with precipitation can cause large losses to the infrastructure and economy. These losses can be significantly reduced if one can improve the lead time and the accuracy of forecasts and hazard warnings, and extreme weather events. Therefore, a meteorologist as a supplier of weather information needs to know what kind of clouds are linked to hazard and extreme weather events and to be able to classify them in terms of their origin conditions and appearance. Only in this case, his/her forecast, advice, or other additional information on the actual or future weather conditions will be useful for the consumer.

Development of the first cloud classification refers to the beginning of the 18th century. Later it was refined and supplemented. It was a morphological feature, that was laid to its foundation, i.e., the appearance of the clouds. According to this approach, clouds are divided into four families on height: the upper tier clouds middle clouds, low clouds, and clouds of vertical development. Below there is a description of these families, as well as their names and designations.

1. **The Family of Upper Tier Clouds:** They are located at altitudes of more than 6 km and, depending on the appearance, are divided into three forms:
    a. **Cirrus (Ci) Clouds:** (Figure 5) Some are white, often silver, fiber thin and transparent without the "shadows" the cloud. When they occur in small quantities in the form of separate clouds, that are called cirrus of good weather. If they occupy a large part of the sky and are associated with Cirrostratus or Altostratus clouds, they are a precursor of bad weather. Closing the sun, they only slightly weaken the sunshine. Cirrus clouds do not reduce daylight exposure so that objects continue to cast shadows. The cirrus clouds around the sun and the moon is often observed halo. Cirrus clouds are composed primarily of ice crystals in the form of columns height of 0.01 - 0.1 mm. Depending on the appearance of the cirrus clouds are divided into six species (Figure 5.): Cirrus filosus (Ci. fil.), Cirrus uncinus (Ci. unc.), Cirrus vertebratus (Ci. vert.), Cirrus spissatus (Ci. sp.), Cirrus incus-genitus (Ci. ing.) and Cirrus radiatus (Ci. rad.).
    b. **Cirrocumulus (Cc) Clouds:** (Figure 6) Thinner clouds without gray shades, consisting of very small waves and ripples flakes, partly with a fibrous structure. As a rule, Cirrocumulus clouds occur in a small amount. There may be observed crowns and halos in these clouds around the sun and the moon. Depending on the appearance of the Cirrocumulus clouds are divided into three varieties (see Figure 6.).: Cirrocumulus floccus (Cc. floc.), Cirrocumulus undulatus (Cc. und.) and Cirrocumulus cumuliform (Cc. cuf.).
    c. **Cirrostratus (Cs) Clouds:** (Figure 7) A thin whitish or bluish veil of the slightly fibrous structure through which the Sun and Moon shine. These clouds differ from Cirrus in that the veil is plain and continuous. In the daytime, cirrostratus clouds usually do not make shadows of ground objects diluted. Elements of cirrostratus clouds are crystals in the form of needles

Figure 5 Cirrus clouds: a) Cirrus filosus (Ci. fil.), b) Cirrus uncinus (Ci. unc.), c) Cirrus vertebratus (Ci. vert.), d) Cirrus spissatus (Ci. sp.), e) Cirrus incus-genitus (Ci. ing.) and f) Cirrus radiatus (Ci. rad.)

Figure 6. Cirrocumulus Clouds: a) Cirrocumulus floccus (Cc. floc.); b) Cirrocumulus undulatus (Cc und.); c) Cirrocumulus cumuliform (Cc. cuf.)

or hexagonal columns. Quite often they are the precursors of deteriorating weather. Depending on their appearance, the cirrostratus clouds are divided into three varieties (see Figure 7.): Cirrostratus (Cs), Cirrostratus filamentous (Cs. fil.), Cirrostratus nebulosus (Cs. neb.).

2. **The Family of Medium-Level Clouds:** These clouds are located at the altitude of 2-6 km. Depending on the appearance, they are divided into two forms:
   a. **Altocumulus (Ac) Clouds:** (Figure 8) White, sometimes bluish clouds, these clouds have the form of waves or ridges, consisting of individual plates or flakes, usually separated by

Figure 7. Cirrostratus clouds: a) Cirrostratus (Cs); b) Cirrostratus filamentous (Cs. fil.); c) Cirrostratus nebulosus (Cs. neb.)

openings of blue sky, but sometimes merging in an almost continuous cover. They differ from Cirrocumulus clouds with more density and greater visibility dimensions of the individual plates or flakes. Altocumulus clouds are usually watery and colloidal-resistant. They consist of overcooled water droplets of 5-7 microns in radius and does not give precipitation. Depending on the appearance of altocumulus are divided into 6 species (Figure 8.): Altocumulus undulatus (Ac. Und.), Altocumulus translucidus (Ac. Trans.), Altocumulus opacus (Ac. op.), Altocumulus lenticular (Ac. lent.), Altocumulus floccus (Ac floc.) and Altocumulus castellanus (Ac cast.).

b. **Altostratus (As) Clouds:** (Figure 9) Gray or bluish, their uniform veil is slightly fibrous structure. As a rule, they gradually cloud the whole sky. Sometimes on the bottom surface, mild waves are visible. Altostratus clouds are mixed (made up of crystals and overcooled water droplets). Thin Altostratus clouds are composed primarily of ice crystals. The light of the sun and the moon, as it passes through Altostratus clouds, is usually weakened or do not shine at all. Altostratus clouds are of colloidal instability. In the cold season, they give light snow, and in the summer, they give rainfall that usually do not reach the ground. With the growing amount of Altostratus, especially in the transition to Cirrostratus, they are usually followed by heavy precipitation. Depending on appearance, they are divided into three varieties (see Figure 9): Altostratus undulatus (As. und.), Altostratus translucent (As.trans.), and Altostratus opacus (As. op.).

3. **A Family of Low Clouds:** These clouds are located within 2 km in height and, depending on appearance, are divided into three forms:

a. **Stratocumulus (Sc) Clouds:** (Figure 10) These are gray clouds composed of waves or ridges, plates or flakes separated by lumens, but sometimes are merged into a continuous wavy cover. The most parts of clouds are arranged often in regular rows in one or two directions. Sometimes they are observed together with cumulus and altocumulus. Stratocumulus clouds are usually watery. They consist of water droplets of 5-7 microns in radius. Almost always, they are colloidal-stable and do not give precipitation. Depending on the appearance of the Stratocumulus clouds are divided into three varieties (see Figure 10.): Stratocumulus undulatus (Sc. und.), Stratocumulus cumuliform (Sc. cuf.), and Stratocumulus opacus (Sc. op.).

b. **Stratus (St) Clouds:** (Figure 11) This is a uniform layer of gray color, similar to the mist, and located not at the very surface of the earth but at a certain height. Often the lower surface of Stratus is torn and ragged. Usually, Stratus clouds cover the entire sky with a solid veil, but they can sometimes take the form of broken cloud masses. They consist of water droplets of

Figure 8. Altocumulus clouds: a) Altocumulus undulatus (Ac. und.); b) Altocumulus translucidus (Ac. trans.); c) Altocumulus opacus (Ac op.); d) Altocumulus lenticular (Ac. lent.); e) Altocumulus floccus (Ac floc.); f) Altocumulus castellanus (Ac. cast.)

Figure 9. Altostratus clouds: a) Altostratus undulatus (As. und.); b) Altostratus translucent (As.trans.); c) Altostratus opacus (As. op.)

2-5 microns in radius. In parts of the clouds with temperatures below 0ºC, the droplets are overcooled. Here also may be ice crystals. Stratus clouds are often observed in the form of rain drizzle. Depending on the appearance, the Stratus clouds are divided into three varieties (see Figure 11): Stratus (St.), Stratus fractus (S.t fr.), and Stratus nebulosus (St. neb.).

c. **Nimbostratus (Ns) Clouds:** (Figure 12) Dark gray, sometimes with a bluish tint, thick cloud layer. When precipitation, appear to be homogeneous. Usually, Nimbostratus clouds cover the sky with a continuous layer with no gaps. In composition, they belong to mixed clouds. At the bottom of Nimbostratus, at subzero temperatures, they are dominated by small drops of water with a dash of snowflakes. With positive temperature, lower part Nimbostratus consists of relatively large droplets. The predominant droplets are of the size of 7-8 microns. Nimbostratus give long, usually at least 2-3 hours, precipitation; in winter – in the form of

Figure 10. Stratocumulus clouds: a) Stratocumulus undulatus (Sc. und.); b) Stratocumulus cumuliform; c) (Sc. cuf.); Stratocumulus opacus (Sc. op.)

Figure 11. Stratus clouds: a) Stratus (St.); b) Stratus fractus (St. fr.); c) Stratus nebulosus (St. neb.)

Figure 12. Nimbostratus clouds: a) Nimbostratus (Ns); b) Nimbostratus giving precipitation (Ns. pr.); c) Nimbostratus with bands of precipitation (Ns. virga)

snow, in the summer –in the form of rain. Depending on the appearance, Nimbostratus are divided into three varieties (see Figure 12): Nimbostratus (Ns), Nimbostratus giving precipitation (Ns. pr.), and Nimbostratus with bands of precipitation (Ns. virga).

4. **The Family of Clouds of Vertical Development:** Their grounds are located at an altitude of low clouds and peaks are located at an altitude of clouds middle or upper tier. Depending on the appearance, they are divided into of the three forms:
   a. **Cumulus (Cu) Clouds:** (Figure 13a) They are dense, developed vertically clouds with white domed tops, often have cauliflower structure and a flat horizontal base. They may have a form of separate contrasting clouds or clusters of clouds that cover the whole sky. The tops of

Figure 13. Cumulus clouds: a) Cumulus (Cu); b) Cumulonimbus (Cb)

Cumulus clouds are composed of fairly large drops of about 10 microns in radius, and smaller drops of predominantly 6 microns of the radius at the base. Sometimes cumulus clouds are further divided into three types: Cumulus humilis (or “fair-weather clouds»), Cumulus mediocris (medium), and Cumulus congestus (powerful). Cumulus clouds are usually colloidal-resistant, do not give precipitation.

b. **Cumulonimbus (Cb) Clouds:** (Figure 13b) They are powerful white cloud masses with dark bases that are the result of further development of Cumulus congestus. They rise in the form of mountains or towers, the upper parts of which have a fibrous structure. The tops often resemble the shape of an anvil or surrounded by dense cirrus clouds. Squalls or thunderstorms accompany cumulonimbus clouds, including torrential rains fall, sometimes with hail, or snow. Cumulonimbus clouds in the upper part are made up of ice crystals. At temperatures above -15ºC, the cloud is composed of lamellar crystals and overcooled water droplets. A lower part contains water droplets with a dash of snowflakes, and sometimes cereals or hail. In winter with severe frosts, Cumulonimbus clouds may consist only of the crystals, and only from water droplets in tropical countries.

Mist is a set of microscopic liquid droplets or ice crystals in the surface layer of the atmosphere (sometimes up to several hundred meters), making the air less transparent. The mists of water droplets are observed primarily at air temperatures above -20°C, but can even occur at temperatures below -40°C. At temperatures below -20°C, the ice mist dominates. The total mass of water droplets per unit volume of mist (water content of the mist) usually does not exceed 0.05-0.1 g/m^3, but can reach 1-1.5 g/m^3 in some dense mist. The radius of mist droplets (mist microstructure) usually ranges from 1 to 60 microns. The majority of the droplets has a radius of 5-15 microns at a positive air temperature and 2-5 microns at temperatures below freezing.

Mists are divided into two types by their origin:

- *Mists of cooling* are formed by the condensation of water vapor while cooling the air below the dew point.
- *Mists of evaporation* are formed by the evaporation of the liquid from the warmer underlying surface (ponds and wet areas of land) in the cold air.

Also, mists differ in synoptic conditions of their forming:

- *Inner mists* are formed in uniform air masses.
- *Front mists* are formed at the boundaries of weather fronts.

Inner mists prevail in nature, as a rule, they are mists of cooling. They also can be divided into several types:

- *Radiation mists* appear because of radiative cooling of the Earth's surface and the mass of wet surface air to the dew point. Typically, the radiation mist occurs in the cloudless night. Often, the radiation mist occurs in conditions of a temperature inversion that prevents lifting of the air mass. After sunrise, the radiation mist usually quickly dispersed.
- *Advection mist* is formed due to the cooling of warm wet air as it moves over the colder underlying surface. Their intensity depends on the temperature difference between the air and the underlying surface and the moisture content of the air. These mists can develop both over the sea and over the land and cover a huge area, in some cases up to hundreds of thousands of square kilometers.

Front mists are formed near atmospheric fronts and move with them. Saturation of air by steam occurs due to the evaporation of precipitation falling in the front area. In strengthening the mists before the fronts, some role plays here observed a drop in atmospheric pressure, which creates a slight decrease in the adiabatic temperature.

Precipitation is water droplets or ice crystals falling from the clouds to the earth's surface. The value of precipitation is determined in the water layer millimeters (mm). The intensity of the precipitation expressed in mm/min. Both these units are practical.

By the nature of precipitation, it can be:

- *Widespread*, falling from Nimbostratus and high tier clouds in the form of medium-sized drops of rain, or in the form of snowflakes.
- *Rainfall*, falling from Cumulonimbus clouds in the form of large drops of rain, snowflakes or hail.
- *Drizzling*, dropping out of Stratus and Stratocumulus clouds in the form of very fine drops of rain.

Hail usually falls in the warm weather and is usually accompanied by thunder, lightning, and rain. Hail is formed when raindrops, on the path to land, pass through the ground layer of cold air and freeze. The size of hailstones can grow if small hailstones, caught up by strong rising air currents, rise up to a level where the overcooled liquid droplets are located. Water envelops hailstones and freezes them, thus increasing the initial size of the hailstones. Lifting and lowering of hail can occur repeatedly. This results in increasing the number of hailstones ice layers. Most hailstones are made up of alternating layers of snow and ice. This occurs when the hailstone rising consistently pass through levels, where the droplet and crystalline particles are located. Increasing the size of hailstones increases their weight so that they overcome the force of the rising air currents and fall to the ground. In this way, hailstones can reach a diameter of 8-9 cm and weight up to 0.5 kg.

Solar Radiation, Long-Wave Radiation of the Earth and Atmosphere

Today the value $S_0 = 1376$ W/m^2 is accepted as the average amount of radiant solar energy that passes through a unit surface orthogonal to rays of the Sun and located at an average distance of the Earth from the Sun (the solar constant).

If we neglect the influence of the atmosphere, the total energy E_s, received by the Earth from the Sun at a time, will be equal to

$$E_s = \pi R^2 S_0, \tag{1.1.14}$$

where R is the radius of the Earth. Since the surface area of the Earth (for simplicity we present it in the form of sphere) is $4\pi R^2$, the average amount of energy e_s obtained by the area unit of the Earth's surface per unit time is of the order

$$e_s = \frac{S_0}{4} = 344\, w / m^2 \tag{1.1.15}$$

If the Earth's axis is not tilted, the average flow of the incoming energy would be changed from the value of S_0/π at the equator to zero at the poles. However, the tilt of Earth's axis (23.5°) leads to seasonal changes in the distribution of the incoming flow of energy.

Not all the energy incident on Earth is absorbed. Part of it α is reflected or scattered, going back into space so that the actual energy absorbed is equal to the average flow

$$\overline{e_s} = \left(1-\alpha\right) e_s = \left(1-\alpha\right) S_0 / 4 \tag{1.1.16}$$

where α is the Earth's albedo.

If we take the average albedo of the Earth $\alpha = 0.3$, it follows from (1.1.16) that the amount of absorbed energy flux medium at all latitudes will be equal to 240 W/m^2, and from (1.1.15) and (1.1.16) it follows that the amount of reflected and scattered energy at all latitudes is about 100 W/m^2.

The total amount of energy produced by any horizontal section of the Earth's surface for different periods of time can be calculated by the formula

$$\frac{dE_s}{dt} = \left(1-\alpha\right) \frac{S_0}{\overline{r}^2} \sin\beta, \tag{1.1.17}$$

where $\overline{r} = h / H_s$ is the distance of the Earth from the Sun, expressed as a fraction of the average distance H_s.

The distribution of the theoretical amount of heat at the Earth's surface is characterized by the so-called solar or *solar climate*. However, these theoretical results are little consistent with the actual data, as they did not consider the effect of the atmosphere on the solar radiation flux.

In the atmosphere, there are three types of solar radiant energy flows:

- *Direct radiation* is that reaching the Earth directly from the solar disk in the form of a beam of parallel rays.
- *Scattered radiation* is part of the solar radiation that passes through the atmosphere and is scattered by molecules of atmospheric gasses and aerosols.
- *Reflected radiation* is solar radiation reflected by the Earth's surface, clouds etc.

In the cloudless sky of clear atmosphere, the basic processes leading to changes in the distribution of the downward flow of solar energy in the lower layers of the atmosphere are:

- Diurnal variation in height of the Sun above the horizon.
- Scattering, absorption, and reflection of radiation, mainly, by water vapor.
- Scattering and reflection of the downward flow of solar energy by underlying surface.

The path traversed by sunlight through the atmosphere is not the same at different heights of the Sun (see. Figure 3). The lower the height of the Sun, the longer path goes a sunbeam in the atmosphere. The unit of the path is adopted the optical mass $m = 1$, of atmosphere traversed by sunlight at the zenith position of the Sun (in the fall of rays at an angle of 90° to the horizontal surface of the Earth). If we do not take into account the curvature of the earth's surface, and the curvature of the beam path due to refraction in the atmosphere, the dependence of the mass m of the atmosphere on the Sun height above the horizon can be defined by the formula:

$$m\left(h\right) = \frac{dh}{dh_0} = \cos\beta, \qquad (1.1.18)$$

which shows that the optical mass of the atmosphere, depending on the Sun zenith distance, is equal to the ratio of the length of the beam path h in an oblique direction to its path h_0. in the vertical direction. For small values of m, zenith distance varies slowly with β and therefore the formula (1.1.18) allows to determine the mass of the atmosphere m with an accuracy of 5% within the interval of 25°≤β≤90°. For β<25° it should consider the curvature of the earth's surface and the beam path.

The longer the path of sunlight in the atmosphere, the more rays of shorter wavelengths (violet, blue, blue) are scattered and the more rays of long wavelength (red, orange) remain. Therefore, the Sun at the horizon has a reddish hue.

Observations show that of the entire incoming solar radiation 46% falls on the visible part of the spectrum, 47% is in the infrared, and 7% is in the ultraviolet at the upper limit of the atmosphere. When passing through the atmosphere of the spectral composition of solar radiation is changed. The ultraviolet radiation of wavelength less than 0.29 microns does not reach the Earth's surface. It is completely absorbed by ozone in the upper atmosphere. Mainly due to scattering, it is strongly attenuated the most short-wavelength region of the visible part of the spectrum (blue and violet rays) and less the long-wavelength region (orange and red rays). The infrared part of the spectrum has several areas of low energy associated with the absorption of its water vapor and carbon dioxide.

In the actual atmosphere, absorption and scattering of solar radiation depend on the amount of water vapor and aerosol particles in the air, i.e. from the transparency of the atmosphere. The transparency of the atmosphere is characterized by the coefficient of transparency, that shows what part of the solar radiation, which came to the top of the atmosphere, comes in the form of direct radiation to the Earth's surface from the Sun at its zenith position. In real terms, the transmittance ranges from 0.60-0.85. At the same transparency of the atmosphere, transparency coefficient χ varies for different wavelengths. Particularly, at a wavelength of 0.30 microns χ =0.36, and at a wavelength of 0.70 microns χ =0.97. If we denote by S the total flux of direct solar radiation at the Earth's surface, where the mass of the atmosphere in the direction of the Sun is equal to m, then averaged on a certain portion of the spectrum, limited by wavelength of λ_1 and λ_2, the transparency coefficient is expressed by

$$\chi_m = \sqrt[m]{S / S_0} \,. \tag{1.1.19}$$

A convenient unit of measurement of the optical properties of the atmosphere is the optical depth r, corresponding to the height z. For plane-parallel optical thickness of the atmosphere it is determined by the expression

$$r = \int_z^{\infty} \sigma_w\left(z\right) dz \,, \tag{1.1.20}$$

where $\sigma_w = \sigma_d + \sigma_a$ is bulk attenuation coefficient (m^{-1}), σ_d is bulk scattering coefficient, σ_a is bulk absorption coefficient.

The scattering and absorption of solar radiation in the atmosphere greatly depend on the presence and properties of cloud layers. Clouds are composed of liquid or solid spherical particles or other shapes, characterized by the size and refractive index. Usually, these polydisperse particles have some medium size that can be determined by various semi-empirical formulas. These particles are located either in the lower atmosphere or concentrated in limited layers. Vertical particle distribution and concentration change with altitude, usually are little known. Therefore, the atmosphere must be considered non-uniform. The optical thickness varies from 0.1 m^{-1} for a very clean atmosphere to 100 min^{-1} for the strong cloud.

A part of the solar radiation, entering the earth's surface, is absorbed by the surface for its heating and air adjacent to it. On the other hand, each portion of the earth's surface and each quantity of air having a temperature different from absolute zero emits thermal radiation. Radiant heat is a secondary radiation factor that plays an important role in the heat balance of the underlying surface and the atmosphere. The heat flux emitted by the earth's surface (*terrestrial radiation*), is directed to the atmosphere. The heat flux emitted by the atmosphere is directed in all directions. That part of it which reaches the earth's surface constitutes *counter radiation* of atmosphere, and that goes into space forming the *outgoing radiation of the atmosphere*. The Earth and atmospheric radiation, as well as solar radiation, is absorbed and reflected.

These different streams of radiant energy are the electromagnetic waves of different lengths and they differ in spectral composition. To estimate the total amount of heat and light, produced by the Sun for the Earth, there have recorded flows of solar radiation that reach the earth's surface. Their continuous spectrum lies in the range of 0.19 to 24 microns. Therefore, when considering the transfer of radiant heat energy from the solar spectrum, as a rule, there are three parts: ultraviolet (λ<40 microns), visible (40≤λ<0.76 microns) and infrared (0.76≤λ<2 microns). Thermal radiation of the earth's surface and the atmosphere is localized in the area of much longer wavelengths, practically, around 2≤λ≤40 microns. On the basis of this difference in spectral content, direct and diffuse solar radiation is referred as the short-wave, and thermal radiation from the Earth and atmosphere as the long-wave radiation (see Figure 14).

Figure 14. Driving the transformation of solar energy in the Earth-atmosphere system

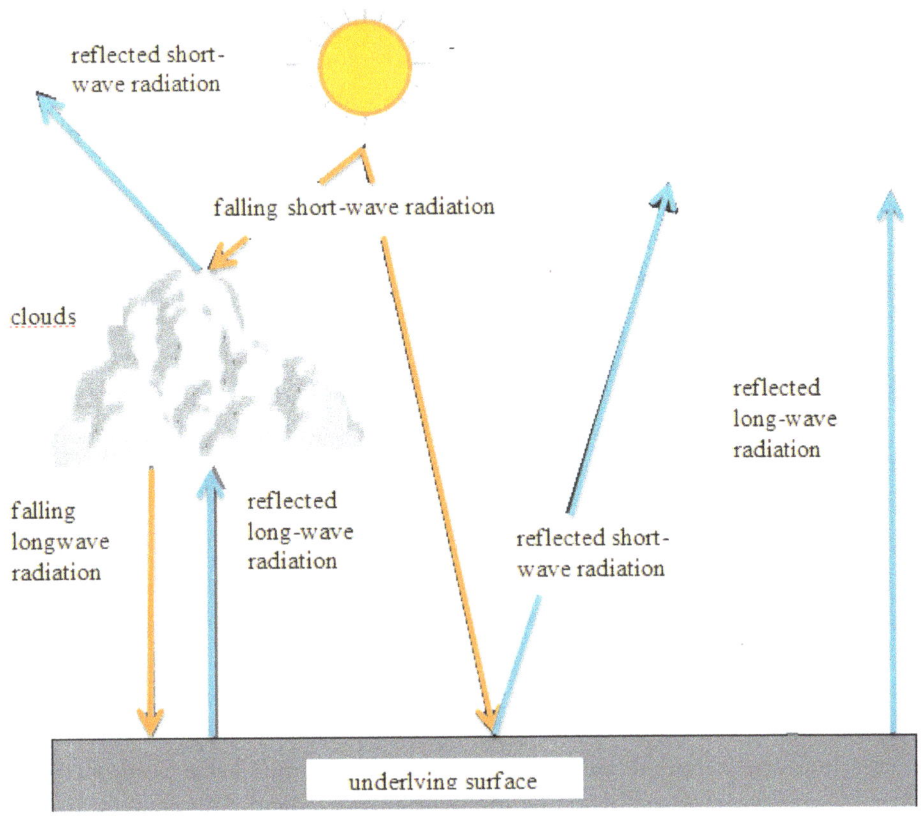

The atmosphere absorbs a comparatively small amount of short-wave radiation, mainly in the infrared part of the spectrum. This absorption is selective: different gasses absorb radiation in different spectral regions and in various degrees. Nitrogen absorbs radiation of short wavelengths in the ultraviolet part of the spectrum. The energy of solar radiation in this region of the spectrum is negligible, and therefore, the absorption of nitrogen has practically no effect on the intensity of solar radiation. Oxygen absorbs solar radiation more, but still very little, in two narrow portions of the visible spectrum and in the ultraviolet portion thereof.

Carbon dioxide and ozone are stronger absorbers of shortwave radiation. The content of carbon dioxide in the atmosphere is negligible, and therefore, the absorption caused by it is generally negligible. The ozone content in the air, even in the stratosphere, is very small, however, it so strongly absorbs ultraviolet radiation, that the amount of radiant energy decreases by a few percent. Thus, tropospheric air is almost transparent for short-wave solar radiation. As direct absorption of solar radiation in the troposphere are few, it may cause an increase in temperature only on the order of 0.5°C per day. The primary radiation absorber in the infrared region of the spectrum in the atmosphere is water vapor, suspended atmospheric aerosols (clouds and solid particles), concentrated in the troposphere, and especially, in the lower part thereof. Especially a lot of radiant energy are absorbed and scattered by the clouds. Therefore, in areas of the world, where for years there is a large number of days with cloudy weather, direct radiation is less than in the regions with the predominantly clear weather.

So, a parallel beam of solar energy a specific spectral composition, uniformly illuminating the upper boundary layer of the atmosphere, is changed in the Earth's atmosphere over the spectrum and is scattered in all directions, in accordance with the laws of absorption and scattering. At the same time, due to the vertical inhomogeneity of the atmosphere, as well as changes in the concentration of the absorbing substance in the way of the spread of radiation, the spectral and angular distribution of the latter depend on the altitude. Regular and random variation of these factors over time cause the corresponding variations of the radiation characteristics. Consequently, the field of short-wave radiation in the Earth's atmosphere has a complex spectral, angular, vertical, spatial, and temporal structure.

The radiant energy of the Sun (solar radiation) is the main and almost the only source of heat for the Earth's surface and its atmosphere. The radiant energy of the Sun is converted into heat partly in the atmosphere itself, but mostly at the Earth's surface. It is here to heat the upper layers of the soil and water, and from there through heat exchange to heat the air. The heated earth's surface and the atmosphere, in turn, emit invisible infrared radiation themselves. Giving this radiation into space, the earth's surface, and the atmosphere cool down. Heat flow from the Earth is negligible compared with the incoming stream from the Sun.

The solar radiation absorption and emission of atmospheric layers have the decisive influence on the heat exchange in the higher layers of the atmosphere. They increase the value and adiabatic temperature changes during the ascending and descending air movements caused by changes in atmospheric pressure.

The radiant energy of the Sun is also the energy source for almost all the natural processes in the atmosphere and on the surface of the Earth, and one of the major weather-factors. The amount of heat accumulated by the earth and atmosphere depends on the influx of solar radiation. Solar radiation heats the surface of land and oceans unevenly. Horizontal temperature changes in the Earth-atmosphere system creates a horizontal pressure gradient, which in turn causes movement of air masses. Even in the same zone of the latitude the solar radiation has different effects. The continents and oceans, forests, deserts, snow-covered and devoid of snow surface of the earth unevenly absorb and reflect solar radiation. The result is a circulation of various sizes, for example, seasonal winds between the continents and the oceans, seas and coasts, mountains and valleys, etc. Thus, the thermal energy of the sun is transformed into kinetic energy of atmospheric motions.

The Earth's surface reflects part of the incoming total radiation, and the rest is absorbed. If we neglect the influence of the atmosphere, the absorbed energy will cause the heating of the Earth's surface unless its radiation into space will be equal to the energy absorbed. The amount of energy E_b emitted by the Earth's surface per unit time is determined by the Stefan-Boltzmann law

$$E_b = \sigma T_b^4, \tag{1.1.21}$$

where T_b is surface temperature σ=5.7×10^{-8} W•m^{-2}•K^{-4} is Stefan-Boltzmann constant.

Calculations show that for the radiation is actually absorbed by the Earth when the assumptions made, the balance will occur when at the equator, the temperature reaches 270°K, the South Pole 150°K and at the North Pole 170°K. In fact, the Earth's surface is much warmer, and the contrast in temperature between the equator and the poles is much smaller. This difference between calculated and the actual surface temperature is due to the presence of the "greenhouse effect", the latent heat flux in the condensation and evaporation of moisture and the water surface of the Earth. The first physical factor in the achievement of the balance includes:

- Absorption of radiation by water vapor in the lower (70% of weight) layer of the atmosphere, as well as carbon dioxide and ozone at higher levels of the atmosphere.
- Reverse the long-wave radiation from the atmosphere.
- The total long-wave radiation of the underlying surface.

The amount of heat given off due to back long-wave radiation from the atmosphere is in an average of 400 cal/cm·day. The second factor requires no special explanation. Heat transfer caused by the evaporation and condensation processes results in the transfer of heat from the surface to the atmosphere in an average amount of about 120 cal/cm·day. The third physical factor affecting the energy balance is the transfer of heat from one area to another by atmospheric winds and ocean currents.

Thermal radiation Earth's surface and atmosphere have the same electromagnetic nature as light and other electromagnetic waves. In general, the three interconnected components can be distinguished in the process of heat transfer by radiation:

- **Radiation:** The volume of the medium heated to a certain temperature emits an electromagnetic field.
- **Dissemination:** The radiated electromagnetic field propagates in a medium apart from the selected volume, being reflected and scattered.
- **Absorption:** The electromagnetic field is absorbed by the medium, giving it an additional energy, part of which is transformed into thermal energy.

In the study of the processes of radiation, scattering and absorption of radiant energy use quantitative characteristics of both the radiation field and the interaction of the radiation field with the material environment in which the distribution of radiant energy occurs. The main ones are the mass absorption coefficient, scattering coefficient of mass, mass emissivity, and reflectance. Determination of these values is based on laws of Kirchhoff, Planck, Stefan-Boltzmann, Wien, and others.

Existing methods of calculating the radiation heat transfer process can be broken down into two broad groups: theoretically derived relationships and empirical formulas. In turn, theoretical methods include two possible approaches to describing the process of heat transfer by radiation: the phenomenological and statistical.

In the phenomenological approach, particular features of the environment and radiation process, as well as their interaction, are not considered. Instead, some general properties are used simulating a particular process, and the physical factors of the medium in which the process is played, are recorded in coefficients determined empirically. Classic examples are the phenomenological approach of (the laws) Newton, Fourier, and Fick, describing the process of friction, heat transfer and thermal conductivity of the impurity diffusion. The properties of the medium in these relationships are reflected using the coefficients of viscosity, thermal conductivity, and diffusion, respectively. In principle, such a consideration is stored in the description of the process of heat transfer by radiation with the aid of Maxwell's equations written for various physical media. But the electromagnetic processes of thermal radiation are random, so Maxwell's equations used to be regarded as some stochastic equations. It should be also noted that the system of Maxwell's equations describing the laws of behavior of a thermal electromagnetic field in the space filled with the medium (substance) is the incomplete system from a mathematical point of view. This system of equations must be supplemented by certain relations that consider the specific properties of the medium, the conditions on the emitting and absorbing surfaces, etc. not derived from the main system.

The objective of the statistical description of the radiation heat transfer process is to obtain a macroscopic or integral characteristics of the system, given its microscopic properties, i.e. the properties of its constituent particles. In this setting, the electromagnetic field is considered as a collection of particles (photons or light quanta), and the radiation heat transfer process is presented as a result of the interaction of these particles with the particles of the substance. This interaction is usually described using the Boltzmann equation written with respect to the specific frequency of the radiation intensity. His right-hand side contains the source-stock terms that describe the processes of absorption, emission, and scattering, which are functions of temperature. Consequently, for the Boltzmann equation is necessary to know the temperature distribution in the region of consideration, because only under this condition the right part of this equation becomes known. Since the temperature distribution is generally not known a priori, and is determined by the process of radiation transfer, it is necessary to supplement Boltzmann equation for total heat exchange by an equation with respect to temperature. In other words, the general problem of heat transfer in view of the radiation should be described by a system of equations.

Long-term monitoring shows that the average annual temperature of the earth's surface and atmosphere changes little anywhere on Earth from year to year. That is, the Earth is in thermal equilibrium: the arrival of heat is balanced by its loss. But as the Earth-atmosphere system receives heat by absorbing solar radiation and loses heat by its own radiation, the Earth-atmosphere system is in radiative equilibrium: the influx of radiation is balanced by the return of radiation into space.

Fundamentals of Atmospheric Thermodynamics

Thermodynamics of the atmosphere is part of dynamic meteorology that studies the changes in the internal state of a moving air parameters and the processes of transition of heat energy into mechanical energy and vice versa, associated with these changes.

The thermodynamics of the atmosphere use the well-known conclusions from the first and the second laws of the general thermodynamics being studied in physics courses.

The main parameters that characterize the physical state of the atmospheric air as a thermodynamic system are the density ρ, pressure p and the absolute temperature T of the pressure, which is determined by the relation (1.1.7) above. At absolute zero, i.e. at t=-273.15°°C, thermal motion of the molecules of an ideal gas is stopped. The absolute temperature T, as the total kinetic energy of the molecules characteristic of the movement, is always positive.

Although air is the gas mixture (see Table 1.1), dry air can be considered as a whole, because its composition is changed in the atmosphere very small compared to the change in the atmosphere of water vapor and various impurities suspended (aerosols). Under the conditions observed in the atmosphere, all the mechanical mixture of gasses forming the dry clean air may be viewed as an ideal gas. As it is known from general thermodynamics, pressure p_d, the absolute temperature T and the volume ϑ or the density ρ_d of an ideal gas for dry air are connected in the state equation

$$p_d = \rho_d RT, \tag{1.2.1}$$

where R is gas constant of dry air:

$$R = \frac{R_0}{m_d} = 287.04\text{J} \cdot \text{kg}^{-1} \cdot \text{K}^{-1},$$

where m_d is the molecular weight of dry air; R_0 is the universal gas constant that is the work of an extension of one mole of gas upon heating by 1 degree. Universal gas constant for dry air has the same value in the International System of Units $R_0 = 8314.36$ J kmol^{-1}·K^{-1}.

The atmosphere always contains a certain amount of water vapor in the mixture with dry air. The equation of state (1.2.1) also reliably holds for wet air, but the specific gas constant of wet air differs from the gas specific constant of dry air and, depending on the amount of water vapor contained in the air, takes different values. For water vapor the equation of state has the form:

$$e = \rho_s R_s T, \tag{1.2.2}$$

where e is water vapor pressure; ρ_s is the mass of water vapor per unit volume; R_s is the gas constant of water vapor.

The molecular mass of the water is $m_h = 18.016$, therefore,

$$R_s = \frac{R_0}{m_h} = 461.50\text{J} \cdot \text{kg}^{-1} \cdot \text{K}^{-1}.$$

According to Dalton's law, an ideal gas is distributed to other gases, not acting on it chemically, as in a vacuum, and the total pressure of a mixture of gases is the sum of the partial pressures produced by individual gases, and the density of the mixture is the sum of the values of the density of these gases

$$p = p_d + e, \rho = \rho_d + \rho_s \tag{1.2.3}$$

Combining the two equations term by term (1.2.1) and (1.2.2), we obtain the equation of state of wet air

$$p = p_d + e = \left(\rho_d R + \rho_s R_s\right) T = \rho R T \left(1 + 0.6078q\right) = \rho R T_v,$$

where $T_v \equiv T(1+0.6078q)$ is virtual temperature, which dry air must have at a predetermined pressure, and its density equal to the density of wet air density under the assumption that it obeys the ideal gas law;.q is specific humidity which is determined by (1.1.11) or (1.1.12).

Virtual temperature of wet air T_v is higher of its actual temperature, so from the equation of state $\rho = p/RT_v$ it follows that the wet air is lighter than dry, at the same pressure and temperature.

Polytrophic Changes in Atmospheric Air

The basis of the thermodynamics theory is the first law of thermodynamics, which is experimentally established the initial position, which expresses the general law of conservation of energy applied to the thermodynamic processes. The essence of the law of conservation of energy is that for all the processes occurring in nature, energy is not destroyed and does not arise out of nothing, but only converted from one form to another.

According to the first law of thermodynamics, heat dQ, put into a unit mass of air, is spent to increase the internal heat energy dE and the work $pd\vartheta$ which makes the air overcoming pressure:

$$dQ = dE + pd\vartheta ,$$

where ϑ is the specific volume of air.

Dry and wet unsaturated air can be regarded as an ideal gas, the internal energy of which is proportional to absolute temperature, $E = c_v T + const$, so

$$dE = c_v dT ,$$

where c_v is the specific heat at constant volume.

Thus, the equation of the first law of thermodynamics for air, considered as an ideal gas, takes the form

$$dQ = c_v dT + pd\vartheta . \tag{1.2.4}$$

For practical purposes, it is advisable to convert the equation so that its right side consisted only of the measured values. For this purpose, we replace the air density in the equation of state (1.2.1) by the specific volume of $\vartheta p = RT$. Differential of the last expression allows to replace the last term $pd\vartheta$ in (1.2.4) by $RdT - \vartheta dp$. This results in:

$$dQ = \left(c_v - R\right) dT - \vartheta dp . \tag{1.2.5}$$

At constant pressure, i.e. when $dp = 0$, it follows from (1.2.5) that

$$dQ = \left(c_v - R\right) dT .$$

$$dQ = \left(c_v - R\right) dT .$$

When isobaric process occurs at constant pressure, the heat consumption dQ is expressed by the formula

$$dQ = c_p dT ,$$

where c_p is a specific heat at constant pressure.

It follows from a comparison of the last two formulas:

$$R = c_p - c_v . \tag{1.2.6}$$

We substitute $c_v - R = c_p$ in equation (1.2.5), replacing the specific volume in his last term of the right side by the expression from the equation of state of $\vartheta = RT / p$. As a result, we obtain the equation of the first law of thermodynamics in the form which is mostly used in meteorology

$$dQ = c_p dT - \frac{RT}{p} dp. \qquad (1.2.7)$$

Moving air mass from one thermodynamic state to another can be done in different ways, depending on the relationship of pressure p, density ρ and temperature T. The simplest process is *a polytrophic process*, which occurs at a constant value of the specific heat c. Consequently, the heat gain is proportional to the increment of temperature for polytrophic processes

$$dQ = cdT. \qquad (1.2.8)$$

From the equality of the left sides of equations (1.2.7) and (1.2.8), it follows that

$$\left(c_p - c\right) dT = \frac{RT}{p} dp.$$

Separating the variables, we obtain

$$\frac{dT}{T} = \frac{R}{c_p - c} \frac{dp}{p}.$$

Integrating between T_0 to T_1 and from P_0 to P_1, we find:

$$\ln \frac{T_1}{T_0} = \frac{R}{c_p - c} \ln \frac{p_1}{p_0}.$$

Potentiating this expression, we obtain the required equation of polytrophic processes

$$\frac{T_1}{T_0} = \left(\frac{p_1}{p_0}\right)^{\frac{R}{c_p - c}} \qquad (1.2.9)$$

Let have the equation of state for cases characterized by parameters P_0, T_0, $\vartheta_0 = 1 / \rho_0$ and P_1, T_1, $\vartheta_1 = 1 / \rho_1$:

$$T_0 = \frac{p_0 \vartheta_0}{R}, T_1 = \frac{p_1 \vartheta_1}{R}.$$

Hence, we obtain:

$$\frac{T_1}{T_0} = \frac{p_1\vartheta_1}{p_0\vartheta_0},$$

and instead of (1.2.9) we have

$$\frac{p_1\vartheta_1}{p_0\vartheta_0} = \left(\frac{p_1}{p_0}\right)^{\frac{R}{c_p - c}}$$

or

$$\frac{\vartheta_1}{\vartheta_0} = \left(\frac{p_1}{p_0}\right)^{\frac{R}{c_p - c} - 1}.$$

Taking into account (1.2.6), we obtain from the last equation:

$$\frac{\vartheta_1}{\vartheta_0} = \left(\frac{p_1}{p_0}\right)^{\frac{c - c_v}{c_p - c}} = \left(\frac{p_0}{p_1}\right)^{\frac{c_v - c}{c_p - c}} = \left(\frac{p_1}{p_0}\right)^{\frac{1}{k}}.$$

Hence, we find the equation of polytrophic:

$$p_0\vartheta_0^k = p_1\vartheta_1^k = \ldots p\vartheta^k = \text{const}, \tag{1.2.10}$$

where k is the so-called polytrophic index.

Polytrophic equation (1.2.10) shows that simple process changes of the thermodynamic state of the air are special cases of the polytropic process. We note the following simple special cases:

1. $c=c_p$: This is a polytrophic process that is characterized by the polytrophic exponent $k = 0$; it follows from (1.2.10) that $p_0 = p_1 = \ldots p = \text{const}$, i.e., in this case there is an *isobaric process.*
2. $c=\infty$: This is a polytrophic process in which all the energy obtained in the process of heat transfer is to spend on the work, and internal energy and temperature remain unchanged. Such a process is determined by the law of Boyle and called *isothermal process.*
3. $c=c_v$: This is a polytrophic process in which all the energy is obtained in the form of heat, and it is only to increase the internal energy of the air $dQ = c_v dT$, i.e., there is *isosteric process* occurring at constant volume.
4. $c=0$: This is the polytrophic process of changing the thermodynamic state of a certain mass of air occurring without input and output of heat, when the air temperature change occur does not at the

process of heat transfer, but through the work of expansion or contraction of air; the process is called *adiabatic process.*

Thermodynamic Properties of Dry Air

When the vertical movement of air in the atmosphere, due to the large pressure differentials, the work on the expansion and contraction of air is much greater than the input of outside heat, that can be neglected. Therefore, as a first approximation, we can assume that the change in the thermodynamic state of the moving air is adiabatic.

In adiabatic processes, heat flow is equal to zero, and the equation of the first law of thermodynamics becomes

$$c_p dT = \frac{RT}{p} dp \cdot \tag{1.2.11}$$

Integrating between T_0 to T_1 and from p_0 to p_1, we obtain the Poisson equation for adiabatic processes:

$$\frac{T_1}{T_0} = \left(\frac{p_1}{p_0}\right)^{\frac{R}{c_p}} . \tag{1.2.12}$$

Using the equation (1.2.11) let us figure out how the temperature of the selected volume of dry air is to change when the external pressure p changes, i.e. when moving the volume vertically.

Designate the temperature and pressure of the vertically moving the elementary volume of air T_i and p_i respectively, and the same values for ambient air T_e and p_e. The equation of the first law of thermodynamics for the adiabatic moving of air portion will be of the form

$$c_p dT_i = \frac{RT_i}{p_i} dp_i \cdot$$

We assume that the movement of air in the atmosphere is relatively slow and the pressure in the air moving portion at any time manages to equalize with the pressure in the surrounding atmosphere at that level. Given this condition quasi-static processes and using the equations of statics and state, we will have

$$\frac{dp_i}{p_i} = \frac{dp_e}{p_e} = -\frac{g}{RT_e} dz . \tag{1.2.13}$$

Then the equation of the first law of thermodynamics for the adiabatic process can be rewritten now as

$$c_p dT_i = -\frac{T_i}{T_e} gdz \;..$$

For small vertical air particle velocities, one can put $T_i/T_e \approx 1$, and so we have

$$\frac{dT_i}{dz} = -\frac{g}{c_p} = -\gamma_a, \tag{1.2.14}$$

The value γ_a in meteorology is called *vertical adiabatic gradient.* For the dry air, which is subject to the laws of an ideal fluid, $\gamma_a = g/c_p \approx 10$ K/km.

It follows from the equation (1.2.12) of the adiabatic processes, that in the vertical movements due to changes in pressure, the temperature of vertically moving air also changes. Therefore, for comparing the thermal state of the different air masses, there used in meteorology the concept of *potential temperature.*

Potential temperature is the temperature, which takes dry air if it would be reconciled to an adiabatic standard pressure p_{00}. If in (1.2.12) one takes $p_0 = p_{00} = 1000$ hPa as the standard pressure, then the temperature T_0corresponding to that pressure will be the potential temperature θ and the Poisson equation (1.2.12) is converted to the form:

$$\frac{T_1}{\theta} = \left(\frac{p_1}{p_{00}}\right)^{\frac{R}{c_p}}. \tag{1.2.15}$$

From this, the expression for the potential temperature can be derived

$$\theta = T\left(\frac{1000}{p}\right)^{\frac{R}{c_p}}. \tag{1.2.16}$$

It is easy to show that the potential temperature does not change at the dry adiabatic processes, and it remains constant when raising or lowering of the air mass. Indeed, the taking logarithm of the expression (1.2.16) and then differentiating the result, we find

$$\frac{d\theta}{\theta} = \frac{1}{c_p T}\left(c_p dT - \frac{RT}{p} dp\right).$$

By comparing the expression in brackets right side of this equation with the equation of the first law of thermodynamics for an adiabatic process (1.2.11), we obtain

$$\frac{d\theta}{\theta} = 0.$$

Consequently, in an adiabatic process ($dQ = 0$), the potential temperature does not change.

The Thermodynamic Properties of Moist Air

We show that thermodynamic processes in the wet unsaturated air do not quantitatively different from the processes in the dry air. If the mass unit of wet air containing steam q gets some heat dQ, then it is consumed for the work $pd\vartheta$ of expansion and the change in internal energy of dry air

$$dQ_d = c_p\left(1-q\right)dT - RT\left(1-q\right)\frac{d\left(p-e\right)}{p-e}, \tag{1.2.17}$$

as well as water vapor

$$dQ_v = c_{ps}qdT - R_vTq\frac{de}{e}, \tag{1.2.18}$$

where $c_{ps} = 1.83c_p$ is specific heat at constant pressure, and $R_v = 1.6R$ is gas constant for water vapor.

Consider the case when the air rises without the exchange of moisture, i.e. q = const. Then from equation (1.1.11), we have:

$$\frac{dq}{q} = \frac{de}{e} - \frac{d\left(p-e\right)}{p-e} = 0$$

or

$$\frac{de}{e} = \frac{d\left(p-e\right)}{p-e} = \frac{dp}{p},$$

as $p \gg e$.

Regarding the latter, let us define the heat dQ got by unit mass of moist air, using expressions (1.2.17) and (1.2.18)

$$dQ = dQ_d + dQ_v = c_p\left(1+0.83q\right)dT - RT\left(1+0.6q\right)\frac{dp}{p}.$$

Remaining within the framework of the adiabatic process in which $dQ = 0$, we find from the last equation:

$$\frac{dT}{T} = \frac{R}{c_p}\frac{1+0.6q}{1+0.83q}\frac{dp}{p}. \tag{1.2.19}$$

Replacing dp / p in the last equation by $-(g/RT)dz$, using a static equation (1.2.13), we have in result:

$$\frac{dT}{dz} = -\frac{g}{c_p}\frac{1+0.6q}{1+0.83q}.$$

Since in the atmosphere $q \sim 10^2$ g/g, one can assume with sufficient degree of accuracy, that $(1+0.6q)/(1+0.83q) \approx 1$. Therefore, the adiabatic temperature gradient of unsaturated moist air does not differ from the adiabatic gradient dry air.

Consider the change of thermodynamic state of the adiabatically rising moist air. Since the process is adiabatic, the ascent is accompanied by a decrease in temperature, and at a certain altitude, water vapor contained in the rising air becomes elastically saturated. Define the level at which the rising air reaches saturation. The height h_k of the condensation level is of interest since it is at a height close to the lower boundary of the convective clouds.

Obviously, at the level of condensation h_k the temperature $T(h_k)$ of rising air is equal to the dew point temperature $\tau(h_k)$. The dew point $\tau(z)$ in the rising air changes due to changes in water vapor elasticity. At the same time, it should be noted that the specific humidity q in the rising air will remain constant as long as the air reaches the saturation state and while no condensation of water vapor occurs.

Expand in Taylor series the functions $T(z)$ and $\tau(z)$ in a neighborhood of h_k, confining ourselves by the terms of the order $O(h_k)$,

$$T(z) = T_0 + h_k \frac{dT}{dz} + \frac{h_k^2}{2}\frac{d^2T}{dz^2} + \ldots + \frac{h_k^n}{n!}\frac{d^nT}{dz^n},$$

$$\tau(z) = \tau_0 + h_k \frac{d\tau}{dz} + \frac{h_k^2}{2}\frac{d^2\tau}{dz^2} + \ldots + \frac{h_k^n}{n!}\frac{d^n\tau}{dz^n}.$$

We use in these expressions the equation $T(z) = \tau(z)$ at the level h_k. This results in:

$$T_0 + h_k \frac{dT}{dz} = \tau_0 + h_k \frac{d\tau}{dz},$$

from where we find the height h_k of the condensation level

$$h_k = \frac{T_0 - \tau_0}{\dfrac{d\tau}{dz} - \dfrac{dT}{dz}}. \qquad (1.2.20)$$

From (1.2.20) it follows that by introducing the average vertical gradients of temperature and dew point, it is possible to obtain an expression for calculating the level of condensation on the surface data. The process of adiabatic air rising to the condensation level allows the use of the formula of dry adiabatic gradient $dT / dz = -g / c_p$. Since the dew point temperature $\tau(z)$ vapor elasticity is equal to its maximum value E, then the equation for the specific humidity q takes the form:

$$q = 0.62197 \frac{E}{p - E}. \tag{1.2.21}$$

Recalling that the specific humidity q in the air rising up to a height h_k remains constant, applying logarithm and then differentiation of (1.2.21) provides:

$$\frac{1}{E}\frac{dE}{d\tau}\frac{d\tau}{dz} = \frac{1}{p}\frac{dp}{dz} = -\frac{g}{RT}.$$

From this, we have

$$\frac{d\tau}{dz} = -\frac{g}{RT}\left(\frac{1}{E}\frac{dE}{d\tau}\right)^{-1}. \tag{1.2.22}$$

We use Clausius-Clapeyron formula in the expression (1.2.22)

$$\frac{1}{E}\frac{dE}{d\tau} = \frac{L}{R_s \tau^2},$$

and the equation (1.2.20) takes the form:

$$h_k = \frac{T_0 - \tau_0}{\frac{g}{c_p}\left(1 - 1.6\frac{c_p \tau^2}{TL}\right)}. \tag{1.2.23}$$

Substituting the numerical values of member variables of the expression (1.2.23), we obtain the formula of Ferrel

$$h_k = 122\left(T_0 - \tau_0\right). \tag{1.2.24}$$

We now consider the change in the thermodynamic state of rising air above the condensation level. In this case, the air becomes saturated with water vapor. Its rising and cooling are accompanied by condensation of water vapor and the release of latent heat of condensation that slow-down individual temperature drop caused by the adiabatic expansion of rising air. Determine the magnitude by which the temperature of the elementary volume of wet saturated air is changed during adiabatic vertically moving at a unit distance.

The specific humidity q is expressed through the maximum water vapor pressure E using the formula (1.2.21). Repeating the process above of logarithm and differentiation of expression (1.2.21), we obtain

$$\frac{1}{q}\frac{dq}{dz} = \frac{1}{E}\frac{dE}{dz} - \frac{1}{p}\frac{dp}{dz} = \frac{1}{E}\frac{dE}{dT}\frac{dT}{dz} + \frac{g}{RT}. \tag{1.2.25}$$

Releasing the latent heat of condensation of water vapor will be exposed to the input of the heat dQ to the elementary volume of adiabatically rising air. The absolute value of this input is equal to $dQ = Ldq$. Herewith dQ (heat input) and dT (change in temperature) will have the opposite signs, as while the temperature increases, the evaporation occurs and latent heat is absorbed. And while the temperature decreases, the steam is condensed and the latent heat released. Consequently, the equation of the first law of thermodynamics for vertically moving portion of the humid air has the form:

$$c_p dT - \frac{RT}{p}dp + Ldq = 0. \tag{1.2.26}$$

We consider the quasi-static condition (1.2.13) in this equation and rewrite it in the form

$$dT = -\frac{g}{c_p}dz - \frac{L}{c_p}dq.$$

Introducing the notation of wet adiabatic gradient $\gamma_{ha}=-dT/dz$, we get:

$$\gamma_{ha} = -\frac{dT}{dz} = \frac{g}{c_p} + \frac{L}{c_p}\frac{dq}{dz}. \tag{1.2.27}$$

Substituting the amount of condensed water in this equation, according to (1.2.25), we find

$$-\frac{dT}{dz} = \frac{g}{c_p} + \frac{Lq}{c_p}\frac{1}{E}\frac{dE}{dT}\frac{dT}{dz} + \frac{Lq}{c_p}\frac{g}{RT},$$

or

$$\gamma_{ha} = -\frac{dT}{dz} = \frac{g}{c_p}\frac{1+\dfrac{Lq}{RT}}{1+\dfrac{Lq}{c_p}\dfrac{1}{E}\dfrac{dE}{dT}}. \tag{1.2.28}$$

The value of wet adiabatic gradient is dependent on vapor elasticity in the atmosphere and, hence, on temperature. Estimates show that the numerator and denominator of the right side of (1.2.28) have different numerical values depending on the temperature and greatly increase with the growing the latter. However, it should be noted that the numerical values of the numerator are always smaller than the corresponding values of the denominator. γ_{ha} at high temperature is small, and as the temperature decreases γ_{ha} increases and approaches the dry adiabatic gradient.

The vertical movement of air in the atmosphere plays an important role in the formation of weather, especially in the formation of clouds and precipitation, i.e. the weather phenomena that are of considerable interest.

One of the main causes of the development of vertical motions in the atmosphere is a horizontal temperature gradient that is the difference between the temperature of the moving elementary volume of air and the temperature of surrounding air. With this temperature difference, there are buoyancy forces that give positive or negative vertical acceleration to the elementary volume of air moving.

Let β is the coefficient of cubic expansion that for an ideal gas is $\beta=1/T$, and $\delta T=T_i-T_e$ is the difference between the temperature T_i of the particles in a moving elementary volume of air and the temperature T_e of the surrounding particles of air. Provided $T_i\neq T_e$ and therefore $\rho_i\neq\rho_e$, the elementary volume of air begins to shift by the force

$$F_a = g\left(\rho_e - \rho_i\right), \tag{1.2.29}$$

acquiring the vertical acceleration

$$\frac{dw}{dt} = \frac{d^2z}{dt^2} = g\frac{\rho_e - \rho_i}{\rho_i} = g\left(\frac{\rho_e}{\rho_i} - 1\right). \tag{1.2.30}$$

From the Poisson equation for the adiabatic processes (1.2.12), one concludes that the $\rho_e/\rho_i=T_i/T_e$, so the equation (1.2.30) can be rewritten as:

$$\frac{d^2z}{dt^2} = g\left(1 - \frac{T_i}{T_e}\right) = g\frac{T_i - T_e}{T_e} \tag{1.2.31}$$

It follows from (1.2.31) that if an elementary volume of air at all levels is warmer than the surrounding atmosphere $T_i > T_e$, then the vertical acceleration is positive. State of the atmosphere in which an elementary volume of air, starting vertically, is accelerated in the same direction, seeking to remove the particle from the initial level, is called an *unstable state* characterized by the development of convection.

If the elementary volume of air at the higher levels becomes colder than the surrounding atmosphere, $T_i < T_e$, then the vertical acceleration is negative. Consequently, an elementary volume of air, which started the vertical offset, is accelerated in the opposite direction of its movement, which tends to return this volume to the original level. This state of the atmosphere is called a *steady state*.

Finally, when $T_i = T_e$, vertical air mass movement depends only on the initial velocity and the vertical acceleration is zero. This state of the atmosphere is called the *indifferent state*.

If the temperature T_{i0} of elemental moving volume of air at baseline is equal to the temperature T_{e0} of ambient air, then after displacement of elementary volume vertically by a distance Δz, the temperature

T_i of considered elementary volume and the temperature T_e of ambient air at a height Δz can be determined using the Taylor series:

$$T_i = T_{i0} + \Delta z\left(\frac{dT}{dz}\right)_i + \frac{\Delta z^2}{2}\left(\frac{d^2T}{dz^2}\right)_i + \ldots + \frac{\Delta z^n}{n!}\left(\frac{d^nT}{dz^n}\right)_i,$$

$$T_e = T_{e0} + \Delta z\left(\frac{dT}{dz}\right)_e + \frac{\Delta z^2}{2}\left(\frac{d^2T}{dz^2}\right)_e + \ldots + \frac{\Delta z^n}{n!}\left(\frac{d^nT}{dz^n}\right)_e.$$

Leaving the terms of order $O[\Delta z]$, we substitute these values of T_i and T_e in (1.2.31). Given equality $T_{i0} = T_{e0}$, we obtain the expression

$$\frac{d^2z}{dt^2} = \frac{g}{T_e}\left[\left(\frac{dT}{dz}\right)_i - \left(\frac{dT}{dz}\right)_e\right]\Delta z,$$

in which we introduce the notation $(dT/dz)_i = -\gamma_a$ for adiabatically transported an elementary volume of air and $(dT/dz)_e = -\gamma$ for ambient air with an arbitrary temperature gradient. This result in:

$$\frac{d^2z}{dt^2} = \frac{g}{T_e}\left(\gamma - \gamma_a\right)\Delta z. \tag{1.2.32}$$

This implies the following criteria vertical stability of the atmosphere:

1. $\gamma>\gamma_a$ that means dry unstable stratification.
2. $\gamma<\gamma_a$ that means dry stable stratification.
3. $\gamma=\gamma_a$ that means dry indifferent stratification.

If we consider the movement of particles in the air, saturated with water vapor, then we need to take the moist-adiabatic gradient γ_{av} in the vertical acceleration formulas instead of dry adiabatic gradient γ_a.

Generally, $T_{i0} \neq T_{e0}$, i.e. the temperature of the moving elementary volume of air at baseline differs from the ambient temperature at this level, and a vertical acceleration, in accordance with formula (1.2.31), is equal to

$$\frac{d^2z}{dt^2} = g\frac{T_{i0} - T_{e0}}{T_e} + g\frac{\gamma - \gamma_a}{T_e}\Delta z. \tag{1.2.33}$$

The height at which the vertical acceleration is zero and the rise of an elementary volume of air stops is called the *level of convection*. Typically, the level of convection is located near the upper boundary of convective clouds.

Taking the vertical acceleration is zero in (1.2.33), we find the height h of the level of convection

$$h = \Delta z = \frac{T_{i0} - T_{e0}}{\gamma_a - \gamma}. \tag{1.2.34}$$

To determine the relationship between the stability of the atmosphere and changes with the height of the potential temperature θ we use the expression (1.2.16) and the procedure used earlier for logarithm and subsequent differentiation in z. This results in:

$$\frac{1}{\theta}\frac{d\theta}{dz} = \frac{1}{T}\frac{dT}{dz} - \frac{R}{c_p}\frac{1}{p}\frac{dp}{dz}.$$

Substituting the value dp / dz in this equation by the value $-(gp) / (RT)$ from the equation of statics, we will have

$$\frac{d\theta}{dz} = \frac{\theta}{T}\left(\frac{dT}{dz} + \frac{g}{c_p}\right). \tag{1.2.35}$$

Since $g/c_p=\gamma_a$, and $dT/dz=-\gamma$, the equation (1.2.35) is converted to the form

$$\frac{d\theta}{dz} = \frac{\theta}{T}\left(\gamma_a - \gamma\right), \tag{1.2.36}$$

and, taking into account (1.2.32), to the form

$$\frac{d^2 z}{dt^2} = -\frac{g}{T}\frac{d\theta}{dz}\Delta z. \tag{1.2.37}$$

At stable state $\gamma_a>\gamma$, $d\theta/dz>0$ and $d^2z/dt^2<0$; otherwise, at unstable state $\gamma_a<\gamma$, $d\theta/dz<0$ and $d^2z/dt^2>0$; at indifferent state $\gamma_a=\gamma$, $d\theta/dz=0$ and $d^2z/dt^2=0$.

Since the atmosphere is stable stratified in the average of its condition, the potential temperature increases with altitude.

From the foregoing, it follows that the stratification of the air mass is characterized by a vertical temperature gradient in the atmosphere. The greater the temperature difference between the elementary rising volume of air and the environment, the greater is the acceleration d^2z / dt^2. However, the vertical temperature gradient in the atmospheric layers can vary significantly with altitude. Therefore, a more user-friendly feature of the stable state of air stratified vertically serves a general power resource of instability, i.e. the work that a buoyancy force can make in the vertical rising mass unit of air.

Elementary work to make by the buoyancy force (1.2.29), acting on a mass unit of air when moving it vertically on an infinitely small distance dz, is

$$dE = \frac{g}{T_e}\left(T_i - T_e\right)dz. \tag{1.2.38}$$

Using the equations of statics and state of the atmosphere, (1.2.38) can be transformed to the following form:

$$dE = -R\left(T_i - T_e\right)d\ln p.$$

It follows that complete the work that will make the buoyancy force in the layer from level of pressure p_1 to level of pressure p_2, is equal to

$$E = -R\int_{p_1}^{p_2}\left(T_i - T_e\right)d\ln p = R\int_{p_2}^{p_1}\left(T_i - T_e\right)d\ln p. \tag{1.2.39}$$

If the elementary volume of rising air is warmer than the environment ($T_i > T_e$), then the energy of instability will be positive, as $p_1 > p_2$. In this case, the state of the atmosphere is unstable. When the energy of instability is negative when elementary volume of rising air becomes cooler than the surrounding environment ($T_i > T_e$), the state of the atmosphere is stable. If the layer is stratified indifferently, then the energy of instability is zero.

Figure 15 shows an aerological diagram, where the X-axis corresponds to the temperature T and the Y-axis, in accordance with formula (2.1.39), corresponds to the value $1\text{n}\,p$, which decreases with altitude and determines the position of a point in the atmosphere. The curve of system state T_i in this figure characterizes the adiabatic temperature change with altitude in an elementary volume of rising air and stratification curve T_e characterizes the vertical distribution of temperature in the atmosphere depending on the $1\text{n}\,p$. According to formula (1.2.39), instability energy is proportional to the area enclosed between the curves and contours of $1\text{n}\,p_1$ and $1\text{n}\,p_2$.

The sign of instability energy is determined in the diagram considering the relative position of the curves status and stratification. If the state curve (adiabatic; dry is below the condensation level, wet is above it) is to the right of the curve of stratification (as shown in Figure 15a and in the upper part of

Figure 15. Aerological diagram

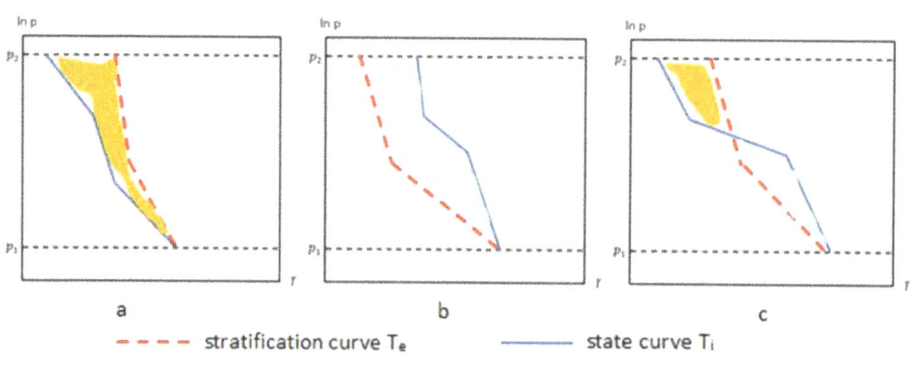

Fig..15c), then the energy of instability will be positive. If the state curve is to the left of stratification curve, as shown in Figure 15b and in the bottom of Figure 15c, the instability energy is negative.

In fact, with the development of convection, rising of some elementary volumes of air is accompanied by a compensative lowering of other elementary volumes and the change in their thermodynamic state. An example is the development of cumulus clouds, in which the powerful upward air movement in the central part is accompanied by a downward movement on the periphery.

Instability energy is released and transformed into the work in turbulence and convection processes, and to a certain extent, in front upward slip processes. For tropical cyclones, it seems to be the primary source of their kinetic energy.

For the convenience of graphic constructions, diagram letterheads have marked by families of following contour lines: isobars, isotherms, dry and wet adiabatic curves, and isograms, i.e. contour isolines of maximum specific humidity. An example of such diagram is shown in Figure 16.

Figure 16. An example of a standard diagram

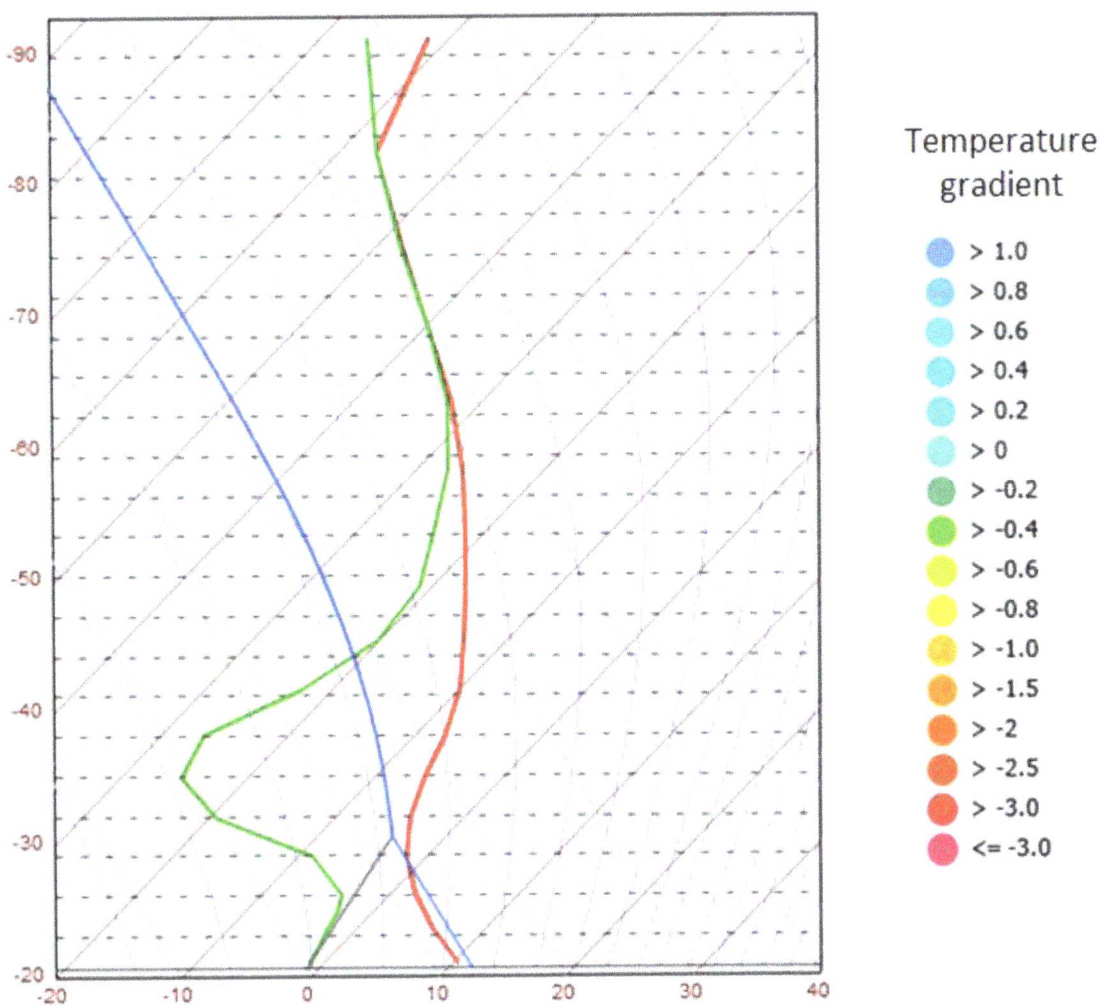

**For a more accurate representation of this figure, please see the electronic version.*

On results of the aerologic upper-air sounding of the atmosphere, there are applied stratification curve and the curve of dew points on the letterhead; the latter represents also the change in specific humidity with height.

The state curve is carried out from the starting point at the stratification curve to the level of condensation on the dry adiabatic, and above – by the wet adiabatic. If there is an inversion of temperature or isothermal in the surface layer of air, then the starting point on the curve of stratification is adopted an upper limit of surface inversion or isothermal. The pressure at the condensation level is determined by the position of the point of intersection of the dry adiabatic passing through the initial point on the curve of stratification, and isogram passing through its corresponding starting point on the curve of dew points.

SUMMARY

The chapter introduces the subject of weather forecasting by giving basic concepts of the science of meteorology, namely, astronomical data, basic information about the composition and structure of the atmosphere, determination of the main meteorological variables, solar radiation, longwave radiation of the Earth and atmosphere. There are given fundamentals of atmospheric thermodynamics and polytrophic changes in atmospheric air that include thermodynamic properties of dry and moist air and conditions for vertical air stability.

This foundation will be used in the following chapters to provide general characteristics of atmospheric processes, their classification and investigation.

Chapter 2
General Characteristics and Classification of Processes in the Earth–Atmosphere System

ABSTRACT

Atmospheric processes affect the heat flows coming from above, from space, and from below, from the earth's surface. The solar radiation that comes to Earth from outer space is the main source of energy of atmospheric processes. It is the radiant energy of the Sun that is converted into heat in the atmosphere and at the Earth's surface, kinetic energy, and other forms of energy. But the Sun's rays heat the earth's surface larger than the air directly, so between the earth's surface and the atmosphere, there is lively exchange of heat as well as moisture. The structure of the earth's surface and its relief are very important for these processes. The chapter presents the picture of heat and the moisture circulation in the atmosphere and gives physical basics of atmospheric general circulation including fundamentals of air mass circulation, local physiographic impact on the atmospheric air movement, in-mass atmospheric processes, and basic laws of pollution spreading in the atmosphere.

THE HEAT AND THE HYDROLOGIC CYCLE IN THE ATMOSPHERE

In addition to heat exchange by radiation, there is heat exchange by *conduction* between the Earth's surface and the atmosphere. Mixing of air in the vertical direction is extremely important for the heat transfer inside the atmosphere. The air that contacts directly with the earth's surface forms a thin laminar boundary sublayer and exchanges with the surface by heat due to molecular conduction. Above the laminar boundary sublayer inside surface atmospheric boundary layer, there acts another, more efficient heat transfer – by the turbulent thermal conductivity. Mixing air due to turbulence promotes the very rapid transfer of heat from one layer of the atmosphere to another. There are here the greatest vertical gradients of meteorological quantities that increase heat transfer from the ground into the air and vice versa. So, if there is a cooling (heating) air from the earth's surface, the turbulent flow continuously deliver the warm (cold) air from the upper layers to the place of cold (warmed) air. This stream will maintain

DOI: 10.4018/978-1-5225-2636-0.ch002

the temperature difference between the air and the surface and, therefore, support the process of heat transfer between the air and the surface. Changing the temperature in the upper layers of the atmospheric boundary layer is not so high (the vertical turbulent exchange reduces the vertical temperature gradient on the outer edge of the boundary layer), but this changing has spread to a more powerful layer of the atmosphere. Thus, the loss (acquisition) of heat by the Earth's surface would be greater than it would be in the absence of turbulence.

Because of diurnal variation, air temperature above the sea surface usually reaches a minimum of 2-3 hours before sunrise, and the maximum at 15-16 o'clock. On land surface, this process has less inertia and is shifted to 1-1.5 hours. This diurnal variation is characteristic only for stable weather. It is violated by heat exchange processes in the atmosphere, for example, when the warm air mass is replaced by the cold one. The night time temperatures may be even higher in such cases than at daytime. In overcast, the temperature difference is considerably less than under clear skies. During precipitation and after, the temperature is usually lowered.

The daily amplitude of temperature changes depends on the time of year and on the cloudiness. In the latter case, it decreases. The daily amplitude in the open region of sea or ocean is about 1.0-1.5°C and in the closed sea, it can reach 10-15°C. In areas with sharply continental climate, for example, in the desert, the daily fluctuations of temperature are maximal and can reach 30°C.

Since the heating value of the earth's surface decreases in an average from the tropics to the poles, the horizontal temperature gradient in the troposphere has the same direction. The dependence of pressure altitude field on the temperature field leads to the fact that the rotating globe holds west air transfer. Zonal movements of air masses over the underlying surface with dramatically different properties (continents or oceans) imply changes in their physical properties, i.e., the process of transformation. Therefore, the main features of the real circulation of the atmosphere in the zonal movement of air masses depend on the transformation caused by the distribution of continents and oceans, the surface of which are heated differently by the action of solar radiation.

Thus, the temperature in a restricted area may vary because of the continuous change of air at a given location, i.e. due to the entrance of air from the atmosphere of other places where it has a different temperature. These temperature changes associated with the influx into this place of new air masses from other parts of the globe are called *advective*. When air flows to this place at a higher temperature, then we talk about heat advection; if lower temperatures, then we talk about cold advection. Horizontal advective heat transfer between the southern and northern latitudes is carried by meridional transfer of air masses and is about 1019 Cal/day.

A significant part of the heat input to the Earth's surface passes in a latent form by expenditure on the evaporation of water. Then, with condensation of steam at atmosphere, this heat is used for heating the air.

In addition to heat exchange between the atmosphere and the earth's surface, there is a constant circulation of water or *hydrologic cycle*. Hydrologic cycle on the Earth is carried out by a continuous process of water movement in the atmosphere, the hydrosphere, and the Earth's crust, accompanied by phase transformations of water. The hydrologic cycle is composed mainly of water vapor, water vapor transport in a distance, its condensation (cloud formation), precipitation, seepage (infiltration) and drains of the precipitated water. The water evaporates into the atmosphere from the surface of the ocean and other bodies of water, moist soil, and vegetation in the form of steam for which it is spent a large amount of heat from the upper layers of soil and water. The steam – water in its gaseous state – is an important part of the air. Water vapor spreads upwards through the turbulent diffusion in an atmosphere and is transferred by air flows (advection) from one place to another. By lowering the temperature of the wet air due to adiabatic process or the heat output, water vapor condenses being transformed into a liquid or solid state. So, clouds and fogs are formed. In the condensation process, there is released into the atmosphere large quantities of latent heat. Partial condensation of water vapor leads to ground hydrometeors. Air currents also carry clouds. The water returns to the surface during precipitation from the clouds, then re-evaporates again, etc. In this single process, there is a continuous passage of water from the surface and back into the atmosphere. A part of the precipitated water flow goes into reservoirs through the land. Along with the heat exchange, the hydrologic cycle is one of the basic climate processes.

The total amount of water on the Earth in the present historical epoch remains constant; the average level of the oceans and the average moisture content of the atmosphere also do not experience significant changes. This means that rainfall and vaporization are equalized for the whole globe over a long period. The average height of the precipitation in the whole world for a year is equal to 1000 mm what corresponds to 511 000 km^3 of water. Herewith, 108 000 km^3 (21% of the total deposit) falls over land and 403,000 km^3 (79%) falls over the oceans. Almost half of all the rainfall occurs in the region between 20° North and 20° South. Average annual precipitation is the greatest at equatorial latitudes (1500-1800 mm). It is down in the subtropics to 800 mm and again increases to 900-1200 mm in temperate latitudes. It is dramatically reduced in the Polar Regions (up to 100 mm or less), thus, both polar zones account for only 4% of the precipitation. It is evaporated from the surface of World Ocean a water layer of the height of 1250 mm (450 000 km^3); 1120 mm of this amount is returned to the ocean in the form of precipitation and 130 mm by runoff from the land. On the land surface on average per year, it is evaporated the water layer of 410 mm (61 000 km3), the precipitation falls on the land of 720 mm. Hydrologic cycle components ratio may differ from average conditions for certain areas of the Earth's. There are areas where the amount of rainfall is much higher or lower than the evaporation.

Physical Foundation of Atmospheric General Circulation

The general circulation of the atmosphere is called the closed flow of air in the hemispheric or across the globe, leading to the latitudinal and meridional transport of momentum, energy, and water. Since the problems of the general circulation of the atmosphere are the main issues of climatology, synoptic and dynamic meteorology, their research is very important and at the same time a very complex problem of meteorology.

The results of long-term ground, upper air, missile, and satellite observations allow scientists to propose several schemes and theoretical models of the general circulation of the atmosphere. Given that these data are still significantly limited and allow only getting a general schematic representation of the transformation of air masses, the problem of creation of the theory of the general circulation of the atmosphere at the moment is still far from complete.

In a number of early proposed schemes of the general circulation of the atmosphere, it was taken into account only one reason for air movement – temperature difference of equator-pole system. Solar energy comes in different latitudes of the globe unevenly: more – at the equator and in the tropics, less – in the high latitudes (Figure 1).

The uneven distribution of heat in the atmosphere results in an uneven distribution of the atmospheric pressure and the pressure distribution influents the movement of air or air currents. These air flow, carried by the large-scale transfer of huge masses of cold and warm air along the meridians, is trying to equalize the temperature difference between high and low latitudes of the Earth. The atmospheric mass is heated at the equator, rises up and moves toward the poles and is replaced by colder masses, coming from the

Figure 1. The radiant energy of the Sun – the main source of heat for the Earth's surface: A and B - high latitudes, B - the equator and tropics

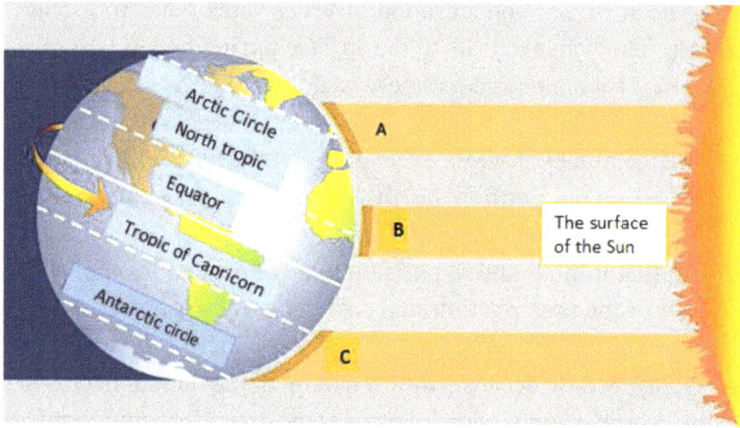

region of high latitudes, thereby creating near the equator a permanent low-pressure zone while cold air masses descend from the poles to the place of the warm air space. Thus, in different areas of the Earth, these oppositely directed flows together with the vertical – ascending and descending – movements would have to create a closed loop of air circulation. For example, in the tropics, it would be the air transport of the equator, its rise above the equator, moving to the top of the subtropical latitudes, where air lowering and reverse movement toward the equator. Likewise, conditions of general circulation in the other zones of the Earth were realized.

Such an idea is too simplistic. It is to a certain extent applicable to the circulation only in the tropics. As it is learned from the long-term observations, in extratropical latitudes, air moves from one latitude to another not only by vertical closed circulation but also with the help of the meridional components whose direction alternates at the same level (horizontal closed circulation). In the northern hemisphere in each horizontal closed loop formation, rotating counterclockwise (cyclone), air shift is created to higher latitudes in the front and to the low latitudes in the rear; in a loop formation, rotating clockwise (anticyclone) – vice versa. This implies that at any given point in time on the same level for some meridians there dominate meridional components directed to the north, and for other, neighboring, – to the south. It should be noted that the meridional components generally cannot be the same direction along the latitudinal range. It would be needed for this a zonal component of the pressure gradient pointing in the same direction on the entire circle of latitude. But this cannot occur since it would imply the existence of the gap in the pressure at some point in the latitudinal range, what is impossible.

Only meridional transfer of air would have been possible if the Earth does not rotate around its axis, and its surface has been smooth (see Figure 1). Under the influence of Earth's rotation, resulting Coriolis forces deflect the flow from the meridional direction. Calculations show that if the contrast in temperature between the equator and the poles was small, the air flow direction would have a symmetrical appearance. Symmetrical circulation can conduct only a relatively small heat flow. When heating (at the equator) and cooling (at the poles) increases, the heat transfer also increase. The symmetrical recirculation mode is replaced by a more powerful – wave mode – for sufficiently strong heating and rapid rotation. Elements of both types can be combined in circulation within one atmosphere, but one of them usually predominates.

Figure 2. The circulation of the atmosphere on the assumption that the Earth has a uniform surface

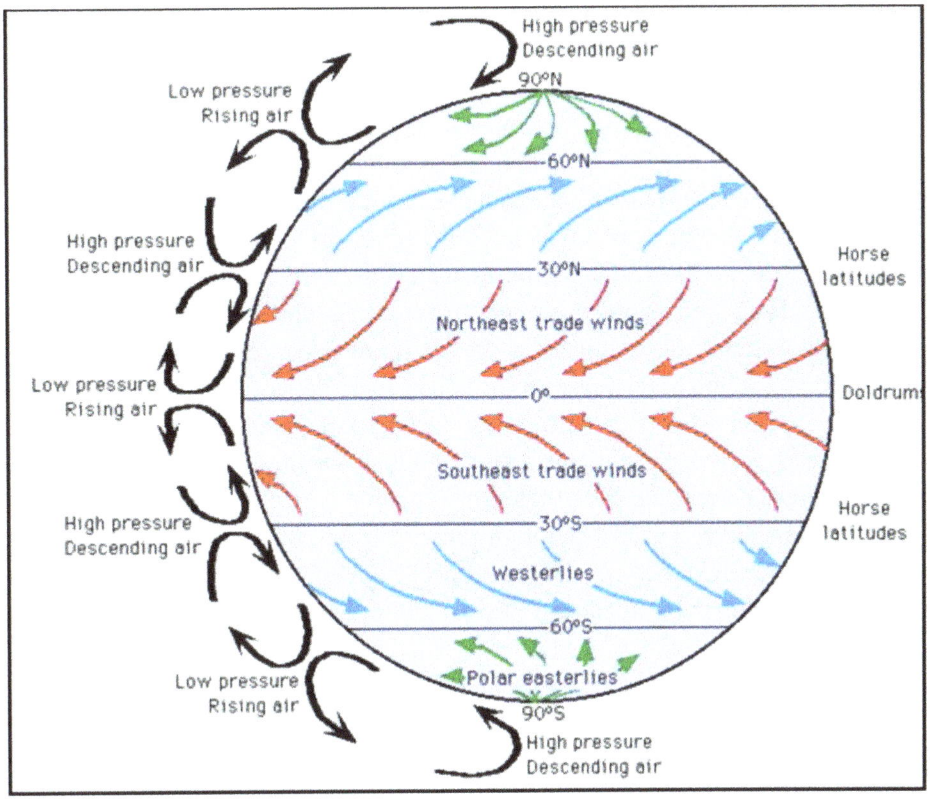

The wave circulation mode dominates on the Earth. Under the influence of Earth's rotation (Figure 2), air currents are deflected to the right in the northern hemisphere, and to the left in the southern, thus forming the winds of the northeast direction in the north hemisphere, and the southeast in the south hemisphere. These winds are called *trade winds*. The heated above the equatorial zone the humid air is cooled while lifting in high latitudes. At the altitude of 4 km and above, the air "spreads" on the North and South. Due to the rotation of the Earth, these flows deviate and form *antitrade* winds – southwest in the northern hemisphere and northwest in the southern hemisphere. In the northern hemisphere antitrade aimed first to the pole, gradually deflected toward the East. When these air masses reach approximately 30-35°North, they change direction almost to the East, preventing further flow of air masses to the pole. Thus, at these latitudes air accumulation occurs, further increasing antitrade when meeting with the air flow from the higher latitudes. Accumulation of air leads to lowering its pressure and increasing the outside near the Earth's surface. Following 30-35° of latitude in both hemispheres forms a high-pressure area, the surrounding land to the South and North of the equator. The air that descends to the earth's surface is dry and warm. That is why near the tropical high-pressure areas there formed huge deserts of Africa, Asia, America, and Australia. One part of the air descended near the earth's surface is directed toward the poles, creating warm southwest winds in the northern hemisphere and northwest – in the south. They are known as "the prevailing west winds." The second part of this air is directed to the equator, forming northeast and southeast trade winds. These trade winds collide near the equator and largely cancel each other. Zone of the meeting called the "equatorial zone of comfort", where the winds are usually weak and unstable in direction.

Thus, at the boundaries of areas of high pressure converted to the poles, southwest and west winds blow in the northern hemisphere, and north-western and western winds blow in the southern hemisphere at the southern boundary of the high-pressure zone. To the South of this zone in Antarctic waters, there is a continuous belt of very low pressure. Thus, there is constantly blowing the very strong northwest wind of 40th and 50th latitudes of the southern hemisphere. In the cold season, they are often of hurricane force. Not by chance, these latitudes are considered the most turbulent of the globe and are called the "roaring forties" and "furious fifties". Continuous ocean ring over which the wind is constantly raging spans the whole globe. Here are gigantic waves up to 26 meters, which has no equal in any place of the world ocean. Over the Antarctic ice continent, there is a permanent area of high atmospheric pressure. The air is moved from the south to the north of the equator and departing due to the Earth's daily rotation to the left, creating over Antarctica (with its banks and part of the Antarctic waters that wash the mainland) the system of very persistent and strong winds of southeast direction. The southern polar winter, the speed of the wind reaches 200-290 km/h.

In the northern hemisphere, north of the high-pressure zone, the wind distribution pattern is less regular. This is due to the asymmetry of the location of high and low-pressure zones with respect to geographical coordinates, and the influence of the warm Gulf Stream which prevents excessive cooling air these latitudes.

This scheme of the general circulation of the atmosphere is largely idealized. The actual direction of the wind in the winter and summer are different from the real wind in the zonal scheme. Most have a clear zonation of winds at the equatorial zone. In the northern hemisphere winter and summer prevailing winds are north-easterly direction, and in the south – the winds of the south-eastern direction, the trade winds. Most clearly expressed are the trade winds over the Pacific Ocean. Over the mainland and near the trade winds are violated by another system of currents – *monsoons,* which occur due to cyclonic activity associated with a large temperature difference between sea and land. In winter monsoon is directed from the continent to the ocean, and in the summer – from the ocean to the continent. Monsoon transfer of air masses occurs in the coastal areas of East Asia.

Monsoon is the system of air currents over large areas in the tropics of West Africa to Southeast Asia and Indonesia, where in single season prevail the wind in one direction, and the other is just the opposite, or close to it. For example, over the Arabian Sea and the Bay of Bengal during the summer months, the winds blow from the southwest, and in the winter – from the northeast. The care that in different parts of the world is paid on monsoons is due to not only to the seasonal change of the prevailing winds but also with the laws of rainfall during the monsoon. The lack of monsoon rains leads to drought, crop failure, shallowing of rivers. At the same time too intense monsoon with stormy, prolonged rains causes floods. A characteristic feature of the monsoon climate is that in the rainiest month of the year rain falls almost 50 times more than in the driest. Specific features of the monsoon are its stability during the season and the change from one-half of a year to another, what is named its seasonality. Reasons of monsoon wind and their direction seasonal change are associated with the arrival of solar radiation on the earth's surface.

Thus, the most stable feature in the distribution of both the wind and the associated atmospheric pressure over the globe is zoning of the distribution. Zonal circulation is manifested in the predominance of the meridional pressure gradient on the latitude and, therefore, the predominance of latitudinal components of the wind (east or west) over the meridional component. This component of the one or the other direction (east or west) prevails at the same time or continuously at a large on latitude Globe area.

The degree of predominance of the meridional zone components can be different. In tropical ocean areas, the predominance of eastern components in air transport in the lower troposphere is expressed very sharply and easily discernible even on separate synoptic maps on some days. In general, the tropical meridional components are at least 10 times lower than zonal. The predominance of westerly winds in the temperate zone of the southern hemisphere is also well expressed. At the same time, in many parts of the temperate latitudes of the northern hemisphere the wind often varies in direction and the predominance of western transfer can be discerned only from the statistical analysis of a large observational material. Finally, there are areas, such as East Asia, where the prevailing wind direction in the lower troposphere is closer to the meridian than zonal.

Air masses are moving at the surface and at high altitudes on Earth, not only horizontally but also vertically. Even though the vertical velocity of air flow is small, they play an important role in the vertical air exchange, cloud formation, precipitation and other weather phenomena. There are other features in the distribution of vertical movements. Analysis of observational data indicates that temperature contrasts of Pole-the Equator are unevenly distributed in latitude. There is a relatively narrow area where concentrated a large part of the energy of the atmospheric circulation. It is marked by the maximum value of the pressure gradient and, consequently, the wind speeds. For these areas, the concept of the high-altitude frontal zone was introduced (HAFZ) and strong westerly winds associated with it became known as jet streams or jets. Typically, the wind velocity along the jet axis is greater than 30 m/s, the vertical wind velocity gradient exceeding 5 m/s per 1 km, and horizontal velocity gradient reaches 10 m/s and over per 100 km. HAFZ covers a large geographical space: its width is 800-1000 km, height is 12-15 km, and the length is of 5-10 thousand.

The real atmosphere is an extremely viscous mobile environment, always in a movement that is the most diverse in size, direction, and speed. The Earth is not a smooth homogeneous sphere. Due to mountains and valleys, additional perturbations are imposed on the general circulation that causes local mesoscale circulation. One can observe elementary air masses in the atmosphere which originate and break away from the mainstream and for some time moves on their own as a holistic formation, and then gradually dissolve and mix with the surrounding air. Vortex structure of these elementary formations assists intense mixing between the different parts of the environment and their strong interaction. In the theoretical fluid mechanics, such movements, in which the trajectory of the individual air particles are irregular, sinuous and rapidly changing in time and space, are called turbulent. Realization of the turbulent nature of the motion in the atmosphere is seen by observing a spread of smoke emitted from industrial enterprises. Smoke is usually distributed as a meandering torch, composed of several puffs of different sizes. The most important consequence of the turbulent nature of atmospheric motions is both vertical and horizontal exchange of momentum, heat, moisture, and other physical substances.

The Air Masses, Atmospheric Fronts, and Frontal Surface

The large volume of air having a horizontal size of several hundred or thousands of kilometers and vertical dimensions about 5 km, characterized by a uniformity of temperature and humidity and moving as a single system in one of the trends of the general circulation of the atmosphere called in meteorology the *air mass*. The homogeneity of the properties of the air mass is achieved by forming it on a uniform underlying surface and the radiation conditions when this air mass is delayed for a long time in the area of the formation. The values of meteorological parameters within the air mass change slightly retaining their continuity, horizontal gradients are small.

Air masses are classified primarily on the hotbed of their formation depending on the location in one of the latitudinal zones – Arctic or Antarctic, polar or temperate latitudes, tropical and equatorial. According to geographical classification of air masses can be divided into the main geographical types for the latitudinal zones, which are located in their hotbeds:

1. Arctic or Antarctic air (AA).
2. Polar or moderate air (PA or MA).
3. Tropical air (TA).

These air masses, moreover, divided into the sea (s) and continental (c) air masses: sAA and cAA, sMA and cMA (or sPA and cPA), sTA and cTA.

Thermodynamic classification divides the air masses in the warm, cold, and neutral. The warm (cold) is called the air mass, which is warmer (colder) than its surrounding environment and is gradually cooled (heated) in the region, trying to get close to thermal equilibrium. The neutral air mass is called the mass, which is in thermal equilibrium with its environment, i.e. for an extended period maintaining their properties without a significant change.

Unstable and stable equilibrium can characterize the air mass. This separation of air masses takes into account one of the most important results of the thermal exchange – the vertical temperature distribution and the corresponding type of vertical equilibrium. With stable (SAM) and unstable (UAM), air masses are associated certain weather conditions.

Stable called air mass, which is dominated by stable vertical equilibrium, i.e. when vertical temperature gradient γ is less than moist-adiabatic gradient γ_{ha} in the bulk of its thick layer. Thermal convection in the SAM does not develop. The average value of the vertical temperature gradient in the SAM typically is less 0.6°C/100 m. Inversion and isothermal layers can be found here.

Unstable is called the air mass in the main body of which wet-instability equilibrium prevails. There are distinguished the absolute instability or dry-instability when $\gamma > \gamma_a$, and the relative instability or wet-instability when $\gamma > \gamma_{ha}$. The average value of the vertical temperature gradient in an unstable air mass is most often more than 0.6°/100m. In an unstable air mass, may develop both a thermal and dynamic convection. The weaker the wind, the more predominant is the thermal convection.

The masses of warm and cold air, when they meet each other, cannot immediately mix, so naturally, the properties of these air masses on the border between the two are very different. It is the transition zone between the two air masses where the sharp increase in horizontal gradients of meteorological quantities occurs approaching the abrupt transition from one value to another. In the meeting place of different air masses, there is a clear strip – the transition zone between air masses of different physical properties that in meteorology called the *front surface* or *zone*. At the intersection of the imaginary surface and the earth's surface, a so-called *atmospheric front line* appears. The horizontal length of the front, as well as air masses, has hundreds or thousands of kilometers, vertically – about 5 km, the width of the front zone of the Earth's surface is of the order of hundreds of kilometers, on the heights – several hundred kilometers. Frontal zones are characterized by significant changes in temperature, humidity, wind direction along the horizontal surface, both on the ground level and above.

Warm air density is lower than the density of cold air. Therefore, the front is an inclined plane, which always has a slope towards the cold air at a very small angle to the ground. Cold air as denser, when to meet a warm air, dives under it, and, like a wedge, picks it up. Imagining front between air masses, one

Figure 3. Schemes structure of main fronts and their symbols

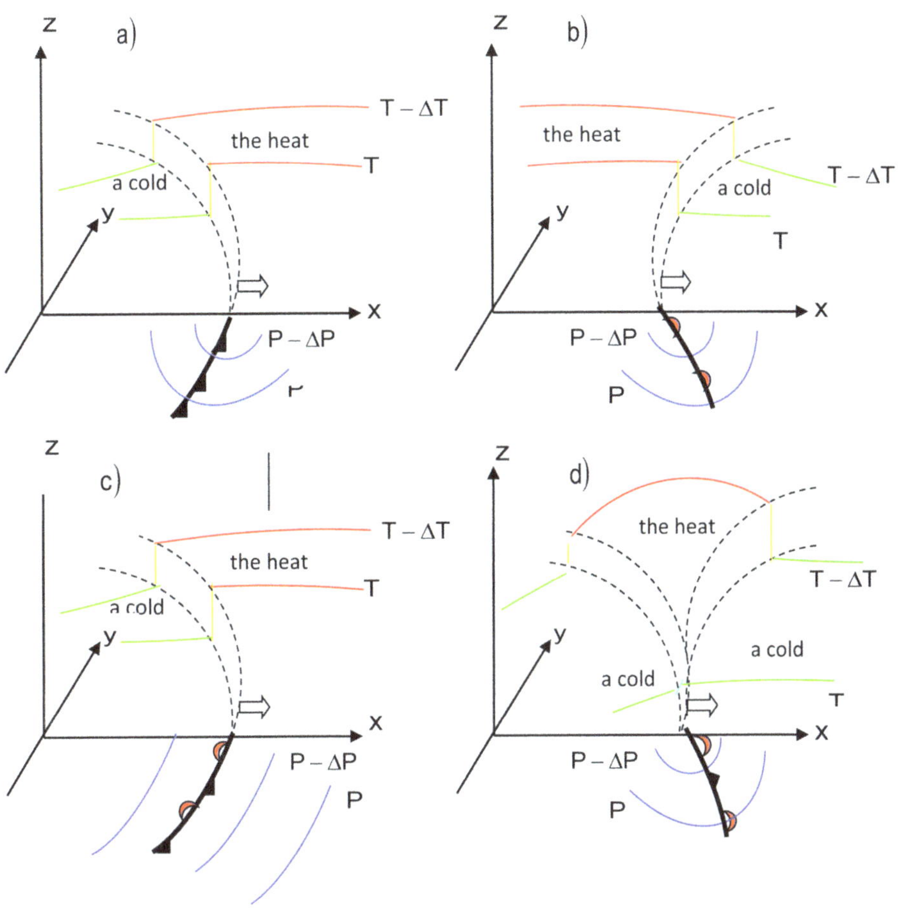

must always keep in mind that this is an imaginary surface inclined above the ground. The front line, which is obtained at the intersection of this surface and the earth's surface, is marked on weather maps.

After the meeting, the cold and warm air mass continues its movement in the direction of the mass movement, which is moving at a higher speed. At the same time the frontal surface that separates the air mass is also moving. Depending on the direction of the front surface movement one distinguishes cold and warm fronts (Figure 3). The cold atmospheric front is that which moves in the direction of higher temperatures, i.e. the upcoming cold air moves faster than the retreating warm (Figure 3a). After the passage of a cold front, comes cooling. The atmospheric front is called the warm when warm air moves toward lower temperatures and move faster the retreating cold (Figure 3b). After the passage of a warm front, comes warming.

Sometimes there are slow-moving fronts (Figure 3c), which are called quasi-stationary. Since the direction of their movements is uncertain, then they are marked on the synoptic maps as both warm and cold. Sometimes a cold front can catch up the warm and lift warm air to a certain height. Such fronts are called occluded or occluded fronts (Figure 3d). On both sides of the front occlusion arranged air masses, more similar in their properties than air masses separated by cold or warm fronts. Their horizontal length does not go beyond a cyclone.

There are some common classifications of fronts based on the importance of the circulation of the fronts and their spatial extent, the peculiarities of movement, the vertical structure and weather conditions. According to the geographical indication in accordance with the geographical classification of air masses one distinguishes the following fronts:

- *The Arctic front* is the front between the Arctic and polar (moderate) air masses of the northern hemisphere. Located on the southern border of the Arctic air mass. Usually, there are several co-existing branches of the Arctic Front, Arctic Front sometimes encircles continuously all the northern hemisphere.
- *Polar front* or the *front of temperate latitudes* is the southern boundary of the temperate air masses separating air masses of temperate latitudes and tropical air.

By circulation importance and the spatial extent, one can identify *main* (tropospheric, high), *secondary* (near the surface, low) and the *upper* atmospheric fronts.

The main concern atmospheric fronts with a large horizontal (several thousand kilometers) and vertical (several kilometers) length. The main fronts divide air masses, which differ significantly in their properties. They are traced on surface and altitude weather maps.

Secondary atmospheric fronts exist only in the lower layers of the atmosphere, near the Earth's surface and not more than AT850 (standard isobar surface of 850 hPa), and are not found in the temperature field on the heights. These are usually fronts in the inhomogeneous air mass, dividing it into two air masses are of the same origin. The most frequent case of secondary fronts is a cold front in a horizontal inhomogeneous cold air mass (of Arctic or polar air), which is followed by a "fresh" and the colder part of the same air mass. The secondary fronts are often observed in the cyclone rear of the main front. Secondary fronts exist no more than 1-2 days.

The top fronts are called edges that are absent near the Earth's surface but are expressed in the high places. They may be formed as a result of erosion of the front surface of the Earth, but being continued at heights. These fronts can also be independently formed at altitudes before reaching the ground.

Typically, atmospheric fronts are associated with characteristic cloudiness. Most often it is a whole system of high-layered and nimbostratus clouds, sometimes cumulonimbus and associated clouds of upper and lower tiers. There are many dangerous and natural phenomena at atmospheric fronts that are characteristic of dirty weather, including significant rainfall, strong winds, sometimes storms, dust storms or blizzards, and so on. To predict the weather, it is necessary to know the characteristics of air masses, the place of their collision, i.e. atmospheric fronts, their direction and speed of movement. It is predicted the oncoming of the front in the whole range of its characteristic signs, and above all – on supposed (or precalculated) position of cyclones and their narrow gullies in which atmospheric fronts are located.

Fundamentals of Circulating Air Movement

The atmosphere is the extremely mobile environment where it is constantly formed and destroyed eddies of various sizes. According to the scheme of Vilhelm Bjerknes (Figure 4), proposed in 1919, cyclones arise and develop at the wave of the front. Cyclone is a grand relatively flat horizontal atmospheric vortex having across hundreds of kilometers, and sometimes 2000-3000 km. The air in this vortex gyrates around the center (the area of low pressure) gradually approaching it. In the northern hemisphere, this rotational movement goes on counterclockwise under the influence of the Earth's rotation. Due to the

Figure 4. The life cycle of mid-latitude cyclone

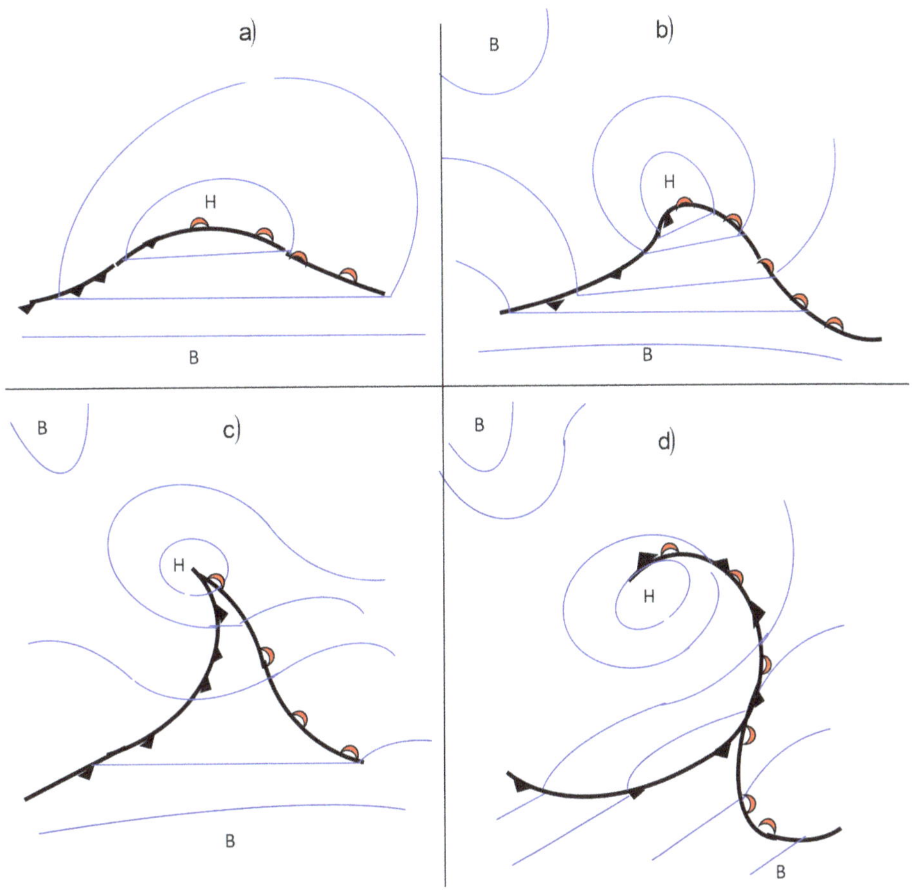

rotation of the Earth about its axis, cyclone winds are not directed to the center of the cyclone and are tangent to a circle described around the center.

The cyclone is formed when a mass of cold polar air meets a mass of warm and moist air. This the latter that rushing into the cold mass forms some blowout to a cold front. These streams of air with different temperatures and densities, slipping relative to one another, create a wave on the front surface and, consequently, on the front lines that like an arc faces by its concavity in the direction of the warm air (Figure 5a). Part of the arc, located at the front, the eastern part of the cyclone is a warm front. The second part of the arc, located in the rear, western part of the cyclone, is a cold front. In between the warm and cold fronts of the cyclone, there may be areas of good weather which usually lasts only a few hours. This deflection of the front line is accompanied by a decrease in pressure near the top of a wave.

The founded wave, usually moving to the east, north, or southeast, is deformed (Figure 4a – 4d). The emergence of a wave at the front is accompanied by deformation of the thermobaric field of the troposphere. The blowout of warm air penetrates further north, forming a well-defined warm sector of the cyclone. In front of the warm sector warm air slides over the colder and denser air. A condensation of water vapor occurs and strong layered rain clouds are formed with following precipitation (rain or snow) of widespread type. The width of the zone of frontal precipitation is about 300 km in the summer, and

400 km in winter. At several hundred kilometers in front of the warm front, updraft can reach a height of 10 km and more. The moisture condenses forming ice crystals. They make the ridge of white cirrus clouds. Thus, the cirrus clouds often herald on forthcoming a warm front of the cyclone. Moist, warm air of the warm sector, passing over a cold surface of the earth, constitutes the low stratus clouds, frost, and fog. After the passage of a warm front comes warm cloudy weather with southern winds, muslin, or mist. Then comes cold front. Along with it, the cold air dives under warm and displaces it upwards. There are formed cumulonimbus clouds are thunderstorms, heavy rains accompanied by strong winds. The width of the cold front precipitation zone is about 70 km away. After that comes the rear part of the cyclone, bringing strong winds and cool weather with cumulus clouds. When the cold air drives away to the east warm, clear weather is set. As the penetration of warm air in the blowout of the mass of cold air, it is increasingly surrounded by cold air and warm air is displaced upward. Cyclone is developing, with the winds in the area covered by them are enhanced.

But the cold air moves in a cyclone at a higher rate than warm air, so cold front overtakes a warm gradually, and the warm sector size is reduced. In the end, the cyclone cold front merges with a warm front, forming the so-called front occlusion (Figure 4d). Warm air rises, where it is cooled and released from the moisture that falls to the ground as rain or snow. Near the earth's surface is only the cold air, the air temperature difference disappears, the cyclone begins to fade. But complete uniformity in the air mass is not. Therefore, at the front, there is a crest of a new wave of the second cyclone. Cyclones are always in series, each following the cyclone passes just south of the previous one.

Thus, in the cyclone life is divided into four stages of development: an initial stage (wave step), a young cyclone stage, the maximum development stage, filling stage of the cyclone.

For the initial stage of the cyclone, which lasts about a day, the typical process is characterized by the deformation of the front surface (wave step) until the first emergence of closed isobars on the surface weather map. The pressure difference between the center and the periphery is less than 5-10 hPa. At altitudes, the vortices in the initial stage are not traced.

In the second stage of development, the duration of which is also usually not more than a day, the cyclones have already at least 2 closed isobars.

The thermobaric field continues to deform, cyclone deepens, it becomes a powerful atmospheric vortex with significant wind speeds. The cyclonic circulation extends into the upper atmosphere.

One can divide a young cyclone into three zones, differing sharply on weather conditions (Figure 5):

- Zone I is the front and center of the cold sector of the cyclone that is before the warm front. The weather in this area has a character determined by the properties of a warm front. The closer to the center of the cyclone and to the front line, the stronger is the system of clouds and the probability of heavy precipitation.
- Zone II is the rear part of the cold sector of the cyclone that is behind the cold front. Here the properties of the cold air mass determine the weather. With sufficient humidity and significant instability in the area, heavy rains fall.
- Zone III is the warm sector between warm and cold front. In winter in the warm sector of the young cyclone it is marked solid clouds St, Sc, and sometimes there are advection fog and drizzle. In summer in the warm sector of the cyclone, depending on the humidity of the air mass, one can observe both overcast and cloudy weather, and sometimes even a lightning storm. The day celebrated mostly cumulus clouds.

Figure 5. Driving a young cyclone

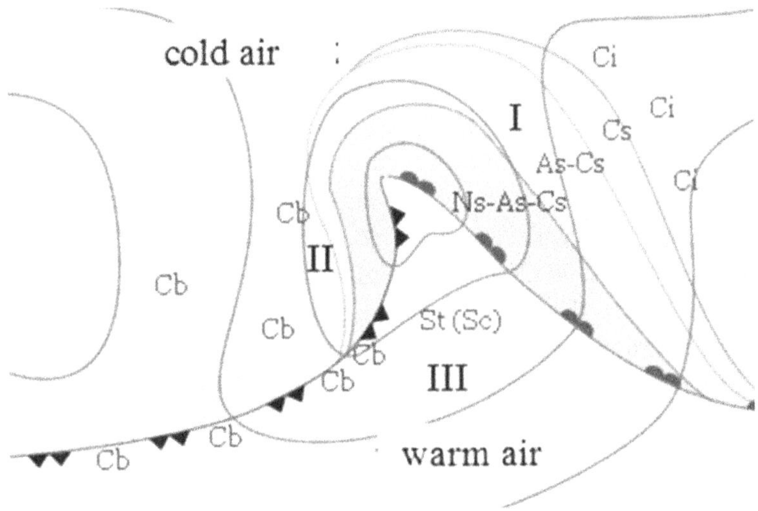

Due to friction in the lower atmosphere, it is observed in the cyclone, in addition to the circular movement of air, even the movement from the periphery to the center, and therefore, there is a constant vertical upward air movement and cooling as it rises. The air cools and becomes saturated with water, it forms clouds, producing precipitation. In the cyclone, especially near their centers, there are always great horizontal gradients of pressure and, therefore, there is constantly observed strong gusty winds (vortices).

The third stage is characterized by the lowest pressure in the cyclone center. The duration of the stage is not more than 12-24 hours. The cyclone is filled at the final stage. Pressure rises on the surface of the Earth at the center of the cyclone. Horizontal pressure gradients and wind speed gradually decrease. This stage is the longest, it is four nights or more.

The second half of the life of the cyclone is characterized by a decrease of its forward speed, a significant transformation thermobaric field of the troposphere. The cyclone becomes high, thermally symmetrical (cold) pressure systems. In the case of the occluded cyclone, the weather is different according to the characteristics of air masses on both sides of the front occlusion. At the front occlusion is a combination of clouds of cold and warm fronts. It is formed a common strip of precipitation of stratus clouds of rising sliding (As-Ns) and convective cumulonimbus (Cb), which will drop the front line and behind it.

In well-developed cyclone (maximum development stage), cloud band on the satellite images is a vast cloud vortex with a powerful cloud-based system. Closure of cloud spirals of cold and warm fronts leads to the formation of a unified helix of occlusion front. The closing process starts in the center of the cyclone and the cyclone is gradually shifted to the periphery.

In its origin cyclones are divided into two main groups: the cyclones of temperate latitudes and tropical cyclones (hurricanes, typhoons).

Cyclones of temperate latitudes occur mainly in the areas of atmospheric fronts, where the two different air masses meet. In the northern hemisphere, the most extensive cyclones usually occur over the waters of the Atlantic and Pacific oceans. Repeatability of them depends on the time of year and the geographical area. On average, in the northern hemisphere cyclones over the European part of the continent, they are more frequent in winter, and on the Asian part – in summer. Cyclones of temperate latitudes have a diameter of about 2-3 thousand kilometers and more.

Weather in temperate latitude cyclone is heterogeneous: there are distinguished the front and rear of the cyclone, the left, and right – relatively to the direction of its movement. In front of the cyclone, there dominate continuous stratiform clouds of the warm front and heavy precipitation with winds of southern quarter of the horizon. At the rear of the cyclone, behind the cold front, the weather is characterized by instability with the loss of storm type of precipitation and gusty winds of northwest and north quarters. Cloudiness may be discontinuous, and even with occasional clearings, and in the summer – of convective type. The left (mostly northern) part of the cyclone is characterized by weather conditions, which can be called as intermediate between the front and rear portions of the cyclone. There prevail winds of east and the north-eastern quadrant, solid clouds, widespread rainfall, falling intermittently and gradually turning into intermittent showers thereof. Right south of the cyclone in a certain period of its life is a "warm sector". It is filled with warm air mass, which eventually displaced upward. Here, depending on the season and type of air mass, the weather can be varied, but mainly without precipitation, with fog or low stratus thin clouds, often cloudless, and always warm, with winds of the south and southwest quarter.

In temperate latitudes, including the greater part of European territory, cyclones often move from west to east, coming from the Atlantic Ocean. The warm moist air of the cyclone comes from the ocean. Cold air is introduced to them in the rear from the most northern part of the ocean or the Arctic lands and ice-covered seas. Often, cyclones move in the temperate latitudes from subtropics. The warm air of the cyclone has a very high temperature in the temperate latitudes and causes significant warming either in winter and summer. In developing cyclone, cold Arctic air can penetrate far south and cause a sharp cooling in the southern latitudes. At the same time, the warm air in front of the cyclone, penetrating far to the north, leading to an increase in temperature in the northern latitudes, and even beyond the Arctic Circle.

Tropical cyclones have as their homeland the oceans in the equatorial region of between 10-15° north and south latitude; their diameter is often a few hundred kilometers, and height is from 5 to 15 km. Tropical cyclones can occur at any time of the year in the tropical parts of all oceans except the southeastern part of the Pacific Ocean and the South Atlantic. The most frequently (in 87% of cases), tropical cyclones occur between latitudes 5° and 20°. At higher latitudes, they arise only in 13% of cases. It never mentioned the occurrence of cyclones north of 35° north latitude and south of 22° of south latitude. Tropical cyclones that have reached considerable intensity in each region have their own name. In the eastern Pacific and in the Atlantic they are called hurricanes (from the Spanish word "Huracan" or the English "Hurricane"), in the countries of the Indian subcontinent they are called cyclones or storms, in the Far East they are called typhoons (the Chinese word "tie", which means a strong wind). For typhoons, there used the four lists of names, for hurricanes, one is set. Each typhoon or hurricane formed in the given calendar year, except the name is assigned a serial number, the two-digit number, for example, 1615, which means the number fifteenth typhoon in 2016.

Most often, they are formed in the northern part of the tropical Pacific: here, on average, about 30 cyclones can be traced for the year. Tropical cyclones go to temperate latitudes in the period from late June to early October and are most active from August to October. A distinctive feature of the cyclone of this group is that they are thermally homogeneous (i.e., no temperature contrasts between different parts of the vortex). They contain a tremendous amount of energy and bring gales and heavy rainfall.

Tropical cyclones are formed where there is a high-water surface temperature (above 26 °), and the temperature difference between the water and air exceed 2°. This leads to increased evaporation and moisture stocks increase in the air, what to a certain extent, determines the accumulation of heat in the atmosphere and contributes to the vertical air recovery. Emerging powerful convection carries volumes of air, warmed and moistened above the water surface. The rotation of the earth gives rise air vortex motion, and the whirlwind becomes like a very energetic giant gyroscope. The central part of the crater called "the eye of the storm." This phenomenon affects by the features of its 'behavior'. When the eye of the storm has well defined the precipitation on its boundary is suddenly stopped, the sky clears up and the wind weakens considerably, sometimes to calm. The form of the eye of the storm can be very different, it is constantly changing. Sometimes there is even a double eye. The average diameter of the eye of the storm in the well-developed cyclone is equal to 10-25 km and in the destructive one it is 60-70 km. Moving toward temperate latitudes, tropical cyclones gradually lose their strength and decay.

Typhoons are the most powerful and destructive tropical cyclones that occur over the ocean to the north-east of the Philippines. The average lifetime of a typhoon is 11 days and the maximum is 18 days. Minimum pressure observed in these tropical cyclones varies widely from 885 to 980 hPa. Maximum daily precipitation totals reached 400 mm, and the wind speeds up to 20-35 m/s. The main season of typhoon output in temperate latitudes is from July to September.

Tornado is strong storms on Earth that can cause an unusual, small, but violent clouds. Tornadoes whirl at a rate of hundreds of kilometers per second, and when they reach the Earth's surface, sweep away almost everything in its path along the long and narrow strip of following. As a rule, a tornado lasts few minutes, but the most powerful and dangerous of them can last for hours.

Cyclones of temperate latitudes are not so terrible, but sometimes they can also bring a lot of destruction by evoking the heavy rains and violent floods. But in general, the role of cyclones in temperate latitudes is more beneficial. It consists primarily in the fact that warm air carries them far to the north, and the mass of cold air moving far to the south. Thus, a series of cyclones have the basic mechanism that ensures the exchange of air between the latitudes. Europe is obliged to cyclones because winters are not as severe, and the humidity in the summer is sufficient for the cultivation of diverse vegetation.

The speed of moving cyclones decreases as they age. Most often, they are moving at a speed of 40-50 km/h, overcoming the distance in the 1000-1500 km per day and more. However, often they can move at a speed of 70-80 km/h and even more, overcoming per day 1800-2000 km. There are occasions when the cyclone, which was moving quite fast, suddenly slows down and becomes inactive. On the territory occupied by the sedentary cyclone, cool weather with precipitation is set which can stand as long as the increase in the atmospheric pressure in the cyclone will not lead to its disappearance (filled).

Wind velocity distribution is uneven in the cyclone. In the moving cyclone, the highest speed of the wind is centered on the right side in the direction of its movement, where the wind in the field of the cyclone and airflow due to the circulation of a large scale, are in the same direction. Wind speed near the center of the cyclone is small either at the earth's surface and the height, only 5 m/s. With increasing distance from the center, the wind speed at a certain distance and increases and after reaching the maximum value falls again. In some cases, the wind speed reaches a maximum at a very short distance from the center and occupies a very narrow zone, in other cases, the maximum wind speed has a wide area at a great distance from the center. The average radius of the storm area with wind speed exceeding 15 m/s is 250 km and with a speed exceeding 25 m/s is 50-100 km. It is therefore not surprising that the cyclone as it moves can devastate a town, leaving intact adjacent one. The winds in the cyclone sometimes reach hurricane force. Most often it is observed in the northern region of the Atlantic and the Pacific, as well as in the northwest of Europe, where cyclones are developing very intensively. The highest wind speed observed in the tropical cyclone. Depending on the wind speed values are divided into the powerful cyclones (speed is over 25 m/s) and superpower – super typhoon or super hurricane (more than 75 m/s).

Usually, inclement weather is associated with the passage of the cyclone with precipitation and strong winds. But cyclone consists of several diverse air masses with different characteristics on the weather. The cyclone may be rainy and sunny weather, depending on the properties of air masses in the front and rear parts of it. Precipitation in the cyclone is concentrated around its center. The most intense rainfall drops from the most powerful thermal towers of cumulonimbus clouds, surrounding the center of the cyclone. Precipitation has the same kind of spiral strips on the radar screen as cloudiness, as they fall out of the same clouds. Data on rainfall in cyclones and their distribution is very diverse, because, firstly, the precipitation measurement due to the sparse observation network does not quite accurate, and secondly, strong wind accompanies rainfall, carrying rain from the place of its loss over long distances. Estimates show that the passage of a tropical cyclone over Virginia (USA) on an area of 8,000 km^2, 700-750 mm of rain fell over 6 hours. The Manila rainfall reaches 1168 mm per day, which is 2-3 times the average annual in temperate latitudes. Almost every year, tropical cyclones carry many human lives, produce devastation, and cause huge losses.

Atmospheric fronts exist mainly in cyclones and in their peripheral portions, so-called trough, where the convergence of air flows is always observed in the lower atmosphere to maintain contrast between the two air masses, i.e. for the existence of the front itself. With the oncoming of a cyclone or its trough, atmospheric pressure usually falls, and it is a necessary featuring the front. Passage through is accompanied by characteristic changes in wind speed and direction and, depending on the type of front (cold, warm), the emergence of a certain sequence of clouds and precipitation. During the passage of the cold front the wind, increasing, turns to the right. If before the front has come wind blew from the west, then after the front he usually blows from the northwest or north. In the center of the cyclone, the “ends” of cold and warm fronts converge: in the east of the center of the cyclone the weather of warm front dominates, and in the west – the weather of cold front. Behind the cold front comes the cold air and cooling occurs. Since the cool air is denser, its propagation is accompanied by an increase in pressure, which indicates the approaching high-pressure zone – anticyclone.

An anticyclone is an area of high pressure in the atmosphere with the maximum at the center (1050-1070 hPa at sea level). The diameter of the anticyclone is of the order of thousands of kilometers. Air flows therein from the central portion to the periphery, in the northern hemisphere twisting clockwise. Thus, in an anticyclone air is spreading from the center to the outskirts, which eliminates the possibility of convergence and interaction of diverse air masses. Herewith, in the place of air which flowed into the surface layer, air is lowered from the higher atmospheric layers. By lowering the air comes under increasing pressure, being compressed and heated. Due to the heating of air clouds, if they were, are scattered. Even cumulus clouds that formed in the afternoon at the updrafts from the heated ground in the evening evaporate and disappear. However, with significant humidity in the cold half of the year in the central part of the anticyclone can be observed continuous clouds, and fogs are observed in both winter and summer. Usually, in the center of the anticyclone, it is established clear (dry and hot) weather with blue sky, on which occasionally easy and fast clouds swim. In summer, the anticyclone at night is cool, clear, and quiet. The day, heated by the Sun the Earth’s surface generates vertical flows of warm moist air (convection) so that light cumulus clouds can emerge. Sometimes, when very strong convection, cumulus clouds may be transformed even to cumulonimbus clouds, leading to thunderstorms. In winter, in clear weather, anticyclones are accompanied by severe frosts.

Depending on the geographical area of anticyclone origin there are distinguished extratropical and subtropical anticyclones. The emergence and development of anticyclones are closely linked with the development of cyclones, it is essentially a single process. If in one area a mass deficit is created, then in the neighbor area it is its excess. Anticyclones cover an area comparable to the size of the continents on which they develop better in the winter, and over the oceans in the summer. The average repeatability of anticyclones is 2.5-3 times smaller than the cyclones.

An anticyclone weather varies significantly in different sectors. Weather conditions on the outskirts of the anticyclone are in general like the weather conditions in adjacent sectors of adjacent cyclones. The northern edge of the anticyclone is usually directly related to the warm sector of an adjacent cyclone. In the cold half of the year, there is often observed overcast, sometimes weak rainfall and often mist occur. In the summer of this sector of the anticyclone, the weather is clear, in the afternoon cumulus clouds may develop. Western outskirts of the anticyclone are adjacent to the front of the area of low pressure. In the cold half of the year in this part of the anticyclone often stratocumulus clouds occur from which weak rain falls. Precipitation area is quite extensive, and moves along the isobars, skirting anticyclone clockwise and undergoing some changes. In summer, on the western edge of the anticyclone at high temperature and humidity often large cumulus clouds develop and thunderstorms occur. South outskirts of the anticyclone are adjacent to the northern part of the cyclone. It is often observed stratus clouds from which precipitation falls in winter. In this part of the anticyclone a large pressure drop is created, so often the wind increases and there are snowstorms. The rear part of the cyclone borders eastern outskirts of the anticyclone. In summer, when the air mass is unstable, in the afternoon cumulus clouds are formed, rainfallы shower and thunderstorms rattle. In winter, there may be cloudless weather or not the solid stratus clouds.

Anticyclones move generally slower than cyclones with an average speed of 25 km/h, and a long time can often stay in one place. Each cyclone or anticyclone lasts 3-5 days on average, moving sometimes during this time many thousands of kilometers and causing sudden sharp changes in weather. A characteristic feature of most of the areas of temperate latitudes is a frequent alternation of cyclones and anticyclones. The latter leads to the climatic peculiarities of these regions – frequent changes in wind direction, temperature fluctuations, alternating clear and cloudy weather, and thus, short-term warming, and cooling, etc. In some years, the conditions are created when in the same areas more often than usual anticyclones come. Here they are delayed; in summer and clear sunny weather air quickly warmed and dried. So, the lack of moisture in the soil is beginning to emerge, which over time can pass into the drought.

Every year over the northern hemisphere of the Earth about a series of hundred cyclones and anticyclones go through. Their movement seems chaotic, nevertheless, if to consider the trajectory of cyclones and anticyclones on the map, in this chaos a certain regularity is manifested – one can select areas (*atmospheric action centers*), in which most of cyclones and anticyclones start out.

There are permanent and seasonal centers of atmospheric action. The permanent centers of atmospheric action include *equatorial depression* – a strip of low atmospheric pressure, covering the globe near the equator. Equatorial depression does not coincide with the geographical equator and moves away from the equator, depending on the season, to the north or the south, in the hemisphere, where the summer at that time takes place. In the equatorial depression located intertropical convergence zone. On both sides of the equatorial depression the subtropical high-pressure zones are located, often breaking up into separate areas – oceanic subtropical anticyclones. In the northern hemisphere, they are:

- Icelandic minimum (center of the cyclones of the Atlantic) that is about two hundred kilometers from Reykjavik.
- The Azores maximum (North Atlantic anticyclone) that is over the subtropical latitudes of the Atlantic Ocean.

- Honolulu maximum (North Pacific anticyclone, Hawaiian maximum) that is over the subtropical latitudes of the Pacific Ocean.

Nevertheless, these centers do not remain in place. In some years, they overcome up to 5000 km along the meridian and even further, they deviate by latitude. Sometimes, very rarely, these areas of low or high pressure disappear altogether. Over the past century, there were up to a dozen of such years altogether.

For seasonal atmospheric centers of action, they are:

- Aleutian winter minimum in the North Pacific Ocean.
- Icelandic winter at least in the northern part of the Atlantic Ocean. These depressions are very deep and extensive in winter and in fact disappear in the summer. Some authors refer them as the permanent centers.

In winter one can find anticyclones over continental areas which in summer give way to depression:

- Siberian (Asian) winter maximum with the center over the Mongolian plateau.
- Canadian winter maximum.
- Asian summer minimum (South Asian, Middle-Asian summer depression) with the center over Afghanistan.

Influence of Local Physiographic Conditions on the Atmospheric Air Movements

We said above on the atmospheric circulation on a large scale. In some areas, it is noted a number of characteristic features of the transformation of the general circulation of the atmosphere due to the influence of local physical and geographical conditions, generating a local system of winds. Some of them can be regarded as a consequence of the Earth's surface topography influence on the overall air flow. Indeed, when the air flows along the rugged earth surface all its irregularities have a mechanical effect on the flow, changing its direction and speed, as well as serve as a reason for the appearance of the vertical velocity component. This is due to the curvature and the convergence or divergence of the flow streamlines at their wrapping large and small irregularities of the relief (ridges, hills, canyons, etc.). At elevations airflow significantly affects the compressibility of air. Over the peaks, lines of flow are raised relative to adjacent areas generally following the relief of the mountain. Conversely, the lines of flow are concave on some high-altitude mountain. However, there are so-called airwave – the wave of airflow generated over the leeward slopes. Around these waves may be significant upward vertical flows associated with intense vortex formation on the lee slopes, often leading to the formation of a multi-tiered cloud (Figure 6).

The general circulation is subjected to some more local mesoscale circulation, occurring in a calm and clear weather. They have a relatively small horizontal extension and may be caused either unequal heating of different areas of the underlying surface or the influence of relief, creating disturbances in the general air flow.

Dryers are dry, warm air masses that descend at high speed from the mountains into the valleys. The main cause of the dryer is overstepping air mass through the mountain barrier. The air rises on the windward slopes and cools at about 10°C per 100 m of rising to the level of condensation of water vapor.

Figure 6. The natural phenomenon of airflow over the mountains

Above the condensation level, the cooling rate is 0,60°C/100 m, i.e. by moist-adiabatic law. This leads to the formation of clouds and precipitation. Crossing the ridge the airflow begins quickly to descend the hill, heated at 10°C for every 100 m. Therefore, the clouds on the lee side of the ridge evaporate and the air reaches the foot of the mountain being quite warm and dry. Figure 7, for clarity, the airflow is shown in the form of mist, cool air as white, and warm air as pink.

The dryers can occur at any time of day and year in all mountain areas. Their duration is usually less than one day, however, in certain cases can be up to 5 days and more. A dryer wind speed is often much higher than 10 m/s. Besides indignation of the general circulation of the atmosphere dryers impact on the local weather conditions. The sharp rise in temperature accelerates the melting of snow in the mountains and sometimes causes severe flooding of mountain rivers. In summer, these winds, drying the soil can harm vegetation.

Breeze is the periodic local wind that occurs in a calm and clear weather on the coasts of seas and lakes during the warm season due to the difference of land and water temperatures (see Figure 8). After sunrise, the land surface and the air above warms up more quickly than the sea. Since in warm air pressure decreases with altitude slower than in the cold, then in the course of heating the air over the land isobaric surfaces at heights tilt down towards the sea.

Therefore, the air outflow starts from the land towards the sea. Due to increasing the air mass over the sea pressure in the lower layers becomes higher than over land, i.e., the isobaric surface is tilted down towards the land. This will cause the movement of air in the same direction (the sea breeze – Figure 8a.).

At night, the land surface cools faster than the sea surface, that is why at altitudes air movement comes from the sea to the land, and in the lower layers of air is directed from land to sea (offshore

Figure 7. The scheme of dryer formation

Figure 8. The scheme of periodic breeze circulation

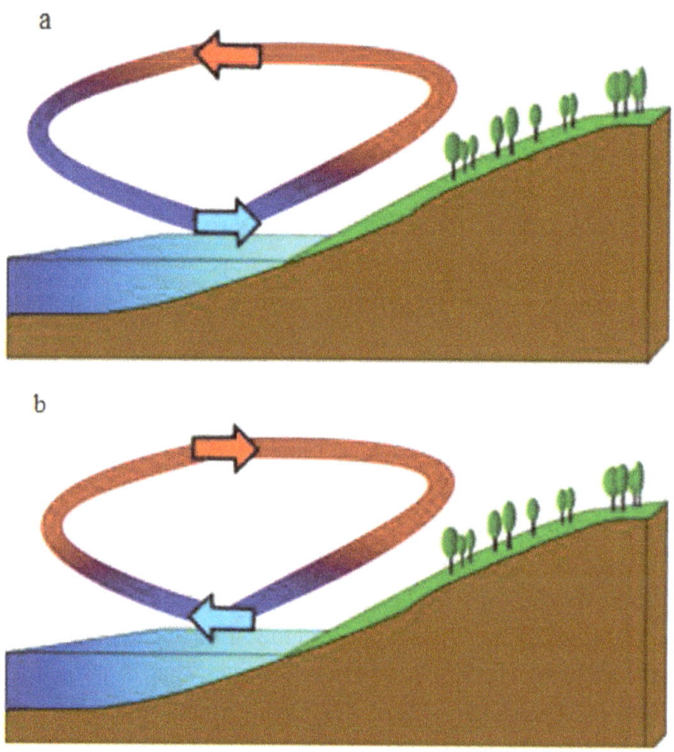

Figure 9. Driving bora

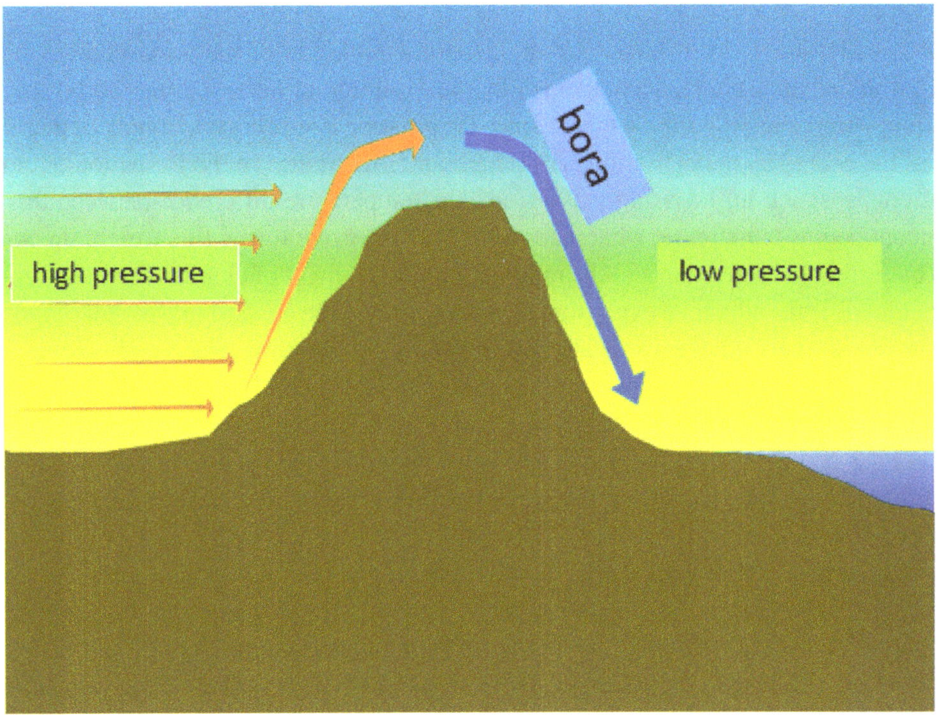

breeze – Figure 8b.). During the day, a breeze penetrates shoreline at 20-30 km, while the wind speed reaches the value of 6.4 m/sec.

The thickness of the vertical layer covered by a breeze is 200-500 m. Above this, passing through the calm zone, the wind blows in the opposite direction. Over the sea during the sea breeze downward movement of air dominates, which can form a temperature inversion. Because of this, the turbulence over the sea continues to decrease. This area is free of clouds. Even the front storm clouds are quickly eroded in moving from land to sea. After sunset, the sea breeze on the coast is changing. It is much less and extends to a height not exceeding 300 m.

Bora is a strong gusty cold wind at a speed of almost 60 m/s, blowing mainly in the winter near the coast hills on the shoreline. Bora formed as a result of lowering the cold denser air down the mountainside of the coastal hills to the sea, like a drop of water in the waterfall (Figure 9). Above the sea, bora extends in the minor distance of about 10 km. The estimated calculations show that the wind speed, only duc to thc cold air falling from a hcight of 400 m, can rcach values of 25-30 m/s. If we add to this the speed of the wind due to the general circulation of the atmosphere we will get those great values of speeds that are cited above.

Besides already considered local disturbances of the general circulation of the atmosphere, there are many other types of disturbances arising because of the impact of physical and geographical irregularities of the underlying surface. These include periodic mountain-valley winds, glacial winds, winds of gorges, passes, etc., but we will not dwell on their description. Interested readers can find references elsewhere.

Inmass Atmospheric Processes

Inmass processes have a great influence on the general circulation of the atmosphere. Condensation of water vapor in the atmosphere, accompanied by the evolution of heat, supplies additional energy to upward airflows supporting their movement. Since evaporation needs large amounts of thermal energy, the release of the energy by condensation is of paramount importance in the formation of clouds. With the onset of condensation in the rising mass of air, the temperature difference between this mass and the environment begins to increase, what leads to a more intensive lifting this weight up. And it is accompanied by further condensation and the rapid growth of the mass of clouds. The main reason for the formation of clouds is the adiabatic expansion that occurs when the upward movement of air. The speed of this movement is small, an average is about 0.03-0.05 m/s. Since the process of raising or lowering the air mass is held for a long time, the upward movement of large volumes of air plays a huge role in the formation of clouds and precipitation. If we assume that the average lifting speed is 0.03 m/s, the mass of air per day could rise by more than 2.5 km, and under normal conditions cool on 20-25°C. With high humidity, such cooling is sufficient for the formation of powerful clouds and heavy precipitation in the form of rain and hail.

The whole system of clouds appears due to the orderly slow rise of warm air, namely: cirrus, cirrostratus, altostratus, and nimbostratus. However, the most disturbing influence on the general circulation of the atmosphere provides the formation of convective cumulus and cumulonimbus clouds, caused by thermal convection and large vertical temperature gradient in the ambient air. They appear in the summer of the continent about 10 o'clock of local time, they reach the maximum development up to 15-16 o'clock, and then begins its "melting" to the complete disappearance of the night. Figure 10 shows the development scheme of the cumulus clouds during the day with a typical well-developed convection in a homogeneous air mass.

At the first stage of development (Figure 10a) cumulus clouds are thin. The horizontal dimensions of the individual clouds can vary from a few hundred meters to 2-3 km. Vertical upward movement in it has a speed of about 1-2 m / s. If the cloud is developed at this stage to 3-4 km altitude and thus remains in positive temperature zone, the vertical velocity in the central portion thereof can be 5 m/s.

Figure 10. Stages of development of cumulus clouds by thermal convection

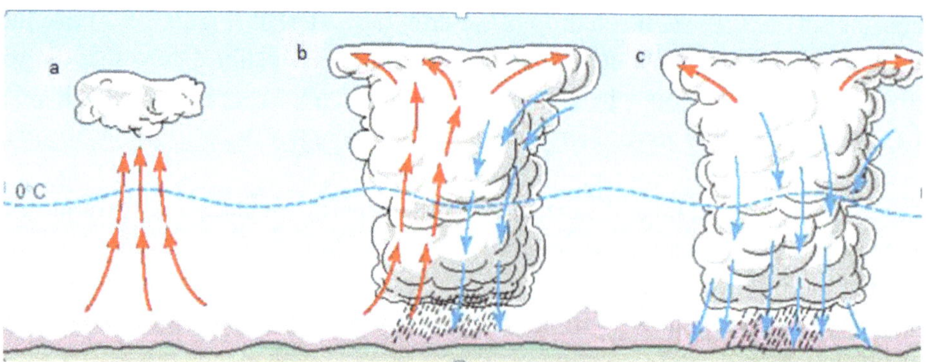

In the second stage, vertical cloud development reaches 5-7 km, i.e., its upper part is in a negative temperature zone. The rounded upper boundary of the cloud shows (Figure 10b) that the cloud consists of overcooled droplets, and the velocity of the ascending air is quite significant. The top of the cloud reaches the vertical velocity of 10-15 m/s. In temperate latitudes, the vertical development of these clouds is about twice the size of the horizontal base.

The third (highest) stage of development of cumulus clouds corresponds to the moment when the upper part of overcooled droplets crystallize, and therefore the upper edge of the cloud leaves the rounded shape and takes the form of the anvil (Figure 10c). At this stage, the upper boundary of the clouds over the continent in summer reaches 8-9 km. Sometimes it comes to the level of the tropopause, i.e. the anvil is formed mostly at the altitude where the air temperature is -25 – -30°C. In the upper part of the greatly developed cloud, there is observed significant upward movement at speeds of 35-40 m/s. At the same time in this part of the cloud, where precipitation forms and falls, vertical movements become descending. The horizontal dimensions of cumulus clouds formed by thermal convection in a uniform air masses, typically are 15-20 km.

At the fourth stage, the convective rise of air is stopped in most parts downdrafts develop, resulting in the slow collapse of cumulonimbus cloud from the bottom up. Sometimes the highest crystalline part of the cloud may persist for a few hours if it is dominated by the upward movement (about 1-3 m/s).

The considered above basic laws of development of the atmospheric processes established by many scientists meteorologists mainly on the basis of theoretical and experimental studies allow today to analyze the current weather conditions, to determine the reasons for the formation of natural and dangerous weather phenomena. In the general case, under the weather one understands the physical state of the atmosphere at given or other location at some time or time range, which is described by a number of characteristics which are called meteorological variables (sometimes, items). The most important of these are pressure, temperature, humidity, wind speed, and direction, cloudiness. One characterizes the weather and the meteorological phenomena, which include precipitation (rain, hail, snow), mist (radiative and advective) inmass and front phenomena (thunderstorm, squall, tornado), and a number of optical and electrical phenomena in the atmosphere. Thus, the weather defines a complex set of meteorological variables and phenomena that are closely related. Observations of them are conducted at meteorological stations around the world. All information received at meteorological stations, using remote measurement techniques (satellites, radiosondes et al.) enters the World Meteorological Centers, located in Moscow, Washington, and Melbourne. These data are processed on powerful computers with the use of statistical, synoptic, and hydrodynamic methods.

Each set of observed meteorological variables and phenomena reflects a certain state of air environment – weather case (Figure 11), which undergoes a change over time. For short periods of time the weather changed little, so monitoring the weather characteristics is not maintained in all cases continuously. Instant weather observation applies to some prolonged period or defines it by averaging a number of observations for this period. The meteorological practice considers one hour for the least amount of time. However, information about the weather in an hour is not always necessary. They are important, for example, for aviation and have no value for agriculture.

Average rating forecast per day is the most often used in meteorology with the specification of certain characteristics, such as the temperature at the time of its maximum and minimum values.

Figure 11. An example of the weather map

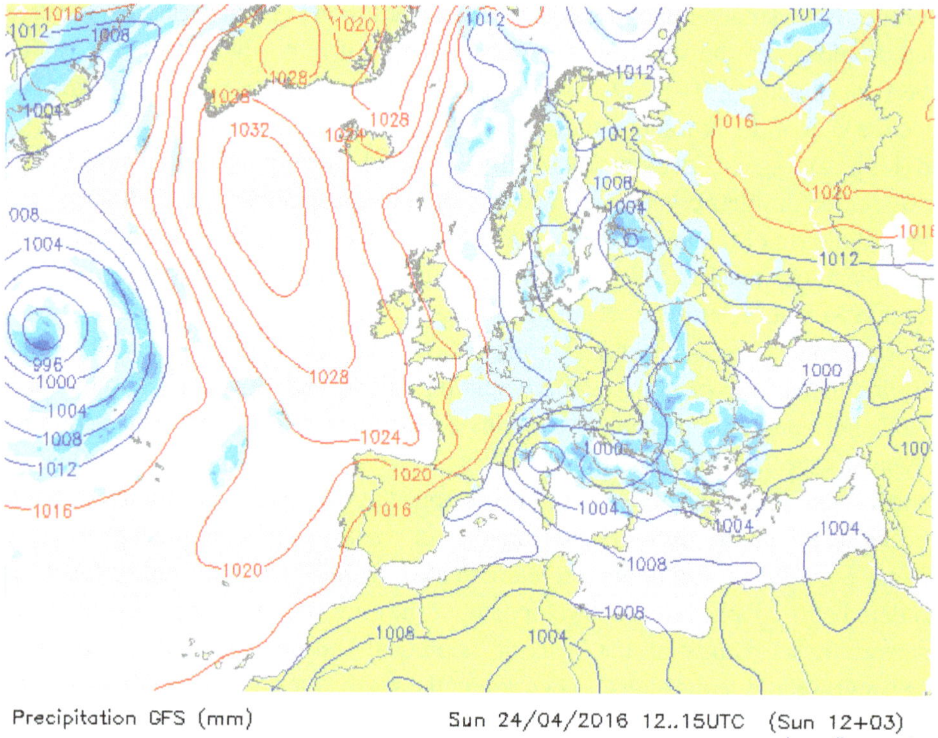

Natural and Dangerous Weather Phenomena

This section describes the conditions for the occurrence of dangerous and natural meteorological phenomena, which are often the cause of the emergency. Specifically, we provide synoptic situations that cause these or other hazardous weather events in the mid-latitudes of the Northern Hemisphere, offers some practical advice on the timely detection of dangerous and natural weather events.

Weather quality essentially depends on the type of synoptic processes which determine the weather in the area of our interest. If the processes that determine the weather occur in a homogeneous air mass, these processes are called inmass. If the weather conditions are defined by the fronts passing through the area under consideration, these processes are called front. Long-term meteorological services experience has shown that the most significant weather deterioration is associated with the passage of fronts. Therefore, the crucial importance for determining the meteorological conditions is the correct recognition of the fronts, their direction, and speed of movement.

However, this is not enough. It is needed the constant thorough analysis of the high-altitude synoptic maps (barometric topography maps). In addition, the high-rise that maps directly indicate wind direction and speed at different altitudes, they also help to better assess the intensity of the various natural phenomena, occurring both at the front and in the homogeneous air masses. The big influence is the correct definition of the location of jet streams, wind shifts, convergence zones and, especially, the divergence of high-altitude air flows.

Besides, evaluation of meteorological conditions will not be complete if it does not consider the effect of the terrain on the changing air currents, the formation of clouds, mists, and precipitation, as well as the formation of increased atmospheric turbulence zones of strong winds and squalls.

Here in presenting issues related to dangerous and natural meteorological phenomena, most of the above circumstances are considered.

Dangerous meteorological phenomena and include:

Heavy rain is a phenomenon when precipitation is 50 mm or more (on the mountain, mud and rainfall areas of 30 mm or more) per 12 hours. Loss of heavy rainfall characterized by a large areal spotting. In 63% of cases, they fall in the territory of one region, and in 27% of instances – in areas of 2-4 regions. Heavy rains have a pronounced annual variation. The highest frequency (about 70%) is in June - August. This is because in the summer months, the air masses coming from the sea, more saturated with moisture and a maximum water vapor; partial pressure is also observed in these months. In the warm season, heavy rains are usually accompanied by hail and thunder, and sometimes squally wind, which increases their adverse effects.

Heavy rains are observed on the cold fronts, sometimes they observed in exacerbation of inactive fronts. These rains may be associated with stationary cyclones. A complex interaction of macro- and mesoscale processes causes precipitation. Rainfall and their frequency depend not only on the nature of the synoptic, physical, and geographical conditions but also on the local features of the territory.

In mountainous areas, heavy rains can cause the rise of groundwater and water levels in rivers and reservoirs.

Large hail is when hailstones are of the diameter of 20 mm. Hail is a local phenomenon, often it falls by individual isolated spots in areas from a few dozen to a few hundred square kilometers, at least – in the form of hail tracks.

Hail is caused by intense convective processes occurring in the atmospheric fronts (primary and secondary cold) and their systems, moving in the hollows of cyclones, as well as on front-occlusion of cold type. Sometimes hail is associated with the development of strong convective clouds on the inactive cold fronts or the fronts with active frontogenesis on a stationary front by increasing two-way advection. Especially hail is observed in the daytime at the front with a wave disturbances.

The damage depends on the size of the hail, their density, intensity of precipitation. The hailstones of more than 30 mm in diameter can destroy crops, damage roofs of buildings, and beat bird and sheep.

Wind and squalls have the maximum speed of 25 m/s and more. They are observed mostly in the cold season, with a maximum in January (23%). Almost the same recurrence is in early spring (March - 22%) and late fall (November - 19%).

Strong winds usually occur during the passage of deep cyclones with the advection of cold and sharp pressure drop (sometimes up to 20 hPa), as well as in storm zones between cyclones and anticyclones.

Strong wind by blowing dust and snow, increases the load on the wire, high-rise buildings, impairs visibility and thereby hinders the traffic. A strong wind causes an intense evaporation, dramatically reducing soil moisture.

Squall situations can be created almost every year (77% probability), but certain patterns have not been identified in their spatial distribution.

Squalls occur mostly during the passage of cold fronts, secondary fronts, and occluded fronts. The necessary conditions, in this case, are sufficiently large reserves of moisture in the warm air mass, its unstable moisture stratification, significant contrasts in air temperatures both on the underlying surface and at heights. Squalls are either inmass and front, the latter is the most dangerous.

Squalls often lead to disastrous consequences, inflicting losses primarily for agricultural production and cause lodging of cereal crops, damage trees, destroy buildings and sometimes terminate transmission lines, and suspended the construction work.

Tornadoes are vortices with a vertical component, seldom recurring phenomena. These natural phenomena are difficult to register, the least investigated and therefore have not been sufficiently studied.

Usually, tornadoes occur only in the warm season. Their formation is associated with the interaction of warm and dry tropical air with cold and humid air masses from the Arctic basin. In the zone of the section of these air masses, there are significant contrasts in temperature of about 10-15°C. Large vertical instability of the warm air mass near the cold front creates the conditions for the formation of a thermodynamic convection and, therefore, a very powerful cumulonimbus cloud in the vertical and horizontal extent. A tornado often accompanies the very powerful storm cloud – a strong whirlwind – with an almost vertical or bent axis. Externally, the vortex has the form of hanging from the cloud and reaching the surface of the earth the spiral pillar of cloud, like a huge trunk. Tornadoes occur both over land and over the sea (waterspout). The diameter of a water tornado is usually 25-100 m, and over the land, its diameter exceeds 100 m. Ravage of tornado causes huge wind speed reaching 90 m/s and more, which corresponds to the speed of movement of fronts and cyclones. The wind speed in a tornado is defined by its destructive power and reaches hundreds of kilometers an hour. Vertical movement in tornadoes also reaches enormous speeds up to 90 m/s. Such powerful upward movement of air is due to a sharp decrease in pressure in the center of the tornado to 40-100 hPa. Tornadoes are very destructive, often causing catastrophic destruction and lead to casualties. The width of destruction band is relatively small, from a few hundred meters up to 5 km. But the consequences, in this case, may be extreme: the victims among the people; fallen trees, poles, and masts; destruction of buildings and structures; destruction of agricultural products; accidents and disasters on transportation and other. The tornado can lift into the air and move a few meters' cattle, logs, and even heavy technique. Small river it can suck to the bottom, as happened with the Moscow River in 1904. As a rule, a tornado is accompanied by heavy rain and large hail, which increases its danger.

These dangerous weather phenomena such as heavy rain, hail, strong winds, squall, and tornado are usually seen during thunderstorms. Thunderstorms and hazards caused by them are the results of atmospheric instability, which manifests itself in the occurrence of significant vertical movements of very moist air at large temperature gradients in the formation of cumulus and cumulonimbus clouds. Provided great moisture instability of the atmosphere, the whole process of the formation of cumulonimbus clouds can last 30-40 minutes. The basic process, predetermining the formation of thunderstorms within the homogeneous air masses, is a thermal convection in a pure form, or in association with dynamic turbulence or influenced by the orographic terrain. This determines inmass showers and thunderstorms formed over the continent mainly in the summer after the midday hours when the air temperature near the ground surface maximum. Accordingly, over the sea, these phenomena are observed more often in winter and at night. Typical synoptic conditions of occurrence of inmass thunderstorms are the rear part of the cyclone and filled the cyclone. In the warm coastal areas with mountainous terrain, air is much warmed and moistened, the vertical power of the cloud reaches 8-10 km, almost to the tropopause. This is sufficiently promoted by the rise of air masses on the mountain slopes. Thunderstorms in these regions are quite intense, especially from 14 to 17 o'clock. On plain territory thunderstorms appear sluggish, as clouds do not develop vertically sufficient. The vertical power of cumulonimbus clouds here reaches 4-5 km, i.e., to delaying (inversion) layers. Most of these storms and showers are observed near the poorly defined fronts, which appear only in the convergence of the surface wind. In addition to thermal convection, cumulonimbus clouds are often formed as a result of the displacement of warm moist air by coming shaft of cold air caused by a cold front. This process contributes to the emergence of the frontal thunderstorms. Unlike inmass thunderstorms, the main feature of frontal storms is their large horizontal extent. Thunderstorm zone is usually located along the front over several hundred kilometers, and across the front of a few tens of kilometers. The higher the temperature of the warm air mass and the greater its moisture content, the more intense is the formation of cumulonimbus clouds, and the higher the probability of strong thunderstorms with hail and squalls. In turn, the development of vertical movements is essentially dependent on the nature of the horizontal air flows at high altitudes in the front area, i.e., on the wind velocity distribution. The most favorable conditions for the development of powerful cumulonimbus clouds with strong thunderstorms and showers are created at divergence altitude air currents, characterized by a decrease in the wind speed. Storm clouds on the cold fronts often reach the height of the tropopause and sometimes exceed it.

Thunderstorms of warm front are much rarer. If warm air stratified with moisture instability, vertical movement in it becomes turbulent, and then on the background of stratus-like clouds arise cumulonimbus clouds, accompanied by thunderstorms. Thunderstorms of warm fronts occur in the mid-latitudes when cyclones moving from the south and southwest, and mass of moist tropical air are submitted in their warm sectors. In this, high-altitude flows are divergent. A characteristic feature of these storms is their greatest activity during evening and night hours.

Particularly noteworthy are severe thunderstorms accompanied by simultaneous showers with hail, squalls, and sometimes tornadoes. Analysis of synoptic conditions shows that the formation of such natural weather phenomena is observed near the point of occlusion of the cyclone, especially near the center of the small wave disturbance. And in both cases, there are very contrasting temperatures. In the warm air mass, daytime temperatures generally exceed 30˚C (to 35˚C), and in the cold air, they are within 16-22˚C. The wave disturbance formed on the main cold front, do not develop into independent cyclones, and very fast, with a speed of 80-100 km/h, move in the direction of the wind in the middle and upper troposphere. The difficulty of identifying these disturbances on the synoptic maps is that from the beginning of their formation, they hardly reveal themselves, especially in the morning. But a few hours later near the center of the disturbance, powerful storm clouds with showers arise unexpectedly, especially after midday.

Vertical development of cumulus clouds occurs until the temperature of rising saturated air is equal to the ambient temperature. During this time, cloud height can reach several kilometers. Thus, the storm clouds are often applied to the entire thickness of the troposphere, reaching the tropopause by their peaks. As soon as the cloud will reach the air temperature of -10˚C – -16˚C, and sometimes even lower values, the upper part of the clouds starts to crystallize. This part becomes a rounded shape and takes the form of vague cirrus clouds, having the form of a gigantic anvil. Since the beginning of freezing of the cloud tops, precipitations start to fall out from them as rain, hail and sometimes snow.

Research has shown that the central part of the storm cloud is generally well reflected in the radar display in the centimeter range. In this intense hail flows, if they exist, do not fall into the center of a storm cloud, but its border regions.

Simultaneously with the rainfall with thunderstorms there observed electrical discharges (lightning), between clouds or between a cloud and the ground, and a sharp increase in wind in the form of vortices with a horizontal (“barrage gate”), and sometimes vertical (tornado) axes.

The barrage gate is in front of the storm cloud, typically 1-2 km to the continuous rain curtain. This vortex is formed in counter air movements – anti-clockwise. In front of him is an upward flow, and in the rear – downward one.

Usually, a vortex moves at an altitude of 500 m but sometimes reduced to 50 m. After the passage of the vortex shaft, there is a sharp increase in the wind to 20-40 m/s, accompanied by a change in its direction. At the same time, there is a significant decrease in temperature and increase in pressure due to the air cooling by precipitation.

The strong dust (sand) storm is a transport of large quantities of dust and sand, lifted from the parched soil surface, by strong wind (speed of 15 m/sec or more).

Most often dust storms are observed in the transition zones of the storm with large barometric and thermal gradients, which are formed by the displacement of the powerful anticyclone and the simultaneous activation of cyclonic activity. In addition, dust storms can occur in passing cold fronts, where there observed large pressure gradients and squall winds in the rear of them. Dust storms of such origin are short and local. A significant influence on the intensity of dust storms has the character of the underlying surface (the structure and the degree of soil moisture, the presence of vegetation and orography).

Agriculture especially suffers from dust storms, layers of arable land are carried away from the fields, crops are damaged. Dust storms impair sanitary conditions of cities, transport operation. Prolonged dust storms cause the greatest loss when they act for a few hours or even days.

A Severe snowstorm is a transfer of an early snowfall during the day or night with the wind speed of 15 m/s and more. They occur almost every year (with a probability of 96%).

For snowstorm are typically strong winds, snowfall, the wind and sometimes rising the snow from the surface of the snow cover. Snowstorms occur mostly around the warm front, associated with actively developing and rapidly moving cyclones. In such cases, a cyclone deepens, pressure gradients in the zone rise, so that the wind increases significantly. In addition, at this stage of the cyclone life in front of him, there is the greatest development of the ascending movement, leading to the formation of powerful clouds and very solid precipitation. The width of the zone of snow before the warm front of coming cyclone usually reaches 500 km. But the area of heavy snow often is 100-200 km.

Snowstorms mode depends on many factors: the amount of snow, the power of its bedding, wind speed and direction, physiographic features, and the state of the underlying surface, temperature, etc.

Snowstorm disrupts the normal operation of transport. In mountain areas, they can create a situation conducive to avalanches, and this can cause great destruction, loss of life, destruction of animals and plants.

Heavy snowfall means the precipitation of 20 mm or more in 12 hours or less. Such snowfalls may continuously add to the danger of this phenomenon. Heavy snowfalls are observed in the period from October to April with an average daily temperature of 0°C to -2°C – -4°C. In mountainous areas, it can drop to -15°C. Heavy snowfall in 50% of cases takes place on the territory of one-two areas, and in 40% of cases – on the territory of the three-five areas.

Heavy snowfall is associated with the exit of cyclones from more warm and humid areas and with the passage of cold fronts with wavy perturbations. They can also be formed in the stormy transition zone between cyclones and anticyclones.

Heavy snowfall hampered the work of transport, break off the wire communication lines and power lines, damage the trees. In mountain areas, they can create a situation that contributes to avalanches. Late snowfall can cause the rise of groundwater and water levels in rivers and reservoirs.

Heavy mist has visibility of 100 m or less. Heavy mists are observed most often in the cold half of the year, when, after a significant cooling, a steady heat loss begins in the pre-cooled underlying surface, i.e., there occurs an increase in the relative humidity to saturation and decrease in air temperature.

Mists in origin are divided into inmass and front, cooling, and evaporation. Mists of cooling, depending on air temperature changes, are divided into advective, radiation and advection-radiation.

Advective mists are formed by advection of the warm air mass on a cold underlying surface, whereby the steam contained in the surface layer is cooled to the dew point and reaches saturation. They occur in the western periphery of the stationing anticyclone ridges, in warm sectors of cyclones, and at the removal of warm air from the Atlantic Ocean on the northern periphery of the Azores anticyclone.

Radiation mist is formed in the central parts of anticyclones, spurs, and ridges of high pressure. In winter, these mists are caused by intense radiative cooling of the lower layers of air at night in clear (cloudless) weather. During warmer months, they arise in the clearings behind the front after a rainfall.

Advection-radiation mist is formed by further cooling at night clearings of warm and moist air moving over the cold underlying surface, i.e., under the combined effect of both advection and radiation factors.

Front mists occur over the continent in the cold part of the year in the zone of warm fronts and occluded fronts on that their sites, which are moved very slowly, up to 10 km/h. Sometimes before these fronts, clouds descend to the underlying surface, turning in the mist in a cross-country terrain.

Cooling mists are formed on the slopes of the mountains because of the cooling of moist air rising up the slope on the windward side of the mountains. Finally, evaporation mist (or steaming) occur during the passage of a very cold moist air over the surface of the nonfreezing water in the cold season.

Mists cause reduced visibility, creating a substantial interference with the work of all types of transport. Given the dense mist, when things are not seen for hundreds or even tens of meters, the functioning of modern transport becomes almost impossible. Mist droplets are deposited on the ground constructions and corrode *metals. Mists contribute to the increase of air pollution in large cities with industrial emissions.*

Strong ice is an ice deposition that is observed on a standard icing machine of 20 mm in diameter or more.

The deposition of heavy ice occurs mainly in the advection of warm humid air caused by the movement of cyclonic formations from the front system. Icing of ground objects and vegetation is usually observed in the cold front area after a freezing rain or freezing of wet snow fallen before the sharp cooling. Icing areas located along the front lines, occupy a considerable area and are very dangerous. Ice can be inmass or front, the latter is more intense. A prerequisite of icing is a presence of overcooled water droplets in a cold surface air with negative temperature. Therefore, ice is usually accompanied by such phenomena as fog or thick haze, loss of drizzle or light rain. Distribution of ice depends not only on the general circulation of the atmosphere but also on local conditions (altitude, ruggedness, type of landscape, openness to air flow).

Strong ice leads to emergency situations in all transport types, to breakage of communication lines and power lines.

On the windward side of the mountain ridges or peaks, air masses add dynamic momentum of recovery to the thermal convection, reinforcing the process of cloud formation. They become stronger and denser near the surface of the ground, obscuring the tops of the mountains, and it is a danger of icing. Mountain ridges and peaks equally affect both frontal clouds, and the clouds of uniform air masses. Even nimbostratus clouds associated with a warm front, not posing the danger of the flat terrain, take the form of rain clouds at the windward ridges, in which icing can be significant even if the rainfall is intensive.

Severe frost is a phenomenon where the temperature is below -30°C to mid-latitudes, -25°C for the southern regions. Therefore, to highlight areas exposed to dangerous and especially dangerous temperature, the latter considered by gradations -25, -30, -40°C and below. Intrusions of Arctic air cause the low temperature, usually traveling in anticyclones from the north-east, north or northwest in the rear of diving cyclones. In most of the territories, the arctic air undergoes additional radiation cooling to -30 – -35°C and below, and the cold anticyclonic weather is installed. Strong winds and low drifting snow accompany extreme cold.

Prolonged low temperatures can lead to death from freezing large areas of winter crops and fruit trees. Deep freezing of the soil can cause an accident of underground utilities. Severe frosts lead to an increase in electricity and fuel costs, complicate the work of transport.

Frost means a lowering the surface temperature below 0°C during the growing season, concluded between the dates of stable transition temperature by 5°C in spring and autumn to early crops. Frosts occur in anticyclones and ridges formed in the Arctic air masses, as well as in light-gradient areas of high pressure and barometric spurs oriented from west to east. In addition, they are possible when the advection of cold air occurs in the rear part of the cyclone. Most of these are advection-radiation frosts.

Late spring and early autumn frosts pose the greatest danger to the herbaceous plants and horticultural crops, as the plants at this time intensively vegetate or ripen, and their resistance is low.

The intense heat is characterized by temperatures of +30°C and more to mid-latitudes. High air temperature is set in the high-pressure areas about the arrival of the dry continental air masses and tropical air over the southern or western periphery of sedentary anticyclones from Central Asia and North Africa. In addition, the intense heat is generated in a slow-moving thermal depression resulting from prolonged heavy warm air, as well as barometric spurs, ridges, and anticyclones of the Azores maximum, shifting slowly to the east. In some cases, the high temperature is observed in front of the slow-moving trough, the axis of which is in the Central Europe. The intense heat with a small amount of precipitation or lack thereof forms the dry weather which adversely affects the flora and fauna.

Droughts is a wind of the speed about 5 m/s for 3-5 days at a temperature of +25°C and more, and a relative humidity of less than 30% during the flowering and ripening grain.

Usually, these winds are observed at the anticyclonic type of weather, or in the spurs on the west or south-western periphery of the anticyclone. In dry conditions, when the air warms up much with intensive transformation, the relative humidity drops sharply, and the saturation deficit is increasing. Dangerous and very dangerous winds are observed every year during the period from April to September, usually in June - August. The diurnal variation of natural hot winds like the direction of air temperature and wind speed has a pronounced maximum in the afternoon.

Droughts lead to lower crop yields and sometimes to its complete loss, harming plants at different stages of their development.

Extreme fire danger takes place when the fire hazard index is greater than 10, 000 ° C (according to the formula of V.G. Nesterov). Fire situation is when anticyclonic pressure field is created. It is determined by high-altitude ridges and centers of the anticyclone of the Azores in origin, oriented to the south, southwest, and east. Under these synoptic processes, there is established clear, dry, hot weather with plenty of sunshine, as a rule, in the period from May to October. High temperatures and prolonged lack of precipitation lead to a significant drying of the soil and the creation of an emergency fire danger.

Strong disturbance in the sea and reservoirs takes place when the height of waves on the reservoir waters is 2 m or more, on small seas (such as the Sea of Azov) – 3.5 m or more, on the inland seas (such as the Black or Caspian) – 6 m and more, and on foreign seas and oceans is 10-12 m and more. Strong waving in the sea and reservoirs occur under the influence of the storm wind. It causes mass accidents in the Navy, flooding, destruction of hydraulic and other structures in the coastal zone and at the coast. Because of these processes annually irrevocably lost more than 100 hectares of coastal areas. Placed in the coastal zone of tangible assets (houses, resorts, utilities, agricultural land) are under constant threat of destruction.

High water levels (floods) take place in achieving and exceeding the level of the critical level of rivers, threatening the area. Floods occur during prolonged heavy rains and the melting of snow. On some rivers (e.g., on the Neva River) floods are the result of surge processes.

This natural phenomenon causes considerable losses to agriculture: flooded agricultural land, granaries, warehouses fertilizers. The appearance of shifts and powerful mudflows complicate the situation. In the case of further intensification, other threats arise – the destruction of a building fall, roads, railways, pipes, power lines, and other communications.

Increased groundwater level (flooding) means the increase of the groundwater level in built up areas to depths above drying construction rules with a significant deterioration in living conditions, or in locations of economic facilities, significantly affecting their work in routine mode. The main causes of flooding are an unorganized runoff, soil deterioration of filtration properties of natural drains (ravines, gullies, channels, and small rivers, etc.), reduction of evaporation due to asphalting, the dam effect of the foundation, trace reservoirs, and tunnels. Construction of artificial reservoirs results in the flooding of residential areas, the destruction of houses and resettlement of their inhabitants. The flooding process itself causes adverse effects of natural and man-made environment.

Sill is mud or mudstone flows, suddenly appearing in the beds of mountain streams. Sill processes are developed in the mountainous areas. The massive mud flows lead to destruction, infliction with mud deposits of buildings, highways etc.

Avalanche is a snowfall from the steep slopes of the mountains. Cause falling highways by snow, and the destruction of homes. Areas of mountain ranges are a dangerous avalanche in February - March and during thaws.

An early freeze-up or fast ice is freezing in the courts and the emergence of mass accidents in the Navy, ships and loss of life due to the appearance of the heavily traveled coastal ice or ice cover when frost is in abnormally early stages.

The gap between the coastal ice means separation of the coastal ice and drift under the influence of wind and warming in the open sea with the possible emergencies in the event of a finding on it of people and equipment or other property.

The intensive ice break-up is formed with a sharp warming. May cause a massive failure of vessels, damage and destruction of hydraulic and other structures in the coastal zone due to heavy ice drift and pressure to the beach.

Fast icing of vessels is the intensive deposits of ice on the surface of the vessels (0.7 cm/h and more). Fast vessels icing occurs during the wet and cold weather and leads to massive accidents in the Navy, loss of ships and the loss of human life.

Basic Laws of Pollution Spreading in the Atmosphere

In the process of air pollution, along with a variety of natural phenomena (volcanic eruptions, forest fires, weathering, and erosion, etc.), human activities associated with the development of natural resources, technological advancing of industry, agriculture, energy, and transport play the increasing role. This activity sometimes leads to undesirable consequences. Due to the imperfection of using technologies, insufficient knowledge, and responsibility, the lack of appropriate comprehensive predictive study, as well as for other reasons some harmful byproducts get the atmosphere. Air pollution causes great, and sometimes irreparable, damage to both the environment and man himself. Harmful substances with different toxic properties, penetrating the human body through inhalation, skin or food pose a real threat to health. These adverse effects may manifest as acute or chronic poisoning, and all sorts of diseases (allergies, cancers, anemia, neuropsychiatric disorders, leukemia, and others). In addition, genetically active substances can cause birth injuries and defects of development.

The specifics of the landscape of industrial centers may exacerbate the negative effect of the pollution. High concentrations of harmful impurities in the atmosphere under certain weather conditions can be maintained over a relatively long period a large territory. This may cause acute massive disease of the population, sometimes with fatal consequences. Such conditions are especially dangerous for children and the elderly and those who suffer from lung and heart diseases or prone to them.

The sickness rate has its regional differentiation. This is especially confirmed by the study of the health of school-age children. The most alarming fact is the presence of children's chronic pathologies. There is a very strong performance of the types of abnormalities that may be associated with atmospheric pollution, namely the incidence of children's tumors, diseases of the respiratory system, skin, and subcutaneous tissue.

Air pollution reduces the productivity and fertility of domestic and wild animals and birds. It causes their disease and even death as a direct consequence of the impact of toxic through inhalation or the skin, and the penetration of food and water. Harmful impurities, falling on the ground and in the water reservoir, contribute to the disease and destruction of vegetation over large areas. The presence in the atmosphere of active chemical additives causes great harm to various building structures and industrial output, leading to premature aging and destruction of buildings, industrial structures, and historical monuments.

The number of air pollutants over the area is defined by two groups of factors – the nature of sources of pollution and the state of the atmosphere. Obviously, the intensity of the sources, as well as physical and chemical properties of the impurities are the major parameters that can be used to classify the type of pollution. There are also other characteristics of the source, such as the shape and duration of emissions of impurities and their effective height.

It is not the aim of the authors to consider in detail the specific characteristics of a variety of sources in terms of air pollution. The natural sources (volcanoes, dust storms, etc.), transport sources (road, air, rail, river and sea transport), and industrial sources are significantly different from one another in the height and shape of the release, as well as the composition and quantity of emitted contaminants. These issues are described in sufficient detail in the specialist literature. Instead, this book gives the advantage of direct study of surface and elevated sources of emissions related to the businesses of certain industries (energy, metallurgy, chemicals, coke, cement, etc.) with their project operation and under accident conditions.

Air Pollution Emissions of Enterprises in Project Mode

When considering the spread of emissions from the chimneys of industrial enterprises operating in project mode, it is generally considered that the diffusion of harmful gasses, liquid, and solids contained in the releases is subject to the same laws of a stationary source. The impurity distribution near steady source is described by a Gaussian law. As you move farther from the place of emission, the pollutant concentration decreases, and its absolute value is higher, the finer fractions are contained at the time of release. The only particles of the diameter of 20 microns or less behave approximately like gas, which is composed of. Larger particles fall out of the aerosol cloud and settle on the ground and near the emission source. The temperature of impurities emitted from the chimneys into the atmosphere, higher compared to the environment, changes the shape of the torch, bends its axis and increases the height of emissions.

The shape of the torch in the wind direction is largely dependent on the structure of turbulence. With a strong instability, large eddies caused by free convection dominate in the boundary layer of the atmosphere. Since the size of the vortices is over the torch diameter, their impact mainly lies in the fact that they either raise or lower the torch, giving its borders the form of waves. These disturbances are moved together with the airflow, increasing in amplitude. With such instability, torch with impurities contained therein reaches the underlying surface near the source, where a large concentration of pollution occurs instantly (Figure 12a). After a source long-acting, due to the constant changes in wind direction, the smoothed distribution of the impurity concentration in the underlying surface takes the form of a fan.

In windy conditions during thermal stratification which is close to neutral, the turbulent exchange is carried out mainly by small eddies of forced convection generated by wind speed shift. Therefore, the extension of the torch from the raised source occurs at approximately the same intensity both vertically and horizontally, and it takes the shape of a cone which is symmetric to the central axis (Figure 12b). Since the diameter of the cone increases in the downstream direction only due to turbulent diffusion and transfer of vertical weakened neutral thermal stratification, the torch reaches the underlying surface at large distances from the source.

In the case of stable atmospheric stratification, turbulence is very weakened, and any incipient updrafts are actively suppressed. Due to the lack of development of vertical diffusion, the torch is thin, but because of the instability of the wind direction, it takes a V-shape (seen as a fan when viewed from above). In the case of constant wind direction, the torch has a ribbon shape and can remain as such without a significant change to about 100 km from the source (Figure 12c). If on the leeward side there are no significant changes in topography, the impurity concentration in the underlying surface at this distance is almost zero, since there is no developed vertical transport.

From the point of view of environmental protection, the most favorable condition for the impurity scattering is the location of the source of the inversion layer. Stably stratified air prevents the spread of impurities in the layers lying below, and moderately unstable stratification of the upper layers assists diffusion of pollution up (Figure 12d). When the inversion layer located over the source and prevents the spread of impurities into the upper layers, the opposite effect occurs. Since the bottom of the stratified air is unstable, the convective mixing assists in the spreading impurities to the underlying surface, and increases the concentration of pollution along the line respective to the torch axis (Figure 12e).

Air Pollution in Accidental Emissions of Enterprises

Define the short-acting sources as instant. There are among them some emergency or salvo emissions of the industrial enterprises, carried out in a relatively short time. A typical type of instant source is the emission of impurities at all possible explosions. Because of the instantaneous power, a cloud of the dust-gas mixture rises to a certain height. Further, it is carried by atmospheric flows as a passive formation and dissipated in the process of turbulent diffusion. The initial lift height is determined from observations or calculations. Sometimes it can reach very high values.

Figure 12. The characteristic shapes of the torch at different atmospheric stratification; temperature gradients: dotted line is for dry-adiabatic; continuous line is for real

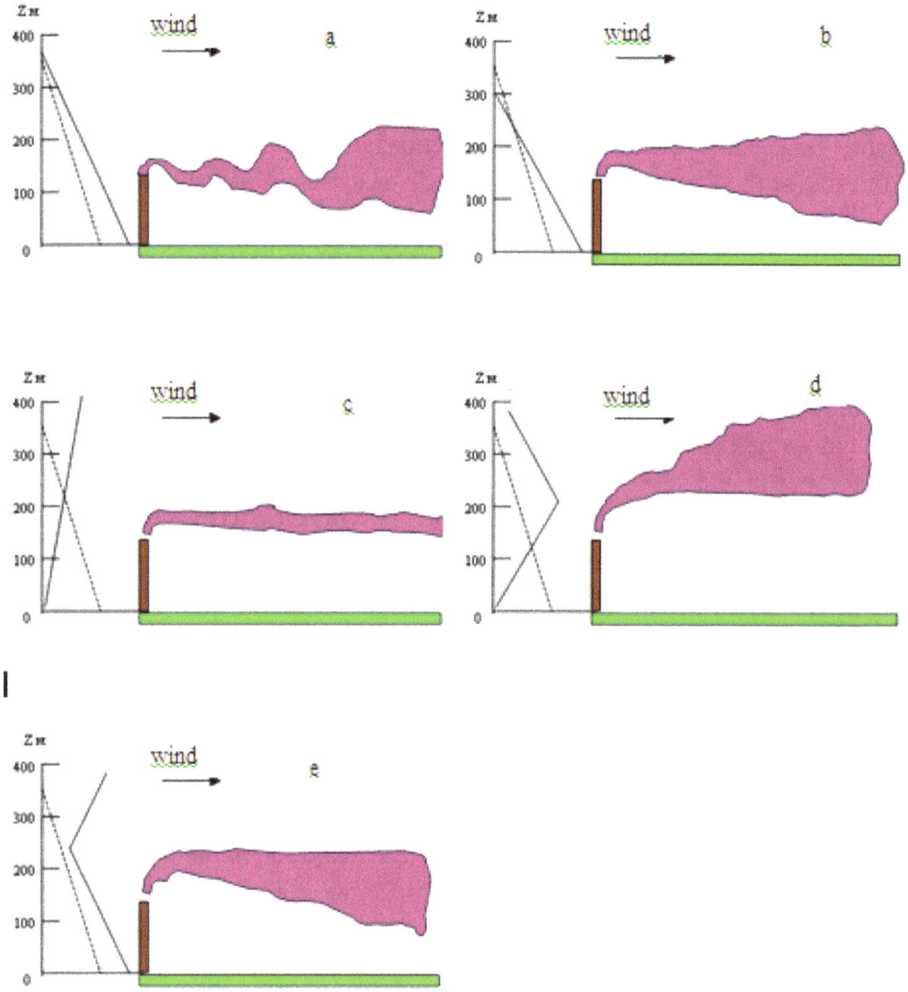

Three groups of factors determine the amount of substance polluted this area after the accident at an industrial facility – emission power, the physical and chemical properties of impurities and atmospheric conditions.

When the emergency release occurred, there is allocated a certain amount of aerosol the *M*, which extends in all directions creating a cloud at a certain height *H*. Cloud first increases, moving in the direction of the wind. Then it becomes lighter at the edges, its dark opaque portion is reduced, the entire cloud becomes lighter and begins to "melt" and eventually disappears. This pattern is especially crisp on a clear day on a background of blue sky.

The process of growth and "blurring" (diffusion) of a cloud is turbulent; it corresponds to a certain effective turbulent diffusion coefficient ν. Wind speed and direction $V = (v_1, v_2, v_3)$ determine the speed and direction of clouds movement (convection) in the atmosphere.

Ceteris paribus, the more powerful the explosion, the higher the impurity cloud will rise and the less will be the concentration of harmful substances in the underlying surface in the downwind sector, as impurities will be longer dissipated by turbulent diffusion. The effective height of the cloud arrangement is increased if the impurity is ejected at high speed and its temperature is much higher than the ambient temperature, i.e., on the emergence of buoyancy.

Accidental releases can be a chaotic mixture of all kinds of solid, liquid, and gaseous substances. The size of solid and liquid particles in diameter may be less than 0.1 microns and a few centimeters. Since the mass of particles 10 microns in diameter is large enough, they fall quickly from the atmosphere after ejection, so they are more likely happen near the accident site. Particles with a diameter smaller than 1 mm longer remain in a suspended state. Their spread more depends on the state of the atmosphere, they can remain in the air for a few days and be transported by the wind over long distances.

Impurities in the cloud can be subjected to erosion by precipitation, chemical and physical transformation due to contact with moisture or other contaminants in the atmosphere. The movement of air causes both the scattering and orderly transfer of impurities in the cloud. If the value of atmospheric vortices is less than the cloud size, the vortices dissipate impurities, else they transport them. The cloud id eroded (deformed) under the action of the wind either in the direction of travel and transverse and vertical directions due to the turbulent transport. Wind direction determines the trajectory of displacement, and the wind speed – the speed of movement of the clouds of impurities from the release point. The vertical temperature gradient in the cloud level is essential, since the unstable stratification assists lifting a cloud, and its inversion one delays it.

Prediction as the Functional Procedure of an Emergency Prevention

The issue of man-made and natural security of actions, states, and processes that directly or indirectly lead to damage to the natural environment, individuals, and humanity in general, can be solved primarily as an effective management system based on reliable forecasts. Probably the first who realized that to anticipate emergencies and prepare for them is cheaper than to eliminate the consequences of them were Napoleon III and Le Verrier. Napoleon III commissioned Le Verrier to investigate the causes of the storms of enormous power that destroyed the Anglo-French squadron in Balaklava bay on November 14, 1854, during the Crimean war. Le Verrier has shown that the movement of the storm to the Black Sea took place with an almost constant rate, and this movement can be traced back to the synoptic charts consistently in the development of the storm. He concluded from this that if the observations were carried out on a large number of stations and their data with sufficient speed transmitted to the central office, the comparing and analyzing weather maps, it was possible to warn of the danger to the fleet.

The objective of environmental protection and rational use of natural resources can be partially solved with the help of a work, revealing the meteorological aspects of air pollution, i.e. associated with the prediction of the conditions in which high concentration of impurities arise in the surface layer of the atmosphere. Both the short-term (within 1-3 days) and long-term forecasts are of practical interest. The interest on short-term forecasts is due to at least two circumstances. The first is that emission of harmful substances into the atmosphere and their concentration in the air is quite large in many cities and industrial centers, and the level of available technologies is not always able to provide the necessary purification of the emission. So, naturally, the question arises of the possibility of termination or reduction of emissions at least in a relatively short period when there is a bad weather situation, leading to dangerous levels of air pollution in residential areas. The second is that at the major accidents in chemical plants, nuclear power plants and other sources of salvo emission of harmful substances into the atmosphere, owing to their surprise, there is no real opportunity to prevent or regulate harmful emissions, and at certain moments of time in a certain area there can be created abnormally dangerous levels of air pollution. In these cases, the operational forecast data on the concentrations of harmful substances in the air are important for timely informing authorities and population about the danger of attack. The results of long-term forecasts that consider the adverse weather conditions in which high concentrations of harmful substances can be observed have a practical interest in the design and construction of new plants as emissions sources and for determining the locations of the evacuation from the contaminated areas after the man-made accidents.

One of the operational procedures to prevent an emergency is the prediction with the purpose of early mapping probability of its occurrence and development on this basis of a comprehensive analysis of the causes and sources that contribute to its occurrence. However, the experience has shown that forecasting remains one of the very complex and insufficiently developed methodical problems. This can be explained by the lack of consistency in predicting the probability of emergencies and the possible situation in which the impact of natural phenomena on various engineering and technical systems should be comprehensively measured, as well as the parameters, or combinations thereof, directly related to these systems and to anthropogenic activities. The quality of these forecasts is low and is determined by the validity and reliability of the procedures used, which does not always allow to calculate the set of all possible emergency scenarios, to estimate the probability of their occurrence and the effectiveness of preventive measures. The magnitude of the risk of the hazards is significantly affected by the element of surprise, intensity, speed, duration, and frequency (recurrence) of their development. There are no clearly expressed practically significant temporal patterns of spontaneous natural phenomena now. There are no such laws also in the development of man-made disasters (though, of course, a certain connection with their natural phenomena does exist).

Given the unexpected occurrence of various natural phenomena, the probability of their manifestation may be provided as follows:

- First, for each type of phenomena, there can be set specific spatial reference.
- Second, most of the phenomena have some repeatability; there is such a tendency that the higher the intensity of the phenomenon, the less often it is repeated with the same force.
- Third, to a greater or lesser reliability, there may be the dependence of the devastating effects of natural phenomena on its scale, duration, and intensity of hydrometeorological processes.

Information on the time of occurrence of a natural phenomenon with serious consequences is an important factor in reducing the risk. Unfortunately, obtaining such information is possible only for certain types of natural phenomena, and it is based on both the data of immediate onset of symptoms (or other phenomena) and the phenomenon statistical repeatability. Regarding the latter, the forecast of phenomena on the basis of statistics is obvious and very unreliable, but the measure of the risk of hazards can be estimated with an accuracy of season and detail. This applies, for example, the possibility of fire. From 40 to 70% of fires are usually in the summer (the period of dry weather and high winds), most of them developed from 11 to 17 hours of the day. Large fires occur less often in the winter – in the freezing dry and windy weather. But the patterns of occurrence of floods and sills in mountainous areas highlight the design and construction of protective structures, the implementation of measures to prevent potential damage to land and economic complex objects.

These provisions demonstrate the possibility of a variety of risk management of catastrophic events, both natural and man-made. Among the elements of the risk management must be:

- Zoning areas according to the degree of danger.
- The risk-adjusted organization of industrial use of the territory (in particular, the construction of chemical plants in the less exposed areas of the population).
- Regular monitoring of hazards.
- The construction of protective structures.
- Operational counteraction to hazards (all available means) by the administration before and during its development.

Today, in all developed countries, the most promising are numerical methods of weather forecasting. This is a physical-mathematical method of modeling the properties of the atmosphere changes at different altitudes for both small areas and for regions that span multiple continents and even the entire globe.

Mathematical methods of forecasting are not new. English meteorologist L. Richardson back in 1922, developed the "method of the previous calculation of the pressure field" and tried to calculate the weather forecast for Europe in 6 hours ahead. Although the numerical results of his calculations were completely unrealistic, yet his proposed principle of calculation through small intervals was the basis for modern numerical forecasting methods.

The idea of numerical prediction method is reduced to a mathematical interpretation of the laws of atmospheric processes. For this purpose, one chooses a system of equations describing these extremely complex patterns as close as possible to reality. The major difficulties of this choice are due to the fact that the equations describing atmospheric circulation processes and phenomena in it, are differential nonlinear equations of the second order in partial derivatives and contain a large number of unknowns. The next step is to replace the mathematical model of atmospheric circulation, presented in the form of differential equations, integrals, logarithmic, exponential, and transcendental functions, by working formulas, which are reduced to the implementation of the four basic arithmetic operations.

The accuracy of numerical weather prediction depends on three main factors:

- Completeness of physical prediction scheme, considering all of those factors affecting the weather.
- A sufficient amount or a reasonable accuracy of observational data at the start of the forecast.
- The method used for solving equations.

In meteorology, there is the notion of accuracy (corroboration) of forecasts. In general, according to the World Meteorological Organization (WMO), accuracy forecasts for the month and the season is 60-65%. And the less the forecast period, the higher are the rates. Numerical predictions for the medium term (up to 10 days), produced by hydro-meteorological centers are characterized by the following results: 75-80% of accuracy in a decade, 95% for a day. Short-term forecasts, for example, for aviation, from several hours to half a day, even approaching the absolute skill. Unfortunate weather forecasts are becoming seldom. Most of the inaccurate forecasts, usually associated with the appearance of any quick, unexpected atmospheric disturbances. The forecast of these rapid, unexpected atmospheric disturbances is far from completion, especially for a long time, and remains the main task of the weather forecasting. Yet it should be noted that scientists have penetrated some of the mysteries of weather, feel the rhythm of the natural weather machine, and even tried to artificially influence the processes that shape the weather.

In numerical forecasts of macroscale processes, the time during which the difference between the two solutions with slightly different initial conditions (of the order of the measurement error) is not greater than the difference between two randomly selected states of the atmosphere, is regarded as the limit of predictability of the atmosphere. The works of many researchers of atmospheric dynamics in the years 1980-1990, were focused on the assessment of the predictability limits. According to (Bouttier, 1994), the predictability border of large-scale atmospheric movements ranges from 10 to 30 days.

Over the past 20 years, a significant progress has made in forecasting macroscale weather changes using hydrodynamic models, and advances in the modeling of mesoscale and regional events are much more modest. In areas with characteristic dimensions 5000×5000 km in a relatively short period (up to 48 hours), one can often see only one or two energetically active weather system. Different systems quite differently react to the initial perturbations, the influence of the underlying surface and the physical processes occurring in the medium itself (the transition of potential energy into kinetic energy). Modern research on estimation of the boundaries of predictability of mesoscale and regional processes show that their perturbations can be predictable in the range 24-72 hours, if the mathematical models will use more accurate initial and boundary conditions and improved parameterization of some physical processes of sub-grid scale.

Modern hydrodynamic atmospheric models are classified according to the size of the territory for which the calculations are made. *Global models* are used to predict values of meteorological variables for the entire globe. They provide the resolution of longitude and latitude of about 0.5° which is sufficient to describe the weather-forming processes of a large scale. Among them, meteorological practices most commonly used US Global Forecasting System (GFS), a model European Center for Medium-Range Forecasting (ECMWF), British Unified Model (UM, UKMET) and the Japanese Global Spectral Model (GSM9603). Global models of the atmosphere, even at high spatial and temporal resolution, do not currently consider all the peculiarities of the local areas for forecasting as the world meteorological centers do not have a complete set of data available to the regional services that can involve additional automatic sensing and surveillance systems.

Contemporary high-performance computing systems provide the ability to use numerical models of *regional models* of high spatiotemporal resolution and more accurate reference to the local geographical and geophysical conditions. So, these models can produce more detailed forecasts. Regional models, by reducing the area of the calculation to one country or region, can increase the horizontal resolution up to 1-2 km and reproduce more accurately the processes on a small scale what is necessary to predict the phenomena related to free and forced convection, gravity waves, and so on. The most known of them are the model Aire Limitée Adaptation Dynamique Dévelopement International (ALADIN, France), Consortium for small-scale modeling (COSMO), Regional spectral model (RSM9630, Japan), Deutscher Wetterdienst (Offenbach, Germany), Mesoscale Model 5 (MM5, USA), Weather Research and Forecasting-Advanced Research WRF (WRF-ARW, USA).

The model WRF is a numerical mesoscale weather forecasting system of the latest generation developed for operational forecasting (Skamarock, 2008) as well as for research purposes (Bagrov, 2007; Guziy et al, 2008). WRF is a free model of open source software that allows to use it not only in a different state, and private weather office with the possibility of fine-tuning for different needs. In many countries, the output of WRF model calculations has been successfully used in the prediction of natural emergency situations of different nature (strong winds, heavy rains, etc.).

WRF can be used in a wide range of applications at scales ranging from a few meters to thousands of kilometers. It can numerically solve the complete system of equations of atmosphere hydrothermodynamics which includes the equation of continuity, momentum transfer (considering non-hydrostatics and compressibility effects), the transfer of moisture and energy. Its distinctive features are several dynamic cores, three-dimensional data assimilation system and software architecture aimed at parallel computing, and extensibility of the system. WRF offers a library of different parametrizations of processes in the atmospheric boundary layer, the formation of clouds and precipitation phenomena of convection and turbulence. The model implements the ability to create sub-region and use computing algorithms assimilation of observational data. The is used the vertical σ-coordinate system in WRF which allows to describe a complex topography and the horizontal Arakawa grid of C-type (i.e., a volumetric cell is given as a rectangular parallelepiped, all scalar values are set in the center, and the components of the vectors are set in the centers of the respective sides). In the integration of the hydrodynamic equations, there is used splitting by directions (the explicit scheme is used in horizontal directions, and implicit one – for calculation of the vertical flow and velocity) and by physical processes. WRF is a regional model, so for its initialization and setting the boundary conditions, results of calculations of global numerical weather prediction models are needed. The model WRF work cycle consists of data preparation, including horizontal and vertical interpolation on a grid pattern, improved interpolated data by including weather stations and radiosonde observations, and numerical integration.

Preventive Measures for the Management of Natural and Technogenic Safety

Yet in the works of Greek historian and writer Plutarch (46-126 AD), there is the record that after a great battle, accompanied by fires, an increase in precipitation was observed. Eyewitnesses of the battles of Borodino, Waterloo, and the bombing of Dresden show that showers also accompany these fiery events. When one or another region was covered by prolonged drought, scientists have thought over the question of the use of this phenomenon to the active influence of the weather. And not just thought. Beginning from the XVIII century, in various countries, including Russia, USA, Ukraine, Australia, and some western European countries, to cause precipitation or dispel the storm clouds they began to shoot at the clouds from the guns. Precipitation sometimes fell, but there was no certainty that artillery processing clouds cause them. Without thorough research, it has been difficult to answer the question, what was the cause of strengthening the coagulation of cloud particles and how to convert them into big drops of rain.

In the first half of the 1950s, the first successful scientific experiments in Russia, Ukraine, and many other countries were carried out and obtained experimental results on the evocation of artificial precipitation. What is the mechanism of formation of artificial rain? In the rapidly developing clouds, the rain often starts due to mechanical enlargement of the drops, which occurs because of their merging. The growth of droplets occurs through the condensation of water vapor on the already formed drop and is particularly intense when developing clouds up, drops fall into the negative zone and freezing temperatures, or if the ice crystals appear in the cloud. In such cases, the droplets grow quickly, simultaneously merge with others, and falls depending on the temperature in the form of rain or snow in the lower atmosphere. If a mature cloud grows very quickly and reaches the area with negative air temperatures, then hail may occur in the center of the cloud due to the rapid freezing of large droplets. This prompted the path of influence on these clouds. For the induced rainfall, it is needed to create a certain amount of ice crystals (nucleation) in the cloud at a height with negative temperatures to "run the reaction" of intensive crystallization cloud droplets.

The particles of solid carbon dioxide are cooled by evaporation to near -80°C and icicles are formed around them. They can grow to a size of 10^{-6}cm, and then can continue to grow due to diffusive transport of water vapor with drops on crystals, regardless of the presence of artificial cooling. This transfer of water vapor is explained because the relative humidity over the pieces of ice is always greater due to their low surface temperature, and hence the pressure of the mixture of air and water vapor is smaller than above droplets. This causes the "pumping" of water vapor from the higher pressure to lower. Thus, the overcooled droplets are reduced due to evaporation, ice crystals grow and they are transformed over time into snowflakes that fall on the ground. Thus, the cloud dissipated (destroyed) forming precipitates. This process is carried out at a temperature of about -10°C (silver iodide particles cause crystallization even at a temperature below -5°C).

The results of long-term studies have shown that in limited areas it is possible to redistribute the rainfall: to increase where they are needed, and to reduce, where there is no need for them. Examples of the destruction of clouds (to prevent possible precipitation) over Moscow during the major holidays and on the Pripyat in the Chernobyl accident are well known. Positive results were also given by the work on the artificial increase of precipitation in the experimental range of Dnipro, Crimea, Bulgaria, and Mongolia, conducted by the research staff of the Ukrainian Research Hydrometeorological Institute. Today, we found ways to increase the amount of moisture in those regions where it is never enough by climate data. To create the necessary moisture reserves in these regions, the work in an artificial increase in the amount of precipitation is carried out in the cold period of the year, creating a permanent layer of snow.

Thus, the serious scientific and experimental research allowed not only to study the physical basis of the formation of clouds and precipitation but also confirmed the possibility of controlling these processes. Currently, the only open issues remain about the economic and political feasibility of these projects. It is obvious that the work on an active influence on the clouds in one country can affect weather conditions in neighboring countries.

Methods of dispersion of overcooled clouds have been used successfully for the "opening" of the airports of low clouds and fogs. Best of all it is possible to spray dry ice in an overcooled fog and in cases where the temperature in the low clouds and fog is below -4°C. Carbon dioxide in these cases is sprayed by special units, located at specific locations of the airport. The effect of the scattering of mist in such a way is so significant that large airports today in many countries around the world are provided with special facilities and devices for the scattering overcooled clouds and fog covering the runways. Seeding fog carbon dioxide is possible with aircraft, but in this case, the aircraft must take off from the nearby airfield, not closed by fog. Conversely, for the opening of airfields of small or warm mist (low clouds) their bottom heating by gas burners located along the runways are used.

Investigation of active influence on the clouds found its implementation in the struggle with forces of nature disaster. It was invented special anti-hail cannon shells and anti-hail rockets filled with the same dry ice, silver iodide, and other chemicals. The explosion of such the shell (rocket) "sows" cloud with a huge amount of very small particles of the reagent. One gram of these substances forms in the cloud at a temperature of -10°C about 10^{12} ice crystals, thus dramatically increasing the number of "nucleation" of hail. The same supply of water in the cloud is distributed among a considerably larger number of hailstones so that they cannot substantially grow in volume. Small snowflakes and hailstones also fall into the lower layers of the clouds, melt, and fall as rain. In addition, because of artificial coagulation of water droplets, the initial volume of hail cloud grows and the associated reduction of horizontal temperature gradients occurs. Therefore, such the natural disturbances as strong winds, squalls, and tornadoes occurring during thunderstorms, showers and hail events almost completely disappear.

The vertical velocity of the air flow is not the same at all altitudes in the hail cloud. There is a height where the vertical speed reaches 20 m/s and even more. Rising flows in a cloud, as it is known, maintain growing in volume drops and hailstones in the hung state or lift. The more vertical speed in hail clouds, the larger size hailstones can reach. Thus, the maximum number of drops (maximum conductivity) of hail cloud, located above the high vertical speed, forming a *convective cell*. Using radar one can not only watch the hail clouds in general but rather pinpoint the position of convective cells in this cloud, i.e., a section in which the hail loss becomes inevitable when sufficient development. It is in this place you need to send a shell filled with a special chemical (or biological and neutral) reagent that instead of large hailstones rain will cause in the worst case snow pellets from which there will be no harm to crops. Now, in many countries, the fields and gardens of the regions with hail danger are surrounded with missile installations through which throughout the season hail almost does not drop with any big hailstones on the territory of the protected area.

At the end of the 20-th century, experiments to tame hurricanes have been held in the United States on the project "Insane storm". Their essence was to sow clouds that accompany hurricanes from aircraft at an altitude of about 10 km. To carry out special bombs filled with silver iodide weighing 180 g were used. Experiments have shown that when confronted with drops of supercooled water, the crystals of silver iodide form icicles, and the heat of crystallization is released. Thus, heating of air occurs, lower layers of which begin to rise. The wall of clouds surrounding the void ("Eye", see. Figure 13) of the hurricane is destroyed and supplanted by more powerful air currents that form a new wall of clouds. It is at a far distance from the center of the hurricane so that the air flow should overcome a long way. Changing the structure of a hurricane causes it to weaken due to the decrease of the centripetal acceleration. For example, in dealing with the hurricane "Debbie" 13 aircraft participated in 1969. After the first treatment of clouds, wind speed has fallen to 30%. The next day, when the hurricane began to gain momentum again, the processing of cloud repeated. This not only reduced the wind speed by 15% but also changed the trajectory of its movement. Instead, to continue its move to the coast, it followed in the open ocean.

Figure 13. The eye of Super Typhoon Maysak from the ISS
Photo by Terry Virts/NASA/ISS); Source: https://commons.wikimedia.org/w/index.php?curid=39345137.

There are methods of dealing with such dangerous natural phenomenon, as the late spring and early autumn frosts, which bring a lot of damage in agriculture and horticulture. To reduce the cooling of the soil on the morning of land, which must be protected from frost, it is necessary to pour the water in the evening vigorously. There is an old and proven method of protecting plants by fire smoke.

SUMMARY

The solar radiation coming to the Earth from outer space is the main source of energy of atmospheric processes. It is converted in the atmosphere and at the Earth's surface into heat, kinetic energy, and other forms of energy. It is going on not directly but mainly through an exchange between the Earth's surface and the atmosphere, including the exchange of heat and moisture. So, atmospheric processes affect the heat flows coming from above, from space, and from below, from the earth's surface, and the structure of the Earth's surface is very important for these processes.

This chapter gives the general picture of heat and the moisture circulation in the atmosphere and provides physical fundamentals of atmospheric circulation including air mass circulation, local physiographic impact on the atmospheric air movement, in-mass atmospheric processes, and basic laws of pollution spreading in the atmosphere. In following chapters, these basics will be applied to mathematical modeling of atmospheric processes.

REFERENCES

Bagrov, A.N. (2007). Comparative evaluation of the success of forecasts weather elements on the basis of a number of domestic and foreign models of the atmosphere of various sizes. *Inform. Coll. Scient.-research. hydrometeorological center of Russia, 35*, 3-20. (in Russian)

Guziy, A.M., Kovalets, I.V., Kushchan, A.A., Zheleznyak M.I. (2008) Numerical weather forecast system WRF-Ukraine. *J. Mathematical Machines and Systems*, (4), 123-131. (in Russian)

Skamarock, W. C., Klemp, J. B., Dudhia, J., Gill, D. O., Barker, D. M., Duda, M. G., & Powers, J. G. et al. (2008). *A descpription of the advanced research WRF version 3. NCAR Technical Note NCAR/TN-475+STR*. Boulder, CO: National Center for Atmospheric Research.

Chapter 3
Equations Satisfied Physical Processes in the Atmosphere

ABSTRACT

The chapter considers theoretical foundation of the modern hydro-mechanics in their application to the atmosphere dynamics problems. General concepts, definitions, and notation needed for mathematical modeling atmospheric processes, particularly, the elements of kinematics and deformation of continuous media, and curved orthogonal coordinates are introduced. The basic equations of mechanics of viscous stratified fluid are considered including the equations expressing the law of conservation of mass, the equations expressing the law of momentum, the equations expressing the law of conservation of energy, the concentration equations for mechanics of heterogeneous medium. At the end of the chapter, the complete list of the basic equations of the model of atmosphere is given and the initial and boundary conditions in the hydrodynamic problems are described.

PRELIMINARY NOTES

Cognition of nature and solving many related urgent tasks require the construction of new and improvement of existing models for deeper and more detailed description of the microscopic and macroscopic physical objects, interactions, and events. The experience and the inherent essence of science indicate that the answers to the questions of meteorology, to various questions about the material properties of the complicated interactions of the inorganic world is largely due to some of our common universal concepts, ideas, laws, and methods.

To date, it has already accumulated huge scientific information, developed the theory and collected observations data on the behavior of meteorological fields on the flow and balance of the atmosphere, homogeneous air masses and heterogeneous formations.

Explicit establishment of common principles and internal connections between different theories and observed effects contribute to a deeper understanding of the actual state of science, the correct assessment of the known and emerging scientific achievements and best orientation in the wealth extracted information. These can serve as a starting basis for further scientific development.

DOI: 10.4018/978-1-5225-2636-0.ch003

In presenting the theoretical foundations of fluid mechanics and thermodynamics it is useful to be guided by the following provision, which must be emphasized clearly. Introduced and accepted concepts and their relations have some precise meaning only within the framework of a set of models that are designed for a scientific description of the domain of real phenomena that we are interested in. This also applies to such fundamental concepts as space, time, power, speed, temperature, entropy, etc. As we know, the idea of a Newtonian force is totally inadequate to describe the mechanical interactions in the modern sophisticated models of continuum mechanics.

Therefore, it is reasonable to talk about the basic concepts and laws in the framework of the aggregate of wide or narrow classes of models that have already been introduced explicitly or may be introduced. The necessary clarity is achieved by determining the correct models where the emerging contradictions can be easy to handle. But, obviously, that all the models reflect reality only approximately and only in a certain region. Refinement, sophistication, a new simulation or in a way simplification of the existing models – is an ongoing process related to scientific progress.

Theoretical foundations of dynamic meteorology are particularly useful not only in terms of the already known applications but in terms of future challenges, which will be the subject of research and applications in the future. In this regard, in recent years it has clearly revealed the need to address in dynamic meteorology as a common basis for the development of thermodynamics, hydrodynamics, circulation of the atmosphere, physical processes of evaporation and condensation of steam. Commonality and the inextricable relation of these sections of meteorology and physics force us to consider them as a whole.

The modern theory of nonlinear continuum mechanics based on approximations of a Newtonian fluid allows representing a highly complex subject of dynamic meteorology in the form of a system of equations that reflect the corresponding physical laws and are well proven in many applied problems of hydrodynamics.

In general, these equations include following:

- Thermodynamic equation.
- Mass balance equations.
- Momentum balance equation.
- Rheological equation.
- The concentration balance equation.
- The energy balance equation.
- The equation of state, which connects the internal energy with temperature, density and the strained state of the medium.
- Fourier (Fick) law, which establishes the proportional relationship between the heat flux (mass) and temperature gradient (concentration).

As far as it is known, these equations are universal; they are applicable to all liquids in a wide range of changes in pressure, temperature, and velocity regardless of the liquid composition, the outer limits, and motion modes. The ability to use them in modelling many important tasks of hydro- and thermodynamics is confirmed by practice. Based on their analytical and numerical solutions, the fields of velocity, pressure, temperature, and concentration of suspended particles can be very successfully predicted.

Until recently, seemed insurmountable technical and mathematical difficulties have prevented the use of a complete system of hydrodynamics equations for a general study of processes and phenomena in meteorology, including local circulation on the background of macroscale movements (convection, orographic and other processes). Significant improvements in this area relate to the advent of modern high-performance computing facilities and the development of numerical methods for solving the equations of hydrodynamics.

Hydrodynamic modeling allows moving from simple estimates and calculations of certain physical processes to a new stage of work – detailed numerical investigation which significantly reduces the need for laboratory measurements and field observations, and in some cases, eliminates them. The use of advanced computers and mathematical methods in dynamic meteorology allows improving the accuracy and quality of the simulation results and ensuring reducing the cost of scientific and operational research.

This chapter outlines the theoretical basis of modern fluid mechanics applied to problems of atmospheric dynamics. Air movement is regarded as a movement of the continuous medium from a position of Euler's method. Here are the basic equations of fluid mechanics, heat, and mass transfer, widely used in the modeling of the physical processes that occur in the atmosphere. In most cases, it is only conventional notation of differential equations for a homogeneous liquid excluding chemical reactions. Comprehensive information about the simulation of flows with chemical reactions can be found in (Dorrance, 1966).

General Concepts, Definitions and Designations

In mathematical modeling of physical processes occurring in the Earth's atmosphere, the most widespread concept that the discrete structure of the medium under normal conditions is not taken into account, i.e., dry and humid air, cloud formations or a balanced mixture of aerosols in the atmosphere is considered as a continuum (continuous medium). The concept of the state of the environment is considered here as an individual sample, which moves when the air is moving. Since the physical properties of the surrounding air particles may vary over time, it is necessary to imagine a single sample in the form of an infinitely small volume, which will retain its individuality. This volume will be called an *elementary volume*.

The movement of air in the atmosphere will be considered as a motion of a continuous medium from the Euler's method point of view. This method suggests the change of parameters of motion of continuous medium particles (velocity, acceleration, density, temperature, pressure) passing through a given fixed point in space over time and changing them when moving from one point in space to another. Thus, the method of Euler considers parameters of the field of air movement as a function of time t and the coordinates $\mathbf{X}\equiv(x,y,z)$.

Elements of Kinematics and Deformation of Continuous Media

Let the projection of the velocity vector $\mathbf{V}$ of some individual elementary (infinitesimal) continuous medium volume in Cartesian coordinates x,y,z by u,v,w. Then the *trajectory* of elementary volume of medium (Figure 1a) can be defined by the equations

$$\frac{dx}{u(x,y,z,t)} = \frac{dy}{v(x,y,z,t)} = \frac{dz}{w(x,y,z,t)} = dt\,. \tag{3.1.1}$$

The *streamline* (Figure 1b) is a line whose tangent at any point coincides with the direction of velocity at a given time

Figure 1. The line of flow and the particle trajectory

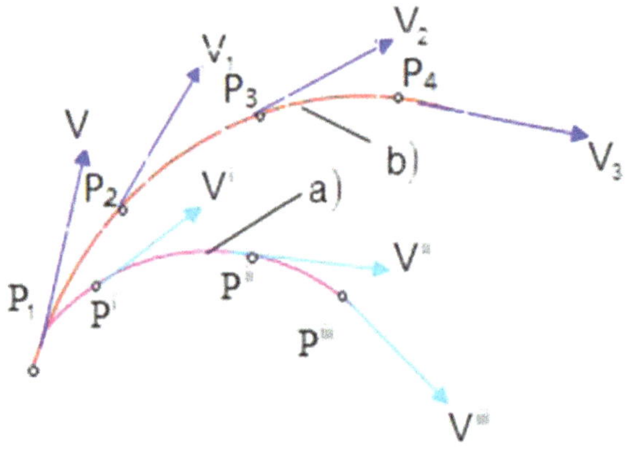

$$\frac{dx}{u\left(x,y,z,t\right)}=\frac{dy}{v\left(x,y,z,t\right)}=\frac{dz}{w\left(x,y,z,t\right)},\qquad(3.1.2)$$

where *t* play the role of a parameter.

If the speed of an elementary volume of the medium moving past a given point in space, changes over time, such a motion of a continuous medium is called *transient*.

In the general case of the nonstationary velocity field, the equation (3.1.1) and (3.1.2) are not the same. In the case of the stationary velocity field time *t* in equation (3.1.1) and (3.1.2) does not appear explicitly, and if we ignore at once the extreme right member of the proportions in (3.1.1), we get the same set of equations for the streamline and for the trajectory. Consequently, in a steady velocity field streamlines and trajectories coincide.

Let some elementary volume of a continuous medium at time *t* has the coordinates *x,y,z* and the speed **V**. During the time Δt this volume is moved to the position $x+u\Delta t$, $y+u\Delta t$, $z+w\Delta t$, and its velocity will acquire the value of

$$\mathbf{V}+\Delta\mathbf{V}=\mathbf{V}\left(x+u\Delta t,y+v\Delta t,z+w\Delta t\right).$$

Then

$$\frac{d\mathbf{V}}{dt}=\lim_{\Delta t\to 0}\frac{\mathbf{V}\left(x+u\Delta t,y+v\Delta t,z+w\Delta t\right)}{\Delta t}=\frac{\partial\mathbf{V}}{\partial t}+u\frac{\partial\mathbf{V}}{\partial x}+v\frac{\partial\mathbf{V}}{\partial y}+w\frac{\partial\mathbf{V}}{\partial z}=\frac{\partial\mathbf{V}}{\partial t}+\left(\mathbf{V}\cdot\nabla\right)\mathbf{V},\qquad(3.1.3)$$

i.e., the absolute acceleration $d\mathbf{V}/dt$ equals the sum of the local component $\partial\mathbf{V}/\partial t$, considering the transient nature of the movement, and advection-convective component $(\mathbf{V}\bullet\nabla)\mathbf{V}$, considering the movement of the elementary volume of a continuous medium in the space. It characterizes the rate of change along the trajectory of motion of a volume element of a continuous medium. Here and below, the operator ∇ is

Hamilton differential operator, it is used with a point (•) to refer to the divergence of a vector function, and without a point – to denote the gradient of a scalar function.

The movement is called *steady* if the speed **V** is a function of only elementary volume coordinates of continuum in space

$$\frac{d\mathbf{V}}{dt} = u\frac{\partial \mathbf{V}}{\partial x} + v\frac{\partial \mathbf{V}}{\partial y} + w\frac{\partial \mathbf{V}}{\partial z} = \left(\mathbf{V}\cdot\nabla\right)\mathbf{V}. \tag{3.1.4}$$

Consequently, the speed of all the elementary volumes of the continuous medium passing through a given point in space is constant.

Below it will be useful to introduce the concepts of discrepancy and vorticity rates. The flow rate Q_v through a given surface S is the surface integral

$$Q_V = \iint_S V_n dS\,,$$

where V_n is a normal component of the velocity vector to the surface element.

The *discrepancy rate* $\nabla \bullet \mathbf{V}$ in the field point (x,y,z) is called the value $\Delta\wp$ of volume flow rate per unit through the closed surface bounding the volume element, which includes this point:

$$\nabla\cdot\mathbf{V} = \lim_{\Delta\wp\to 0}\frac{\iint_S V_n dS}{\Delta\wp} = \frac{\partial u}{\partial x} + \frac{\partial v}{\partial y} + \frac{\partial w}{\partial z}. \tag{3.1.5}$$

It is easy to obtain from this a mathematical expression of Ostrogradskii – Gauss theorem

$$\iint_S V_n dS = \iiint_\wp \nabla\cdot\mathbf{V}d\wp\,\cdot \tag{3.1.6}$$

The *circulation rate* Γ in a closed curve L with a certain traversal direction can be determined by the formula

$$\Gamma = \oint_L \mathbf{V}dr = \oint_L \left(udx + vdy + wdz\right),$$

where dr is an element of the curve L.

Based on Stokes' theorem, we have:

$$\Gamma = \oint_L \mathbf{V}dr = \iint_S \left(\mathrm{rot}\mathbf{V}\right)_n dS\,,$$

where S is the surface bounded by the contour of L, rot**V** is the *vorticity* determined by the relation

$$\mathrm{rot}\mathbf{V} = \mathbf{i}\left(\frac{\partial w}{\partial y} - \frac{\partial v}{\partial z}\right) + \mathbf{j}\left(\frac{\partial u}{\partial z} - \frac{\partial w}{\partial x}\right) + \mathbf{k}\left(\frac{\partial v}{\partial x} - \frac{\partial u}{\partial y}\right), \tag{3.1.7}$$

where **i,j,k** are the unit vectors that define the directions of the axes x,y,z.

Vector lines of the vorticity field rot**V** are called the *vortex lines* (Figure 2a). The part of the continuum bounded by the vortex surface formed by vortex lines is a *vortex tube* (Figure 2b).

The flow of the vorticity vector through an arbitrary cross section of the vortex tube is the same at any given time along the tube (Helmholtz theorem). Indeed, if to apply the Ostrogradskii-Gauss formula (3.1.6) to the vortex tube bounded by two arbitrary cross section S_1,S_2 and its lateral surface S_s, and consider the fact that

$$\nabla \cdot \left(\mathrm{rot}\mathbf{V}\right) = \frac{\partial}{\partial x}\left(\frac{\partial w}{\partial y} - \frac{\partial v}{\partial z}\right) + \frac{\partial}{\partial y}\left(\frac{\partial u}{\partial z} - \frac{\partial w}{\partial x}\right) + \frac{\partial}{\partial z}\left(\frac{\partial v}{\partial x} - \frac{\partial u}{\partial y}\right) \equiv 0\,,$$

we will have

$$\iint_S V_n dS = \iint_{S_1} \mathrm{rot}_n \mathbf{V} dS + \iint_{S_2} \mathrm{rot}_n \mathbf{V} dS = 0\,.$$

For elementary vortex tube, you can choose a small section S_1 and S_2, that are normal planes to the surface of the tube. Then, up to the small higher order, you will have $S_1 rotV_1 = S_2 rotV_2$, where V_1 is the speed in the section S_1 and V_2 is the speed in the section S_2. From this equation, it follows that the square of the cross-sectional area of this tube cannot be zero, as this would lead to an increase of rot**V** in this section to infinity. Thus, the vortex tube can not end in a continuous medium. They either form a closed ring or base on the solid or free surface.

Irrotational or *potential motion* of a continuous medium is called a movement with rot**V** = 0. For such a movement **V** can be represented as the gradient of a function, called the *potential speed*:

Figure 2. The vortex line and the vortex tube

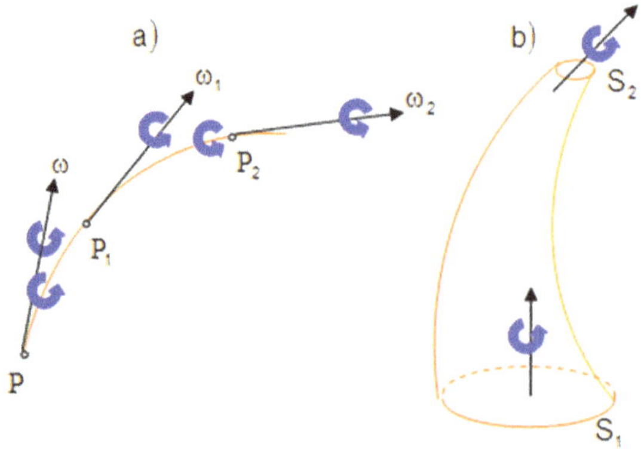

$$\mathbf{V} = \nabla\varphi = \mathbf{i}\frac{\partial\varphi}{\partial x} + \mathbf{j}\frac{\partial\varphi}{\partial y} + \mathbf{k}\frac{\partial\varphi}{\partial z},$$

In accordance with the rules of vector analysis we have

$$\mathbf{V}\times\mathrm{rot}\mathbf{V} = \nabla\left(\frac{V^2}{2}\right) - \left(\mathbf{V}\cdot\nabla\right)\mathbf{V}.$$

Therefore, equation (3.1.4) can be written as

$$\frac{d\mathbf{V}}{dt} = \frac{\partial\mathbf{V}}{\partial t} + \nabla\left(\frac{V^2}{2}\right) - \mathbf{V}\times\mathrm{rot}\mathbf{V}. \qquad (3.1.8)$$

Until now, all the calculations were carried out for the so-called non-accelerated or inertial coordinate system $\mathbf{X} = (x,y,z)$. In meteorology, the motion of the atmosphere is convenient to refer to a coordinate system fixedly associated with the Earth. Relative to the Sun, the Earth carries 365 revolutions around its polar axis for the year, that is the time of one revolution around the Sun, moving with uniform acceleration in its orbit around the Sun. Thus, the coordinate system fixedly associated with the Earth, is not an inertial system, even with respect to the Sun, which in turn moves with acceleration with respect to the fixed stars. Therefore, it is necessary to express the acceleration in an inertial frame of reference by moving the characteristics relative to the Earth. Given that the greatest acceleration of the Earth relative to the immovable stars is due to its rotation around the polar axis, the absolute direction of which is constant, the Earth can be seen as a body that performs a pure rotation with respect to an inertial coordinate frame moving with the Earth.

Suppose $\mathfrak{R}$ is the vector function of time and the three spatial coordinates. If $\mathbf{X} = (x,y,z)$ is Cartesian coordinates taken in inertial system whose origin is on the rotation axis of the Earth, then the following relation holds:

$$\mathfrak{R} = \mathbf{i}\Re_x + \mathbf{j}\Re_y + \mathbf{k}\Re_z,$$

where $\Re_x, \Re_y, \Re_z$ are the components along the axes x,y,z, respectively. Now let $\mathbf{X}' = \left(x', y', z'\right)$ is a Cartesian coordinate system fixedly connected with the Earth revolving around the Earth's polar axis with angular speed Ω and the origin of which coincides with the origin of the inertial system given in Figure 3.

Then, if $\mathbf{i}', \mathbf{j}', \mathbf{k}'$ are the unit vectors corresponding to x', y', z', then $\mathfrak{R}$ can also be written as:

$$\mathfrak{R} = \mathbf{i}'\Re_{x'} + \mathbf{j}'\Re_{y'} + \mathbf{k}'\Re_{z'},$$

where $\Re_{x'}, \Re_{y'}, \Re_{z'}$ are the components of $\mathfrak{R}$ along the axes x', y', z'. Since the directions of the unit vectors $\mathbf{i}', \mathbf{j}', \mathbf{k}'$ are non-permanent in time, after differentiating the above expression we have

Figure 3. The relation of relative and absolute speed

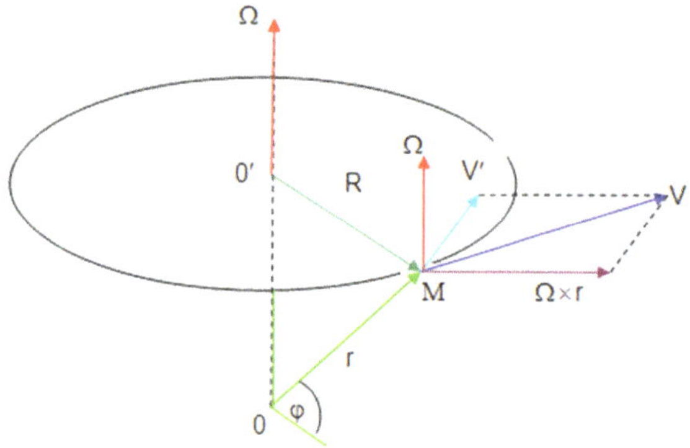

$$\frac{d\mathfrak{R}}{dt} = \mathbf{i}'\frac{d\Re_{x'}}{dt} + \mathbf{j}'\frac{d\Re_{y'}}{dt} + \mathbf{k}'\frac{d\Re_{z'}}{dt} + \Re_{x'}\frac{d\mathbf{i}'}{dt} + \Re_{y'}\frac{d\mathbf{j}'}{dt} + \Re_{z'}\frac{d\mathbf{k}'}{dt}. \tag{3.1.9}$$

The first three terms on the right side of this equation are the total derivative $d'\mathfrak{R} / dt$ taken with respect to the rotating coordinate system x', y', z'. As the system of coordinates $\mathbf{X}' = \left(x', y', z'\right)$ rotates with a constant angular velocity Ω, then

$$\frac{d\mathbf{i}'}{dt} = \mathbf{\Omega} \times \mathbf{i}', \frac{d\mathbf{j}'}{dt} = \mathbf{\Omega} \times \mathbf{j}', \frac{d\mathbf{k}'}{dt} = \mathbf{\Omega} \times \mathbf{k}'.$$

Thus, the equation (3.1.9) can be rewritten as:

$$\frac{d\mathfrak{R}}{dt} = \frac{d'\mathfrak{R}}{dt} + \mathbf{\Omega} \times \mathfrak{R}. \tag{3.1.10}$$

This equation establishes that the difference between the total derivatives of any vector field $\mathfrak{R}$ measured in rotational and inertial coordinate systems is equal to the vector product of $\mathfrak{R}$ and the angular speed $\mathbf{\Omega}$ of the rotational system.

If we identify $\mathfrak{R}$ with the radius vector **r** that is defined by the equation:

$$\mathbf{r} = \mathbf{i}x + \mathbf{j}y + \mathbf{k}z = \mathbf{i}'x' + \mathbf{j}'y' + \mathbf{k}'z',$$

then similar to (3.1.10), we obtain

$$\frac{d\mathbf{r}}{dt} = \frac{d'\mathbf{r}}{dt} + \mathbf{\Omega} \times \mathbf{r}. \tag{3.1.11}$$

Since $d\mathbf{r} / dt$ is the velocity vector, what follows from (3.1.11) should be

$$\mathbf{V} = \mathbf{V}' + \mathbf{\Omega} \times \mathbf{r}\,, \qquad (3.1.12)$$

where **V** and $\mathbf{V}'$ are the speed of the elementary volumes of the continuous medium, measured relative to the inertial and rotating systems, respectively. Identifying further arbitrary vector of $\mathfrak{R}$ with the **V**, we find that

$$\frac{d\mathbf{V}}{dt} = \frac{d'\mathbf{V}}{dt} + \mathbf{\Omega} \times \mathbf{V}\,.$$

Substituting (3.1.12) in this equation and considering **Ω** the constant, we obtain

$$\frac{d\mathbf{V}}{dt} = \frac{d'}{dt}\left(\mathbf{V}' + \mathbf{\Omega} \times \mathbf{r}\right) + \mathbf{\Omega} \times \left(\mathbf{V}' + \mathbf{\Omega} \times \mathbf{r}\right) = \frac{d'\mathbf{V}'}{dt} + 2\mathbf{\Omega} \times \mathbf{V}' + \mathbf{\Omega} \times \left(\mathbf{\Omega} \times \mathbf{r}\right). \qquad (3.1.13)$$

In this equation, $d\mathbf{V} / dt$ is the acceleration relative to the inertial system, whereas $d'\mathbf{V}' / dt$ is the acceleration relative to the rotating system. The difference between these accelerations can be seen as result of virtual forces per unit mass, what should be taken in account in the rotating system, even if no real power does act. The term $2\mathbf{\Omega} \times \mathbf{V}'$ is the so-called Coriolis acceleration, and the term **Ω** ×(**Ω** ×r) is the well-known centrifugal acceleration.

Note that different continuous media respond differently to the action applied to them outside. This distinction is so important that the best way to make classification of continuous media by their physical properties is considering the changes that will cause applied to them external forces. Any external force acting on the solid medium always belongs to one of two categories: volume (mass) and surface forces.

Volume forces act on all the points of the continuous medium, regardless of whether these points lie on the surface bounding the medium or are inside of it. Such forces are considered per mass unit. If we denote **F** the force acting on a mass unit of a continuous medium, and its components on the coordinate axes as F_x, F_y and F_z, then the volume force acting on a volume element $d\wp$ of continuum media is obviously determined by the expression $\rho\mathbf{F}\,d\wp$, and its components – by expressions $\rho F_x d\wp$, $\rho F_y d\wp$ and $\rho F_z d\wp$. The main vector of volume forces acting on the finite volume of a continuous medium is expressed by the triple integral over this volume

$$\iiint\limits_{\wp} \rho\mathbf{F}\,d\wp\,,$$

or, in constituents,

$$\iiint\limits_{\wp} \rho F_x d\wp,\ \iiint\limits_{\wp} \rho F_y d\wp,\ \iiint\limits_{\wp} \rho F_z d\wp\,.$$

The dimension of the volume force **F** acting on the element volume is $ML^{-1}T^{-2}$ (mass, length^{-1}, time^{-2}).

Note the main difference between the vectors **F** and $\mathfrak{R}$. The vector **F** is a single-valued function of a vector space and time points that constitutes a vector field. The vector $\mathfrak{R}$ gains at each point of an infinite set of values, depending on the orientation of the site to which a voltage is applied, and therefore does not constitute a vector field.

Let us dwell on the exact concept of the pressure at a given point of a continuous medium. To do this, select in a continuous medium arbitrary small area ΔS and define the force Δf, with which fluid particles act on one side of this area. Taking the ratio of the force Δf to the area size, we find the average value of the force with which the particles act in the area. Moving on to the limit of this ratio when the area size tends to zero, we find the new value of p, called the *pressure* at a given point of a continuous medium, i.e.,

$$p = \lim_{\Delta S \to 0} \frac{\Delta f}{\Delta S}..$$

Pressure observed in separate points of a unmovable continuous medium is called *hydrostatic*, and in movable continuous medium – it is called *hydrodynamic*. Hydrodynamic pressure is a function of position and time, that is $p=p(x,y,z,t)$.

Curved Orthogonal Coordinates

In the simulation of atmospheric circulation, we must use a variety of coordinate systems: Cartesian in different versions (stereographic, Mercator and other map projections) and spherical. Let us recall briefly the basic formula of the theory of orthogonal curvilinear coordinates. The position of point P (Fig. 4) in three-dimensional space can be determined by specifying its three Cartesian coordinates x,y,z and the radius vector **r** with the projections x,y,z, and by any other three numbers x_1,x_2,x_3 of curvilinear coordinates associated with the coordinates x,y,z by unique correspondence:

$$\begin{aligned} x &= x\left(x_1, x_2, x_3\right), \\ y &= y\left(x_1, x_2, x_3\right), \\ z &= z\left(x_1, x_2, x_3\right). \end{aligned} \tag{3.1.14}$$

The Jacobian $\partial(x,y,z)/\partial(x_1,x_2,x_3)$ is different from zero, then

$$x_1 = x_1\left(x, y, z\right), x_2 = x_2\left(x, y, z\right), x_3 = x_3\left(x, y, z\right).$$

The length of the arc element in Cartesian coordinates is determined

$$\left(ds\right)^2 = \left(dx\right)^2 + \left(dy\right)^2 + \left(dz\right)^2. \tag{3.1.15}$$

If to differentiate the equation (3.1.14) and substitute the result into the equation (3.1.15), we obtain $(ds)^2 = (h_1dx_1)^2 + (h_2dx_2)^2 + (h_3dx_3)^2$, where

$$\left(h_1\right)^2 = \left(\frac{\partial x}{\partial x_1}\right)^2 + \left(\frac{\partial y}{\partial x_1}\right)^2 + \left(\frac{\partial z}{\partial x_1}\right)^2,$$

$$\left(h_2\right)^2 = \left(\frac{\partial x}{\partial x_2}\right)^2 + \left(\frac{\partial y}{\partial x_2}\right)^2 + \left(\frac{\partial z}{\partial x_2}\right)^2,$$

$$\left(h_3\right)^2 = \left(\frac{\partial x}{\partial x_3}\right)^2 + \left(\frac{\partial y}{\partial x_3}\right)^2 + \left(\frac{\partial z}{\partial x_3}\right)^2.$$

Let i_1,i,i_3 are corresponding unit vectors of the curvilinear orthogonal coordinate system; $\mathfrak{R}$ is an arbitrary vector, and $\mathfrak{I}$ is an arbitrary scalar. Then we have the following expressions for the gradient, divergence, vorticity, and the Laplace operator in curvilinear coordinates of the general form:

$$\nabla\mathfrak{I} = \mathbf{i}_1\frac{1}{h_1}\frac{\partial\mathfrak{I}}{\partial x_1} + \mathbf{i}_2\frac{1}{h_2}\frac{\partial\mathfrak{I}}{\partial x_2} + \mathbf{i}_3\frac{1}{h_3}\frac{\partial\mathfrak{I}}{\partial x_3},$$

$$\nabla\cdot\mathfrak{R} = \frac{1}{h_1h_2h_3}\left[\frac{\partial}{\partial x_1}\left(h_2h_3\mathfrak{R}_1\right) + \frac{\partial}{\partial x_2}\left(h_1h_3\mathfrak{R}_2\right) + \frac{\partial}{\partial x_3}\left(h_1h_2\mathfrak{R}_3\right)\right],$$

$$\mathrm{rot}\mathfrak{R} \equiv \nabla\times\mathfrak{R} = \frac{1}{h_1h_2h_3}\left\{\begin{array}{l} h_1\left[\frac{\partial}{\partial x_2}\left(h_3\mathfrak{R}_3\right) - \frac{\partial}{\partial x_3}\left(h_2\mathfrak{R}_2\right)\right]\mathbf{i}_1 + h_2\left[\frac{\partial}{\partial x_3}\left(h_1\mathfrak{R}_1\right) - \frac{\partial}{\partial x_1}\left(h_3\mathfrak{R}_3\right)\right]\mathbf{i}_2 \\ +h_3\left[\frac{\partial}{\partial x_1}\left(h_2\mathfrak{R}_2\right) - \frac{\partial}{\partial x_2}\left(h_1\mathfrak{R}_1\right)\right]\mathbf{i}_3 \end{array}\right\},$$

$$\nabla^2\mathfrak{R} = \frac{1}{h_1h_2h_3}\left[\frac{\partial}{\partial x_1}\left(\frac{h_2h_3}{h_1}\frac{\partial\mathfrak{I}}{\partial x_1}\right) + \frac{\partial}{\partial x_2}\left(\frac{h_1h_3}{h_2}\frac{\partial\mathfrak{I}}{\partial x_2}\right) + \frac{\partial}{\partial x_3}\left(\frac{h_1h_2}{h_3}\frac{\partial\mathfrak{I}}{\partial x_3}\right)\right]. \quad (3.1.16)$$

The expression $(\mathbf{V}\bullet\nabla)\mathbf{V}$ included in the absolute (the substantial) acceleration (3.1.3) can be written as

$$\left(\mathbf{V}\cdot\nabla\right)\mathbf{V} = \left(\frac{v_1}{h_1}\frac{\partial}{\partial x_1} + \frac{v_2}{h_2}\frac{\partial}{\partial x_2} + \frac{v_3}{h_3}\frac{\partial}{\partial x_3}\right)\left(\mathbf{i}_1v_1 + \mathbf{i}_2v_2 + \mathbf{i}_3v_3\right),$$

where v_1,v_2,v_3 are the components of vector **V** in the directions x_1,x_2,x_3. Given that the unit vectors are essentially functions of the coordinates, we get

$$(\mathbf{V}\cdot\nabla)\mathbf{V}=\mathbf{i}_1\left(\frac{v_1}{h_1}\frac{\partial v_1}{\partial x_1}+\frac{v_2}{h_2}\frac{\partial v_1}{\partial x_2}+\frac{v_3}{h_3}\frac{\partial v_1}{\partial x_3}+\frac{v_1v_2}{h_1h_2}\frac{\partial h_1}{\partial x_2}+\frac{v_1v_3}{h_1h_3}\frac{\partial h_1}{\partial x_3}-\frac{v_2^2}{h_1h_2}\frac{\partial h_2}{\partial x_1}-\frac{v_3^2}{h_1h_3}\frac{\partial h_3}{\partial x_1}\right)$$
$$+\mathbf{i}_2\left(\frac{v_1}{h_1}\frac{\partial v_2}{\partial x_1}+\frac{v_2}{h_2}\frac{\partial v_2}{\partial x_2}+\frac{v_3}{h_3}\frac{\partial v_2}{\partial x_3}+\frac{v_1v_2}{h_1h_2}\frac{\partial h_2}{\partial x_2}+\frac{v_2v_3}{h_2h_3}\frac{\partial h_2}{\partial x_3}-\frac{v_1^2}{h_1h_2}\frac{\partial h_1}{\partial x_2}-\frac{v_3^2}{h_2h_3}\frac{\partial h_3}{\partial x_2}\right)$$
$$+\mathbf{i}_3\left(\frac{v_1}{h_1}\frac{\partial v_3}{\partial x_1}+\frac{v_2}{h_2}\frac{\partial v_3}{\partial x_2}+\frac{v_3}{h_3}\frac{\partial v_3}{\partial x_3}+\frac{v_1v_3}{h_1h_3}\frac{\partial h_3}{\partial x_1}+\frac{v_2v_3}{h_2h_3}\frac{\partial h_3}{\partial x_2}-\frac{v_1^2}{h_1h_3}\frac{\partial h_1}{\partial x_3}-\frac{v_2^2}{h_2h_3}\frac{\partial h_2}{\partial x_3}\right)\tag{3.1.17}$$

All formulas written above can be used to obtain the equations of atmospheric circulation patterns in any curvilinear orthogonal coordinate system. The following are examples of two types of coordinates:

- Cartesian coordinate system:

$$\begin{aligned}&x_1=x,\quad h_1=1,\quad v_1=u,\\&x_2=y,\quad h_2=1,\quad v_2=v,\\&x_3=z,\quad h_1=1,\quad v_3=w.\end{aligned}\tag{3.1.18}$$

- Spherical coordinate system $0\le\lambda\le2\pi$, $-\pi/2\le\varphi\le+\pi/2$, $0\le r<\infty$, $z=r-r_0$.

$$\begin{aligned}&x_1=\lambda,\quad h_1=r\sin\varphi,\quad v_1=u,\\&x_2=\varphi,\quad h_2=r,\quad v_2=v,\\&x_3=z,\quad 3=1,\quad v_3=w,\end{aligned}\tag{3.1.19}$$

where λ is longitude, φ is latitude, r is radial distance, $r_0 = 6367.456$ km is radius of the Earth, z is height from the underlying surface.

The Basic Equations of Mechanics of a Viscous Stratified Fluid

The equations of motion of air will be displayed based on conservation laws for mass, momentum, and energy as well as equations defining the strained and thermodynamic state of a viscous liquid.

Although the physical viscosity in itself is not important for hydrodynamic atmospheric models that are discussed in this book, it is important indirectly, as a means of removing the mechanical energy of the system. The viscosity results in strain on the borders, which can be related to the intensity of strain. It turns out that within the scale where the viscosity is important, changes in the most viscosity and compressibility effects can be neglected. And then its influence on the momentum is the same as the influence of diffusion processes, so the momentum may be considered by adding the diffusion fluxes to the equations of motion.

In the derivation of the equations of motion of air, it is necessary to distinguish motion relative to a fixed coordinate system in space and movement relative to a fixed world coordinate system moving with the Earth. In what following, the first of these movements will be called absolute, and the second – relative. Where it is necessary, we will use special symbols to distinguish them.

The Equations Expressing the Law of Conservation of Mass

In general, of the spatial motion of the moving medium, flow field is defined by the velocity vector $\mathbf{V} = (u,v,w)$, a pressure p, and density ρ. The mass of the selected small material element that retains its identity remains constant in the moving continuum, but the volume of this element can be changed, and consequently, its density may change. Obviously, the rate of change of the density of the selected item is associated with a field of velocities of the continuous medium. The equation establishing this connection is called the *equation of continuity* and is derived from the *law of conservation of mass*.

If we denote by ρ the density of air, located at time t in a randomly selected from the entire continuum volume $\wp$ of mass m, then, by definition of density, we can write

$$m = \iiint_{\wp} \rho d\wp .$$

According to the law of conservation of mass, $dm / dt = 0$. Change of mass m per time unit during the motion of a continuous medium can occur both due to changes in the density ρ

$$\frac{dm}{dt} = \iiint_{\wp} \frac{\partial \rho}{\partial t} d\wp ,$$

and due to changes of the volume $\wp$, which may take the considered mass continuum in subsequent time

$$\frac{dm}{dt} = \iint_{S} \rho V_n dS ,$$

where S is the surface bounding the volume $\wp$, V_n is the speed of movement of the element dS of the surface in the direction of the normal. Consequently,

$$\frac{dm}{dt} = \iiint_{\wp} \frac{\partial \rho}{\partial t} d\wp + \iint_{S} \rho V_n dS . \qquad (3.2.1)$$

Equation (3.2.1) expresses the law of conservation of mass in the integral form. To give the differential form of this law we transform the surface integral into a volume one. Using the formula Ostrogradskii-Gauss we substitute its result into (3.2.1), combining the two into a single integral. Thus, we obtain

$$\iiint_{\wp}\left(\frac{\partial \rho}{\partial t}+\nabla\cdot\rho\mathbf{V}\right)d\wp=0\,.$$

In view of the arbitrariness of the selected volume to ensure the last equality, it is necessary

$$\frac{\partial \rho}{\partial t}+\nabla\cdot\left(\rho\mathbf{V}\right)=0\,. \tag{3.2.2}$$

Equation (3.2.2) is the differential form of the law of conservation of mass and is called the *continuity equation.*

Let write in expanded form the formulae of spatial differentiation (divergence) for the two main systems of coordinates:

- In Cartesian system of coordinates (x,y,z)

$$\nabla\cdot\left(\rho\mathbf{V}\right)=\frac{\partial \rho u}{\partial x}+\frac{\partial \rho v}{\partial y}+\frac{\partial \rho w}{\partial z}\,,$$

- In spherical system of coordinates (λ,φ,z)

$$\nabla\cdot\left(\rho\mathbf{V}\right)=\frac{1}{r\cos\varphi}\frac{\partial}{\partial\lambda}\left(\rho u\right)+\frac{1}{r\cos\varphi}\frac{\partial}{\partial\varphi}\left(\rho v\cos\varphi\right)+\frac{1}{r^2}\frac{\partial}{\partial z}\left(r^2\rho w\right).$$

In all the problems associated with the movement of the continuous medium, the continuity equation (3.2.2) has a fundamental role. It can be represented in another form. As it takes place

$$\nabla\cdot\left(\rho\mathbf{V}\right)=\rho\nabla\cdot\boldsymbol{V}+\left(\boldsymbol{V}\cdot\nabla\right)\rho\,,$$

and

$$\frac{d\rho}{dt}=\frac{\partial\rho}{\partial t}+\left(\boldsymbol{V}\cdot\nabla\right)\rho\,,$$

the equation (3.2.2) can be written as

$$\frac{d\rho}{dt}+\rho\nabla\cdot\boldsymbol{V}=0\,. \tag{3.2.3}$$

If a continuous medium can be represented as an incompressible fluid, for which ρ=const, the continuity equation becomes the equation of incompressibility

$$\nabla \cdot V = 0 \tag{3.2.4}$$

It is clear from this equation that for an incompressible fluid there may not be an arbitrary movement, but only one in which the fluid velocity particles satisfy the condition of equation (3.2.4). If the motion of an incompressible fluid has the potential speed ϕ, that is, **V**=∇ϕ, then

$$\nabla \cdot V = \nabla\left(\nabla \phi\right) = \Delta \phi .$$

Thus, the motion of an incompressible fluid velocity potential satisfies the Laplace equation.

A common form of (3.2.2) or (3.2.3) of the representation of the equation of continuity in our opinion is not convenient for its numerical solution. It is much easier to realize difference schemes for the continuity equation written in the form:

$$\frac{d\mathcal{R}}{dt} \equiv \frac{\partial \mathcal{R}}{\partial t} + \left(\mathbf{V} \cdot \nabla\right)\mathcal{R} = -\nabla \cdot \mathbf{V}, \tag{3.2.5}$$

where

$$\mathcal{R} = \ln\left(\rho\right). \tag{3.2.6}$$

The Equations Expressing the Law of Momentum

According to the basic law of mechanics, the change of momentum vector K,

$$K = \iiint_{\wp} \rho \mathbf{V}\, d\wp ,$$

given the constant fluid mass m, located in the volume $\wp$ at a time t, per unit of time is equal to the geometric sum of external forces F_i acting on the given mass, i.e.

$$\frac{dK}{dt} = \frac{d}{dt}\iiint_{\wp} \rho \mathbf{V}\, d\wp = \sum_i F_i \tag{3.2.7}$$

In contrast to the dynamics of a solid body or system of bodies, the dynamics of continuous media considers not the forces itselves but the density of their distribution. External forces F_i consist of volume $F_{\wp}$ and surface F_s forces. Volume force has a magnitude that is proportional to the mass

$$\iiint\limits_{\wp} \rho\left(\mathbf{g} - 2\mathbf{\Omega} \times \mathbf{V}\right) d\wp \tag{3.2.8}$$

and is the sum of the vectors of gravity **g** and deflecting Coriolis force 2**Ω**×**V** in the system rotating with an angular velocity **Ω**. Components of the integrand on the axes have the following meanings:

$$\begin{aligned} &\rho\left[g_1 - 2\left(\Omega_2 w - \Omega_3 v\right)\right], \\ &\rho\left[g_2 - 2\left(\Omega_3 u - \Omega_1 w\right)\right], \\ &\rho\left[g_3 - 2\left(\Omega_1 v - \Omega_2 u\right)\right], \end{aligned} \tag{3.2.8a}$$

where g_1, g_2, g_3 and $\Omega_1, \Omega_2, \Omega_3$ are the components of the gravity vector **g** and the angular velocity **Ω** in these systems of coordinates (x,y,z) and (r,λ,φ).

To specify expressions for g_1, g_2, g_3 and $\Omega_1, \Omega_2, \Omega_3$, limit ourselves by the Earth's northern hemisphere, rout the axis 0y (0φ) along the meridian to the North, the axis 0x (0λ) along the parallel to the East, the axis 0z (0r) along the radius of the Earth and note that the vector of the angular velocity of the Earth's rotation **Ω** is parallel to the Earth's axis to the North. Then we get

$$\begin{aligned} g_1 &= 0, \\ g_2 &= 0, \quad , \\ g_3 &= -g \end{aligned}$$

$$\Omega_1 = 0, \Omega_2 = \mathbf{\Omega} \cos\varphi, \Omega_3 = \mathbf{\Omega} \sin\varphi,$$

and the components of the forces of gravity and Coriolis (3.2.8a) will take the form

$$-2\rho\mathbf{\Omega}\left(\mathrm{w}\cos\varphi - v\sin\varphi\right), -2\rho\mathbf{\Omega} u \sin\varphi, -\rho\left(g + 2\mathbf{\Omega} u \cos\varphi\right). \tag{3.2.8b}$$

Surface forces act on the surface S bounding the volume $\wp$ from the surrounding liquid. They are caused by normal and shear strains and are set by components of the strain tensor

$$F_s = \iint\limits_S \left(\tau_n - p\mathbf{n}\right) dS, \tag{3.2.9}$$

where τ_n is given by vectors τ_x, τ_y, τ_z, standing for the frictional force acting on an area unit orthogonal to axes, respectively x,y,z, using the Cauchy formula

$$\tau_n = \tau_x \cos\left(n, x\right) + \tau_y \cos\left(n, y\right) + \tau_z \cos\left(n, z\right).$$

Vectors τ_x, τ_y, τ_z are recorded via their projections on the coordinate axes *x,y,z* as follows:

$$\tau_x = \tau_{xx}\mathbf{i} + \tau_{xy}\mathbf{j} + \tau_{xz}\mathbf{k},$$
$$\tau_y = \tau_{yx}\mathbf{i} + \tau_{yy}\mathbf{j} + \tau_{yz}\mathbf{k},$$
$$\tau_z = \tau_{zx}\mathbf{i} + \tau_{zy}\mathbf{j} + \tau_{zz}\mathbf{k},$$

In our notation, the first subscript in projections of the strain τ denotes the axis orthogonal to the area, the second index denotes the axis, on which this strain is projected. For example, τ_{xy} denotes projection on the *y*-axis of the strain applied to the area orthogonal to the *x*-axis. The values of the identical indices $\tau_{xx}, \tau_{yy}, \tau_{zz}$, representing the projections of the strains τ_x, τ_y, τ_z on the normals to the respective areas are called normal strains. Projections $\tau_{xy}, \tau_{xz}, \tau_{yz}, \ldots$ on an axis's lying in the plane of areas are called shear strains. The set of normal and shear strains form the strain tensor of the 2nd rank

$$\begin{pmatrix} \tau_{xx} & \tau_{xy} & \tau_{xz} \\ \tau_{yx} & \tau_{yy} & \tau_{yz} \\ \tau_{zx} & \tau_{zy} & \tau_{zz} \end{pmatrix}.$$

This tensor is symmetric, because $\tau_{xy} = \tau_{yx}$, $\tau_{xz} = \tau_{zx}$, $\tau_{yz} = \tau_{zy}$. Some of its components depend on the choice of directions of the coordinate axes, but the tensor is a physical quantity that expresses a certain state of the continuum media – its tension – and does not depend on the choice of the directions of the coordinate axes, i.e., it forms the field.

Substituting (3.2.8) and (3.2.9) to (3.2.7) and using the formula Ostrogradskii-Gauss we can unite the integrals. Due to the arbitrariness of the volume $\wp$, in which the integration is carrying out, thus, we obtain the equation of motion in differential vector form

$$\frac{d\boldsymbol{V}}{dt} = -\frac{1}{\rho}\nabla p + \mathbf{g} + 2\mathbf{V} \times \boldsymbol{\Omega} + \nabla \cdot \vec{\tau}_n, \tag{3.2.10}$$

where, as in (3.1.3), $d\mathbf{V}/dt$ – the substantial acceleration that is the sum of the local component of $\partial\mathbf{V}/\partial t$, taking into account the transient nature of the flow, and advection-convective component $(\mathbf{V}\bullet\nabla)\mathbf{V}$, taking into account the movement of the liquid particles in space. The equality $\left(\mathbf{V} \cdot \nabla\right)\mathbf{V} = \mathrm{rot}\mathbf{V} \times \mathbf{V} + \nabla V^2 / 2$ may be used for submission of the projections of the acceleration on the direction in of axis of the curvilinear coordinate system. All the needed for these expressions of vortex and gradient projections in curvilinear coordinates can be found in the course of vector analysis, for example, (Laptev, 1975). Omitting simple calculations, we give the final form of the projections of the right side of this equation:

- In rectangular Cartesian coordinate system (x,y,z)

$$\frac{\partial u}{\partial t}+u\frac{\partial u}{\partial x}+v\frac{\partial u}{\partial y}+w\frac{\partial u}{\partial z},$$
$$\frac{\partial v}{\partial t}+u\frac{\partial v}{\partial x}+v\frac{\partial v}{\partial y}+w\frac{\partial v}{\partial z},$$
$$\frac{\partial w}{\partial t}+u\frac{\partial w}{\partial x}+v\frac{\partial w}{\partial y}+w\frac{\partial w}{\partial z}.$$

- In the spherical coordinate system (z,λ,φ)

$$\frac{\partial u}{\partial t}+\frac{u}{r\cos\varphi}\frac{\partial u}{\partial\lambda}+\frac{v}{r}\frac{\partial u}{\partial\varphi}+w\frac{\partial u}{\partial z}+\frac{uw}{r}-\frac{uv}{r}\mathrm{tg}\varphi,$$
$$\frac{\partial v}{\partial t}+\frac{u}{r\cos\varphi}\frac{\partial v}{\partial\lambda}+\frac{v}{r}\frac{\partial v}{\partial\varphi}+w\frac{\partial v}{\partial z}+\frac{vw}{r}+\frac{u^2}{r}\mathrm{tg}\varphi,$$
$$\frac{\partial w}{\partial t}+\frac{u}{r\cos\varphi}\frac{\partial w}{\partial\lambda}+\frac{v}{r}\frac{\partial w}{\partial\varphi}+w\frac{\partial w}{\partial z}-\frac{u^2+v^2}{r}.$$

The strain tensor for a viscous solid medium (liquid or gas) in component form is (Landau, 1954; Sedov, 1976):

- In rectangular Cartesian coordinate system (x,y,z)

$$\tau_{xx}=2\nu\frac{\partial u}{\partial x},\tau_{yy}=2\nu\frac{\partial v}{\partial y},\tau_{zz}=2\nu\frac{\partial w}{\partial z},$$
$$\tau_{xy}=\tau_{yx}=\nu\left(\frac{\partial v}{\partial x}+\frac{\partial u}{\partial y}\right),\tau_{xz}=\tau_{zx}=\nu\left(\frac{\partial u}{\partial z}+\frac{\partial w}{\partial x}\right),$$
$$\tau_{yz}=\tau_{zy}=\nu\left(\frac{\partial w}{\partial y}+\frac{\partial v}{\partial z}\right).$$

- In the spherical coordinate system (z,λ,φ)

$$\tau_{\lambda\lambda}=2\nu\left(\frac{1}{r\cos\varphi}\frac{\partial u}{\partial\lambda}+\frac{w}{r}+\frac{v}{r}\mathrm{tg}\varphi\right),$$
$$\tau_{\varphi\varphi}=2\nu\left(\frac{1}{r}\frac{\partial v}{\partial\varphi}+\frac{w}{r}\right),\tau_{rr}=2\nu\left(\frac{\partial w}{\partial z}\right),$$

$$\tau_{\lambda\varphi} = \tau_{\varphi\lambda} = \nu\left(\frac{1}{r\cos\varphi}\frac{\partial v}{\partial\lambda} + \frac{1}{r}\frac{\partial u}{\partial\varphi} - \frac{u}{r}\mathrm{tg}\varphi\right),$$
$$\tau_{\varphi r} = \tau_{r\varphi} = \nu\left(\frac{\partial v}{\partial z} + \frac{1}{r}\frac{\partial w}{\partial\varphi} - \frac{v}{r}\right),$$
$$\tau_{\lambda r} = \tau_{r\lambda} = \nu\left(\frac{1}{r\cos\varphi}\frac{\partial w}{\partial\lambda} + \frac{\partial u}{\partial z} - \frac{u}{r}\right).$$

Here ν is the kinematic viscosity of the continuous media with the• dimension m^2•c^{-1} (below ν denotes the coefficient of "turbulent viscosity").

The equation (3.2.10) above with the expressions for the volume and surface forces forms the basis on which a complete theory of dynamic meteorology has developed. Now modify somewhat this equation, using the expression (3.2.6) for the coefficient 1/ρ. The result is:

$$\frac{d\boldsymbol{V}}{dt} = -e^{-\mathcal{R}}\nabla p + \mathbf{g} + 2\mathbf{V}\times\boldsymbol{\Omega} + \nabla\cdot\vec{\tau}_n\,. \tag{3.2.10a}$$

It follows from the differential equations (3.2.5) and (3.2.10a) that for the determination of the velocity of motion u,v,w, the density ρ (or, in the conventional notation, $\mathcal{R} = \ln(\rho)$ and pressure p as a function of coordinates (x,y,z) or (r,λ,φ) and the time t given on mass force component $F_\wp$ there are only four equations. Consequently, the problem is uncertain (for the five unknown functions there are only four equations). Usually, the equation of state (1.2.4) is selected as the fifth equation of the system.

However, the equation (1.2.4) introduces a new (sixth) unknown function (the temperature T). So, to complete the system of equations it is necessary to introduce the sixth equation, which is the equation of conservation of energy. It should be noted that temperature is no less important meteorological prognostic value than wind or pressure for the problems of weather forecast. Therefore, in cases where the atmosphere is considered as incompressible medium (at a large distance from the surface of the Earth), the addition of the conservation of energy equation is of important independent significance.

The Equations Expressing the Law of Conservation of Energy

The configuration of the temperature field greatly depends on the configuration of the velocity field in non-isothermal flows. On the other hand, the temperature field is the cause of a disturbance of air medium uniformity. The density of the medium with the higher temperature is reduced, and there appears an unstable density distribution (it is stable only if there is inversion at the upper limit, given the absence of the possibility of circulation). In this connection, they distinguish forced convection when the motion is due to external mechanical (advective) exposure, and a free convection caused by the motion of the medium during the actual heat transfer process.

In general, the presence of a temperature field results in a change of fluid viscosity, which also affects the profile of the velocity field. Thus, the velocity and temperature field in the moving medium together are a consequence of thermal and mechanical interactions and cannot be separated from each other.

The basic differential equation of heat propagation in a moving environment can be derived from the energy conservation law and the Fourier heat conduction law. The law of conservation of energy states that the change of the total energy (the sum of the internal U and kinetic $V^2 / 2$ energy) of a mass of air over a period of time equal to the work of all the forces applied to a given mass of air in a given period of time plus the amount of heat that will receive this air mass over the same period of time as a result of heat conduction, radiation or chemical reactions, i.e.

$$\iiint_{\wp} \frac{\partial}{\partial t}\left[\rho\left(U+V^2/2\right)\right]d\wp + \iint_{S} \rho\left(U+V^2/2\right)V_n dS = A_p + A_s + A_m + E_K + E_R \cdot \tag{3.2.11}$$

The first integral on the left side of the equation (3.2.11) expresses the change in air mass energy contained in the volume $\wp$ at a time given immutability of this volume, and the second reflects the energy change under consideration the air mass per unit time due to the variability of volume $\wp$. The following notation is used in the right-hand side of this equation:

- $A_p = -\iint_S p\left(n\mathbf{V}\right)dS$ is the work of pressure forces per unit of time,
- $A_s = \iint_S \left(\tau_n \mathbf{V}\right)dS$ is the work of the forces of friction per unit of time,
- $A_m = \iiint_{\wp} \rho\left(F_{\wp}\mathbf{V}\right)d\wp$ is the work of mass force per unit of time,
- $E_K = \iint_S \mathbf{q}_n dS$ is the amount of heat received by a given mass of air through the unit surface area per unit volume inside,
- $E_R = \iiint_{\wp} \rho Q d\wp$ is the amount of heat generated by given mass of air per unit of time caused by radiation or chemical reactions.

In the formulas above, the operation of round brackets (* *) is the scalar product, the vector with index n is the projection of this vector to the inner normal to the element dS. Substituting them in (3.2.11) and transforming the surface integrals into volume ones using the formula Ostrogradskii-Gauss, one can rewrite (3.2.11) in differential form since the volume of integration is chosen arbitrarily:

$$\begin{aligned}&\frac{\partial}{\partial t}\left[\rho\left(U+V^2/2\right)\right] + \nabla\cdot\left[\rho\left(U+V^2/2\right)\mathbf{V}\right] = -\nabla\cdot\left(p\mathbf{V}\right)\\ &+\rho F_{\wp}\mathbf{V} + \nabla\cdot\mathbf{q} + \rho Q + \frac{\partial}{\partial x}\left(\tau_x\mathbf{V}\right) + \frac{\partial}{\partial y}\left(\tau_y\mathbf{V}\right) + \frac{\partial}{\partial x}\left(\tau_z\mathbf{V}\right).\end{aligned} \tag{3.2.12}$$

The left side of equation (3.2.12) can be presented in the form of the substantial derivative by the following simple transformation:

$$\rho\frac{\partial}{\partial t}\left(U+V^2/2\right)+\rho u\frac{\partial}{\partial x}\left(U+V^2/2\right)+\rho v\frac{\partial}{\partial y}\left(U+V^2/2\right)$$
$$+\rho w\frac{\partial}{\partial z}\left(U+V^2/2\right)+\left(U+V^2/2\right)\left(\frac{\partial\rho}{\partial t}+\nabla\cdot\rho\mathbf{V}\right)=\rho\frac{d}{dt}\left(U+V^2/2\right).$$

Let apply scalar multiplication to the equation (3.2.10) and **V** and take into account the obtained result in (3.2.12). In view of (3.2.6) for the coefficient 1/ρ, we will have:

$$\frac{dU}{dt}=-e^{-\mathcal{R}}p\nabla\cdot\mathbf{V}+\left(\tau_x\frac{\partial\mathbf{V}}{\partial x}+\tau_y\frac{\partial\mathbf{V}}{\partial y}+\tau_z\frac{\partial\mathbf{V}}{\partial z}\right)+\nabla\cdot\mathbf{q}+Q. \tag{3.2.13}$$

The term in brackets on the right side of this equation is usually called the dissipative function and denoted as

$$\tau_x\frac{\partial\mathbf{V}}{\partial x}+\tau_y\frac{\partial\mathbf{V}}{\partial y}+\tau_z\frac{\partial\mathbf{V}}{\partial z}=\mathrm{diss}\Phi\left(\mathbf{V}\right). \tag{3.2.14}$$

Calculation of scalar products of vectors

$$\tau_x=\tau_{xx}\mathbf{i}+\tau_{xy}\mathbf{j}+\tau_{xz}\mathbf{k},$$
$$\tau_y=\tau_{yx}\mathbf{i}+\tau_{yy}\mathbf{j}+\tau_{yz}\mathbf{k},$$
$$\tau_z=\tau_{zx}\mathbf{i}+\tau_{zy}\mathbf{j}+\tau_{zz}\mathbf{k},$$

and

$$\frac{\partial\mathbf{V}}{\partial x}=\frac{\partial u}{\partial x}\mathbf{i}+\frac{\partial v}{\partial x}\mathbf{j}+\frac{\partial w}{\partial x}\mathbf{k},$$
$$\frac{\partial\mathbf{V}}{\partial y}=\frac{\partial u}{\partial y}\mathbf{i}+\frac{\partial v}{\partial y}\mathbf{j}+\frac{\partial w}{\partial y}\mathbf{k},$$
$$\frac{\partial\mathbf{V}}{\partial z}=\frac{\partial u}{\partial z}\mathbf{i}+\frac{\partial v}{\partial z}\mathbf{j}+\frac{\partial w}{\partial z}\mathbf{k}.$$

will be carried out in accordance with the scalar multiplication table (Table 1).

Table 1.

	i	**j**	**k**
i	1	0	0
j	0	1	0
k	0	0	1

Then, if it is assumed that the components of the strain rate tensor τ_x, τ_y, τ_z are given by the formulas in Sec. 3.2.2, then, for example, in a Cartesian coordinate system (*x,y,z*) after simple transformations we obtain

$$\text{diss}\Phi\left(\mathbf{V}\right) = \nu\left\{2\left[\left(\frac{\partial u}{\partial x}\right)^2 + \left(\frac{\partial v}{\partial y}\right)^2 + \left(\frac{\partial w}{\partial z}\right)^2\right] + \left(\frac{\partial u}{\partial y} + \frac{\partial v}{\partial x}\right)^2 + \left(\frac{\partial v}{\partial z} + \frac{\partial w}{\partial y}\right)^2 + \left(\frac{\partial w}{\partial x} + \frac{\partial u}{\partial z}\right)^2\right\}.$$

It is clear from this expression that the dissipation of mechanical energy has a positive value. There always happens influx of heat, i.e., converting the mechanical energy into heat. However, this influx is very small, given the wind speeds that occur in the Earth's atmosphere, so it is usually neglected.

According to Fourier's law, the amount of heat passing through a unit area per unit time is proportional to the temperature gradient in a direction that is orthogonal to the area under consideration. So, we have

$$\frac{1}{S}\frac{dU}{dt} = \mathbf{q} = -\rho c_p k_T \frac{\partial T}{\partial n}, \tag{3.2.15}$$

where k_T is the thermal diffusivity, its dimension $m^2 \bullet c^{-1}$; T is the absolute temperature.

The change in internal energy for an ideal gas is equal to $dU = c_\wp dT$, where $c_\wp$ is the specific heat capacity of the medium at a constant volume per unit mass.

Considering the continuity equation (3.2.2a) in the form

$$\nabla \cdot \mathbf{V} = -\frac{1}{\rho}\frac{d\rho}{dt},$$

and taking into account that $c_p dT = c_\wp dT + d\left(p / \rho\right)$, we have the equation (3.2.13) after some transformations in the form

$$\rho c_p \frac{dT}{dt} = \frac{dp}{dt} + \nabla \cdot \left(\rho c_p k_T \ \nabla T\right) + Q, \tag{3.2.16}$$

where c_p is the specific heat of the medium at constant pressure per mass unit, Q is the density of the internal source (sink) of heat.

In dynamic meteorology, they represent the gradient of the vertical component of the heat flow temperature (3.2.15) as the sum of dry adiabatic gradient $\gamma_a = g/c_p$ and real temperature gradient in the environment $\gamma = -\partial T/\partial z$, i.e. the vertical heat flow is presented in the form $F_{hz} = -\rho c_p k_T \left(\partial T / \partial z + \gamma_a\right)$.

Regarding the latter, we write the final formula for the expression $\nabla \bullet (k_T \nabla T)$ in projections:

- In rectangular Cartesian coordinate system (*x,y,z*)

$$\frac{\partial}{\partial x}\left(\rho c_p k_T \frac{\partial T}{\partial x}\right) + \frac{\partial}{\partial y}\left(\rho c_p k_T \frac{\partial T}{\partial y}\right) + \frac{\partial}{\partial z}\rho\left[\acute{A}c_p k_T \left(\frac{\partial T}{\partial z} + \gamma_a\right)\right].$$

- In the spherical coordinate system (z,λ,φ)

$$\frac{1}{r\cos\varphi}\frac{\partial}{\partial\lambda}\left(\frac{\rho c_p k_T}{r\cos\varphi}\frac{\partial T}{\partial\lambda}\right) + \frac{1}{r}\frac{\partial}{\partial\varphi}\left(\frac{\rho c_p k_T}{r}\frac{\partial T}{\partial\varphi}\right) + \frac{\partial}{\partial z}\left[\rho c_p k_T \left(\frac{\partial T}{\partial z} + \gamma_a\right)\right].$$

Continuity equation (3.2.2), the dynamics of the medium (3.2.6), condition (1.2.4) and energy balance (3.2.16) make up the system of equations of continuum mechanics. This is a closed system as the number of variables (u,v,w,p,ρ,T) in it equals to the number of equations if we add communication between variables based on the different physical assumptions. Some of these assumptions about the physical characteristics of v, kT we justify and accept below.

There is another useful form of the equation (3.2.12), often used in dynamic meteorology. We move for this its first term on the right side to the left side of the equation (3.2.12), and run the following chain of transformations with the help of the equation of state below:

$$\rho c_p \frac{dT}{dt} - \frac{dp}{dt} = \frac{c_p p}{RT}\frac{dT}{dt} - \frac{dp}{dt} = \frac{c_p p}{R}\left(\frac{1}{T}\frac{dT}{dt} - \frac{R}{c_p p}\frac{dp}{dt}\right) = \frac{c_p p}{R}\frac{d}{dt}\ln\left(\frac{T}{p^{R/c_p}}\right) = \frac{\rho c_p T}{\theta}\frac{d\theta}{dt},$$

where $\theta = T\left(p_{00} / p\right)^{R/c_p}$ is a potential temperature (1.2.16), obtained by a particle of the air, provided adiabatic raise or lower it to the level with the pressure p_{00}, starting with the level of pressure p and the absolute temperature T; p_{00} is a characteristic pressure chosen arbitrarily.

As (see (1.2.35))

$$\frac{\partial T}{\partial z} + \gamma_a = \gamma_a - \gamma = \frac{T}{\theta}\frac{\partial\theta}{\partial z},$$

the vertical heat flow is equal to

$$F_{hz} = -\rho c_p k_T \left(\frac{\partial T}{\partial z} + \gamma_a\right) = -\rho c_p k_T \frac{T}{\theta}\frac{\partial\theta}{\partial z}.$$

Indeed, taking the logarithmic derivative of the expression for the potential temperature $\theta = T(p_{00}/p)^{\gamma}$, we have

$$\frac{1}{\theta}\frac{\partial\theta}{\partial z} = \frac{1}{T}\frac{\partial T}{\partial z} - \frac{\gamma-1}{\gamma}\frac{1}{p}\frac{\partial p}{\partial z},$$

from which based on the equations of statics and the state after simple transformations we obtain:

$$\frac{T}{\theta}\frac{\partial \theta}{\partial z}=\frac{\partial T}{\partial z}+\frac{\gamma-1}{\gamma}\frac{g\rho T}{p}=\frac{\partial T}{\partial z}+\gamma_a,$$

where it is considered that

$$\frac{\gamma-1}{\gamma}\frac{g}{R}=\gamma_a, \frac{gRT}{p}=1.$$

Considering the fact that the adiabatic temperature changes at the possible compression and expansion of the elementary volumes of air during their horizontal movements are very small, expressions for horizontal heat flow can be represented as:

$$F_{hx}=-\rho c_p k_T\frac{T}{\theta}\frac{\partial \theta}{\partial x}, F_{hy}=-\rho c_p k_T\frac{T}{\theta}\frac{\partial \theta}{\partial y}.$$

After these transformations, the equation (3.2.16) takes the form:

$$\frac{d\theta}{dt}=\nabla\cdot\left(k_T\nabla,\right)+Q, \tag{3.2.16a}$$

If the motion of the medium is non-polytropic, the term Q of equations (3.2.16) and (3.2.16a) specifies the corresponding effects. These effects are of two main types:

1. Radiation exchange with the environment that requires being aware of the radiation energy flux density $Q_R=\nabla\bullet F_R$, which can be calculated if we know the distribution and status of the absorption factors, radiation, reflection, and dispersion.
2. Heating due to the phase transitions (the release of latent heat) and chemical reactions, the effect of which may be represented by a member of Q_F determining the heat intensity per volume unit. This term is nonzero in two cases: when the humidity q reaches saturation q_S and when the speed of changes in pressure and temperature of the material particles are such that the saturation humidity q_S (p,T) decreases. The latter condition is usually equivalent to the requirement that the movement is upward $((v_3 = w > 0))$. Thus, when the internal heating caused Q_F release latent heat at a pseudo-adiabatic process, then

$$Q_F=\begin{cases}0 & if\ q<q_S\ or\ dq_S/dt\le 0,\\ -\dfrac{L}{1-q_S}\dfrac{dq_S}{dt} & otherwise.\end{cases} \tag{3.2.17}$$

Another way to consider the heat of condensation, if any, is to put $Q_F = 0$i n (3.2.16*a*) and to replace θ by the equivalent potential temperature θ_e, which an elementary volume would have if all the moisture contained in it condenses and the latent heat released, went to the volume of heating. If instead of water ice is formed, then q_S is identified with the q_i – the value of saturation corresponding to the ice. Note that in practice q_S small, and hence the factor $(1 - q_S)$ in (3.2.17) can be approximated by unity. Then (3.2.17) can be written as

$$Q_F \approx -L\frac{dq_S}{dt},$$

and the value $Q=\nabla \bullet F_R + Q_F$ of the right-hand side of equations (3.2.16) and (3.2.16*a*), can be called a of source-sink member.

Balances of Scalar Quantities in the Atmosphere Model

The water due to the originality of its physical and chemical properties has a unique position and plays a special role in nature among all the substances on the Earth. The transition of any substance from the gaseous in the liquid state is possible only at temperatures below the critical value ($T < T_{cr}$). The temperature occurring in the atmosphere, is always below of the water critical temperature, and often even below the melting point of ice. So, the water vapor is practically a unique in the atmosphere that can move from one aggregate state to another.

Water in the atmosphere is constantly in the process of phase transformations between three natural states. These are a gaseous state, i.e. vapor, liquid state – water, rain drops, and solid state – ice. Studying water vapor condensation resulting in the formation of clouds, the evolution of the particle size in the clouds under the influence of gravitational coalescence of drops is an essentially important problem as the droplet size influences on the radiative properties of clouds, determining the flow of solar and thermal radiation to the underlying surface, and the intensity of precipitation.

The accuracy of the physical description of the cloud formation process is one of the important tasks in the simulation of the atmospheric processes. The choice of a method to describe the process of cloud formation determines the level of estimation of the moisture distribution in the lower atmosphere, and the magnitude of the water balance of the atmosphere and the underlying surface.

Cloud is a system of particles suspended in an air-vapor ionized environment. Each cloud particle can be characterized by a certain set of parameters (the particle mass, its geometrical sizes, charge, the coordinates of the center of gravity, etc.). Thus, the clouds are a special case of atmospheric aerosols. Therefore, all the problems arising in the study of aerosols and clouds are the problems of physics. However, due to the major role of the liquid phase of water in clouds, the study of clouds mainly relies on the study of the properties of water, its phase transitions, and processes of enlargement of water droplets and crystals, what makes the task specific enough, though no less complicated.

The value of the different components of the atmosphere in the atmospheric processes is not only determined by their relative content in the air mass. Aerosol particles of dust, smoke and the like are in weight a negligible fraction of atmospheric air. Even in a strongly dusty air, the fraction of aerosol particles is not greater than 10^{-6} of the air mass in which they are contained and does not exceed the value 10^{-9} for the whole atmosphere. In other words, it is the three or four orders of magnitude smaller than the mass fraction of water vapor. And despite this, the role of aerosols in atmospheric processes is very important, especially in context of the problem of the formation of clouds of aerosol particles and reaction with steam. In the absence of nuclei of condensation, the clouds could occur in the atmosphere only at high altitudes due to the ion based condensation of water vapor. The fact that the mass of water vapor per volume unit of air is the several orders of magnitude greater than the mass of the aerosol particles has a significant impact on the variability of the optical characteristics of aerosols. Thus, even at a constant concentration of solid aerosol particles, it is enough to change the external conditions (e.g., ambient temperature, and hence relative humidity, intensity, and spectral composition of the incident solar radiation), that immediately the transformation may occur in size and composition of the aerosol particles due to transfer of some proportion of water from the gaseous state into aerosol one. And now there is a still not always clear mechanism of growth of aerosol particles with the relative humidity, especially when it is quite far from saturation. In some cases, special physical and chemical properties of some aerosol particles determine it.

It should be noted also the fact that the presence of aerosols and moisture in the atmosphere plays an extremely important role in shaping the climate of the entire Earth. Their presence significantly affects the thermal system of the atmosphere and Earth's surface. It is known that the amount of incoming solar radiation depends on the composition of the atmosphere primarily by the presence of aerosols. Aerosols absorb and scatter short wavelength solar radiation, thereby reducing the amount of radiation flow coming to the underlying surface (according to modern data albedo a of the system the Earth and its atmosphere is $\alpha \approx 0.3 \div 0.4$). This lowers the temperature of the underlying surface. In turn, the water vapor of the atmosphere captures a significant part of the earth's surface longwave radiation and directs it back. The surface absorbs it, resulting in an additional heat reemitted into the atmosphere. This phenomenon is called the *greenhouse effect*. According to current estimates, the water vapor determines about 60% of the greenhouse effect. It is calculated that in the absence of the greenhouse effect, the Earth's temperature would be $T = 255^{\circ}K$, but with his view of it is equal to $T = 288^{\circ}K$. The phase transitions of water in the atmosphere and at the Earth's surface can also affect the energy balance of the system Earth-atmosphere - space.

Thus, in the general problem of atmospheric circulation modeling, it is necessary to consider the processes, the main components of which are diffusive transport of a scalar value (the concentration of water vapor, cloud droplets or crystals, suspended particles of various impurities, etc.) and convective transport due to the motion the medium itself.

The main factor determining the features of the motion of aerosol particles in the air flow is their considerable weight in comparison with the molecules of the medium (the particles is more inertial than the gas molecules). The mechanics of the aerosol particles is characterized by the fact that of the three forces acting on the particle: the external, environmental resistance and the interaction forces between the particles, the latter is much less than the first two, so they are in most cases neglected and the motion of the particles are regarded as independent from each other. However, in the clouds, it is often necessary to consider the forces of interaction between cloud particles.

There are known methods for simulating the movement of aerosols with two interpenetrating continuous media with different speeds (the *pseudo-liquid method*). As one of the two media is selected gas, another medium is constructed of the aerosol particles and called *aerosol liquid*.

Sometimes there is introduced a correction to the aerodynamics equation of particles for the effect of their interaction. The simplest case of aerosol dynamics is a uniform motion of spherical particles under a constant force. The equations of motion of particles, in general, can be obtained from the known results of classical hydrodynamics. For (almost) monodisperse aerosols, as well as in cases where it is important to know the behavior of particles near the obstacles and the intensity of the inertial deposition, the equation of motion of the aerosol liquid is reduced to an equation of motion of a single aerosol particle in an average speed field of a gaseous medium. Although these methods are not widely used in dynamic meteorology problems, because of its bulkiness, we will briefly touch on the principles of aerosol fluid motion simulation in the next section.

As a rule, in the simulation of the propagation of atmospheric substances, the medium under consideration is believed to be continuous, that means a continuous distribution of matter and the physical characteristics of its state. They do not specify in many cases what kind of environment is in question, as dry and moist air and suspended particles in it possess similar properties (continuity and fluidity). The similarity of properties allows using the same equations as base models describing two processes: convection (i.e., transfer of substance due to the motion of the medium) and diffusion (i.e., the viscosity properties of the medium). Physical and chemical properties of aerosol particles are shown in these equations by the terms that describe the uniqueness of the physical and chemical properties of these aerosol particles.

Let $\mathfrak{H}$ be a scalar. For example, it can be the intensity (concentration) of the aerosol substance that is transferred by wind flow, having advective and convective components. If along the path of movement of the aerosol substance its intensity (concentration) retained, the transfer equation is:

$$\frac{\mathrm{d}\mathfrak{H}}{\mathrm{dt}} \equiv \frac{\partial \mathfrak{H}}{\partial \mathrm{t}} + \left(\mathrm{V} \cdot \nabla\right) \mathfrak{H} = 0 .$$

If the substance reacts with the aerosol medium or disintegrates, the transfer equation becomes

$$\frac{\partial \mathfrak{H}}{\partial t} + \left(\boldsymbol{V} \cdot \nabla\right) \mathfrak{H} + \phi \mathfrak{H} = 0 .$$

where ϕ is a factor determining the speed of response of substance to the environment.

Indeed, when $\boldsymbol{V} = 0$ (no wind), the concentration will vary according to

$$\frac{\partial \mathfrak{H}}{\partial t} + \phi \mathfrak{H} = 0 ,$$

from where we have $\mathfrak{H} = \mathfrak{H}_0 \exp\left(-\phi t\right)$.

In the presence of a source of the substance, which creates (generates) the concentration of the considered scalar values according to the function $f(\mathbf{X},t)$, the transport equation takes the form:

$$\frac{\partial \mathfrak{H}}{\partial t} + \left(\boldsymbol{V} \cdot \nabla\right) \mathfrak{H} + \phi \mathfrak{H} = f . \tag{3.2.18}$$

This equation describes, for example, the wet adiabatic process occurring during cooling of moist air mass after its temperature decreases to the dew point τ, corresponding to the onset of condensation of water vapor. The resulting condensation products in the form of water droplets or ice crystals do not change the total moisture content of the air, but only alter the ratio between the gas phase and the condensed water. Obviously, the wet adiabatic process is reversible, as participating in the movement considered the volume of air, water droplets or ice crystals, may evaporate if the volume of air moved into the area with high temperature.

Completing the construction of the model of spreading a scalar quantity in the air, it is necessary to describe another feature of the behavior of weighted substance, since even in calm weather cloud formed by the considered substance dissolves by the diffusion process.

Just as the use of the Fourier heat conduction law has led us to the heat balance equation, likely use of the Fick's law for material flow $D = -d_T \nabla \mathfrak{H}$ after substitution in the equation (3.2.18) allows to set the equation of propagation of a scalar quantity $\mathfrak{H}$ by not only convection but diffusion also.

$$\frac{\partial \mathfrak{H}}{\partial t} + \left(V \cdot \nabla\right) \mathfrak{H} + \phi \mathfrak{H} = \nabla \cdot \left(d_T \nabla \mathfrak{H}\right) + f , \tag{3.2.19}$$

where d_T is the diffusion coefficient of dimension $m^2 \bullet c^{-1}$.

Transfer and diffusion of heavy aerosols are described by an equation that takes into account the settling of aerosols in the air, which is subject to the Stokes' law

$$\frac{\partial \mathfrak{H}}{\partial t} + u\frac{\partial \mathfrak{H}}{\partial x} + v\frac{\partial \mathfrak{H}}{\partial y} + \left(w - w_g\right)\frac{\partial \mathfrak{H}}{\partial z} + \phi \mathfrak{H} = \nabla \cdot \left(d_T \nabla \mathfrak{H}\right) + f . \tag{3.2.20}$$

Let us specify the source-sink member $f(\mathbf{X},t)$ for simulated scalar quantities; f_θ – for temperature θ (see 1.2.16), f_q – for the specific humidity q (see 1.1.12), f_δ – for the specific water content δ (water vapor condensation product), f_s – for the aerosol concentration s:

$$f\left(\mathbf{X}, t\right) \equiv \left\{\begin{matrix} f_\theta \\ f_q \\ f_\delta \\ f_s \end{matrix}\right\} = \left\{\begin{matrix} -\nabla \cdot F_R - L\left(Q_H - S_k + S_u\right) \\ Q_H - S_k + S_u \\ Q_H + S_u + \dfrac{\partial F_g}{\partial x_3} \\ -S_s - J_s \end{matrix}\right\}.$$

Here Q_H is the rate of change of the specific humidity in the moving volume unit of air in those portions of the atmosphere where the humidity reaches saturation q_S; S_k,S_u are sub-grid scale processes associated with the speed of cloud and precipitation formation; $F_g = w_k \delta$ is the value of drops flux as a result of their gravitational deposition; w_k is an average rate an of gravitational deposition of droplets; S_s is the rate of change of the impurity concentration in a volume unit of moving air of that part of the atmosphere where the chemical reactions take place; J_s is the speed of erosion impurities by precipitation.

Mechanics of Inhomogeneous Medium Concentration Equations

It was assumed in the previous section that the air is a homogeneous continuous medium that has the properties of Newtonian fluids. However, experimental investigations have established that the natural mists and clouds resulting from the sublimation and condensation of water vapor in the atmosphere consist of suspended water droplets and/or ice crystals of different size, i.e. they are polydisperse. Concentration (number) of the particles varies over a wide range from tens to thousands of particles per 1 cm^3. Furthermore, the air contains the plurality of suspended solid and liquid particles of natural and industrial origin. Therefore, atmospheric air is a mechanical mixture of various ingredients. If all the ingredients are identical in their phase composition, it is called *homogeneous* mixture, otherwise, it is called *heterogeneous* one.

Theoretical description of inhomogeneous flows, by their physical or chemical composition, regardless of whether it is a homogeneous or heterogeneous medium, requires the adoption of basic assumptions about the continuity of all the sets of particles moving together as a single ingredient or a mixture of them. Just as it was adopted in the mechanics of a homogeneous continuous medium, we assume that in the elementary volume of the mixture and in the elementary volumes of ingredients, despite the smallness of these volumes, there is contained a sufficiently large number of particles, in order to be able to apply a statistical averaging of the physical parameters of these particles on their set.

The dynamic state of an inhomogeneous air that is the mechanical mixture of different chemical composition and phase state of matter (which can be both neutral and react with each other) can be determined by the assignment of coordinates and velocities (pulses) of a mixture of particles. Let at some moment each particle of the mixture has a certain position in the velocity space characterized by the vector $\mathbf{V}_i$. In addition, the position of the particle in the physical space we define by the vector $\mathbf{r}$. The properties of each ingredient of the mixture can be expressed in terms of the distribuon function $f_i(\mathbf{r},\mathbf{V}_i,t)$, i.e., we assume that the mixture is in a non-equilibrium state. The total number of particles of the i-th ingredient located in the elementary volume unit of the physical space at time t, is obtained by integrating the distribution function over all possible values of velocities $\mathbf{V}_i = (u_i,v_i,w_i)$:

$$n_i\left(\mathbf{r},t\right) = \iiint_{-\infty}^{+\infty} f_i du_i dv_i dw_i \,. \tag{3.2.21}$$

Then, by definition of the density of the i-th ingredient, considered as a continuous medium, we have

$$\rho_i = n_i m_i \,,$$

where m_i is the mass of a single particle.

Summing over all ingredients of the mixture in accordance with the density determination as the mass per volume unit we get the following expression for the density of the mixture at some point $(\mathbf{r},t)$:

$$\rho = \sum_i n_i m_i = \sum_i \rho_i \,.$$

The relation

$$\frac{\rho_i}{\rho} = \frac{n_i m_i}{\sum_i n_i m_i} = q_i, \sum_i q_i = 1 \tag{3.2.22}$$

is called the mass concentration of particles of the i-th ingredient.

Knowing the distribution function $f_i(\mathbf{r},\mathbf{V}_i,t)$, we can calculate the average value $\bar{\mathbf{V}}_i$ of particles velocity of the i-th component.

$$\bar{\mathbf{V}}_i = \frac{1}{n_i} \cdot \mathbf{V}_i f_i \left(\mathbf{r}, \mathbf{V}_i, t\right) d\mathbf{V}_i .$$

The average mass flow rate, which is a weighted average value, as each particle contributes to this rate, is determined by the formula

$$\mathbf{V}\left(\mathbf{r},t\right) = \frac{\sum_i n_i m_i \bar{\mathbf{V}}_i}{\sum_i n_i m_i}. \tag{3.2.23}$$

The difference of velocities $\bar{\mathbf{V}}_i - \mathbf{V}$, representing an average rate of propagation $\mathbf{V}_i^*$ of the i-th component (ingredient) through the continuous inhomogeneous medium at a given point, is called the diffusion rate of the i-th component.

The continuity equation of the i-th component is given by

$$\frac{\partial \rho_i}{\partial t} + \nabla \cdot \left(\rho_i \bar{\mathbf{V}}_i\right) = Q_i = \sum_j Q_{ij}, \tag{3.2.24}$$

where Q_{ij} characterizes the rate of weight gain per volume unit of the i-th component due to the reaction of the transition from the j-th component in the i-th one at speeds Q_{ij}.

Summing both sides of equation (3.2.24) for all components (ingredients), we obtain the continuity equation for the mixture as a whole:

$$\frac{\partial \rho}{\partial t} + \nabla \cdot \left(\rho \mathbf{V}\right) = -\phi\rho, \tag{3.2.25}$$

which considers the law of conservation of mass given chemical reactions in a mixture:

$$Q_i = \sum_j Q_{ij} = -\phi\rho .$$

In the equation (3.2.24), the average velocity of the mixture $\bar{\mathbf{V}}_i$ is unknown. This is a significant and characteristic difference between the velocities of the individual phases and the mixture as a whole in the mechanics of multiphase systems, which is the result of a mismatch of mixture movement and its individual constituent phases. The observed discrepancy leads to considerable complication of the system of motion equations of an inhomogeneous medium, which requires for their definition in-depth ideas about the mechanism of interfacial exchange of mass, momentum, and total energy, as well as the chemical and physical kinetics of phase transformations. Despite the extreme complexity of these phenomena requiring for their study to attract data from diverse fields of mechanics, thermodynamics, and thermochemistry, there received now the results of the motion studies of multiphase media, that quite satisfy the practice, by simplifying the tasks associated with the abandonment of the study of some details of the phenomenon.

In the problems of dynamic meteorology, it is used so-called "diffusion" approach in which all processes are in thermodynamic equilibrium and hydrodynamic equations for a mixture are the same as that for a homogeneous medium with a density ρ and viscosity coefficient ν the corresponding mixture.

Let express the average speed $\bar{\mathbf{V}}_i$ of movement of an individual component in equation (3.2.24), according to (3.2.23), by the average speed $\mathbf{V}$ of the environment in general and the rate of diffusion $\mathbf{V}_i^*$ of the component. We get

$$\frac{\partial \rho_i}{\partial t} + \nabla \cdot \left[\rho\left(\mathbf{V}_i^* + \mathbf{V}\right)\right] = -\phi\rho_i,$$

or, by introducing from (3.2.22) the concentration $s_i = \rho_i/\rho$, we can write

$$\frac{\partial\left(\rho s_i\right)}{\partial t} + \nabla \cdot \left[\rho s_i\left(\mathbf{V}_i^* + \mathbf{V}\right)\right] = -\phi\rho s_i.$$

Applying the well-known rules of differentiation, after simple transformations, considering the continuity equation mixture (3.2.24), we finally have

$$\rho\frac{\partial s_i}{\partial t} + \rho\mathbf{V}\nabla s_i = -\phi\rho s_i - \nabla \cdot \left(\rho s_i \mathbf{V}_i^*\right), \tag{3.2.26}$$

This equation is called the "equation of concentration" of the i-th ingredient (component). The left side of the equation is a collection of local, advection and convective changes in the concentration of i-th ingredient in the mixture flow. The right side characterizes the diffusion (spreading) of the i-th ingredient in the mixture $\nabla \cdot \left(\rho s_i \mathbf{V}_i^*\right)$ and the occurrence of the i-th ingredient due to physical is the chemical transformation of the j-th ingredient into i-th one ($\sigma\rho q_i$).

Let us take the scheme of a binary mixture (i=1,2) and consider the temperature T_i of the individual components to be the same and equal to the total equilibrium temperature T of the mixture at this point. Let us also take the Fick's law of diffusion in the form, according to which the rate of diffusion of the i-th component is determined by the concentration gradient of the formula

$$\mathbf{V}_i^* = -\frac{d}{s_i}\nabla s_i \,. \qquad (3.2.27)$$

The right part in it expresses the speed of mass diffusion, i.e. the rate of propagation i-th ingredient relative to the mixture, which is due to heterogeneity of the fields of this component concentration. The diffusion flow is directed against the gradient ∇s_i of the value s_i, i.e. it carries particles of the i-th ingredient from a high concentration area to a low concentration one. The value of $d=\rho\nu/Sc$ (where ν is the kinematic viscosity, Sc is the dimensionless Schmidt test) is a factor of mass diffusion; it may depend on the state of the atmosphere, i.e. temperature and pressure. Values of d for common substances can be found in special reference books; for example, it is equal to 2.4 x 10^{-5} m^2s^{-1} for water vapor in air at 281°K. Diffusion of an ingredient can also be caused by gradients of temperature and pressure, but these effects are insignificant on a very small scale, on which the diffusion occurs.

Substituting (3.2.27) in the right side of (3.2.26), we obtain the following form of concentration equation of the i-th ingredient

$$\rho\frac{\partial s_i}{\partial t} + \rho\mathbf{V}\nabla s_i = -\phi\rho s_i + \nabla\left(\mathrm{d}\nabla s_i\right), \qquad (3.2.28)$$

The so-called "diffusion" approach, where each component (ingredient) is passive (coincidence of motions of the mixture and its individual constituent components), a mixture speed in the last equation can be determined from the equations of the dynamics of a homogeneous medium (3.2.5) with the values of physical constants, appropriate for mixture.

A Complete List of the Basic Equations of the Model of Atmosphere

The problem of atmospheric circulation modeling includes a system of equations:

- Continuity:

$$\frac{d\mathcal{R}}{dt} \equiv \frac{\partial\mathcal{R}}{\partial t} + \left(\mathbf{V}\cdot\nabla\right)\mathcal{R} = \nabla\cdot\mathbf{V}\,. \qquad (3.2.29)$$

- Hydrodynamics in vector form:

$$\frac{d\boldsymbol{V}}{dt} = -e^{-\mathcal{R}}\nabla\, p + \mathbf{g} + 2\mathbf{V}\times\mathbf{\Omega} + \nabla\cdot\vec{\tau}_n\,. \qquad (3.2.30)$$

- Distribution of heat in the atmosphere:

$$\frac{d\theta}{dt} = \nabla\cdot\left(k_T\nabla, - F_R\right) - \frac{L}{c_p}\left(Q_H - S_k + S_u\right). \qquad (3.2.31)$$

- Distribution of specific humidity:

$$\frac{dq}{dt} = \nabla \cdot \left(k_d \nabla \mathrm{q}\right) + Q_H - S_k + S_u \,. \tag{3.2.32}$$

- Distribution of condensed moisture (water content):

$$\frac{d\delta}{dt} = \nabla \cdot \left(k_d \nabla'\right) - Q_H + S_u + \frac{\partial F_g}{\partial x_3} \,. \tag{3.2.33}$$

- Distribution of weighted impurity concentrations:

$$\frac{ds}{dt} = \nabla \cdot \left(k_d \nabla \mathrm{s}\right) - S_s - J_s \,. \tag{3.2.34}$$

- State:

$$p = e^{-\mathcal{R}} RT \,. \tag{3.2.35}$$

In the system of equations (3.2.29) – (3.2.35), the following notation is used: ρ is the air density, $\mathcal{R} = \ln\left(\rho\right)$; $T=t+273.15°\mathrm{K}$ is the absolute temperature; $\theta = T\left(p_{00} / p\right)^{R/c_p}$ is the potential air temperature; p is the air pressure in millibars, p_{00} is the pressure on the underlying surface; R=287.04 J•kg•K^{-1} is the universal gas constant for dry air; c_p=3.5R is the specific heat at constant pressure; q is the specific humidity (the mass of water vapor per unit of air mass); δ is the specific water content (the weight of condensed water per unit of air mass); s is the impurity concentration in the air; **Ω** is the angular speed of rotation of the Earth; g=(0,0,-g) is the acceleration of gravity; $\vec{\tau}_m$ is the strain tensor of the 2nd rank; k_T is the coefficient of turbulent temperature conductivity; k_d is the turbulent diffusion coefficient; F_R is the radiation energy flow density, which depends on the distribution and factors of the absorption, radiation, reflection and dispersion; $L$is latent heat of evaporation; Q_H is the rate of change of the specific humidity in the moving volume unit of air in those portions of the atmosphere where the humidity reaches saturation q_S; S_k,S_u are sub-grid scale processes associated with the rate of cloud and precipitation formation; $F_g=w_k\delta$ is the value of drops flow as a result of their gravitational sedimentation; w_k is the average rate of gravitational deposition of drops; S_s is the rate of change of the impurity concentration in a moving volume unit of air in that part of the atmosphere where the chemical reactions take place; J_s is the rate of leaching contaminants by precipitation.

When integrating the system of equations (3.2.29) - (3.2.35), the vector equation (3.2.30) for $\mathbf{V}=(v_1,v_2,v_3)$ should be broken down into three scalar equations, corresponding components in the three coordinate directions x_1,x_2,x_3. Since the equations (3.2.29) - (4.2.35) are implemented for arbitrary coordinate directions, they can be chosen arbitrarily. Usually two axes x_1,x_2 are selected to be parallel to hypothetical (smooth) surface of the Earth (e.g., calm sea surface). At the same time, one of them x_1 is directed to the

East, and the second one x_2 – to the North. The third coordinate axis x_3 is directed upwards, orthogonally to the hypothetical smooth earth's surface.

The system of equations (3.2.29) - (3.2.35) containing the turbulent flows of momentum, heat, and mass, as well as terms characterizing sources (sinks) of mass and energy, due to the phase transformation of atmospheric moisture and radiative processes in the system of the atmosphere – the Earth. Direct inclusion of this micro- and mesoscale processes in the atmospheric circulation model is inappropriate for two main reasons.

Firstly, these processes are nondeterministic and cannot be described by means of accurate differential equations. The actual picture of the turbulent flows of momentum, heat and mass in the boundary layer, the phase, and radiative processes in the real atmosphere is quite complicated. It depends on the geometric and optical characteristics of the underlying surface, the vertical gradients of heat and humidity, convective motions in layers with moist-adiabatic instability and in inversion layers. It also depends on cloud structure, many gaseous and fine-dispersed components of air with different spectral absorption properties of radiation, scattering and others. The mathematical models of the atmosphere now cannot adequately reproduce the whole picture of the turbulent, multiparameter, stochastic, phase and radiative processes.

Secondly, the numerical implementation of the model would require reducing the mesh size of less than 1/4 of the characteristic scale of the phenomenon. This would lead to an increase in the total number of grid points in the solution area and, consequently, the problem of numerical integration of the model would be unrealistic even for the most powerful modern computers. Therefore, the processes of interaction of the atmosphere with the underlying surface, the phase of the radiant heat exchange, convection and cloud formation are referred to as sub-grid scale. Given these features, the problem of creating semi-empirical (parametric) model of subgrid-scale processes has been and remains relevant.

The complexity of the physical mechanism of movement processes in the atmosphere causes the structural complexity of the equations of hydrodynamics (3.2.29), (3.2.30). Even under the assumption of the constancy of the physical properties of the environment, they are non-linear partial differential equations of the second order. The presence of small parameters in the highest derivatives in the equations of hydrodynamics (3.2.30) and the non-linear nature of these equations leads to a complicated structure of the solutions, and today the problem of solving these equations is still open for the following reasons:

- There is no not only sufficiently broad class of analytic solutions, but also any proof of existence theorems of solutions for the entire range of regime parameters.
- The structure of classical and generalized solutions is insufficiently studied.
- There is no well-developed techniques and computational tools for applications in a wide range of parameters if moving medium in three-dimensional space.

It is necessary to note that there are specific atmospheric processes directly related to the partial differential equations of various modifications, depending on the size of the computational domain. Each type of partial differential equations has its own general theory and methods of their solutions.

If the system of equations (3.2.29) - (3.2.35) is selected on the surface of the integrated system of parabolic equations at all points of the surface, there is one family of characteristics at each point of which the following system of equations (3.2.29) – (4.2.35) holds. If the derivative with respect to t is equal to zero, i.e., we consider the stationary process, the system of equations (3.2.29) – (4.2.35) is of the elliptic type and the actual characteristics do not exist. In the cases when the coefficients of the second derivatives are zero, the system (3.2.29) – (3.2.35) are of the hyperbolic type that describes two different families of characteristics.

Therefore, different types of equations lead to different solutions. In the case of variable coefficients on the highest derivatives, the type of system of equations can vary from point to point. This should be considered in the formulation and numerical solution of hydrodynamic prediction problem because the equations form a system of mixed elliptic-parabolic type in mesoscale models, and hyperbolic-parabolic type in the macroscale models. Therefore, it is necessary to use different numerical methods for solving equations in the two cases.

There are some opportunities for simplification in the equations of hydrodynamics (3.2.29), (3.2.30), as the relative role of different nature of the forces that determine the movement of an elementary volume of air, significantly varies depending on the physical environment. Based on additional, justified by practices assumptions, relating both to the general properties of the medium and the various approximate approaches to describe the mechanical and physical processes that accompany its movement, it is possible, the first, to simplify the initial equations and boundary conditions, and, the second, to identify those solutions that reflect the main physical features of the movement, ignoring minor details.

One approach to the study of mesoscale dynamics of the atmosphere is the task decomposition of the computational domain in the region of non-viscous flow and the boundary layer region, in each of which the solution is obtained in the framework of the simplified equations (3.2.29) - (3.2.35). The theory of the boundary layer is a special form of synthesis of mathematical fluid dynamics and real-liquid theory. A characteristic feature of the boundary layer theory is that its mathematical apparatus is based on a physical model, which with great clarity and fullness is focused on real properties of a viscous liquid. However, the system of physical representations, which is the basis of the boundary layer theory, is valid only to the extent when, the first, the Reynolds number is very large value and, the second, the pressure at some point of the outer edge of the boundary layer extends down without any distortion until the underlying surface. These assumptions lead to significant simplification of the original equations (3.2.29) - (3.2.35). However, not all characteristic types of flows encountered in applications can be calculated with the help of the boundary layer equations. The study of complex atmospheric flows in confined areas, in zones of return flow (boundary layer separation, traces, secondary vortices and so on), and atmospheric convection flows is possible only on the basis of the solving the full hydrodynamics equations, despite the fact that the mathematical model based on the equations (3.2.29) - (3.2.35), it is not an adequate reflection of the real physical process.

In addition, it should be noted that the integration of the equations in the approximation of the boundary layer theory for well-posed boundary conditions still causes considerable difficulties. The main methods of their solution are numerical methods. A variety of numerical methods and physical problems that can be solved based on the boundary layer equations for viscous incompressible fluid and gas is not considered here. Their detailed review can be found in the monographs (Anderson, Tannehill and Pletcher, 1990; Belov, Borisenkov and Panin, 1989; Fletcher, 1991).

The Initial and Boundary Conditions in the Hydrodynamic Problems

Differential equations with partial derivatives (3.2.29) – (3.2.35), obtained from the common conservation laws and applied to the phenomenon under study of stratified fluid mechanics, have an infinite number of solutions. To highlight from this set any unique solution, it is necessary to formulate additional conditions characterizing the desired solution. Central to the theory of equations with partial derivatives is proving the existence and uniqueness of precisely those solutions that satisfy these conditions. The problem is called well-formed if small changes in the input data lead to a small change in the solution they meet, or in other words, if desired solutions are sustainable.

To complete the physical definition of the solutions of equations of hydrodynamics and heat and mass transfer the initial and boundary conditions must be specified. Setting these ancillary data is a prerequisite for a well-formed problem, and in a certain sense, the starting point for its specific solution.

Let there is given finite domain $\wp$ with boundary $\mathcal{B}$ in the space $\mathbf{X} = x_1, x_2, x_3$. Assume that the system of equations (3.2.29) – (3.2.35), which includes the closing relations, holds in the domain $\wp$. Air movement, described by this system is made either through the action of mass forces, or the action of external forces, creating a pressure drop p. The last is formally manifested in the formulation of appropriate boundary conditions. The system of equations (3.2.29) – (3.2.35) allows to determine the values field of the unknown functions $\Re = \left(\mathbf{V}, p, T, q, s, \delta\right)$ in the computational domain $\wp$ at time t_n for given initial conditions at time $t = t_0$ and boundary conditions on the boundary $\mathcal{B}$.

In general, at $t = t_0$, an initial condition is given by

$$\Re\left(\mathbf{X}, t_0\right) = \Re_0\left(\mathbf{X}\right), \mathbf{X} \in \left(\wp + \mathcal{B}\right). \tag{3.2.36}$$

The boundary conditions on the border $\mathcal{B}$ may be related to one of three classes:

- Dirichlet condition (the first boundary value problem)

$$\Re = C_1\left(\mathbf{X}, t\right), \mathbf{X} \in \mathcal{B}, t > 0, \tag{3.2.37}$$

- Neumann condition (the second boundary value problem)

$$\frac{d\Re}{dn} = C_2\left(\mathbf{X}, t\right), \mathbf{X} \in \mathcal{B}, t > 0, \tag{3.2.38}$$

- Mixed condition of Robin (the third boundary value problem)

$$\frac{d\Re}{dn} + C_3\left(\mathbf{X}, t\right) + C_4\left(\mathbf{X}, t\right), \mathbf{X} \in \mathcal{B}, t > 0, \tag{3.2.39}$$

where $\Re_0, C_1, C_2, C_3, C_4$ are given functions of arguments $\mathbf{X}$ and t, $d\Re / dn$ is the derivative with respect to the outward normal to $\mathcal{B}$.

The specific boundary conditions are stated depending on the type of boundary, its surface properties, dynamic and thermal conditions at the boundaries, and finally, the physical properties of the continuous medium. We distinguish the following types of lateral limits and the corresponding boundary conditions for viscous inhomogeneous liquid:

1. Boundary of inflow of media (defined by Dirichlet boundary conditions (3.2.37)).
2. Boundary of outflow of media (defined by Neumann boundary conditions (3.2.38)).
3. Solid impenetrable boundary (by definition, no particle of the medium does not cross the boundary, and as a result of adhesion boundary condition for the velocity has the form

$$\mathbf{V} = V_s, \mathbf{X} \in \mathcal{B},$$

where V_s is the speed of surface movement. When wrapping fixed solid boundaries $V_s = 0$. For the purposes of heat and mass transfer exchange various modes can be set (3.2.37) – (3.2.39).
4. The phase boundary between media with different densities. This boundary is not precise, so the molecules of both the continuous medium may cross it (for example, by evaporation from the surface of one of them). The boundary between the two media it is not just the bordering surface. Experimental studies show that it has a finite thickness, within which the properties of the medium are different from the properties of the two media, i.e., it is a "transitional" zone. Therefore, while setting the boundary conditions at the phase boundary, it is necessary to first describe the transition film (buffer zone) using thermodynamic equations and state equations that consider the diffusion processes in the interface.
5. The free boundary. It is defined by the condition of uniform flow at infinity that has an asymptotic nature for all the desired functions

$$\Re \to \Re_B, \mathbf{X} \to \infty. \tag{3.2.40}$$

Although this condition has a relatively simple form, its quite accurate numerical implementation, usually in the field of finite size, is fraught with many difficulties. Limited by the finite control volume $\wp$, we have to move the boundary conditions from infinity (3.2.40) on its boundary $\mathcal{B}$, why it is necessary to know the asymptotic behavior of the solutions. In the absence of prior ideas about the structure of the flow in the outer remote area, the transfer of boundary conditions on the boundary of $\mathcal{B}$ is usually made by the iterative process of solving the problem for the two test volumes.

The above list of possible boundary conditions reflects only simple common types of boundary conditions. Other forms of boundary conditions will be discussed below when describing the performances of specific meteorological problems.

The quality of numerical weather prediction depends essentially on several the following factors. These include: the development of data base integrating information from different measuring systems and the satellite; coordination in space and time irregularly distributed synoptic and meteorological data obtained from national meteorological services; the quality of the initial fields of meteorological and related boundary conditions; theoretical principles of hydrodynamic model and numerical method of its implementation. At this time, two traditionally distinct tasks – an objective numerical analysis and prediction using numerical models – are combined and called four-dimensional analysis (FDA). FDA is appointed to get the sequence of fields of meteorological elements, which are aligned not only with the results of observations but also with the physical laws of air flows in the atmosphere. FDA includes four components: observation data; objective analysis based generally on methods of optimum interpolation; initialization algorithms and predictive models.

According to the formal requirements of the system of field observations of mid-latitude should provide a horizontal atmospheric sounding in increments of no more than 500 km for wind speed, temperature and humidity at standard levels (1000, 850, 700, 500, 300, 200, 100, 50, ... hPa), and the measurement of surface pressure and temperature of the water surface. However, the high cost of modern surveillance systems limits the fulfillment of these requirements.

An alternative to the expansion of the observation network is an objective analysis procedure, consisting of primary treatment, preliminary validation, consistent correction, and optimal interpolation. The most important factor in the objective analysis is its logical basis. It allows to consider such factors as the spatial distribution of meteorological observation points relative to each other and the grid points; a variety of errors of observing systems; data for previous periods and climate information; the consistency of the pressure field and wind speed. Controlling data and error evaluation is also the positive feature of this analysis. It is best when the covariance taken in the interpolation errors of observations are adequate real covariates these errors. However, in practice, it appears that the applied optimal interpolation is insufficiently sensitive to small changes in the adopted covariance structure. The synoptic measurement data obtained from orbital satellites even complicate the task of objective analysis. At present, the use of algorithms for objective analysis is limited mainly by the assimilation of data in planetary or hemispherical scale, and this, in turn, dictates the problem of special data assimilation techniques for regional numerical weather prediction models.

Obviously, the increase in data assimilation quality gets closer to the reality of predictive models. The system of observation assimilation is based on the principle of statistical linear estimation, which is based on the least squares method minimizing the difference between the available observations and assesses the state of the atmosphere. The term "observation" is used here in its broadest sense. It refers to all the information that is available in the quantitative form, as well as that contained in the equations of the model data assimilation.

SUMMARY

The main goal of this chapter is to give theoretical foundation of the modern hydro-mechanics in their application to the atmosphere dynamics problems. Therefore, general concepts, definitions, and notation needed for mathematical modeling atmospheric processes are introduced and basic equations of mechanics of viscous stratified fluid are considered. They include the equations expressing the law of conservation of mass, the equations expressing the law of momentum, the equations expressing the law of conservation of energy, the concentration equations for mechanics of heterogeneous medium. The complete list of the basic equations of the model of atmosphere and the initial and boundary conditions for the hydrodynamic problems are given at the end of the chapter.

The basic knowledge of this chapter on mathematical modeling atmospheric processes is used in following chapters to develop methods of mathematical modeling atmospheric circulation.

REFERENCES

Anderson, D., Tannehill, J., & Pletcher, R. (1984). *Computational fluid mechanics and heat transfer* (Vols. 1–2). New York: Hemisphere Publishing Corporation.

Belov, P. N., Borisenkov, E. P., & Panin, B. D. (1989). *Numerical methods of weather forecasting*. Leningrad: Gidrometeoizdat Publisher. (in Russian)

Dorrance, W.H. (1966). *Hypersonic viscous flow*. Moscow: Mir. (in Russian)

Fletcher, K. (1991). *Computational Techniques for Fluid Dynamics (vols. 1-2)*. Moscow: Mir Publisher. (in Russian)

Landau, L. D., & Lifshitz, E. M. (1954). *Continuum Mechanics*. Moscow: GITTL.

Laptev, G. (1975). *Elements of vector calculus*. Moscow: Nauka. (in Russian)

Sedov, L. I. (1976). *Continuum Mechanics (vols. 1-2)*. Moscow: Nauka. (in Russian)

Chapter 4
Methods of Mathematical Modeling Atmospheric Circulation

ABSTRACT

In this chapter, atmospheric processes are considered on a planetary scale that is impossible to imagine without a detailed three-dimensional modeling based on geophysical fluid dynamics equations of thermodynamics, radiation transfer, the kinetic equations describing the actual movement of air masses, conversion and transfer of energy and matter, the formation of clouds, aerosols, and rainfall and other atmospheric processes.

PLANETARY AND HEMISPHERIC ATMOSPHERE MODEL: ISOBARIC COORDINATE SYSTEM

All the processes occurring in the Earth's atmosphere and associated with the movement of air masses are inherently heterogeneous on their properties in the space and variability in time. Spatial scales are determined by the size or length of the disturbance wave, and the time scales – by the time of their existence, or the period of oscillation. The range of these phenomena varies significantly: from very small short-term turbulence to the jet stream (giant waves). Globally, these movements determine the transfer of momentum, heat, passive contaminants of water, dust, and minor constituents.

Unequal streams of solar radiation and the heterogeneity of the Earth's surface in different geographical areas cause uneven heating of the air. This results in the movements in the atmosphere, which promote the redistribution of heat. The water vapor and impurities of various chemical substances also affect the processes in the atmosphere. Under certain conditions, the vapor condenses, creating a mist and clouds. Clouds are the source of many atmospheric phenomena – rain, thunderstorms, and a variety of optical effects. In addition, clouds significantly alter the energy resources in the atmosphere, because the extra heat is released during condensation of water vapor and clouds themselves sensitively reduce the influx of solar radiation to the Earth's surface and reduce its heat loss. Although the amount of internal (po-

DOI: 10.4018/978-1-5225-2636-0.ch004

tential) energy transformed into kinetic energy is relatively small, this process has a significant impact on the energy of the atmospheric circulation. The manifestation of this energy is felt in the convective zones, jet streams, winds associated with the trough, and others. They are subject to weather forecasts. A small amount of the energy converts shows a subtle imbalance, whereby kinetic energy is generated, what also affects to the complexity of theoretical studies of such processes.

Let us recall and analyze the system of equations (3.2.29) – (3.2.35) with respect to the modeling of large-scale atmospheric circulation. Since in the large-scale atmospheric processes vertical acceleration is negligibly small, the equation (3.2.30), which expresses the law of conservation of momentum, is only used to describe the horizontal movement, and the equation of motion in the projection on the vertical axis, is simply reduced to the hydrostatic equation

$$\frac{\partial p}{\partial z} = -\rho g = -e^{\mathcal{R}} g\,. \tag{4.1.1}$$

This equation is the fundamental relation that connects the height z and the pressure p. The validity of (4.1.1) for the Earth's atmosphere is confirmed by both theoretical and field observations. It was found that in the process of adapting the atmosphere is always committed to the state of hydrostatic equilibrium. This feature and the need to address the underlying surface orography make appropriate to use the pressure p as the independent variable instead of height z. When constructing barometric topography maps, hydrometeorological services adopt geopotential of the isobaric surface $\Phi=gz$ as the function in question, so this dictates the transition to the isobaric coordinates. The pressure p in this case plays the role of the independent variable.

Before proceeding to adapt the system of equations (3.2.29) – (3.2.35) to the isobaric coordinate system, we must agree that the orographic field hydrodynamic model (3.2.29) – (3.2.35) has the same horizontal resolution as meteorological fields that are included in the model. Therefore, as the field of orography is a function of the horizontal coordinates, we a priori leave aside the issue of the horizontal finite-difference approximation errors influence on the pressure gradient in the case of orographic subgrid scale, i.e. when the orographic field is due to the very steep mountains.

The velocity field $\mathbf{V}$ = (u,v,ϖ) at an isobaric coordinate system is a rate of change of isobaric coordinates ofan elemental air volume

$$\left(u, v, \varpi\right) = \frac{d}{dt}\left(x, y, p\right)$$

and allows obtaining the equations of motion for the variables (u,v,ϖ). To do this, we transform the full (the substantial) derivative (3.3.1), considering the transition from the old variables (x,y,z) to the new variables (x,y,p)

$$\frac{d}{dt} \equiv \frac{\partial}{\partial t} + \frac{\partial}{\partial x}\frac{dx}{dt} + \frac{\partial}{\partial y}\frac{dy}{dt} + \frac{\partial}{\partial p}\frac{dp}{dt} \equiv \frac{\partial}{\partial t} + u\frac{\partial}{\partial x} + v\frac{\partial}{\partial y} + \varpi\frac{\partial}{\partial p}, \tag{4.1.2}$$

where $\varpi=dp/dt$ is the vertical velocity.

The equations of motion (3.2.30) expressing the law of conservation of momentum for describing horizontal movements include horizontal pressure gradients, for which the relations of the isobaric coordinate system must be modified. For generality, we denote by ξ any of the coordinates x or y. As

$$\left(\frac{\partial p}{\partial \xi}\right)_p = \left(\frac{\partial p}{\partial \xi}\right)_z + \frac{\partial p}{\partial z}\left(\frac{\partial z}{\partial \xi}\right)_p = 0,$$

it follows from the statics equation (4.1.1) that

$$\left(\frac{\partial p}{\partial \xi}\right)_z = g\rho\left(\frac{\partial z}{\partial \xi}\right)_p = \rho\left(\frac{\partial \Phi}{\partial \xi}\right)_z = e^{\mathcal{R}}\left(\frac{\partial \Phi}{\partial \xi}\right)_z. \tag{4.1.3}$$

The divergence operator (3.1.5) in the isobaric coordinate system is also undergoing change. To derive it, let denote a mass elementary volume by $\rho\delta x\delta y\delta z$. It follows from the hydrostatic equation (4.1.4) that.δz=-$\delta p/(\rho g)$. Further, the rate of the relative change of weight of $-\delta x\delta y\delta p / g$ must be zero:

$$\frac{1}{\delta x\delta y\delta p}\frac{d}{dt}\left(\delta x\delta y\delta p\right) = \frac{1}{\delta x}\frac{d}{dt}\left(\delta x\right) + \frac{1}{\delta y}\frac{d}{dt}\left(\delta y\right) + \frac{1}{\delta p}\frac{d}{dt}\left(\delta p\right) = 0.$$

Letting $\delta x\delta y\delta p \to 0$, we obtain

$$\frac{\partial u}{\partial x} + \frac{\partial v}{\partial y} + \frac{\partial \varpi}{\partial p} = 0. \tag{4.1.4}$$

Its simplicity is another reason for using the isobaric coordinate system.

The most accurate circulation of the atmosphere can be described by drawing a spherical coordinate system (3.1.19). Let denote atmospheric wind velocity components by u,v,w along the coordinate axes: λ ($u > 0$ in the East direction); φ ($v > 0$ in the direction to the North) and z ($w > 0$ in the direction of “up” opposite to gravity). The z coordinate is determined as the difference $= r - r_0$ between the radial distance r and the distance r_0 to the geopotential at sea level, i.e. the distance measured up from the geopotential surface at sea level. Considering (4.1.1) - (4.1.4) the system (3.2.29) – (3.2.35) takes the form:

- The continuity equation

$$\frac{1}{r\cos\varphi}\frac{\partial u}{\partial \lambda} + \frac{1}{r}\frac{\partial v}{\partial \varphi} + \frac{\partial \varpi}{\partial p} - \frac{v}{r}\operatorname{tg}\varphi = 0. \tag{4.1.5}$$

- The equation of atmospheric motion

$$\frac{\partial u}{\partial t}+\frac{u}{r\cos\varphi}\frac{\partial u}{\partial\lambda}+\frac{v}{r}\frac{\partial u}{\partial\varphi}+\varpi\frac{\partial u}{\partial z}+\frac{u\varpi}{r}-\frac{uv}{r}\mathrm{tg}\varphi$$
$$=-\frac{1}{r\cos\varphi}\frac{\partial\Phi}{\partial\lambda}-2\Omega\left(\mathrm{w}\cos\varphi-v\sin\varphi\right)+F_u ,$$

$$\frac{\partial v}{\partial t}+\frac{u}{r\cos\varphi}\frac{\partial v}{\partial\lambda}+\frac{v}{r}\frac{\partial v}{\partial\varphi}+\varpi\frac{\partial v}{\partial p}+\frac{v\varpi}{r}+\frac{u^2}{r}\mathrm{tg}\varphi=-\frac{1}{r}\frac{\partial\Phi}{\partial\varphi}-2\Omega u\sin\varphi+F_v . \tag{4.1.6}$$

- The equation of propagation of scalars $\mathfrak{H}=\left(\theta,q,\delta,s\right)$ in the atmosphere

$$\frac{\partial\mathfrak{H}}{\partial t}+\frac{u}{r\cos\varphi}\frac{\partial\mathfrak{H}}{\partial\lambda}+\frac{v}{r}\frac{\partial\mathfrak{H}}{\partial\varphi}+\left(\varpi-w_g\right)\frac{\partial\mathfrak{H}}{\partial p}+\phi\mathfrak{H}=f+F_{\mathfrak{H}} . \tag{4.1.7}$$

The underlying surface is solid and impermeable and at the same time serves as the origin of the coordinate system $p = p_B$. The relevant kinematic condition can be written as

$$\varpi=0 \text{ on condition } p=p_B \tag{4.1.8}$$

A similar condition is stated for the top of the atmosphere $p = p_T$:

$$\bar{\sigma}=0 \text{ on condition } p=p_T . \tag{4.1.9}$$

Integrating the continuity equation (4.1.5) with respect to p and using the condition at the upper limit (4.1.9), we obtain an explicit expression for the rate of p (analogous to vertical speed ϖ):

$$\varpi=-\int_{p_T}^{p}\left(\frac{1}{r\cos\varphi}\frac{\partial u}{\partial\lambda}+\frac{1}{r}\frac{\partial v}{\partial\varphi}-\frac{v}{r}\mathrm{tg}\varphi\right)dp . \tag{4.1.10}$$

Using the conditions (4.1.8), (4.1.9), we can create an equation for determining the pressure p_H at the surface. To this end, we integrate the continuity equation (4.1.5) with respect to p from p_H до p_T. Thus, we obtain an equation for the pressure trend at the Earth's surface

$$\frac{\partial p_B}{\partial t}=-\left(\frac{u}{r\cos\varphi}\frac{\partial p_H}{\partial\lambda}+\frac{v}{r}\frac{\partial p_H}{\partial\varphi}\right)-\int_{p_T}^{p_B}\left(\frac{1}{r\cos\varphi}\frac{\partial u}{\partial\lambda}+\frac{1}{r}\frac{\partial v}{\partial\varphi}-\frac{v}{r}\mathrm{tg}\varphi\right)dp . \tag{4.1.11}$$

Consequently, the lower limit pressure trend is calculated from the formula (4.1.11) but is not specified. This fact shows the duality of this coordinate system as the pressure p_H must be the dependent variable from the a physical point of view.

While selecting the spherical coordinate system due to its wide use one must be aware that this system has the drawbacks noted above, and the problem of constructing an optimal, in some sense, the coordinate system for the simulation of the general circulation of the atmosphere is still waiting to be solved.

No less popular among meteorologists is the σ-system of coordinates in which the following value is used as the vertical coordinate

$$\sigma = \frac{p}{p_B}, \tag{4.1.12}$$

where $p_B = p_B\left(\lambda, \varphi, t\right)$ is the pressure at the underlying surface, and $\sigma \in [0,1]$.

The analog of vertical speed in the coordinate σ -system is dependent variable

$$\dot{\sigma} = d\sigma / dt = d\left(p / p_B\right) / dt,$$

for which corresponding kinematic conditions are written in the form

$$\dot{\sigma} = 0 \text{ on condition } \sigma = 0 \text{ and } \sigma = 1. \tag{4.1.13}$$

After carrying out some simple transformations,

$$\frac{dp}{d\sigma} = p_B, \frac{1}{r\cos\varphi}\frac{\partial p}{\partial \lambda} = \frac{\sigma}{r\cos\varphi}\frac{\partial p_B}{\partial \lambda}, \frac{1}{r}\frac{\partial p}{\partial \varphi} = \frac{\sigma}{r}\frac{\partial p_B}{\partial \varphi}$$

the system of equations (3.2.29) – (3.2.35) takes the form

- The continuity equation

$$\frac{1}{p_B}\left(\frac{\partial p_B}{\partial t} + \frac{u}{r\cos\varphi}\frac{\partial p_B}{\partial \lambda} + \frac{v}{r}\frac{\partial p_B}{\partial \varphi} + \dot{\sigma}\frac{\partial p_B}{\partial \sigma}\right) + \frac{1}{r\cos\varphi}\frac{\partial u}{\partial \lambda} + \frac{1}{r}\frac{\partial v}{\partial \varphi} + \frac{\partial \dot{\sigma}}{\partial \sigma} - \frac{v}{r}\mathrm{tg}\varphi = 0. \tag{4.1.14}$$

- The equation of atmospheric motion

$$\frac{\partial u}{\partial t} + \frac{u}{r\cos\varphi}\frac{\partial u}{\partial \lambda} + \frac{v}{r}\frac{\partial u}{\partial \varphi} + \dot{\sigma}\frac{\partial u}{\partial \sigma} + \frac{u\dot{\sigma}}{r} - \frac{uv}{r}\mathrm{tg}\varphi$$
$$= -\frac{1}{r\cos\varphi}\frac{\partial \Phi}{\partial \lambda} + \frac{1}{r\cos\varphi}\frac{\sigma}{p_B}\frac{\partial \Phi}{\partial \sigma}\frac{\partial p_B}{\partial \lambda} - 2\Omega\left(\mathrm{w}\cos\varphi - v\sin\varphi\right) + F_u,$$

$$\frac{\partial v}{\partial t}+\frac{u}{r\cos\varphi}\frac{\partial v}{\partial \lambda}+\frac{v}{r}\frac{\partial v}{\partial \varphi}+\dot{\sigma}\frac{\partial v}{\partial \sigma}+\frac{v\dot{\sigma}}{r}+\frac{u^2}{r}\mathrm{tg}\varphi = -\frac{1}{r}\frac{\partial \Phi}{\partial \varphi}+\frac{1}{r}\frac{\sigma}{p_B}\frac{\partial \Phi}{\partial \sigma}\frac{\partial p_B}{\partial \varphi}-2\Omega u\sin\varphi+F_v \quad (4.1.15)$$

- The equation of propagation of scalars $\mathfrak{H}=\left(\theta,q,\delta,s\right)$ in the atmosphere

$$\frac{\partial \mathfrak{H}}{\partial t}+\frac{u}{r\cos\varphi}\frac{\partial \mathfrak{H}}{\partial \lambda}+\frac{v}{r}\frac{\partial \mathfrak{H}}{\partial \varphi}+\left(\dot{\sigma}-w_g\right)\frac{\partial \mathfrak{H}}{\partial \sigma}+\phi\mathfrak{H}=f+F_{\mathfrak{H}}. \quad (4.1.16)$$

From equation (4.1.14), we can obtain expressions for the pressure trend at the Earth's surface

$$\frac{\partial p_B}{\partial t}=-\int_0^1\left[\frac{1}{r\cos\varphi}\frac{\partial\left(up_B\right)}{\partial \lambda}+\frac{1}{r}\frac{\partial\left(vp_B\right)}{\partial \varphi}\right]d\sigma \quad (4.1.17)$$

and for the analog of the vertical speed

$$\dot{\sigma}\left(\sigma\right)=\frac{\sigma}{p_B}\int_0^1\left[\frac{1}{r\cos\varphi}\frac{\partial\left(up_B\right)}{\partial \lambda}+\frac{1}{r}\frac{\partial\left(vp_B\right)}{\partial \varphi}\right]d\sigma-\frac{1}{p_B}\int_0^{\sigma}\left[\frac{1}{r\cos\varphi}\frac{\partial\left(up_B\right)}{\partial \lambda}+\frac{1}{r}\frac{\partial\left(vp_B\right)}{\partial \varphi}\right]d\varsigma, \quad (4.1.18)$$

where ς is the variable of integration.

There are known also some more attempts, besides the above coordinate systems, to use other types of the vertical coordinate in solving the problems of numerical weather prediction, as well as generally in synoptic meteorology (Phillips, 1957). Particularly, map projections in the Cartesian coordinate system are widely spread. The need to use them arises since the earth's surface cannot be projected onto a plane without distortion. Map Projections allow maintaining well-defined characteristics: area, shape (angles between intersecting curves), scale (relation between the distance on the map and the distance on the Earth), the bearing. All major map projections and variants thereof (stereographic, Mercator and others.) that are used in meteorology are described elsewhere (see, e.g. (Manabe, Smagorinsky & Strickler, 1965). Unfortunately, there are no uniform conformal projections, which would display the entire sphere on the finite area of the plane. This problem is usually solved by combining several projections and interpolation between them in those areas where the projections intersect. This interpolation method increases the complexity and creates additional difficulties in the formulation of the numerical schemes having desired properties, such as the conservation of mass and energy. Therefore, in this book, we do not describe the hydrodynamics equations related to cartographic projections.

The atmospheric processes in boundary layer vary with height mainly influenced by the earth's surface and the turbulent exchange. The wind velocity (magnitude and direction) in the free atmosphere also changes with altitude, but the major role here is no longer played by the friction force, but the pressure gradient change under the influence of the horizontal temperature gradient. Due to the large horizontal scale weather disturbances, horizontal force components of intrinsic viscosity caused by the molecular and turbulent transfer of momentum are small too compared with the Coriolis force and horizontal pressure gradient. Consequently, the free atmosphere model with a high degree of accuracy can serve as a model of an ideal medium (without friction). To establish quantitative relationships, let us turn to the equations (5.4.24). If we neglect the terms describing the molecular and turbulent transfer, we get

$$\frac{1}{p_B}\left(\frac{\partial p_B}{\partial t}+\frac{u}{r\cos\varphi}\frac{\partial p_B}{\partial\lambda}+\frac{v}{r}\frac{\partial p_B}{\partial\varphi}+\dot{\sigma}\frac{\partial p_B}{\partial\sigma}\right)+\frac{1}{r\cos\varphi}\frac{\partial u}{\partial\lambda}+\frac{1}{r}\frac{\partial v}{\partial\varphi}+\frac{\partial\dot{\sigma}}{\partial\sigma}-\frac{v}{r}\mathrm{tg}\varphi=0\,;$$

$$\frac{\partial u}{\partial t}+\frac{u}{r\cos\varphi}\frac{\partial u}{\partial\lambda}+\frac{v}{r}\frac{\partial u}{\partial\varphi}+\dot{\sigma}\frac{\partial u}{\partial\sigma}+\frac{u\dot{\sigma}}{r}-\frac{uv}{r}\mathrm{tg}\varphi$$
$$=-\frac{1}{r\cos\varphi}\frac{\partial\Phi}{\partial\lambda}+\frac{1}{r\cos\varphi}\frac{\sigma}{p_B}\frac{\partial\Phi}{\partial\sigma}\frac{\partial p_B}{\partial\lambda}-2\Omega\left(\mathrm{w}\cos\varphi-v\sin\varphi\right),$$

$$\frac{\partial v}{\partial t}+\frac{u}{r\cos\varphi}\frac{\partial v}{\partial\lambda}+\frac{v}{r}\frac{\partial v}{\partial\varphi}+\dot{\sigma}\frac{\partial v}{\partial\sigma}+\frac{v\dot{\sigma}}{r}+\frac{u^2}{r}\mathrm{tg}\varphi=-\frac{1}{r}\frac{\partial\Phi}{\partial\varphi}+\frac{1}{r}\frac{\sigma}{p_B}\frac{\partial\Phi}{\partial\sigma}\frac{\partial p_B}{\partial\varphi}-2\Omega u\sin\varphi.$$

$$\frac{\partial\mathfrak{H}}{\partial t}+\frac{u}{r\cos\varphi}\frac{\partial\mathfrak{H}}{\partial\lambda}+\frac{v}{r}\frac{\partial\mathfrak{H}}{\partial\varphi}+\left(\dot{\sigma}-w_g\right)\frac{\partial\mathfrak{H}}{\partial\sigma}+\phi\mathfrak{H}=f.$$

(4.1.19)

In conclusion, the approximate forms (4.1.5) – (4.1.7) and (4.1.14) – (4.1.16) of the equations (3.2.29) – (3.2.35) are most widely used as a large-scale model atmospheric circulation.

THE INITIAL CONDITIONS FOR MACRO-SCALE PROBLEMS

To address a meteorological problem, we need the total physical definition of the solutions of equations of hydrodynamics and heat and mass transfer, as well as specification of initial and boundary conditions. Setting these ancillary data is a prerequisite for a well-defined problem, and in a certain sense, the starting point for a specific addressing it. The system of equations (4.1.5) – (4.1.7) or (4.1.14) – (4.1.16) allows for given initial conditions (3.2.36) at time $t = t_0$ and boundary conditions (3.2.37) – (3.2.39) at the boundary $\mathcal{B}$ to determine the field values of the unknown functions at time $t = t_0$.

However, the modern system of meteorological observations cannot fully meet the requirements of the initial conditions on which the success of macroscale hydrodynamic prognosis depends. Even negligible inconsistency between air mass fields and wind velocity leads to the generation of fast-growing high-frequency gravitational waves with amplitudes significantly higher than observed in the real atmosphere. Ideally, one needs values of the main meteorological parameters (surface pressure, wind velocity vector, temperature, and humidity) in each node of the computational grid, covering the area of solving problems of the weather forecast. In addition, the data on the temperature of bodies of water surfaces, soil moisture and snow cover are also required.

The most serious obstacles to the numerical solution of hydrodynamic equations of the macroscale model are properties of the atmosphere. Large-scale horizontal acceleration of the air is about an order of magnitude smaller than any of the forces (i.e. the Coriolis force or the horizontal pressure gradient force) per mass unit, taken separately. Minor errors in the acceleration valuation generate high-speed and high-frequency perturbations; the amplitude of these disturbances significantly more than high-amplitude perturbations observed in the Earth's atmosphere. Consequently, the horizontal wind velocity and pressure gradient components must be measured and transmitted by communication channels up to 1% in order the horizontal accelerations included in equation (4.1.6) can be calculated with accuracy up to 10%. However, the wind velocity components and the horizontal pressure gradient in practice are not measured with an accuracy of 1%. Moreover, their measurement at individual weather stations is not representative of the large-scale atmospheric motions, if you focus on the accuracy of such limits. Therefore, horizontal accelerations calculated directly from the measured Coriolis forces and the horizontal pressure gradient will contain a 100% error and will be much more than acceleration corresponding to the large-scale atmospheric circulation. Such erroneous acceleration will generate high-speed and high-frequency perturbations; the amplitude of these perturbations are much more than high-amplitude perturbations observed in the Earth's atmosphere.

This conclusion suggests that the atmosphere itself is close to a state of equilibrium, and is much closer than it follows from direct measurements. Therefore, for sufficiently accurate calculation of horizontal accelerations included in the equation (4.1.6), one must first find out to what extent the large-scale atmospheric circulations are in the balance. In this case, it may be possible to impose artificial equilibrium conditions on the initial state of the atmosphere that reduce erroneous great acceleration. The task of formulating the conditions of the natural state of equilibrium between the pressure distribution and large-scale motions of the atmosphere has been among the main tasks of dynamic weather prediction during the entire period of development of numerical methods of weather forecasting. This problem is closely related to the problem of filtering solutions of equations (4.1.6) related to sound and gravity waves (Prusov & Doroshenko, 2006).

For similar reasons, the horizontal divergence of the wind speed cannot be accurately calculated based on the results of direct measurements of wind speed. It is possible to establish indirectly that the value of

$$D_g = \frac{1}{r\cos\varphi}\frac{\partial u}{\partial\lambda} + \frac{1}{r}\frac{\partial v}{\partial\varphi}, \tag{4.2.1}$$

typically, is an order of magnitude less than the value of the individual terms $r^{-1}\cos^{-1}\varphi\partial u/\partial\lambda$ and $r^{-1}\partial v/\partial\varphi$, as the latter tend to almost completely offset each other. Therefore, to calculate the horizontal divergence D_g up to 10%, wind velocity components must be measured to an accuracy of 1%. However, wind speed measurement error exceeds 10%. Therefore, the horizontal divergence and horizontal wind acceleration calculated based on direct measurement of the wind will contain the error of about 100%. The calculated horizontal divergence D_g is usually much larger than divergence corresponding to the large-scale motions of the atmosphere. Therefore, the trend in the surface pressure and the analog of the vertical air speed cannot be accurately calculated using equations (4.1.5) and (4.1.6) until the erroneous perturbations divergence are repaid in any way.

The initial fields of meteorological variables obtained by measurements must nevertheless satisfy the equations of hydrodynamics constituting circulation model of the atmosphere. This is due, on the one hand, with the above drawbacks in the background information, and on the other hand, the inadequacy of the system used. Any practically acceptable system of macro-scale model equations cannot and should not display all the details of the meteorological fields included in the initial data. The system of hydrodynamic equations of the macroscale model is always based on a simplified model of the real processes. But if we assume that the accepted model of atmospheric circulation provides almost satisfactory accuracy, it is useful to carry out some "fit" the source data, so that these approximate data could better satisfy the system of equations. In such a fitting process, commonly called "matching the initial fields" raw data error can be largely reduced. The procedure for effective preparation of the initial information is widely used in modern practice, but not all the possibilities are exhausted.

Almost until 1970, most research has been devoted exclusively to the problem of studying the characteristics of fields of meteorological variables, calculated using models with precise and perturbed input data, in order to develop requirements for the grid density of meteorological stations and data assimilation techniques. Studies have shown that the main difficulties in the assimilation of data (alignment of fields of pressure and wind speed, control of the occurrence of collateral gravitational waves), can be overcome by means of a multi-element approach. Modern data assimilation method has three components: the collection of observational data; an objective analysis; initialization (initial approval fields of meteorological variables). The objective analysis is to use a multiple-optimal interpolation in which the covariance statistical properties of observational errors and background meteorological fields to minimize the mean square error of interpolation. By initializing suppressed high-frequency vibrations resulting from the inconsistency of fields of pressure and wind speed. There are several ways to initialize developed, based on the solution of both the linearized and complete the equations of motion.

Harmonization of the Wind Field, Temperature, and Geopotential on the Basis of Geostrophic Relations and Equations of Statics

In the free atmosphere, the actual wind is close to the geostrophic wind, and the static equation is performed with great accuracy. Therefore, the measured values of the temperature T, the horizontal component of the wind velocity vector v_1,v_2 and values calculated on the basis of the geopotential field $\tilde{\Phi}$

$$\tilde{v}_1 = -\frac{1}{\ell}\frac{\partial \tilde{\Phi}}{\partial x_2},$$
$$\tilde{v}_2 = -\frac{1}{\ell}\frac{\partial \tilde{\Phi}}{\partial x_1}, \quad (4.2.2)$$
$$\tilde{T} = -\frac{1}{R}\frac{\partial \tilde{\Phi}}{\partial \ln p}$$

should differ by a relatively small amount $\varepsilon' = \left(v_1', v_2', T', \Phi'\right)$. In (4.2.2) ℓ=2ωsinφ and values, marked on top by a wavy line, indicate the value of the corresponding variable after the correction (the model value). Obviously

$$\Phi' = \tilde{\Phi} - \Phi,$$

$$\begin{aligned} v_1' &= \tilde{v}_1 - v_1 = -\frac{1}{\ell}\left(\frac{\partial \Phi'}{\partial x_2} + \frac{\partial \Phi}{\partial x_2}\right) - v_1, \\ v_2' &= \tilde{v}_2 - v_2 = -\frac{1}{\ell}\left(\frac{\partial \Phi'}{\partial x_1} + \frac{\partial \Phi}{\partial x_1}\right) - v_2, \\ T' &= \tilde{T} - T = -\frac{1}{R}\left(\frac{\partial \Phi'}{\partial \ln p} + \frac{\partial \Phi}{\partial \ln p}\right) - T, \end{aligned} \tag{4.2.3}$$

and there is an objective function

$$J = \iiint_Q \left[\alpha_1^2\left(v_1'^2 + v_2'^2\right) + \alpha_2^2 T'^2 + \alpha_3^2 \Phi'^2\right] dx_1 dx_2 d\ln p\,, \tag{4.2.4}$$

that is an integral sum of squared differences between the observed values and the corresponding model values. In (4.2.4), the weighting factors $\alpha_1^2, \alpha_2^2, \alpha_3^2$ are chosen different, so that, firstly, to take into account the likely closeness of measured values and calculated ones by formulas (4.2.2), and, second, to display the contrast of the errors of measurement of wind and temperature.

In order to minimize the objective function (4.2.4), set it's the first variation δJ equal to zero

$$\iiint_Q \left[\alpha_1^2\left(v_1'^2 dv_1' + v_2'^2 dv_2'\right) + \alpha_2^2 T'^2 dT' + \alpha_3^2 \Phi'^2\, \mathrm{d}\,\Phi'\right] dx_1 dx_2 d\ln p = 0\,.$$

Using arbitrariness of the volume Q and equating the integrand, we obtain

$$\alpha_1^2\left(v_1'^2 dv_1' + v_2'^2 dv_2'\right) + \alpha_2^2 T'^2 dT' + \alpha_3^2 \Phi'^2 d\Phi' = 0 \tag{4.2.5}$$

By the content of the problem, functions v_1, v_2, T, Φ are given in relations (4.2.3) for Φ, which determine the values of $\Im' = \left(v_1', v_2', T', \Phi'\right)$. One can choose and, therefore, vary only the field Φ'. On this basis, we get

$$\begin{aligned} \delta v_1' &= -\frac{1}{\ell}\frac{\partial \delta \Phi'}{\partial x_2}, \\ \delta v_2' &- \frac{1}{\ell}\frac{\partial \delta \Phi'}{\partial x_1}, \\ \delta T' &= -\frac{1}{R}\frac{\partial \delta \Phi'}{\partial \ln p} \end{aligned}$$

and equation (4.2.5) takes the form

$$\alpha_1^2\left[\frac{\partial}{\partial x_2}\left(\frac{v_1'}{\ell}\right)-\frac{\partial}{\partial x_1}\left(\frac{v_2'}{\ell}\right)\right]+\alpha_2^2\frac{\partial}{\partial \ln p}\left(\frac{T'}{R}\right)+\alpha_3^2\Phi', \tag{4.2.6}$$

Equation (4.2.6) together with the last three equations in (4.2.3) constitute a system of four equations for the four functions v_1', v_2', T', Φ'. The solution of this system does not cause difficulties.

The Dynamic Negotiations

A natural generalization of the approach discussed above is the use of complete equations in the form (4.1.5) - (4.1.7) or (4.1.14) - (4.1.16) instead of the geostrophic balance. We pose the problem for any of the systems of equations (4.1.5) - (4.1.7) or (4.1.14) - (4.1.16) in the following general form

$$\begin{gathered}\frac{\partial \Im(r,t)}{\partial t}=F(r,t), r\in Q, 0\le t<\infty,\\ \Im(r,t)\big|_r=\varpi(r), \Im(r,0)=\Im_0(r),\end{gathered} \tag{4.2.7}$$

where $\Im=(v_1,v_2,T,\Phi)$.

The simplest method of implementation of (4.2.7) is a "step over method" for solving the Cauchy problem. It amounts to the calculation of the function $\Im(r,t)$, first, for the time $t+\Delta t$, and then for the time $t-\Delta t$. The non-linear terms on the right side $F(r,t)$ for time $t-\Delta t$ are calculated not with the initial data (4.2.7), but the results obtained for time $t+\Delta t$. Naturally, the calculated initial field will thus differ from the original fields. Repetition of this procedure leads, ultimately, to the mutual agreement of the fields, which they obtain within the selected closed system of equations (4.2.7).

One can offer a more efficient algorithm for solving the problem (4.2.7) by the establishment algorithms with nested grids. To implement it in $Q(r)$ there is constructed a sequence of condensing twice grids. The initial conditions are given by the coarsest grid (4.2.7). Since this grid has limited number of nodes, the amount of computation in the solution of the problem (4.2.7) on a time step is small. It is easy to show that the establishment of a stationary solution of the problem (4.2.7) is fast enough because of the exponential nature of the decay of perturbations at the initial conditions. After the establishment, the resulting solution is interpolated on a finer grid and we choose it as the initial conditions. Repeating this procedure more on finer fractional meshes, one can then obtain the refined solution of (4.2.7), based on the Runge method using all the grids.

Modern Methods of Analysis of Primary Meteorological Fields

With the development of numerical methods of weather forecasting, there are increasing demands to identify appropriate initial conditions for the model. In the early stages, more than forty years ago, meteorological observations were synchronous in time and carried out for the same physical variables (pressure, temperature, humidity, horizontal wind components) that are used in predictive models to describe the state of the atmospheric flow. Long before the advent of computers, meteorologists fulfill what has been called the analysis of the weather situation, which consisted of the daily correcting forecast of the previous day by new observations. The result of this analysis was available in the form of weather maps. With the development of numerical models, this task was assigned to the computer. Fundamentals of two-dimensional interpolation algorithms were formulated in (Bergthorsson & Doos, 1955; Cressman, 1959). They persist essentially unchanged to this day. Background fields defined in the nodes of a regular grid of predictive models are interpolated in the observation points. The difference between the observations and interpolated values is then used for correction of the background field and produce a first approximation. In the works (Eliassen, 1954; Gandin, 1963) that laid the foundations of statistical analysis based on the functions of the covariance of meteorological fields, there was defined what became known as the optimal interpolation. After many improvements, and with many variations, the optimal interpolation to this day is the basis of most of the procedures for determining the initial conditions for numerical weather prediction.

One of the first problems encountered when using the optimal interpolation to determine the initial conditions, it was the unrealistic appearance of high-frequency oscillations, which were initially interpreted as gravitational waves. The solution to this problem was found in the field of practice of primary initialization performed after proper analysis. Initialization is to impose requirements of the balance of the initial mass and wind fields a priori, that does not depend on the time. However, the requirement to balance the initial fields may conflict with the demand of closeness of the initial fields to the available observations. In fact, as is well known (Lorenz, 1963), atmospheric processes are extremely high sensitive to initial conditions. The latter, in addition to the birth of a new discipline of dissipative deterministic chaos, led meteorologists to understand the fact that the deterministic forecast atmospheric circulation severely limited in time, and efforts that they have made have been in vain. On the other hand, scientists have added incentive to conduct research aimed at determining, as accurately as possible, the initial conditions for numerical weather prediction. By the end of the sixties of the last century, the development of satellite observing systems has increased the amount of information produced by continuous in time synoptic observations. This allowed studying more deeply the dynamic evolution of air flows and the influence of the initial conditions assignment errors on the quality of forecasts. The term "data assimilation" was introduced at that time to determine the approach in which observation, distributed over time, are merged together with the dynamic numerical flow model to determine as precisely as possible the state of the atmosphere.

Since then, continuous progress in the theory of physics of the atmosphere, in improving the efficiency of numerical methods and algorithms that increase the performance of computer systems have led to a slow but steady progress in methods of data assimilation. This progress, together with the development of own numerical weather prediction models and advances in observational systems, significantly contributes to the continuous improvement of the quality of numerical weather prediction. It is worth to note that the share of the resources of the entire process of numerical weather prediction, directed to data assimilation, is constantly increasing. If at the initial stage of numerical methods for the analysis of the estimated cost forecast was negligible compared to the computational cost of the 24-hour forecast, today the cost of calculations required for data assimilation (in addition to the cost of the integration model for the assimilation period) is usually equal to the cost of 24-hour forecast. And in the world's leading meteorological centers, considering this problem daily, 24-hours data assimilation is equivalent at a cost to 10-day or more forecast.

Today, several meteorological centers involved in the project of re-analysis of the collected observations using modern computer capabilities and learning algorithms. Such projects, being carried out over a period of several decades, must give a uniform description of the atmospheric circulation over a longer period. Long sequences of restored states of the atmosphere, thus obtained, will be very useful for climatological studies of various kinds. This is particularly important now, due to the increased interest in the potential impact of human activity on the climate and the need to define possible climate changes timely and reliably. One may mention the main difference between the assimilation of data aimed at identifying the initial conditions of numerical weather prediction, and a posteriori re-analysis of previous observations. In the first case, it is possible to use only the observations made before the time at which it is desired to evaluate the state of the flow. In the case of re-analysis, there is no reason to ignore the observations made after the evaluation time. In this case, it is natural to wish to use algorithms that can process the information contained in the observations back and forth in time.

The goal of data assimilation can be described in general terms as follows: using all the available information to determine as far as possible as accurately as possible, the state of the atmospheric flow. Depending on the goal you may want to determine the flow state at a given time or the development of the flow for a period. Two sources determine available information: observations and dynamical model. Different types of observations have significant differences in nature, the degree of accuracy, and irregularities in the temporal and spatial distribution. Furthermore, observations can be "direct" when they give the same physical quantities to be used in the description of the flow (typically the speed, temperature, atmospheric humidity) or "indirect", giving values that are complex function (usually some forms of integral) values, selected to describe the flow. Satellite measurements of the infrared radiation flow emitted by the atmosphere are examples of indirect measurements.

The second source of information used in data assimilation is presented by a dynamic model, and more generally, by the physical laws that govern the evolution of the flow. These fundamental physical laws are the laws of conservation of mass, energy, and momentum, and the numerical model is nothing more than numerically convenient (and approximate) state of these laws. Obviously, it should be possible to clarify certain concepts of meteorological fields through an appropriate combination of observations and the physical laws governing the flow. For example, the time derivative of the surface pressure is, when the hydrostatic approximation, the integral over the vertical divergence of the horizontal wind. Surface pressure observations made at the given point in comfortable times, therefore, contain information about the wind field. One can give another simple example: information on wind field can also be obtained from the observed motion of tracers, such as humidity.

One would assume that the observations on the one hand, and the physical laws that govern the flow, on the other hand, together contain all the necessary information, and there is no need to add anything else to the two main sources of it. This is true in principle, but in practice, the situation is somewhat different. It may be useful, for example, the clear introduction of climate assessments in data assimilation. It seems important also to use geostrophic balance. It is known that, at least in mid-latitudes, the large-scale atmospheric flows are geostrophically balanced. We have already mentioned that the numerical weather prediction, made using the initial conditions not properly balanced, will give at least some time, high-frequency unrealistic motion. Geostrophic balance should be the necessary consequence of physical laws governing the flow. However, this is only an asymptotic property, in the sense that any solution of the corresponding equations will asymptotically tend to the approximation of the geostrophic balance, but its presence is not required at the initial time. Thus, it was found that the numerical models with nonlinear forms advection, dissipation and power generation by integrating on a long range provide solutions that seek to geostrophic balance, even if the initial conditions were non-geostrophic. However, this property is still not enough to guarantee the availability of the geostrophic balance in the fields, obtained by assimilation of meteorological observations over a period of, say, 24 hours. One must explicitly introduce the corresponding requirement of the geostrophic balance of the atmospheric flow into the process of data assimilation.

Today there is no such information, which could be considered absolutely accurate, whether it comes from the observation, from the physical laws supplied (especially if these physical laws are expressed in the form of the discrete numerical model) or from any other sources. And thus, some uncertainty always remains in the results of data assimilation, proceeding from the uncertainties in the accuracy of the various sources of information used for assimilation. The ideal data assimilation system must, therefore, give not only an assessment of the state of the flow but also assess the associated uncertainty.

The first difficulty while performing calculations on the assimilation of meteorological observations is the dimension of the problem. The number of independent scalar meteorological observations, performed in a 24-hour interval, now is about 10^5. Regarding the dimension of the largest numerical prediction models (i.e., the number of independent parameters that define the models in a state of flow at a given time), it is today of the order of 10^6-10^7. Numerical models require approximately about an hour of computer time to perform calculations on a 24-hour period of modeling time. In addition to at least one such 24-hour integration, the most effective modern data assimilation algorithms require solutions of one or more linear systems of equations, the dimension of which is equal to the number of observations or model dimension. The corresponding computational load within the operating limitations of numerical models imposes very strong restrictions on data assimilation. These restrictions are critical to the selection and application of data assimilation algorithms. Another difficulty arises from the non-linear (chaotic) nature of atmospheric flows. It imposes severe restrictions on the data assimilation, as well as on the predictability. However, it should be mentioned that most of the work performed in the data assimilation is implemented in the framework of a corresponding local linear approximation so that all non-linearity effects may not be taken into account in the process of data assimilation.

The problem of determining as accurately as possible the atmospheric flow conditions together with the corresponding uncertainty is rather complicated. Nevertheless, it can be resolved, at least to a reasonable degree of accuracy, by using a generalized least squares approach. Description of existing methods of assimilation of meteorological observations can be found in (Daley, 1991; Bennet, 1992; Thiebaux & Redder, 1987; Wahba, 1990).

These techniques are reduced to the estimation problem, in which it is desirable to find a state of the physical system based on information that is extremely heterogeneous in origin, nature, and accuracy. They use the probabilistic approach and aim to get a reasonable approximation of the distribution function of the conditional probabilities of the system state on this available information (Tarantola, 1987; Lorenc, 1986). The implementation of numerical algorithms is based on equations that define the physical system under consideration.

Consistent and Variational Assimilation

From a purely algorithmic point of view (regardless of the underlying theory) data assimilation now exists in two forms – consistent and variational.

The sequential data assimilation, which has been used in fact in all operational numerical weather forecasts prior to the 90s of the last century, suggests that the model data assimilation is integrated into the time interval in which the distributed input monitoring. Whenever the model reaches the end of the segment on which observations are available, the state of the predicted model is used as a background, is "updated" or "adjusted" by new observations. The integration of the model then starts again from the updated state, and the process repeats. The operational numerical weather prediction model has the state of the atmosphere, obtained at the end of the assimilation period, as the initial state for the prediction. Correction of the background state by newly received observations is called analysis. The sequential data assimilation, therefore, is a series of tests, carried out at the moment of observation, and the integration of the model between the moments of analysis. All sequential data assimilation algorithms, developed to date, use only one passage that is performed over a period of assimilation so that each individual observation is used only once.

An attractive feature of sequential data assimilation is that it provides a constant update your status forecasted by the model: each new data set is used to correct the last assessment of atmospheric flow conditions. This feature makes the sequential data assimilation is well adapted to numerical weather prediction. Numerical prediction is usually performed once a day. To define the initial conditions for the new forecast for the time t_0 it starts with the initial conditions of the previous forecast at t_0 - 24 and carries out 24-hour sequential assimilation. This approach is used for operational purposes in many forecasting centers and is carrying out in some of them for decades without interruption (but with the continuous improvement of both the model and the assimilation of the algorithm).

However, sequential data assimilation has a serious drawback. Due to the sequential nature of each separate set of observations affects the state of estimated flow only in subsequent times, but not in the previous. As such, spreading information contained in the observations is made only from the past to the future, but not from the future into the past.

Variational assimilation, for its side, aims to combine global model solution with all the observations available in the period of assimilation. Today, the existing algorithms for variational data assimilation define a scalar function that for any model solutions to the range of assimilation measures the "distance" or "error" between the decision and the available observations. So-called objective function (or cost function) is the sum of squares of the differences between the observations and the corresponding model values, for example,

$$J = \sum_j \alpha_j \left(\mathcal{B}_j - \mathcal{B}_j^0 \right)^2$$

where $\mathcal{B}_j^0$ is observation, $\mathcal{B}_j$ is corresponding model values, and α_j is a weighting factor reflecting the accuracy of the various observations. After that, a model solution is sought, which will minimize the objective function. Since the model solution solely determined by the relevant initial conditions at the start of data assimilation, these initial conditions are taken as control variables, i.e., as variables, in relation to which the minimization is performed. Minimizing the initial state is performed through an iterative process, where each step requires an explicit knowledge of the local values of the partial derivatives (or gradient vector) of the objective function in the initial state. This gradient can be determined with little additional computational cost through the use of conjugate equations of the assimilation model.

Some Generalizations of the Linear Statistical Evaluations of the Method of Least Squares

Most (but not all) data assimilation algorithms of the sequential or variational type used to date for analytical or operational objectives, can be described as more or less simplified forms of the statistical linear least squares estimation method. Many "non-linear" applications actually correspond to the very similar linear cases. Statistical evaluation of the linear least squares method is a classic tool. The basic principle is very simple, even if the practical application of the systems of large sizes, such as meteorological fields, may give rise to many problems. Before turning to the modern meteorological applications, let invite the reader to consider the basic principles of statistical estimation of the linear least squares method using the simplest examples.

Suppose we want to determine the following unknown scalar ξ^t for two known measurements $\mathcal{B}_1$ and $\mathcal{B}_2$ like

$$\mathcal{B}_1 = \xi^t + \varepsilon_1, \tag{4.2.8}$$

$$\mathcal{B}_2 = \xi^t + \varepsilon_2, \tag{4.2.9}$$

where ε_1 and ε_2 are errors of observation. These errors are not known, but it is assumed that the statistical error characteristics of instruments used are known. More correctly, it is expected that these tools have unbiased estimation errors, i.e.,

$$E(\varepsilon_1) = E(\varepsilon_2) = 0, \tag{4.2.10}$$

where E(*)denotes statistical averaging, and that variances of ε_1 and ε_2 are known

$$E\left(\varepsilon_1^2\right) = \sigma_1^2, E\left(\varepsilon_2^2\right) = \sigma_2^2, \tag{4.2.11}$$

For simplicity, we also assume that the errors of the observations are uncorrelated

$$E(\varepsilon_1, \varepsilon_2) = 0 \tag{4.2.12}$$

This is the case if, for example, two observations are made with different tools.

Assessment ξ^t can be obtained as a linear combination of two observations $\mathcal{B}_1$ and $\mathcal{B}_2$

$$\xi^a = a_1\mathcal{B}_1 + a_2\mathcal{B}_2, \tag{4.2.13}$$

where weight a_1 and a_2 are defined. To provide estimation ξ^a to be statistically unbiased, it is required to comply with the conditions $E(\xi^a - \xi^t) = 0$. It holds if

$$a_1 + a_2 = 1. \tag{4.2.14}$$

A minimum of statistical estimation error deviations among all unbiased estimates ξ^a can be found through

$$\sigma^2 = E[(\xi^a - \xi^t)^2]. \tag{4.2.15}$$

Solution to the problem of minimizing (4.2.15) under the condition (4.2.14) is easily found for the respective weights a_1 and a_2, which are inversely proportional to the variance of the corresponding observational error,

$$a_1 = \frac{\sigma_2^2}{\sigma_1^2 + \sigma_2^2}, a_2 = \frac{\sigma_1^2}{\sigma_1^2 + \sigma_2^2} \tag{4.2.16}$$

and the corresponding minimum of estimation variance σ^2 is given by the relation

$$\frac{1}{\sigma^2} = \frac{1}{\sigma_1^2} + \frac{1}{\sigma_2^2}, \tag{4.2.17}$$

which has a simple interpretation: if we call a "precision" the reciprocal of the error variance, then the accuracy of the estimate ξ^a is the sum of the accuracy of observations.

The same estimate ξ^a can be found through a different approach: an acceptable estimate the exact value of ξ^t should be close to the observations, at least within the accuracy of the latter. For any value of ξ, the "distance" between the ξ and observations can be measured through the following quadratic value

$$J(\xi) = \frac{(\xi - B_1)^2}{\sigma_1^2} + \frac{(\xi - B_2)^2}{\sigma_2^2}, \tag{4.2.18}$$

where the observation error variance σ_1^2 and σ_2^2 take into account the accuracy of the observations. Now, the value of ξ, which minimizes $J(\xi)$, is precisely estimated by ξ^a, obtained by the equation (4.2.13) - (4.2.16). So, minimization (4.2.15) provides another way to determine the estimation ξ^a, based on a variational principle.

Formula (4.2.13) - (4.2.16) can be generalized for the case of any number of m observations $B_i = \xi^t + \epsilon_k$, $I = 1,2,\ldots, m$. This generalization is obvious if the observation errors are not correlated. However, more realistic is to consider the observation, for which there is no need to consider the quality of assessments. The observations are rarely performed at the times and at the points of the space in which the estimations are carried out. Moreover, as already mentioned, many observations are "indirect" and do not bring physical properties of the estimation. For example, satellite radiometer measures the radiative flows emitted by the Earth into space at different wavelengths, while it is desirable to have an assessment of atmospheric temperature and moisture fields. The measured flows are functions of these fields (as well as other variables such as the amount of cloudiness and pressure at the top level, the emissivity of the surface, etc.) expressed as the equation of radiative transfer. Indirect measurements will increase every year, and data assimilation techniques must take into account the need for measurement assessments that are "complex" function of the physical parameters. Statistical linear estimation can be applied to such indirect measurement, of course, within the framework of linearity.

Suppose one needs to evaluate the full vector ξ^t of n dimensions and components ¾ $(i = 1 \ldots n)$. Such a vector may be considered as comprising, for example, values of one or more of meteorological fields (temperature, wind velocity components, humidity) in the given time points in a two- or three-dimensional regular space. The vector ξ^t, which is necessary to assess, is called the state vector, as it generally describes the state of a physical system such as the atmosphere. As for the available observations, they form a vector $\mathcal{B}_j$ (j=1,…,m). It is assumed that the vector $\mathcal{B}$ can be written as

$$\mathcal{B} = \mathcal{M}\xi^t + \epsilon\,, \tag{4.2.19}$$

where $\mathcal{M}$ is the matrix of $m \times n$, which defines the relationship between the estimated parameters and the values of the observations. Matrix $\mathcal{M}$ is called the matrix of observations. If observations are made on estimated physical fields but obtained at points other than the time and space of the points for which the evaluation was found, then $\mathcal{M}$ is a spatial-temporal interpolation. If the observations are obtained from "indirect" functional parameters that are measured, the $\mathcal{M}$ will represent the corresponding linearization of physical and (or) the statistical relation connecting ξ^t and $\mathcal{B}$. In the simple case of (4.2.8), (4.2.9), we have $n = 1$ and $m = 2$, the matrix $\mathcal{M}$ is a matrix $(2\times1)^T$ (where the superscript 'T' denotes transposition). With regard to the dimension m of the vector ϵ, its components are errors due to observations. This error vector, of course, unknown, but we assume that it is not statistically biased, i.e., $E(\epsilon)=0$ and that the covariance of variance of its components, forming a matrix $E(\epsilon\epsilon^T)$ are known.

$$E(\epsilon\epsilon^T)=\Sigma.$$

Thus, the exactly known quantities in the equation (4.2.19) are the vector of observations $\mathcal{B}$ and the matrix of observations $\mathcal{M}$, whereas the vector of errors of observation is known only through its statistical properties and unknown vector ξ^t is completely unknown. In analogy to example (4.2.8), (4.2.9), the estimation of the state vector ξ^a is linear function of $\mathcal{B}$, i.e. it can be represented as

$$\xi^a = A\mathcal{B}\,, \tag{4.2.20}$$

where the matrix A of dimension $n \times m$ is defined. The evaluation ξ^a must be unbiased, i.e., $E(\xi^a - \xi^t)=0$.

This condition is satisfied if

$$A\mathcal{M} = I_n\,, \tag{4.2.21}$$

where I_n is the identity matrix of order n. Furthermore, among all matrices satisfying (4.2.21), it is needed to choose the one that minimizes the variance of the estimation error norm, i.e. matrix that minimizes the trace (sum of the diagonal terms) of the covariance matrix

$$P^a = E\left[\left(\xi^a - \xi^t\right)\left(\xi^a - \xi^t\right)^T\right]$$

of estimation error. The solution of this problem is,

$$A = \left(\mathcal{M}^T \Sigma^{-1} \mathcal{M}\right)^{-1} \mathcal{M}^T \Sigma^{-1}. \quad (4.2.22)$$

Then the matrix

$$P^a = \left(\mathcal{M}^T \Sigma^{-1} \mathcal{M}\right)^{-1} \quad (4.2.23)$$

It contains a variance and covariance estimation errors for the components of ξ^t. In particular, the diagonal members are all variance of estimation errors.

Formulas (4.2.22), (4.2.23) are a generalization of the formulas (4.2.13) - (4.2.16) and represent the variational formulation of the estimation ξ^a, which now corresponds to the minimization of the objective function

$$J(\xi) = \left(\mathcal{M}\xi - \mathcal{B}\right)^T \Sigma^{-1} \left(\mathcal{M}\xi - \mathcal{B}\right), \quad (4.2.24)$$

where ξ is a vector of dimension n. The value of the expression (4.2.24), which is a generalization of (4.2.18), has the following meaning. For every ξ^a, the vector $\mathcal{M}\xi - \mathcal{B}$ is the vector difference between what could occur if the estimated vector would be equal to ξ (if observations would be perfect) and actual observations $\mathcal{B}$. $J(\xi)$ is the quadratic norm of this vector difference weighted so as to take into account through the accuracy of various observations (and possible correlations between different observational errors). Thee evaluation ξ^a is the value of ξ, for which this quadratic rate is minimal, i.e., a value x, which can be obtained by performing an absolutely precise observation in nearby locations.

It must be emphasized that the problem of minimizing the statistical estimation of error variance for ξ^t, and the minimization of the objective function (4.2.24) are a priori quite different problems, despite the fact that they can lead to algebraically identical results in simple cases. The first problem concerns minimizing the value determined in the n-dimensional space of the state vector, whereas the second problem is about the minimization of the value defined on the m-dimensional space of observations.

The vector ξ^a, defined by equations (4.2.20) - (4.2.22) is called the best linear unbiased estimation (Best Linear Unbiased Estimate, or BLUE) of ξ^t in $\mathcal{B}$. General theory of BLUE underlies variational data assimilation methods. In addition to evaluating the ξ^a, statistical linear estimation gives the covariance matrix P^a of the corresponding estimation error, fulfilling thus the above goal: to get the perfect algorithm for data assimilation. It may be noted that the matrix P^a is independent of the observation vector $\mathcal{B}$, but depends only on the observations of the operator $\mathcal{M}$ and the matrix Σ, i.e., the nature and the accuracy of observation. In particular, the statistical theory of linear estimation can be used to determine the perfection of any hypothetical observation system, defined by the fact what such a system can observe and with what accuracy, regardless of the actual observations themselves.

Application of statistical linear estimation requires, in addition to knowledge of the matrices $\mathcal{M}$ and Σ, that the error ξ of observations is unbiased. In fact, if such an error is offset, but the offset is known, the offset can be subtracted from the pre-observation vector to obtain a new, unbiased observation vector. The requirement that the bias is zero, therefore, is limited to the requirement that the statistical average observation error is known.

Thus, before using the statistical theory of estimation one need to know what has been observed (this is expressed through the matrix of observations), and with what the accuracy (it is expressed in terms of the mean and covariance matrix of the vector of observation errors). These requirements may be too stringent, especially regarding new types of devices for which it is difficult to establish reliable values of observational errors. Therefore, the absence of the required information can be compensated, within reasonable limits, using the possible hypotheses on the matrix $\mathcal{M}$ and the corresponding observation error covariance matrix Σ.

The linear form (4.2.20) of a priori assessment ξ^a is limited by the requirement of knowledge of statistical moments only the first two orders. Identification of the most general evaluation of the variance minimum would require full knowledge of the probability distribution function of observation errors. However, in the case where the error vector has a Gaussian distribution, the corresponding conditional probability distribution function for ξ^t is also Gaussian with expectation defined by (4.2.20) - (4.2.22), and covariance defined by (4.2.23). Therefore, in the case of Gaussian the BLUE completely solves the problem of determining the function of the conditional probability distribution of the state vector ξ^t.

The Consistent Form of Statistical Linear Estimation. Kalman's Filtering

The specific form of formulas (4.2.20), (4.2.22) and (4.2.23) are widely used in many applications where the "observation vector" $\mathcal{B}$, given by equation (4.2.20), can be decomposed into two components

$$\mathcal{B} = \left(\xi^{bT}, \mathcal{B}^{0T}\right)^T,$$

where ξ^b is a background estimation of the vector ξ^t, or the background estimation, and $\mathcal{B}^0$ additional set of of observations of the dimension p. Background estimation can be written as (4.2.15)

$$\xi^b = \xi^t + \varepsilon^b, \tag{4.2.25}$$

where ε^b is respective background estimation error. Regarding the additional set of observations $\mathcal{B}^0$, it is assumed to be associated with the observation matrix H of dimension $n \times m$, that can be written as

$$\mathcal{B}^0 = H\xi^t + \varepsilon, \tag{4.2.26}$$

where ε is respective observation error. Formulas (4.2.25) and (4.2.26) are analogs of formula (4.2.19) with the only difference that the dimension of the vector of observations is now to be $m = n + p$, the observation matrix is $\mathcal{M} = \left(I_n, H^T\right)^T$, and the observation error vector is $\varepsilon = \left(\varepsilon^{bT}, \varepsilon^T\right)^T$. Observation error covariance matrix in this case is defined as

$$\Sigma = \begin{cases} \mathrm{E}(\varepsilon^b, \varepsilon^{bT}) & \mathrm{E}(\varepsilon^b, \varepsilon^T) \\ \mathrm{E}(\varepsilon, \varepsilon^{bT}) & \mathrm{E}(\varepsilon, \varepsilon^T) \end{cases} \tag{4.2.27}$$

It is important to emphasize that the background estimation ξ^b is not necessary, and is not a member of the "observations" in the truest sense of the word. Background estimation, for example, may already be known as statistical or "climatological" average of the vector ξ^t, or any other available quantity chosen, for example, from theoretical considerations or as a result of the preceding integration of the numerical weather prediction model. The important point is only that ξ^b must be known numerically, together with relevant statistical covariance error ε^b.

To obtain the corresponding BLUE of ξ^a and covariance matrix P^a of the corresponding estimation error, you can use the formula (4.2.22), (4.2.23). The results can be presented in a form that is very convenient to both points of view, theoretical and numerical. For simplicity, it is assumed that the error ε^b. and ε are statistically uncorrelated so that the off-diagonal terms in (4.2.27) are equal to zero. Matrix $\mathrm{E}\left(\varepsilon^b, \varepsilon^{bT}\right)$ is denoted by P^b, and the matrix $\mathrm{E}\left(\varepsilon, \varepsilon^T\right)$ by R. Then the equation (4.2.20), (4.2.22) and (4.2.23) can be represented as:

$$\xi^a = \xi^b + P^b H^T \left(HP^b H^T + R\right)^{-1} \left(\mathcal{B}^0 - H\xi^b\right), \tag{4.2.28}$$

$$P^a = P^b - P^b H^T \left(HP^b H^T + R\right)^{-1} HP^b. \tag{4.2.29}$$

Equation (4.2.28) determines the state to analyze ξ^a as the sum of background evaluation ξ^b and a correction member. The last is proportional to the vector $\mathcal{B}^0 - H$¾, i.e., the difference between the additional vector of observations and the that one which the operator of observations H could give, if it was applied to the background of ξ^b. This difference substantially represents the degree of discrepancy of the background and the new observations estimates. Obviously, if such difference is exactly zero, i.e. if the background estimation is consistent with new observations, then there is no point at which you want to assess the background correction. In the linear case, the correction applied to the background estimation will appear as a linear function of the difference vector $\mathcal{B}^0 - H\xi^b$. Matrix $I = P^b H^T \left(HP^b H^T + R\right)^{-1}$, which is called the matrix of "winning", considering the relevant precision of the background estimations and observations, as defined by the covariance matrix P^b and the R, gives the best estimate, in the sense of minimum variance state vector ξ^t.

The vector $\mathcal{B}^0 - H\xi^b$ is called the vector of residues or innovative vector. The second name comes from the general theory of estimation and is very informative because this vector effectively describes all the new information contained in the vector of additional observations $\mathcal{B}^0$.

Another point relates to the equation (4.2.28). Its numerical solution requires inverting the matrix HP^bH^T+R, which has the dimension p. Direct use of the equation (4.2.22) requires (at least) the inversion of the matrix $\mathcal{M}^T\Sigma^{-1}\mathcal{M}$, which has dimension n. If $p \ll n$ (that corresponds to the numerical weather forecasts, where the number of observations available at a given time, two orders of magnitude smaller than the dimension of the model state), using the formula (4.2.28) is much more economical in terms of computational cost. In addition, formula (4.2.28) does not require conditions of inversion of covariance matrix Σ, whereas the equation (4.2.22) obviously requires it.

Equation (4.2.29) has the following meaning. It determines the covariance matrix of error analysis P^a as the covariance matrix of background state error P^b minus correction matrix. The latter is symmetrical, with nonzero eigenvalues and determines how the variance error analysis for any of the parameters is close to the corresponding variance error of the background state. The second term on the right side of the equation (4.2.29), therefore, is a gain which is obtained from introducing the vector of additional observations $\mathcal{B}^0$.

There is another way to express the P^a directly from (4.2.23)

$$\left(P^a\right)^{-1} = \left(P^b\right)^{-1} + H^T R^{-1} H .$$

This expression is often used in numerical applications, but its analogy to (4.2.17) is evident: it also represents the condition that "accuracy" of the analysis is determined by background estimation accuracy, on one hand, and additional observation accuracy, on the other.

Thus, from (4.2.24) that the variational problem with estimating ξ^t based on the background state ξ^b and the additional observations of the vector $\mathcal{B}^0$ corresponds to minimizing the objective function

$$J(\xi) = \left[\xi - \xi^b\right]^T \left(P^b\right)^{-1} \left[\xi - \xi^b\right] + \left[H\xi - \mathcal{B}^0\right]^T R^{-1} \left[H\xi - \mathcal{B}^0\right], \tag{4.2.30}$$

where ξ, as above, is a vector of dimension n. The objective function is the sum of two terms. One determines the distance to the background state ξ^b, the other is the distance to the additional observations of the vector $\mathcal{B}^0$. These two terms are weighted by the inverse covariance matrix corresponding to the error.

Formula (4.2.28), (4.2.29) express the most general form of the optimal interpolation used for operational analysis. In practice, this is implemented as follows. The background estimation ξ^b, obtained by the model of data assimilation at the time of the analysis, is combined with the observation vector $\mathcal{B}^0$ at the same time using any of the approximate forms of (4.2.28). After that, the integration model is performed by analysis from the state ξ^a until the following observations, for which the new analysis is performed. This establishes the basis of sequential data assimilation in the form in which it is used today in a variety of options for operational numerical weather forecasts. The use of equation (4.2.28) requires prior knowledge of the covariance matrices P^b and R of forecast and observation errors respectively. Reliable determination of these matrices, especially P^b, generates a lot of problems. Therefore P^b is usually introduced as a model matrix, based on some simple hypotheses about the shape and spatial scales of relevant covariance functions. Also, it is usually assumed that the prediction errors of the geopotential and wind are associated by geostrophic relation for the middle and high latitudes, what allows the determining all the required covariances from the knowledge only of the error covariance of geopotential forecast. Several simplifications are also introduced to reduce the computational cost of (4.2.28) and, especially, (4.2.29). For example, the calculation according to the equation (4.2.28) requires solving at least one linear system of dimension p. In usual meteorological practice, p can have the dimension of the order of 10^4-10^5. To reduce the calculation cost in the analysis, one uses observations only close to each model node. This "selection" of observation is quite "by the law" in the sense that the observations made at a great distance from a given point, should have less influence on the analyzed field at this point. However, experience has shown that such "selection" nevertheless introduces spatial noise in the field of analysis that must be filtered out, followed by smoothing. Regarding equation (4.2.29), it is not used in its complete form and, usually, only the diagonal terms of the matrix P^a are calculated, i.e., the analysis error variance.

Experience also shows that, despite the alleged link between the errors of the forecast wind and geopotential, the fields derived from the optimal interpolation, still contain unrealistic non-geostrophic noise. This noise should be filtered corresponding additional initialization procedures. This question is set out in (Temperton, 1988).

Optimal interpolation in the form in which it is currently used in the operational numerical weather prediction gives satisfactory results. Description of its implementation can be found in many articles or reports. For example, the work (Lorenc, 1981), though it is not new, contain a clear description of the main provisions of the use of optimal interpolation in operational practice, and the results of some of the features that it provides. US National Meteorological Center has introduced into operational practice in the early 90s using the optimal interpolation, built on the equation (4.2.28), but based on iterative minimization of the objective function (4.2.30) (Parrish & Derber, 1992). These results showed improvement of the quality of analysis at a lower computational cost. To a large extent, this has been achieved by eliminating consideration of observations near each of the analyzed node. In addition, an appropriate definition of the matrix P^b eliminates the need for initialization. The results thus show that the optimal interpolation is more effectively to carry out numerically, not through the direct use of the equation (4.2.28) and explicit exact solution of one (or more) of linear systems of equations, but through iterative (and approximate) minimization of the corresponding objective function (4.2.30). After that in a few years, the European Centre for Medium-Range Weather Forecast also put into operational practice such a "three-dimensional analysis of variance" and is currently successfully completing the test "four-dimensional variational analysis."

In addition to the above, there is a much more sophisticated approach than the analysis of the state at a given time, based on the formulas (4.2.28), (4.2.29). This approach is based on the Kalman filter and, considering the linear context, corresponds to the sequential approach. For example, consider the following scenario. Vector observations $\mathcal{B}$ in the form of (4.2.19) are obtained from the equations (4.2.20) - (4.2.22). Appropriate BLUE will be denoted as ξ^a–, and the covariance matrix of the corresponding estimation error as P^a-. At this point, an additional observation vector $\mathcal{B}^0$ in the form of (4.2.26) is available and that allows to determine BLUE ξ^a for ξ^t from combined observations vector $\beta = \left(\mathcal{B}^T, \mathcal{B}^{0T}\right)^T$. Is it necessary to repeat the calculation again, and is there a possibility to use the results of the calculations, which have already carried out and led to ξ^a– and P^a-? From the following discussion, it will be clear that if the vector ε of error observation associated with $\mathcal{B}^0$, is uncorrelated with the vector estimation error ($\xi^a - \xi^t$), then the BLUE ξ^a and the corresponding covariance matrix of the estimation error obtained by formulas (4.2.28) and (4.2.29), will be replaced by ξ^a– and P^a–, respectively. There is no need, therefore, in re-starting the calculations from the original $\mathcal{B}$, and one can take the results of the calculations already done. If the dimension of the vector p additional observations $\mathcal{B}^0$ is small compared with the dimension n of the vector ξ^t, it is obvious that the numerical gain from using formulas (4.2.28) and (4.2.29) is greater than the repetition of all calculations.

Formulas (4.2.28), (4.2.29), therefore, provide a way to update the status of the last evaluation system considering the new data. If an error related to the quality of observations, mutually uncorrelated, then the estimate obtained at any stage of the process will be always BLUE by ξ^t observed in the process earlier, and there is no loss of accuracy of the estimate due to the sequential nature of the procedure.

With the assimilation of observation spread over a range of time in which the state of the system itself is evolving, there is an additional complication. To account for the fourth independent coordinates, the time, again in a linear approach, we consider a system whose state evolves in time according to a linear law

$$\xi_{k+1}^t = M\xi_k^t + \eta_k \,, \tag{4.2.31}$$

where ξ_k^t denotes the state of the system at the time k, while M is a known transition matrix, which expresses the change in the state vector of time between time points of k and $k+1$. The member η_k represents the contribution of a temporary change of the state vector, which does not reflect the transition matrix. We can assume that the transition matrix is a numerical model available to carry out the assimilation of observations, and the member of η_k represents the accumulated effect of the change of the state vector between time points k and k+1 from various processes that are not represented in M. The η_k member will be referred to the model error. It will be an unbiased random vector, uncorrelated in time, with a known covariance matrix Q

$$\mathrm{E}\left(\eta_k, \eta_l^T\right) = Q\delta_{kl} \,, \tag{4.2.32}$$

where δ_{ld} is the classical Kronecker symbol (for simplicity, there is not considered a clear dependence M and Q on the time).

It is assumed that the observation of the general form (4.2.26) are available at successive time instants $k = 0, 1, ..., N$:

$$\mathcal{B}^0 = H\xi_k^t + \varepsilon_k \,, \tag{4.2.33}$$

Appropriate errors ε_k of observation are assumed, as before, unbiased, and uncorrelated in time and have a covariance matrix R (again without the explicit dependence of H and R on time). This leads to the expression

$$\mathrm{E}\left(\varepsilon_k, \varepsilon_l^T\right) = R\delta_{kl} \,, \tag{4.2.34}$$

In addition, the model and the observational errors are assumed to be mutually uncorrelated

$$\mathrm{E}\left(\varepsilon_k, \eta_l^T\right) = 0 \,.$$

It is also assumed that the BLUE ξ_k^a of the state ξ_k^t of the system at the time k is defined by all previous observations until k inclusive, with the covariance matrix P_k^a of the corresponding estimation error. It can be shown that the BLUE system state at the time until $k + 1$ for all observations will be equal to

$$\xi^b_{k+1} = M\xi^a_k . \tag{4.2.35}$$

The covariance matrix of the corresponding estimation error $\xi^b_{k+1} - \xi^t_{k+1}$, is equal to

$$\begin{aligned} P^b_{k+1} &= \mathrm{E}\left[\left(\xi^b_{k+1} - \xi^t_{k+1}\right)\left(\xi^b_{k+1} - \xi^t_{k+1}\right)^T\right] \\ &= \mathrm{E}\left[\left(M\xi^a_k - M\xi^t_k - \eta_k\right)\left(M\xi^a_k - M\xi^t_k - \eta_k\right)^T\right], \\ &= \mathrm{E}\left\{\left[M\left(\xi^a_k - \xi^t_k\right) - \eta_k\right]\left[M\left(\xi^a_k - \xi^t_k\right) - \eta_k\right]^T\right\} \\ &= MP^a_k M^T + Q \end{aligned} \tag{4.2.36}$$

where different hypothesises of no correlation are used. The first term in the last line represents the error at the time $k + 1$, which follows from the error at the time k, while the second term represents the contribution η_k of the model error between the instants k and $k + 1$.

At this stage, easy to introduce the vector of observations $\mathcal{B}^0_{k+1}$ at the time $k + 1$ simply by the use of formulas (4.2.28) and (4.2.29). Then ξ^b and P^b are replaced by ξ^b_{k+1} and P^b_{k+1}, respectively, and $\mathcal{B}^0$ will be replaced by $\mathcal{B}^0_{k+1}$. This leads to BLUE ξ^a_{k+1} system state at time $k + 1$ for all observations, including $k + 1$

$$\xi^a_{k+1} = \xi^b_{k+1} + P^b_{k+1}H^T\left(HP^b_{k+1}H^T + R\right)^{-1}\left(\mathcal{B}^0_{k+1} - H\xi^b_{k+1}\right), \tag{4.2.37}$$

and for the covariance matrix of the corresponding estimation error

$$P^a_{k+1} = P^b_{k+1} - P^b_{k+1}H^T\left(HP^b_{k+1}H^T + R\right)^{-1} HP^b_{k+1} . \tag{4.2.38}$$

The sequential process defined by (4.2.35) - (4.2.35) is called the *Kalman filtering* (Kalman, 1960). General description of the theory of the Kalman filtering and its properties can be found in (Jazwinski, 1970). At any step of the Kalman filter gives the BLUE system state, using all observations before the evaluation. It also gives the corresponding covariance matrix of the estimation error.

In the case of assimilation of meteorological observations, one can see that if the adoption of linear hypotheses underlying equations (4.2.31) and (4.2.33) is permissible, the Kalman filter satisfies the definition of an ideal data assimilation system, namely, the use of all available information for the purpose of getting the most accurate possible description of flow conditions along with the uncertainty arising from the uncertainties of different sources of information. In this case, the available information consists of not only observations $\mathcal{B}^0_k$ of the equation (4.2.31), but also of the model (4.2.33) and initial state and ξ_0^b, from which the entire process must start. With regard to the associated uncertainties, they are determined covariance matrices R and Q, and the initial covariance matrix P_0^b. Kalman filtering combines sequentially all these elements in the searching the BLUE of (4.2.37) and the corresponding covariance matrix (4.2.38).

Application of Kalman filtering to the assimilation of meteorological observations has been studied by several authors, such as (Ghill, 1989; Ghill & Malanotte-Rizzoli, 1991). Experiments performed with different linear systems gave acceptable results and confirmed the possibility of using the method for efficient selecting information contained in the observations and models. The main obstacle to the application of Kalman filtering to the problem of assimilation of meteorological observations is high computational cost. For example, the first term of the last line of the equation (4.2.36) has the form $MP_k^a M^T$, which requires the implementation of the relevant calculation of two successive multiplications on M. At the same time, the multiplication of a vector by M corresponds to integration of the model between time instances of k and $k+1$ in the equation (4.2.35). Using equation (4.2.36) will require $2n$ integrations of the model where n is, as before, the dimension of the model state vector. With n of the order of 10^6-10^7 in meteorological data assimilation problem, it goes far beyond the capabilities of operational applications of Kalman filtering in numerical weather prediction, or even a posteriori assimilation of meteorological observations. Therefore, the calculation by the formula (4.2.36) in operational weather forecasting is replaced by a simple multiplication of analysis error variances with a priori established factor (usually 1.5 for 6-hour forecast), and the corresponding correlations are modeled independently. The corresponding computational cost is negligible in this case, but such a procedure does not consider the effect of a specific weather situation, and especially the instability that may develop on the evolution of the forecast error.

This is the main weakness of submitted operational data assimilation techniques, which can be described as a simplified but efficient form of Kalman filtering. Currently, intensive research is being conducted on the development of algorithms to describe the temporal evolution of the prediction error, which, on the one hand, would be cost-effective enough for practical use, and on the other hand, would be accurate enough to improve these operational methods of sequential assimilation. These studies hold in following directions. It is studying to what extent may be a significant correlation between the forecast errors at the points of a long distance. If it is negligible, then far from the diagonal of the matrix member P_k^a must be zero or a small value and therefore removed. This reduces the cost calculation (4.2.36). In addition, it was found that the greatest impact on the evolution of the forecast errors has geostrophic propagating modes (Lacarra & Talagran, 1988). In this case, it is possible to reduce the cost calculations to a subset of all model modes (Cohn & Parrish, 1991; Dee, 1991; Bouttier, 1994).

Variational Form of Statistical Linear Estimation

When considering only the case of variational estimation, we consider only perfect model, i.e., $\eta_k = 0$ in the equation (4.2.35), so that the exact evolution of the flow is reduced to condition

$$\xi_{k+1}^t = M\xi_k^t, \tag{4.2.39}$$

Variational form (4.2.24) of estimation problem is defined by equations (4.2.33) and (4.2.39), leading to an objective function

$$J(\xi) = \Sigma_{0<k<N} \left[H\xi_k - \mathcal{B}_k^0\right]^T R^{-1} \left[H\xi_k - \mathcal{B}_k^0\right], \tag{4.2.40}$$

where $\xi = \left(\xi_k^T\right)^T$ is a sequence of model states at times related to the model equation (4.2.39). In other words, $J(\xi)$ is the sum of the squares of the differences values model - minus - observation, weighted inversely proportional to the values of the covariance matrix of observation errors. Minimizing the objective function (4.2.40) under the condition (4.2.39) will give at any time k the real state of the BLUE ξ_k^t of the system for all available observations up to the time k. Data assimilation through variational approach will be the same as the result obtained by the Kalman filter (assuming an accurate model, i.e. $Q = 0$ in the equation (4.2.36)). Variational form of statistical estimation, therefore, provides an adaptation of the model to the timing of observations on a global scale.

The assumption of linearity needed to establish a connection with the theory of linear statistical estimation and Kalman filtering is in fact not necessary for the variational problem of the type (4.2.39), (4.2.40). One can easily turn to the problem of minimizing the objective function of the form (4.2.40) under the condition of the form (4.2.39), in which the matrix M and H are replaced with non-linear operators. Indeed, numerical models of atmospheric flows are non-linear, and many observations non-linearly related to atmospheric variables are desirable to evaluate. For example, infrared radiation is associated with satellite measurements of temperature and humidity profiles in the atmospheric column through the equation of radiative transfer in the strongly nonlinear way. Consider the problem of minimizing the non-linear objective function (4.2.40) (i.e., the objective function with non-linear operators of observation) in the nonlinear condition of the form (4.2.39). To emphasize the non-linearity of operators, we will use the notation M' and H' instead of M and H, respectively.

There are mainly two methods for solving the limited minimization problem. The principle of the first method consists in reducing the constraints, i.e., in eliminating some restricting variable so that the transition from the original problem to the problem without any restrictions becomes feasible. In this case, it may be noted that the model solution (4.2.39) uniquely determines the corresponding initial condition ξ_0. The objective function J, thus, can be considered as a function only of ξ_0, for which there are no limitations so that the initial problem is reduced to the problem of minimization without restrictions on ξ_0. The second method, which is based on the principle is much less obvious, consists in finding the unknown coefficients, called Lagrange multipliers, and restricting the problem in the form of the corresponding Lagrangian. In this case, there are N limitations (4.2.39), each dimension n, and a set Λ of Lagrange multipliers consisting of N vectors Λ_k ($k = 0,1, ..., N$-1), each of dimension n. Relevant Lagrangian looks

$$L(\xi,\Lambda) = J(\xi) + \Sigma_{0<k<N} \Lambda_k^T \left[\xi_{k+1} - M\xi_k\right].$$

Minimum for limited minimization problem (4.2.39), (4.2.40) corresponds to fixed-point Lagrangian $L(\xi,\Lambda)$, regarded as a function of the independent variables ξ and Λ.

The method of conjugate equations (Lions, 1971) is the most effective way to minimize the numerical solution of the problem (4.2.39), (4.2.40). It is interesting that the method of conjugate equations can be used in two ways: either by relaxing constraints (4.2.39) so that to use only the initial state ξ_0 as an independent variable (LeDimet & Talagran, 1986; Talagran & Courtier, 1987), or alternatively, using Lagrange multipliers (Thacker & Long, 1988). For example, to solve the problem (4.2.39), (4.2.40) as a problem of minimization without restrictions on the initial state x_0, it is necessary for beginning to be able to relate the variation of the initial state ξ_0 to appropriate variations of the objective function J. For a given initial state, these variations are connected via a local vector of partial derivatives, or the gradient vector, of the objective function by components of the initial state. If it is possible to calculate numerically the gradient vector for the initial state, it is possible to obtain a gradient in the standard algorithm of minimization which determines the minimized initial state through successive iterations. In most cases, there is no possibility to establish analytical expressions for the gradient. But it is possible to define numerically (and approximately) the gradient through the initial state disturbances. This will require an additional calculation of the objective function, i.e., integrating the model for data assimilation period as many times as there are independent components of the initial state. The method of conjugate equations provides a way to calculate the gradient, the computing cost of which is several times higher than the calculation of the distance function.

The basis of the method is very simple. The sequence of operations starts with a certain input vector ω with components $\omega_i(i=1,\ldots,q)$ and provides output vector ϖ with components ϖ_j $(j=1,\ldots,r)$. The equation can describe this sequence

$$\varpi = G\left(\omega\right), \tag{4.2.41}$$

where G specifies the operator to transition from ω to ϖ. For a given perturbation $\delta\omega$ at the input, the resulting disturbance $\delta\varpi$ is of first order

$$\delta\varpi = G'\left(\delta\omega\right), \tag{4.2.42}$$

where G' is the matrix of local partial derivatives, or Jacobian, of ϖ components by ω components. The equation (4.2.42) is called the equation of the linear tangent to (4.2.41). Now let $J\left(\varpi\right)$ is a scalar function of the output ϖ. The gradient of J by ω is determined by the chain rule

$$\frac{\partial J}{\partial \omega_i} = \sum_{j=1}^{r} \frac{\partial \varpi_j}{\partial \omega_i} \frac{\partial J}{\partial \varpi_j}, i = 1,2,\ldots,q,$$

or in matrix notation

$$\forall_\omega J = G'^T \forall_\varpi J, \tag{4.2.43}$$

where the superscript 'T' still represents transposition.

The conjugate method is based on the consistent use of the formula (4.2.43). More precisely, it is assumed that the process G is a superposition of a series of elementary processes, namely

$$G = G_c \circ \ldots \circ G_2 \circ G_1,$$

The Jacobian G' is the product of elementary Jacobians

$$G' = G'_M \circ \ldots \circ G'_2 \circ G'_1$$

and transposition G'^T is the result of the elementary transpositions performed in reverse order

$$G'^T = G_1'^T \circ G_2'^T \circ \ldots \circ G_M'^T.$$

This shows that defining the gradient $\forall_\omega J$ on w numerically, it is enough to return back through direct calculation and to perform the appropriate transposition or associated calculations at every step. The total cost of the dual settlement of (4.2.43) is usually of the same order of magnitude as the cost of a direct calculation (4.2.41). In (Morgenstern, 1984) it is shown that the total number of operations of the conjugate calculation is 4 times less than the total number of direct calculation of the corresponding operations.

In the case of determining the gradient of the nonlinear objective function (4.2.40) corresponding to the initial state ξ_0 of nonlinear assimilation model (4.2.39), associated calculations are reduced to the integration of the equation (Talagran & Courtier, 1987)

$$\delta'\xi_k = M'^T \delta'\xi_{k+1} + H'^T R^{-1} \left[H\xi_k - \mathcal{B}_k^0 \right] \tag{4.2.44}$$

back in time, starting with the "final" state of $\delta'\xi_{N+1} = 0$. In this equation, M' and H' are Jacobians of corresponding nonlinear operators of the model and observations M and H. The gradient of the objective function of the initial state ξ_0 is equal to $2\delta'\xi_0$.

The basic model solution ξ_k is clearly shown in the equation (4.2.44) in the member $H\xi_k - \mathcal{B}_k^0$, that, except for its sign, is updated (innovative) vector of equations (4.2.28) and (4.2.37). This means that the basic solution is calculated and stored before the conjugate integration is performed.

The method of conjugate equations to perform a variational assimilation of meteorological observations was first proposed in (Renenko & Obraztsov, 1976) for a simple line of small problems. Since then, a large number of experiments have been carried out with non-linear models with increasing complexity and dimension, and with different types of observations (Lewis & Derber, 1985; Talagran & Courtier, 1987; Derber, 1987; Courtier & Talagran, 1987; Lorenc, 1988; Navon et al., 1992; Rabier & Courtier, 1992; Thacker & Long, 1988; Sheinbaum & Anderson, 1990; Greiner & Perigaud, 1994). These experiments showed that the variational data assimilation is capable of minimizing an objective function. And, unlike the sequential data assimilation, the information is transmitted in both directions, forward and backward. However, when the objective function contains only terms that measure the difference between the individual observations and the corresponding model counterparts, minimization solution contains unrealistic "noise", often in the form of small-scale fluctuations and/or non-geostrophic gravitational waves. Therefore, the minimization solution is physically justified only when such noise waves correspond to movements of small scale with low energy (Courtier & Talagran, 1990; Thepaut & Courtier, 1991). It should not be assumed that the need to introduce additional members to remove unrealistic oscillations in the estimated fields is not meant to limit the application of variational methods to weather problems. Such members are often called "smoothing", "penalty-making" or "regularizing". They are also used in statistical estimation using a Kalman filter, which, as already mentioned, must lead to the same results as the variational algorithms. In today's operational optimum interpolation, the requirement of the relevant smoothing is satisfied, on the one hand, by the presence of background state ξ^b in data coverage, and on the other hand, by the process of "initialization", which filters unrealistic gravitational wave oscillations.

There are known experiments in the variational assimilation of meteorological observations that have been made on multilevel models. They are based on the initial equations that are similar, but not identical to equations of numerical weather prediction models (Thepaut, Hoffman & Courtier, 1993). There are differences in the resolution, which is still coarser in experiments with variational data assimilation (the number of nodes in a data assimilation model, typically, is one order of magnitude smaller than in the operational models), and in the description of several "natural" processes, such as convection and phase transitions, which are still lacking in explicit form for variational data assimilation. These experiments have shown that the variational data assimilation due to using a system of evolutionary equations is able to transmit the information contained in the observations much more closely than the operational optimal interpolation. However, as in the case of the Kalman filter, even though the conjugate equations offer a more efficient way of calculating the gradient of the objective function, the computational cost is still high. In addition to the need to store model solutions obtained by direct integration that must be performed before each conjugate integration, minimization of the objective function usually requires an additional 10 to 30 iterations of the minimization algorithm. And each iteration itself requires one integration model assimilation period, followed by a conjugate integration. The cost of one conjugate integration is about twice of a direct integration. Thus, one equivalent minimization usually requires from 30 to 100 integrations of model assimilation period. It is of course very expensive. But it is possible to efficiently apply the simplified form of the variational assimilation of observations in which assimilation is carried out on a grid with a lower resolution than a full numerical weather prediction model (Courtier, Thepaut & Hollingsworth, 1994).

The current situation in the methods of data assimilation and meteorological weather analysis thus is as follows. In addition to the relatively simple, economical and, in general, satisfactory operational algorithms, there are two broad classes of algorithms that are able, within the general framework of statistical linear estimation, to allocate reliably the information contained, on the one hand, in the observations and, on the other hand, in the physical laws expressed the model, namely, Kalman filtering and variational data assimilation. To apply these algorithms, it is necessary to express the observations in general form (4.2.19), i.e., it needs to know what is being measured, and with what accuracy. Direct application of any of these two classes of algorithms for numerically expensive and most of the research carried out today on data assimilation is aimed to the search for effective direction in terms of computational cost and simplifications. This problem may persist for a long time as there is no end an increase in the power of computers and the deepening of the modifications of their structures nor the development of new observing systems and more realistic models. New developments in any of these areas can easily disprove any conclusion reached as a result of the most effective ways to perform data assimilation.

Some of the Main Advantages and Disadvantages of Kalman Filtering and Variational Data Assimilation

As mentioned above, both algorithms lead to the same evaluation of the state at the end of the assimilation period for the linear operator *H* of observations and perfect ($\eta_k = 0$) of the linear model (4.2.39). The main difference between the two algorithms is that the Kalman filter operates only on the information from the past to the future, while the variational data assimilation information operates in both directions. We do not cover the procedure, called smoothing Kalman (Anderson & Moore, 1979), which in the first stage can perform calculations on the period of assimilation with the possibility of returning in time for any intermediate time *k* to obtain a BLUE state of the system for all available observations made before, during and after time *k*. A simple and instructive example of the use of such smoothing Kalman is given in (Gaspar & Wunsh, 1989). On the other hand, the variational data assimilation, as opposed to the Kalman filter, does not take account of the fact that the assimilation model, as well as observation, will never be perfect and will always contain errors. It can combine different kinds of model errors. It's enough to consider the model equation (4.2.39) not as a limitation, just satisfying sequence ξ_k of learned states, but as (if to use the definition from (Sasaki, 1970) "weak constraint" satisfied only approximately. This can be accomplished by modifying the objective function (4.2.40) to

$$J(\xi) = \Sigma_{0<k<N}\left[H\xi_k - \mathcal{B}_k^0\right]^T R^{-1}\left[H\xi_k - \mathcal{B}_k^0\right] + \Sigma_{0<k<N}\left[\xi_{k+1} - M\xi_k\right]^T Q^{-1}\left[\xi_{k+1} - M\xi_k\right] \tag{4.2.45}$$

where *Q*, as in (4.2.36), is model error covariance matrix. The second term on the right side of (4.2.45) expresses the fact that the difference $\xi_{k+1} - M\xi_k$ should not be regarded as exactly equal to zero as if it was in the case of the exact model, but the zero in the range of uncertainty established through matrix *Q*. Therefore, we consider the sequence of states with no restrictions, and minimization of *J* is done taking into account the entire sequence $\xi = \left(\xi_k^T\right)^T$. In this case, the sequence of minimizing conditions (4.2.45) gives a BLUE state of the system at all times *k* for all observations for the entire period of assimilation. This leads to the fact that the minimization of (4.2.45) should give the same sequence of states learned as Kalman smoothing. For more information on the variational data assimilation with a weak constraint can be found in (Bennet, 1992; Bennet et al., 1993).

Another difference between the Kalman filtering and variational data assimilation is that the former gives, in addition to the BLUE state of the system, the corresponding covariance matrix of the estimation error, while the latter gives only BLUE. In this sense, only the Kalman filter ensures solving the task set before on perfect data assimilation system. This is achieved through a much higher computational cost of Kalman filtering. From the equations (4.2.23) and (4.2.24) that the covariance matrix of the estimation error is the inverse Hessian (matrix of second derivatives) of the objective function. Some algorithms to minimize the so-called quasi-Newton type (Gill, Murray & Wright, 1992) enable calculating the inverse Hessian while minimizing reliably and at a lower computational cost than Kalman filtering.

After the consideration of theoretical and methodological aspects of the assimilation of meteorological observations and analysis of the field, let us revisit the validity of the adoption of the linear hypothesis, which underlies the theory of filtration Kalman and, though not necessary for variational data assimilation, has a certain value in order to facilitate understanding analysis of the obtained results. It is important that the equations governing atmospheric flows are not only strongly nonlinear but also that their non-linearity arises from one of the most important features of these flows – their chaotic nature. This character imposes strong restrictions on the predictability of the flows, and naturally, wonder how justified the linear hypothesis in the context of data assimilation.

Consider first the Kalman filter equations. Suppose that the operators of the models and observations are non-linear and denote them of *M* and *H*, respectively. If the difference $\xi_k^a - \xi_k^t$ is sufficiently small, the value of $M\xi_k^a - M\xi_k^t$ (second line of equation (4.2.36)) can be approximated by $M'\left(\xi_k^a - \xi_k^t\right)$, where M' is Jacobian taken at the point ξ_k^a. The equation (4.2.36) is, respectively, converted to

$$P_{k+1}^b = M'P_k^a M'^T + Q\,, \tag{4.2.46}$$

Similarly, if the difference $\xi_{k+1}^t - \xi_{k+1}^b$ is small enough so that the innovation vector

$$\mathcal{B}_{k+1}^0 - H\xi_{k+1}^b = H\xi_{k+1}^t - H\xi_{k+1}^b + \varepsilon_{k+1}$$

yok can be approximated by $H'\left(\xi_{k+1}^t - \xi_{k+1}^b\right) + \varepsilon_{k+1}$, then the equation (4.2.37) and (4.2.38), may be replaced by

$$\xi_k^a = \xi_{k+1}^b + P_{k+1}^b H'^T\left[H'P_{k+1}^b H'^T + R\right]^{-1}\left[\mathcal{B}_{k+1}^0 - H\xi_{k+1}^b\right] \tag{4.2.47}$$

and

$$P_{k+1}^a = P_{k+1}^b - P_{k+1}^b H'^T\left[H'P_{k+1}^b H'^T + R\right]^{-1} H'P_{k+1}^b\,. \tag{4.2.48}$$

In these equations, the operator *H* is replaced by the Jacobian H', except in the expression vector for innovation. The algorithm defined by (4.2.46) - (4.2.48), to which a non-linear analog of (4.2.36) should be added, is called the extended Kalman filtering (Jazwinski, 1970). It is based on the fact that the difference between the real and estimated the state of the system is small enough and allows local linearization, as just described above.

Similar arguments play also for variational data assimilation. For linear model and observation operators, the objective function (4.2.40) is a quadratic function of the initial state ξ_0; for non-linear ones, the objective function is not quadratic, but will be approximately quadratic near its minimum. If the initial uncertainty of the system, e.g. at the point where the minimization process begins, is small enough to ensure the preservation of the estimated state of the system throughout the period of data assimilation within this uncertainty, then the theoretical non-linearity of the objective function will have no practical significance. The solution of minimization will be BLUE of flow state.

So, the hypothesis of linearity will always be justified in those cases where the difference between the real and estimated the state of the system is small enough and allows local linearization described above. The validity of this approximation, called the linear tangent, has been tested for many situations. For example, authors of the (Lacarra & Talagran, 1988), have shown that using a linear approximation for the evolution of atmospheric flow in the barotropic model is justified on the periods of 48 or even 72 hours. In (Thepaut & Moll, 1990) the hypothesis of linear tangent is justified within the existing in practice uncertainty for the temperature and humidity in the atmosphere and for radiation observations made TOVS probe system installed on satellites NOAA series. These restrictions are still not allocated in full, and the study of non-linear data assimilation associated with processes able to stimulate drastic variations in model fields began yet in (Miller, Ghil & Gauthiez, 1994; Pires, Vautard & Talagran, 1996). These studies cover mainly the atmosphere.

The Assimilation of Indirect Observations

Until recently, the use of "indirect" satellite observations required prior conversion of them into "geophysical variables" such as for example, temperature and humidity, followed by introducing inverted fields in the data assimilation algorithms. However, such inversion is not necessary for considered methods of evaluation, though does not eliminate the need to define the relevant observations of the operator and the determination of the appropriate errors of observation. In addition, some knowledge on the background condition is required that depends on other available observations itself. This results in interdependence of errors, associated with different "observations" used in the data assimilation. The problems arising from such interdependence can be resolved in the context of statistical estimation, but it seems preferable to avoid them in the first stage. Today, preference should be given to a systematic approach to all types of observations: to enter all the observations in the data assimilation algorithm through an appropriate statement of observations without pretreatment. In the case of the satellite radiation measurements, relevant issues have been studied in (Eyre, 1989; Thepaut & Moll, 1990). Several National Meteorological Services is taking steps to implement the direct entry satellite radiation measurements in operational data assimilation algorithms. The same general trend is observed for other types of observations: for example, the wind speed component measured based on the Doppler effect, ground-based radars and satellite lidar (Sun, Flicker & Lilly, 1991; Courtier et al, 1992).

Besides the quality of the initial data assimilation, to improve the skill of numerical forecasts there are needed: the higher the horizontal and vertical resolution, improving the parameterization of physical processes, development of efficient numerical schemes and improvement of computer technology. Also on the stage there is a solution for improving analysis and forecast on non-homogeneous underlying surface.

BOUNDARY CONDITIONS FOR MACRO-SCALE PROBLEMS

Setting the boundary conditions on the lateral boundaries is not a problem for macroscale models. Indeed, if a forecast based on global forecasting model needs only the boundary conditions on the underlying surface and the upper boundary of the solution domain, then the forecast on the basis of hemispheric model formulates conditions on the equator in the form of no-flow conditions, i.e.:

$$\frac{\partial \Re}{\partial n} = 0\,, \tag{4.3.1}$$

where n is the normal to the equator; $\Re$ is any weather variable.

Basic boundary conditions on the underlying surface are reduced to their impermeability (when the normal component of the velocity w is equal to zero) and adhesion

$$u = u_b\left(\lambda,\varphi\right), v = v_b\left(\lambda,\varphi\right) \text{ on condition } z = 0. \tag{4.3.2}$$

There u_b and v_b are surface movement velocity projections (in particular, for the continent u_b=v_b=0.)

The temperature and humidity on the underlying surface may be defined on the basis of measurements of meteorological stations

$$T = T_b\left(\lambda,\varphi\right), q = q_b\left(\lambda,\varphi\right) \text{ on condition } z = 0. \tag{4.3.3}$$

In the case of setting the boundary conditions in the forecast period, and subject to a sparse network of weather stations, temperature and humidity on the underlying surface are determined from the heat balance equation

$$h = h_1 + h_2 + h_3 + \Delta h \tag{4.3.4}$$

and moisture

$$m = m_1 + m_2 + \Delta m\,, \tag{4.3.5}$$

where h is the heat flow due to radiation exposure; h_1 is the turbulent heat transfer; h_2 is the heat of phase transitions; h_3 is the heat flow at the underlying surface; Δh is absorption or heat loss reserves; m is the influx of water caused mainly by precipitation; m_1 is the moisture loss due to evaporation (evapotranspiration); m_2 is the horizontal moisture transfer to the underlying surface and the inside; Δm is the changes in the moisture content of the underlying surface. The quantities involved in (4.3.4) and (4.3.5) may be obtained, for example, based on climatic data.

The underlying surface in the real world consists of different types of areas. They are characterized by their own combination of relief, radiation, thermal, moisture and aerodynamic characteristics, such as roughness, albedo, thermal conductivity, moisture content, etc. Therefore, each of the surfaces regulates and redistributes its own way incoming heat and moisture, thereby creating individual heat and water balance. These balances in the system of the surface-atmosphere, in addition to the underlying surface, are significantly affected by the state of the atmosphere – its transparency and sustainability.

All this complicates the integration of (4.3.4) and (4.3.5) for theoretical relations obtained by the heat and water balance, since calculation results are highly dependent on the value and reliability of thermal conductivity, absorption, reflection, radiation, and others. Among the existing methods of prediction of flow, the most widely known methods are those using inherently empirical data.

The radiation balance *h*, including shortwave constituents ϱ_s

$$\varrho_s = \varrho_{s\downarrow} - \varrho_{s\uparrow}, \tag{4.3.6}$$

and longwave constituents

$$\varrho_l = \varrho_{l\downarrow} - \varrho_{l\uparrow}$$

(arrows indicate the upward and downward flow of radiation), is the most important energy characteristic of the available tributaries and heat sinks in (4.3.4).

The reflected shortwave radiation $\varrho_{s\uparrow}$ depends on the $\varrho_{s\downarrow}$ and the albedo α of the underlying surface:

$$\varrho_{s\uparrow} = \alpha \varrho_{s\downarrow}.$$

Since the Earth's surface is opaque to shortwave radiation, the no reflected part of the surface is absorbed. Thus, the shortwave radiation balance h_s looks

$$\varrho_s = \varrho_{s\downarrow} - \varrho_{s\uparrow} = \left(1-\alpha\right)\varrho_{s\downarrow}. \tag{4.3.7}$$

The illumination of the horizontal surface is proportional to $\cos\beta$, where β is the angle between the incident solar rays and the direction normal to the surface. Sunny beam at the same time will be regarded as consisting of parallel rays. So,

$$\varrho_{s\downarrow}\left(\beta\right) = \varrho_{s\downarrow}\left(0\right)\cos\beta. \tag{4.3.8}$$

Here, the solar constant $\varrho_{s\downarrow}\left(0\right)$ is taken to be $\varrho_{s\downarrow}\left(0\right) = 2.0$, and the value of $\sin\beta$ is determined by the well-known formula of spherical astronomy

$$\sin\beta = \sin\varphi\cdot\sin\delta + \cos\varphi\cdot\cos\delta\cdot\cos\tau,$$

where φ is the area's latitude; δ is the declination of the Sun; τ is watch angle of the Sun, $\tau = \pi\left(t-12\right)/12$; *t* is the local time in hours, measured from midnight.

For mountain areas, expression (4.3.8) requires clarification:

$$\varrho_{s\downarrow}\left(\beta\right) = \varrho_{s\downarrow}\left(0\right)\cos\beta\cos\xi_\beta.$$

where ξ_β is the angle between the ray and the normal to the surface.

Note that the angle ξ_β is determined not only by the angle of inclination to the horizontal surface, but also the azimuth and zenith angle of the Sun:

$$\cos\xi_\beta = \sin\xi_\beta \cos\xi_w + \sin\phi_a \cos\phi_\beta \sin\omega_\beta \sin\xi_w +$$

$$\cos\phi_a \left(\sin\phi_m \cos\phi_\beta \cos\omega_\beta - \sin\phi_\beta \cos\phi_a \cos\phi_\beta\right)\sin\xi_w ,$$

where ξ_w is the angle of incidence of sunlight on the surface; ϕ_a is the azimuth of the normal projection to the slope on a horizontal surface, measured from the plane of the meridian.

The amount of shortwave radiation $\varrho_{s\downarrow}$, reaching the Earth's surface, is strongly influenced by clouds. The main reasons for reducing $\varrho_{s\downarrow}$ are the refraction of light in the cloud elements (water droplets or ice crystals) followed by reflection and scattering of light by cloud elements.

All existing parameterization methods of shortwave radiation balance of the underlying surface give the calculated results in cloudy sky with large errors. This is due to either an error of setting the radiative properties of clouds and radiation to multiple reflection effects. Since the intensity of radiation varies depending on the optical density of the cloud and their distribution across the sky and position relative to the solar disk (these changes occur very quickly and over a wide range), the estimated individual values of the radiation intensity are beyond the permissible limits.

Numerical experiments show (Feigel'son, 1981) that the complex formulas, that consider the absorption of shortwave radiation by ozone in the stratosphere, Rayleigh scattering and multiple reflections between the clouds, the atmosphere, and the underlying surface, does not have significant advantages in terms of accuracy to more simple models. Apparently, more promising in the numerical schemes is the use of the method of calculating the radiation characteristics on the relative duration of sunshine.

If to denote by *n* the number of hours of the absence of cloudiness in all atmospheric column relating to any grid point on the underlying surface (*i*,*j*), the duration of the day at this point, i.e. the possible maximum of *n*, denote by *N*, then the ratio *n* / *N* will characterize the duration of sunshine in the design point. Assuming that the relationship between $h_{s\downarrow}$ and *n* / *N* is quadratic, we obtain (Feigel'son, 1981)

$$\varrho_{s\downarrow}(\beta) = \varrho_{s\downarrow}(0)\left[C_1 + C_2(n / N) + C_3(n / N)^2\right]\cos\beta\cos\xi_\beta . \tag{4.3.9}$$

where $C_1 = C_2 = 0.2$, $C_3 = 0.5$ are constants determined by statistics. Naturally, the formulas (4.3.7) and (4.3.9) are taken as a first approximation. Further research is needed to verify and clarify the numerical results, in particular, when used "effective "cloud albedo and land surface α_{eff} instead of the surface albedo α in (4.3.7), taking into account the multiple reflection and absorption.

In determining the contribution ϱ_l of longwave radiation in the heat balance of the underlying surface downward $\varrho_{l\downarrow}$ and upward $h_{l\uparrow}$ flows in a cloudless sky can be written as (Zadok, 1982)

$$\varrho_{l\downarrow} = \varepsilon\sigma T_b^4 \Im(m),$$

$$\varrho_{l\uparrow} = \varepsilon\sigma T_t^4, \tag{4.3.10}$$

where $\varepsilon \cong 1$ is the radiation absorption (emission) coefficient; σ is Stefan-Boltzmann constant; T_t is air temperature at the top level of the settlement; $\Im(m)$ is the integral transmission function, which takes into account the effect of water vapor and carbon dioxide, which is calculated according to the formula

$$\Im(s) = 0.471\exp\left(-0.695\sqrt{m}\right) + 0.529\exp\left(-8.94\sqrt{m}\right).$$

Then, according to (4.3.10), we have

$$\varrho_l = \varepsilon\sigma\left[T_t^4 - T_b^4\Im(m)\right]. \tag{4.3.11}$$

With the help of this formula, one can calculate longwave radiation flows in cloudless conditions. In the case of overcast, downdraft $\varrho_{l\downarrow}$ is calculated according to the formula

$$\varrho_{l\downarrow} = \varepsilon\sigma\sum_{i=1}^{2} T_i^4\Im\left(m_{i+1} - m_1\right), \tag{4.3.12}$$

where T_i is the temperature in the upper level calculated for i=1 and at the lower limit temperature of the cloud for i=2.

Part of the excess radiant heat received by the underlying surface is given to the atmosphere in the form of a turbulent flow of heat h_1, which is determined by the relation

$$h_1 = -\rho c_p k_T\left(\frac{\partial T}{\partial z} + \gamma_a\right). \tag{4.3.13}$$

Here, c_p is the specific heat at constant pressure, k_T is the turbulent exchange coefficient for heat turbulent Prandtl number, γ_a is the dry adiabatic temperature gradient.

The heat balance equation (4.3.4) includes heat of phase transition, since it defines the relationship between radiation, molecular and turbulent heat flows to the underlying surface. Increased soil moisture, for example, can alter the albedo. Moisture variations can affect the thermal characteristics of the soil, which in turn will change the heat transfer and its inventory in the soil. The greatest impact on the spending heat is from evaporation. It is this fact that explains the presence of a member describing the process in equations (4.3.4) of thermal and (4.3.5) of water balances. Water loss due to evaporation and the energy expended on evaporation, are related by

$$h_2 = Lm_1, \tag{4.3.14}$$

where L is the specific heat of vaporization or condensation.

The flow of water vapor from the underlying surface as it passes through the turbulent surface layer is given by

$$m_1 = -\rho k_d \frac{\partial q}{\partial z}, \qquad (4.3.15)$$

where k_d is coefficient of turbulent exchange of water vapor.

Substituting (4.3.15) to (4.3.14), we obtain the dependence of the heat flow from the phase transitions:

$$h_2 = -\rho L k_d \frac{\partial q}{\partial z}, \qquad (4.3.16)$$

The value of heat flow into the underlying surface depends on the temperature gradient and the thermal conductivity of this type of surface, i.e.

$$h_3 = -\lambda_g \frac{\partial T}{\partial z}. \qquad (4.3.17)$$

Unfortunately, the thermal conductivity of soil λ_g is not constant for a given type of soil. It varies with time and depth, and depends on the thermal conductivity of individual soil particles, its porosity and moisture content. This must be considered due to the fact that with the same soil inflow temperature heat treatment depends on the magnitude of its conduction. Diurnal variation of temperature changes in soils with low thermal conductivity is more pronounced because the heat exchange occurs in the thin surface layers, and in soils with greater thermal conductivity, it is smooth since heat exchange takes place in a thicker layer.

The last term in equation (4.3.4), absorption or heat loss reserves Δh, combines the enthalpy change of the layer, the advection of heat by horizontal currents in the case of water surface, the heat of phase transitions in the case of snow or ice cover, and others. Therefore, the value of Δh should be set for each individual case.

Combining expressions (4.3.6) - (4.3.17) and substituting them into (4.3.4), we obtain the equation

$$\rho c_p k_T \left(\frac{\partial T}{\partial z} + \gamma_a \right) = f_T \text{ on condition } z = 0, \qquad (4.3.18)$$

where f_T is algebraic sum of heat flows (from radiation, phase transitions in the underlying surface and absorption). One can determine the temperature of the underlying surface from (4.3.18).

Now let us consider the water balance equation. Assuming in (4.3.4) m_1 and Δm known functions of the horizontal coordinates λ, φ for each specific type of the underlying surface and expressing the entry of moisture m into the soil with

$$m_1 = -\rho k_d \frac{\partial q}{\partial z},$$

we will have the moisture balance equation (4.3.5) into the form

$$\rho k_d \frac{\partial q}{\partial z} = f_q \text{ on condition } z = 0, \tag{4.3.19}$$

where q_p is conductivity of precipitation, V_p is the rate of precipitation, k_d is turbulent exchange coefficient for water vapor, f_q is algebraic sum of moisture flows (rainfall, horizontal transfer, change of moisture content).

In many practical cases, without lowering the accuracy of calculations, it is possible to abandon the moisture balance equation (4.3.19) and set the boundary conditions for q directly, using the partial pressure of water vapor E:

$$q = \frac{0.622E}{p_{00} - 0.378E} \approx 0.622 \frac{E}{p_{00}}, \tag{4.3.20}$$

where p_{00} is atmospheric pressure at the $z = 0$.

The equation (4.3.20) is acceptable to set the boundary conditions on the water surface. Introducing the f as relative humidity, being known function of the horizontal coordinates, the expression (4.3.20) can be adapted to the boundary conditions over the land in the form of

$$q = 0.622 \frac{fE}{p_{00}} \text{ on condition } z = 0. \tag{4.3.21}$$

The described methods for setting the boundary conditions at z = 0 based on the account of heat and moisture exchange between the atmosphere and underlying surface are the most stringent. However, due to the fact that the necessary details of the type of soil, moisture, roughness, albedo and surface temperature is often not available or not representative, the practical implementation of these methods is an extremely cumbersome task. Therefore, hydrodynamic weather forecasts, to simplify registration of interaction of the atmosphere with the Earth's surface, there are often recorded the heat transfer processes in the soil moisture active layer (about 1 m depth), where the temperature and humidity are constant.

SUMMARY

In this chapter, the general theoretical framework developed in the previous chapter is applied to modeling atmospheric processes at a planetary scale. This suggests a detailed three-dimensional modeling based on geophysical fluid dynamics equations of thermodynamics, radiation transfer, the kinetic equations describing the actual movement of air masses, conversion and transfer of energy and matter. The material of this chapter also considers the formation of clouds, aerosols, and rainfall and other atmospheric processes.

The planetary model of atmospheric circulation developed here is a necessary step to construct a dynamic atmospheric model at smaller scales – regional and mesoscale – what will be done in following chapters.

REFERENCES

Anderson, B. D. O., & Moore, J. B. (1979). *Optimal filtering*. Englewood Cliffs, NJ: Prentice Hall.

Bennet, A. F. (1992). *Inverse methods in physical oceanography*. Cambridge, UK: Cambridge University Press. doi:10.1017/CBO9780511600807

Bennet, A. F., Leslie, L. M., Hagelberg, C. R., & Powers, P. E. (1993). Tropical cyclone prediction using a barotropic model initialized by a generalized inverse method. *Monthly Weather Review*, *121*(6), 1714–1729. doi:10.1175/1520-0493(1993)121<1714:TCPUAB>2.0.CO;2

Bergthorsson, P., & Doos, B. (1955). Numerical weather map analysis. *Tellus*, *7*(3), 329–340. doi:10.3402/tellusa.v7i3.8902

Bouttier, F. (1994). Dynamical estimation of forecast error covariances in an assimilation system. *Monthly Weather Review*, *122*(10), 2376–2390. doi:10.1175/1520-0493(1994)122<2376:ADEOFE>2.0.CO;2

Cohn, S. E., & Parrish, D. F. (1991). The behavior of forecast error covariances for a Kalman filter in two dimensions. *Monthly Weather Review*, *119*(8), 1757–1785. doi:10.1175/1520-0493(1991)119<1757:TBOFEC>2.0.CO;2

Courtier, P., Gauthier, P., & Rabier, F. (1992). Study of preparation for the use of Doppler wind lidar information in meteorological assimilation systems. Paris: ESA Contract Rep. 3453.

Courtier, P., & Talagran, O. (1987). Variational assimilation of meteorological observations with the adjoint vorticity equation. Part II. Numerical results. *Quarterly Journal of the Royal Meteorological Society*, *113*(478), 1329–1347. doi:10.1002/qj.49711347813

Courtier, P., & Talagran, O. (1990). Variational assimilation of meteorological observations with the direct and adjoint shallow-water equations. *Tellus*, *42A*(5), 531–549. doi:10.1034/j.1600-0870.1990.t01-4-00004.x

Courtier, P., Thepaut, J. N., & Hollingsworth, A. (1994). A strategy for operational implementation of 4D-Var, using an incremental approach. *Quarterly Journal of the Royal Meteorological Society*, *120*(519), 1367–1388. doi:10.1002/qj.49712051912

Cressman, G. P. (1959). An operational objective analysis scheme. *Monthly Weather Review*, *87*(10), 367–374. doi:10.1175/1520-0493(1959)087<0367:AOOAS>2.0.CO;2

Daley, R. (1991). *Atmospheric data analysis*. Cambridge, UK: Cambridge University Press.

Dee, D. P. (1991). Simplification of the Kalman filter for meteorological data assimilation. *Quarterly Journal of the Royal Meteorological Society*, *117*(498), 365–384. doi:10.1002/qj.49711749806

Derber, J. C. (1987). Variational four-dimensional analysis using quasi-geostrophic constraints. *Monthly Weather Review*, *115*(5), 998–1008. doi:10.1175/1520-0493(1987)115<0998:VFDAUQ>2.0.CO;2

Eliassen, A. (1954). *Provisional report on calculation of spatial covariance and autocorrelation of the pressure field.* Report No. 5. Oslo: Institute for Weather and Climate Research, The Norwegian Academy qf Science and Letters.

Eyre, J. (1989). Inversion of cloudy satellite sounding radiances by nonlinear optimal estimation. Part 1 and 2. Theory and simulation for TOVS. *Quarterly Journal of the Royal Meteorological Society*, *115*(489), 1001–1026. doi:10.1002/qj.49711548902

Feigel'son, E. M. (1981). *Radiation in the cloudy atmosphere.* Leningrad: Gidrometeoizdat. (in Russian)

Gandin, L. S. (1963). *Objective analysis of meteorological fields.* Leningrad: Gidrometeoizdat. (in Russian)

Gaspar, P., & Wunsh, C. (1989). Estimates from altimeter data of barotropic Rossby waves in the northwestern Atlantic Ocean. *Journal of Physical Oceanography*, *19*(12), 1821–1825. doi:10.1175/1520-0485(1989)019<1821:EFADOB>2.0.CO;2

Ghill, M. (1989). Meteorological data assimilation for oceanographers. Part 1. Description and theoretical framework. *Dynamics of Atmospheres and Oceans*, *13*(3-4), 171–218. doi:10.1016/0377-0265(89)90040-7

Ghill, M., & Malanotte-Rizzoli, P. (1991). Data assimilation in meteorology and oceanography. *Advances in Geophysics*, *33*, 141–266. doi:10.1016/S0065-2687(08)60442-2

Greiner, E., & Perigaud, C. (1994). Assimilation of Geosat altimetric data in a nonlinear reduced-gravity model of the Indian Ocean. Part 1. Adjoint approach and model -data consistency. *Journal of Physical Oceanography*, *24*(8), 1783–1803. doi:10.1175/1520-0485(1994)024<1783:AOGADI>2.0.CO;2

Jazwinski, A. H. (1970). *Stochastic processes and filtering theory.* Academic Press.

Kalman, R. E. (1960). A new approach to linear filtering and prediction problems. *Journal of Basic Engineering*, *82D*(1), 35–45. doi:10.1115/1.3662552

Lacarra, J. F., & Talagran, O. (1988). Short-range evolution of small perturbations in a barotropic model. *Tellus*, *40A*(2), 81–95. doi:10.1111/j.1600-0870.1988.tb00408.x

LeDimet, F. X., & Talagran, O. (1986). Variational algorithms for analysis and assimilation of meteorological observations: Theoretical aspects. *Tellus*, *38A*(2), 97–110. doi:10.1111/j.1600-0870.1986.tb00459.x

Lewis, J. M., & Derber, J. C. (1985). The use of adjoint equations to solve a variational adjustment problem with advective constraints. *Tellus*, *37A*(4), 309–322. doi:10.1111/j.1600-0870.1985.tb00430.x

Lions, J. L. (1971). *Optimal control of systems governed by partial differential equations.* New York: Springer. doi:10.1007/978-3-642-65024-6

Lorenc, A. C. (1981). A global three-dimensional multivariate statistical interpolation scheme. *Monthly Weather Review*, *109*(4), 701–721. doi:10.1175/1520-0493(1981)109<0701:AGTDMS>2.0.CO;2

Lorenc, A. C. (1986). Analysis methods for numerical weather prediction. *Quarterly Journal of the Royal Meteorological Society*, *112*(474), 1177–1194. doi:10.1002/qj.49711247414

Lorenc, A. C. (1988). Optimal nonlinear objective analysis. *Quarterly Journal of the Royal Meteorological Society*, *114*(479), 205–240. doi:10.1002/qj.49711447911

Lorenz, E. N. (1963). Deterministic nonperiodic flow. *Journal of the Atmospheric Sciences*, *20*(2), 130–141. doi:10.1175/1520-0469(1963)020<0130:DNF>2.0.CO;2

Manabe, S., Smagorinsky, J., & Strickler, R. F. (1965). Simulated climatology of a general circulation model with a hydrologic cycle. *Monthly Weather Review*, *93*(12), 769–798. doi:10.1175/1520-0493(1965)093<0769:SCOAGC>2.3.CO;2

Miller, R. N., Ghil, M., & Gauthiez, F. (1994). Advanced data assimilation in strongly nonlinear dynamical system. *Journal of the Atmospheric Sciences*, *51*(8), 1037–1056. doi:10.1175/1520-0469(1994)051<1037:ADAISN>2.0.CO;2

Morgenstern, J. (1984). *How to compute fast a function and all its derivatives. A variation on the theorem of Baur-Stressen*. Nice: University de Nice.

Navon, I. M., Zou, X., Derber, J., & Sela, J. (1992). Variational data assimilation with an adiabatic version of the NMC spectral model. *Monthly Weather Review*, *120*(7), 1433–1446. doi:10.1175/1520-0493(1992)120<1433:VDAWAA>2.0.CO;2

Parrish, D., & Derber, D. (1992). The National Meteorological Centers spectral statistical interpolation analysis system. *Monthly Weather Review*, *120*(8), 1747–1763. doi:10.1175/1520-0493(1992)120<1747:TNMCSS>2.0.CO;2

Phillips, N. A. (1957). A coordinate system having some special advantages for numerical forecasting. *Journal of Meteorology*, *14*(2), 184–185. doi:10.1175/1520-0469(1957)014<0184:ACSHSS>2.0.CO;2

Pires, C., Vautard, R., & Talagran, O. (1996). On extending the limits of variational assimilation in nonlinear chaotic systems. *Tellus*, *48A*(1), 96–121. doi:10.1034/j.1600-0870.1996.00006.x

Prusov, V. A., & Doroshenko, A. Yu. (2006). *Modeling Natural and Technogenic Atmospheric Processes*. Kyiv: Naukova Dumka. (in Ukrainian)

Rabier, F., & Courtier, P. (1992). Four-dimensional assimilation in the presence of baroclinic instability. *Quarterly Journal of the Royal Meteorological Society*, *118*(506), 649–672. doi:10.1002/qj.49711850604

Renenko, V. V., & Obraztsov, N. N. (1976). A variational method to initialize fields of meteorological elements [in Russian]. *Meteorology and Hydrology.*, *11*, 1–11.

Sasaki, Y. (1970). Some basic formalism in numerical variational analysis. *Monthly Weather Review*, *98*(12), 875–883. doi:10.1175/1520-0493(1970)098<0875:SBFINV>2.3.CO;2

Sheinbaum, J., & Anderson, D. L. T. (1990). Variational assimilation of XBT data. Part 2. Sensitivity studies and use of smoothing constrains. *Journal of Physical Oceanography*, *20*(5), 689–704. doi:10.1175/1520-0485(1990)020<0689:VAOXDP>2.0.CO;2

Sun, J., Flicker, D. W., & Lilly, D. K. (1991). Recovery of three-dimensional wind and temperature fields from simulated single-Doppler radar data. *Journal of the Atmospheric Sciences*, *48*(6), 876–890. doi:10.1175/1520-0469(1991)048<0876:ROTDWA>2.0.CO;2

Talagran, O., & Courtier, P. (1987). Variational assimilation of meteorological observations with the adjoint vorticity equation. Part 1. Theory. *Quarterly Journal of the Royal Meteorological Society*, *113*(478), 1311–1328. doi:10.1002/qj.49711347812

Tarantola, A. (1987). *Inverse Problem Theory*. Amsterdam: Elsevier.

Temperton, C. (1988). Implicit normal mode initialization. *Monthly Weather Review*, *116*(5), 1013–1031. doi:10.1175/1520-0493(1988)116<1013:INMI>2.0.CO;2

Thacker, W. C., & Long, R. B. (1988). Fitting dynamics to data. *Journal of Geophysical Research*, *93*(C2), 1227–1240. doi:10.1029/JC093iC02p01227

Thepaut, J. N., & Courtier, P. (1991). Four-dimensional variational data assimilation using the adjoint of a multilevel primitive-equation model. *Quarterly Journal of the Royal Meteorological Society*, *117*(502), 1225–1254. doi:10.1002/qj.49711750206

Thepaut, J. N., Hoffman, R. N., & Courtier, P. (1993). Interactions of Dynamics and observations in a four-dimensional variational assimilation. *Monthly Weather Review*, *121*(12), 2396–2408. doi:10.1175/1520-0493(1993)121<3393:IODAOI>2.0.CO;2

Thepaut, J. N., & Moll, P. (1990). Variational inversion of simulated TOVS radiances using the adjoint technique. *Quarterly Journal of the Royal Meteorological Society*, *116*(496), 1425–1448. doi:10.1002/qj.49711649609

Thiebaux, H. J., & Redder, M. A. (1987). *Spatial objective analysis, with applications in atmospheric science*. New York: Academic Press.

Wahba, G. (1990). *Spline models for observational data*. Philadelphia: Society for Industrial and Applied Mathematics. doi:10.1137/1.9781611970128

Zadok, V. P. (1982). *Numerical methods used in atmospheric models*. Leningrad: Gidrometeoizdat. (in Russian)

Chapter 5
Equations of Hydrodynamics, Heat, and Mass Transfer in Models of Regional and Mesoscale Atmospheric Processes

ABSTRACT

The formation of meteorological processes and phenomena in the atmosphere in each specific area is influenced by the processes of different scales. However, the significance of each process in different cases is different. Therefore, there are certain priorities for the inclusion of certain processes in the hydrodynamic model. Depending on the size of the territory taken for modeling of processes or individual phenomena in the atmosphere, hydrodynamic models are divided into local and regional. In this chapter, we consider these features in terms of mathematical models.

MESOSCALE AND REGIONAL ATMOSPHERIC MODELS

For mesoscale processes described by local and regional models, the basic equations of hydrodynamics are formulated in Cartesian and spherical coordinates, respectively. The vertical coordinate takes into the account the relief of the underlying surface.

Equations (3.2.10) have a number of specific features, which are exercised in the numerical implementation. One of the essential features of a space-elliptical nature of equations is due to the influence of the strain tensor divergence $\nabla \cdot \vec{\tau}_n$ in the flow field. In this regard, for the solution of equations (3.2.10), typical methods for solving initial-boundary problems and the formulation of conditions on all the boundaries of the computational domain must be used.

DOI: 10.4018/978-1-5225-2636-0.ch005

The system of equations (3.2.10) at the term $\nabla \cdot \vec{\tau}_n$ has a parameter ν, changing the value of which corresponds to a significant change in the smoothness of the solution due to the large vertical gradients of the meteorological variables at the underlying surface in the boundary layer.

Finally, the system of equations (3.2.10) is nonlinear. This non-linearity, typical of many systems of hydrodynamic type, is caused by the inertia components in the equations of momentum. In conjunction with the two features mentioned above, the nonlinear equations (3.2.10) leads to the formation of highly complex spatial-temporal structures. The numerical solution, in this case, is not only irregular in time, but it becomes much more complicated its spatial structure, in particular, it becomes unstable and breaks up the boundary layer, the secondary flows appear in the external flow (in the free atmosphere), etc. In some cases, it is difficult to distinguish the effects of computer instability of the possible solutions from complex behavior.

Thus, the complete system of equations (3.2.29) – (3.2.35) has great generality and universality, but this system of equations is complicated enough for practical applications. Therefore, to simplify (in some cases, very significantly) the model above, while conducting the hydrodynamic modeling of processes and phenomena in the atmosphere of all sizes, they usually use, where possible, known hypotheses and assumptions to take into account specific physical and thermodynamic properties of the atmosphere.

There are processes and phenomena of various types in the atmosphere. Some processes are developed on a very limited area in a short period, other – over large areas for a long time. The modern classification of meso-processes, depending on the spatial extent, distinguishes following three processes of the intermediate scale:

- 2-20 km – it is a α-*mesoscale* process (individual cumulus clouds, certain types of gravitational waves, and others.).
- 20-200 km – it is a β-*mesoscale* process (orographic perturbation convection zone, and others.).
- 200-2000 km – it is a γ-*mesoscale* process (cyclones, atmospheric fronts, the weather in a particular area, and others).

Obviously, the processes of all the scale described above influence the formation of meteorological processes and phenomena in the atmosphere in each specific area. For example, the macro-scale processes mainly influence the formation of the general circulation over large areas, and meso-processes make here their definite but less significant contribution. Therefore, there are certain priorities for the inclusion of certain processes in the hydrodynamic model. Depending on the size of the territory, which is implemented for modeling of processes or individual phenomena in the atmosphere, hydrodynamic models are divided into *local* (α- and β-mesoscale) and *regional* (γ-mesoscale). Consider the features of their formalization.

The horizontal and vertical gradients of wind speed, temperature, and humidity in the atmosphere promote the growth of hydro and thermodynamic instability and nonlinear interaction between perturbations of different temporal and spatial scales. Consequently, the order of the individual terms of the equations in the local models depends on local field conditions, and solutions may vary within wide limits. Of course, simplifying the original system of equations by neglecting some included members in the equation requires a proof of their smallness within the entire region under consideration and solution implementation time, what is today an insurmountable task. Therefore, α- and β- mesoscale models require the use of full hydrodynamics equations. We consider the mesoscale atmospheric model, defined by equations (3.2.29) –(3.2.35) in the Cartesian coordinate system (*x*,*y*,*z*):

- Continuity equation:

$$\frac{\partial \mathcal{R}}{\partial t}+u\frac{\partial \mathcal{R}}{\partial x}+v\frac{\partial \mathcal{R}}{\partial y}+w\frac{\partial \mathcal{R}}{\partial z}=-\left(\frac{\partial u}{\partial x}+\frac{\partial v}{\partial y}+\frac{\partial w}{\partial z}\right). \tag{5.1.1}$$

- Equations of atmospheric motion:

$$\begin{aligned}
&\frac{\partial u}{\partial t}+u\frac{\partial u}{\partial x}+v\frac{\partial u}{\partial y}+w\frac{\partial u}{\partial z}=-e^{-\mathcal{R}}\frac{\partial p}{\partial x}+2\frac{\partial}{\partial x}\left(\nu\frac{\partial u}{\partial x}\right)+\frac{\partial}{\partial y}\left[\nu\left(\frac{\partial v}{\partial x}+\frac{\partial u}{\partial y}\right)\right]\\
&+\frac{\partial}{\partial z}\left[\nu\left(\frac{\partial u}{\partial z}+\frac{\partial w}{\partial x}\right)\right]-2\boldsymbol{\Omega}\left(\mathrm{w}\cos\varphi-v\sin\varphi\right),\\
&\frac{\partial v}{\partial t}+u\frac{\partial v}{\partial x}+v\frac{\partial v}{\partial y}+w\frac{\partial v}{\partial z}=-e^{-\mathcal{R}}\frac{\partial p}{\partial y}+\frac{\partial}{\partial x}\left[\nu\left(\frac{\partial v}{\partial x}+\frac{\partial u}{\partial y}\right)\right]+\\
&+2\frac{\partial}{\partial y}\left(\nu\frac{\partial v}{\partial y}\right)+\frac{\partial}{\partial z}\left[\nu\left(\frac{\partial w}{\partial y}+\frac{\partial v}{\partial z}\right)\right]-2\boldsymbol{\Omega}u\sin\varphi,
\end{aligned} \tag{5.1.2}$$

$$\begin{aligned}
&\frac{\partial w}{\partial t}+u\frac{\partial w}{\partial x}+v\frac{\partial w}{\partial y}+w\frac{\partial w}{\partial z}=-e^{-\mathcal{R}}\frac{\partial p}{\partial z}-g+\frac{\partial}{\partial x}\left[\nu\left(\frac{\partial u}{\partial z}+\frac{\partial w}{\partial x}\right)\right].\\
&+\frac{\partial}{\partial y}\left[\nu\left(\frac{\partial v}{\partial z}+\frac{\partial w}{\partial y}\right)\right]+2\frac{\partial}{\partial z}\left(\nu\frac{\partial w}{\partial z}\right)+2\boldsymbol{\Omega}u\cos\varphi.
\end{aligned}$$

- The equation of propagation of scalars in the atmosphere $\mathfrak{H}=\left(\theta,q,\delta,s\right)$:

$$\frac{\partial \mathfrak{H}}{\partial t}+u\frac{\partial \mathfrak{H}}{\partial x}+v\frac{\partial \mathfrak{H}}{\partial y}+\left(w-w_g\right)\frac{\partial \mathfrak{H}}{\partial z}+\sigma\mathfrak{H}=f+\frac{\partial}{\partial x}\left(d_T\frac{\partial \mathfrak{H}}{\partial x}\right)+\frac{\partial}{\partial y}\left(d_T\frac{\partial \mathfrak{H}}{\partial y}\right)+\frac{\partial}{\partial z}\left(d_T\frac{\partial \mathfrak{H}}{\partial z}\right) \tag{5.1.3}$$

The system of equations (5.1.1) – (5.1.3) allows for the given values of the fields of the functions $u,v,w,\theta,p,q,\delta,s$, at some time instance *t* and the corresponding boundary conditions to determine the value of these functions in the fields some point in time $t+\Delta t$, where $\Delta t>0$. Mesoscale atmospheric processes described by the system, is carried out either by the action of mass forces, or by the action of external forces, creating a drop of pressure *p*. The last is formally manifested in setting the appropriate boundary conditions.

THE CHOICE OF THE COORDINATE SYSTEM OF EQUATIONS FOR MESOSCALE AND REGIONAL ATMOSPHERIC MODELS

In solving the problem of forecasting mesoscale atmospheric processes in the region characterized by the heterogeneity of the underlying surface topography, we have to consider settlement boundaries do not coincide with the physical space of the coordinate lines, what makes difficult to formulate the boundary conditions on the underlying surface and leads to loss of precision solutions. This fact serves as a cause for the introduction of mapping or transformation of physical space (x,y,z) to the space (x_1,x_2,x_3) of curvilinear coordinates. The field of curvilinear coordinates is constructed so that the boundaries of the physical space coincide with the lines in the coordinate space of curvilinear coordinates. Using curvilinear coordinates assumes a curved region in a physical space is converted into a rectangular parallelepiped.

Let us record our equations using curvilinear coordinates as independent variables. We introduce a non-degenerate coordinate transformation associated with the computational domain geometry

$$x_1 = x, x_2 = y, x_3 = H\frac{z - F(x,y)}{H - F(x,y)}, \tag{5.1.4}$$

where $F(x,y)$ is the equations underlying the bottom surface, $H = H(x,y) = \text{const}$ is an altitude adopted as the upper boundary of the computational domain. Coordinate transformation (5.1.4) displays the calculated area in the cuboid with the height of H (Figure 1). The inverse transformation is given by the relations:

$$x = x_1, y = x_2, z = F(x,y) + x_3\left[1 - \frac{F(x,y)}{H}\right].$$

When replacing the variables x,y,z by new variables x_1,x_2,x_3, associated with previous ones by the relation (5.1.4), first order partial derivatives are expressed by well-known formulas (Prusov & Doroshenko, 2006):

$$\frac{\partial}{\partial x} = \frac{\partial}{\partial x_1} - \frac{H}{H - F(x_1,x_2)}\left(1 - \frac{x_3}{H}\right)\frac{\partial F(x_1,x_2)}{\partial x_1}\frac{\partial}{\partial x_3},$$
$$\frac{\partial}{\partial y} = \frac{\partial}{\partial x_2} - \frac{H}{H - F(x_1,x_2)}\left(1 - \frac{x_3}{H}\right)\frac{\partial F(x_1,x_2)}{\partial x_2}\frac{\partial}{\partial x_3},$$
$$\frac{\partial}{\partial z} = \frac{H}{H - F(x_1,x_2)}\frac{\partial}{\partial x_3}.$$

The operator of the total derivative in the new coordinate system takes the form:

Figure 1. Displaying the calculated area in the cuboid

$$\frac{d}{dt}=\frac{\partial}{\partial t}+u\frac{\partial}{\partial x_1}-\frac{Hu}{H-F\left(x_1,x_2\right)}\left(1-\frac{x_3}{H}\right)\frac{\partial F\left(x_1,x_2\right)}{\partial x_1}\frac{\partial}{\partial x_3}$$
$$+v\frac{\partial}{\partial x_2}-\frac{Hv}{H-F\left(x_1,x_2\right)}\left(1-\frac{x_3}{H}\right)\frac{\partial F\left(x_1,x_2\right)}{\partial x_2}\frac{\partial}{\partial x_3} \qquad (5.1.5)$$
$$+\frac{Hw}{H-F\left(x_1,x_2\right)}\frac{\partial}{\partial x_3}=\frac{\partial}{\partial t}+u\frac{\partial}{\partial x_1}+v\frac{\partial}{\partial x_2}+\tilde{w}\frac{\partial}{\partial x_3},$$

where

$$\tilde{w}=\frac{H}{H-F\left(x_1,x_2\right)}\left\{w-\left(1-\frac{x_3}{H}\right)\left[u\frac{\partial F\left(x_1,x_2\right)}{\partial x_1}+v\frac{\partial F\left(x_1,x_2\right)}{\partial x_2}\right]\right\}.$$

The members of equations (5.1.2), containing the components of the pressure gradient $-e^{-\mathcal{R}}\nabla p$ also get a new view (5.1.6):

$$-e^{-\mathcal{R}}\frac{\partial p}{\partial x}=-e^{-\mathcal{R}}\frac{\partial}{\partial x_1}+\frac{He^{-\mathcal{R}}}{H-F\left(x_1,x_2\right)}\left(1-\frac{x_3}{H}\right)\frac{\partial F\left(x_1,x_2\right)}{\partial x_1}\frac{\partial p}{\partial x_3}$$
$$=-e^{-\mathcal{R}}\frac{\partial}{\partial x_1}-\frac{Hg}{H-F\left(x_1,x_2\right)}\left(1-\frac{x_3}{H}\right)\frac{\partial F\left(x_1,x_2\right)}{\partial x_1},$$

$$-e^{-\mathcal{R}}\frac{\partial p}{\partial y}=-e^{-\mathcal{R}}\frac{\partial}{\partial x_2}+\frac{He^{-\mathcal{R}}}{H-F\left(x_1,x_2\right)}\left(1-\frac{x_3}{H}\right)\frac{\partial F\left(x_1,x_2\right)}{\partial x_2}\frac{\partial p}{\partial x_3}$$
$$=-e^{-\mathcal{R}}\frac{\partial}{\partial x_2}-\frac{Hg}{H-F\left(x_1,x_2\right)}\left(1-\frac{x_3}{H}\right)\frac{\partial F\left(x_1,x_2\right)}{\partial x_2},$$

$$-e^{-\mathcal{R}}\frac{\partial p}{\partial z}-g=-\frac{He^{-\mathcal{R}}}{H-F\left(x_1,x_2\right)}\frac{\partial p}{\partial x_3}-g=-g\left[1-\frac{H}{H-F\left(x_1,x_2\right)}\right], \tag{5.1.6}$$

where the transformation $e^{-\mathcal{R}}\partial p/\partial x_3=-g$ is made on the basis of the equation of statics.

After the transition to the coordinates (5.1.4), taking into account (5.1.5) and (5.1.6), the system of equations (5.1.1) - (5.1.3) takes the form:

- Continuity equation:

$$\frac{\partial\mathcal{R}}{\partial t}+u\frac{\partial\mathcal{R}}{\partial x_1}+v\frac{\partial\mathcal{R}}{\partial x_2}+\tilde{w}\frac{\partial\mathcal{R}}{\partial x_3}=-\left(\frac{\partial u}{\partial x_1}+\frac{\partial v}{\partial x_2}+\frac{\partial\tilde{w}}{\partial x_3}\right). \tag{5.1.7}$$

- Equations of atmospheric motion:

$$\frac{\partial u}{\partial t}+u\frac{\partial u}{\partial x_1}+v\frac{\partial u}{\partial x_2}+\tilde{w}\frac{\partial u}{\partial x_3}=$$
$$-e^{-\mathcal{R}}\frac{\partial p}{\partial x_1}-\frac{H\,g}{H-F\left(x_1,x_2\right)}\left(1-\frac{x_3}{H}\right)\frac{\partial F\left(x_1,x_2\right)}{\partial x_1}$$
$$+2\frac{\partial}{\partial x_1}\left(\nu\frac{\partial u}{\partial x_1}\right)+\frac{\partial}{\partial x_2}\left[\nu\left(\frac{\partial v}{\partial x_1}+\frac{\partial u}{\partial x_2}\right)\right]$$
$$+\frac{H^2}{\left[H-F\left(x_1,x_2\right)\right]^2}\frac{\partial}{\partial x_3}\left\{\nu\left[\frac{\partial u}{\partial x_3}+\left(1-\frac{F\left(x_1,x_2\right)}{H}\right)\frac{\partial w}{\partial x_1}\right]\right\}$$
$$-2\Omega\left(\mathrm{w}\cos\varphi-v\sin\varphi\right),$$

$$\frac{\partial v}{\partial t}+u\frac{\partial v}{\partial x_1}+v\frac{\partial v}{\partial x_2}+\tilde{w}\frac{\partial v}{\partial x_3}=$$
$$-e^{-\mathcal{R}}\frac{\partial p}{\partial x_2}-\frac{H g}{H-F\left(x_1,x_2\right)}\left(1-\frac{x_3}{H}\right)\frac{\partial F\left(x_1,x_2\right)}{\partial x_2}$$
$$+\frac{\partial}{\partial x_1}\left[\nu\left(\frac{\partial v}{\partial x_1}+\frac{\partial u}{\partial x_2}\right)\right]+2\frac{\partial}{\partial x_2}\left(\nu\frac{\partial v}{\partial x_2}\right)$$
$$+\frac{\partial}{\partial x_3}\left\{\nu\left[\left(1-\frac{F\left(x_1,x_2\right)}{H}\right)\frac{\partial w}{\partial x_2}+\frac{\partial v}{\partial x_3}\right]\right\}-2\boldsymbol{\Omega}u\sin\varphi,$$

$$\frac{\partial w}{\partial t}+u\frac{\partial w}{\partial x_1}+v\frac{\partial w}{\partial x_2}+w\frac{\partial w}{\partial x_3}=-g\left[1-\frac{H}{H-F\left(x_1,x_2\right)}\right]$$
$$+\frac{\partial}{\partial x_1}\left[\nu\left(\frac{H}{H-F\left(x_1,x_2\right)}\frac{\partial u}{\partial x_3}+\frac{\partial w}{\partial x_1}\right)\right]$$
$$+\frac{\partial}{\partial x_2}\left[\nu\left(\frac{H}{H-F\left(x_1,x_2\right)}\frac{\partial v}{\partial x_3}+\frac{\partial w}{\partial x_2}\right)\right] \qquad (5.1.8)$$
$$+\frac{2H^2}{\left[H-F\left(x_1,x_2\right)\right]^2}\frac{\partial}{\partial x_3}\left(\nu\frac{\partial w}{\partial x_3}\right)+2\boldsymbol{\Omega}u\cos\varphi.$$

- The equation of propagation of scalars in the atmosphere $\mathfrak{H}=\left(\theta,q,\delta,s\right)$:

$$\frac{\partial\mathfrak{H}}{\partial t}+u\frac{\partial\mathfrak{H}}{\partial x_1}+v\frac{\partial\mathfrak{H}}{\partial x_2}+\left(\tilde{w}-w_g\right)\frac{\partial\mathfrak{H}}{\partial x_3}+\sigma\mathfrak{H}$$
$$=f+\frac{\partial}{\partial x_1}\left(d_T\frac{\partial\mathfrak{H}}{\partial x_1}\right)+\frac{\partial}{\partial x_2}\left(d_T\frac{\partial\mathfrak{H}}{\partial x_2}\right)+\frac{H^2}{\left[H-F\left(x_1,x_2\right)\right]^2}\frac{\partial}{\partial x_3}\left(d_T\frac{\partial\mathfrak{H}}{\partial x_3}\right)\cdot \qquad (5.1.9)$$

It is well known that, if not to take into account the active convection zone, the horizontal components of velocity *u* and *v* in the atmospheric movements are several orders of magnitude greater than the vertical component *w*. Therefore, the values of members of $ge^{\mathcal{R}}$ and $\partial p/\partial z$ ∂p/∂z in the third equation of the system (5.1.2) are far exceeds the value of the remaining members of this equation, and the equation of statics holds. The assumption of hydrostatic equilibrium in this form leads to the essential features of the system of equations of hydrodynamics. In particular, in the assumption of hydrostatic equilibrium, pressure is determined up to some arbitrary fixed constant $\overline{p}=\text{const}$. The system of complete equations (5.1.2) adequately describes all three types of atmospheric vibrations – large-scale, gravitational and acoustic, and the use of expressions (5.1.6) leads to the equation of hydrostatic filtering out of acoustic waves. Therefore, the system of equations (5.1.8) in the new coordinates describes, on one side, only large-scale and gravitational vibrations, and the other, the upward movement of the active convection zones, as the third equation is retained in the system for the vertical velocity w.

The practical significance of the system of equations (5.1.7) – (5.1.9) essentially depends on several factors, such as the type of the atmospheric phenomenon or process, applied a numerical method of its decision, the requirement for the accuracy of the result, the type of computer, qualified researchers, restrictions on the time to get the result, serial calculations, etc. These requirements are usually conflicting; their simultaneous and complete implementation leads to the need of developing effective methods and algorithms for the modeling mesoscale atmospheric processes, greatly reducing the cost of computer time and its memory.

The Initial and Boundary Conditions for Mesoscale Models

Mesoscale models describing meteorological processes and phenomena that determine the formation of weather in some areas and regions are designed to provide more reasonable and detailed weather forecasts both in space and time. The quality of these forecasts depends strongly on the accuracy of the set of initial and boundary conditions on the lateral boundaries of the area of solutions, which can pass through areas with large horizontal gradients and high temporal variability of meteorological variables. Formulation of approximate initial and boundary conditions in such cases usually leads to significant errors in weather forecasts.

Initial Conditions

In general, the initial conditions for $t = t_0$ have the form

$$\Re\left(\mathbf{X}, t_0\right) = \Re_0\left(\mathbf{X}\right), \mathbf{X} \in \left(\wp + \mathcal{B}\right), \tag{5.2.1}$$

where $\Re_0$ defines meteorological variables on the basis of an objective analysis of the data.

Sometimes, to improve the accuracy of weather forecasting tasks, in addition to the initial data (5.2.1), solutions use the background of atmospheric processes in the form of:

$$\Re\left(\mathbf{X}, t_k\right) = \Re_k\left(\mathbf{X}\right), \mathbf{X} \in \left(\wp + \mathcal{B}\right), \tag{5.2.2}$$

where $t_k = t_0 - k\tau$ (k=0,1,2,…,K) is the time for which additional information, given on the base of the history of atmospheric processes; τ is time step.

Border Conditions

There are several options for the formulation of the boundary conditions on the lateral boundaries G of the solution field. One of them is based on the fact that the boundary conditions are obtained according to the original synoptic observations for predicting time and considered constant throughout the time interval of calculation

$$\frac{\partial u}{\partial t} = \frac{\partial v}{\partial t} = \frac{\partial w}{\partial t} = \frac{\partial \mathfrak{H}}{\partial t} = 0 \, ,$$

or is based on the assumption of the possibility of extrapolating the values of meteorological variables on the lateral boundary of G from the inside of the field of solutions at each time step with the help of relations of the form

$$\frac{\partial \Re}{\partial t} + v \frac{\partial \Re}{\partial n} = 0 \, ,$$

where $\Re$ is a calculated meteorological parameter, n is the normal to the lateral border of the field, v is the value having a sense of the rate of the transfer $\Re$ and determined by values of $\Re$ in the inner part of the solution field at different times.

Stated thus lateral boundary conditions, as a rule, containing errors, generate fictitious waves of large amplitudes at the borders and in the vicinity, which are distributed in the interior of the domain of the model equations and lead to large errors and an unstable numerical solution. Therefore, to reduce the forecast error in the computational domain is necessary to expand the boundaries of the area, adding a buffer zone, to such an extent that generated disturbance at the boundaries does not reach the control region. Increasing the field of solutions entails increasing requirements for computer performance and thus leads to difficulties of practical realization of such models.

Another approach to the formulation of the lateral boundary conditions, considering the meteorological processes and phenomena of the territory bordering the modeled area, is based on the use of nested grids. This takes into account that meteorological values in the field of mesoscale region are formed on the background of macro-scale circulation of the atmosphere, and the forecast of meteorological task in a limited area can be formulated by combining the two models of different completeness:

- The mesoscale model described by equations of hydrodynamics (5.1.7) or (5.1.9) and implemented on the fine grid.
- The global hydrodynamic model described by the simplified equations (4.1.5) – (4.1.7) or (4.1.14) – (4.1.16) and implemented on the coarse grid.

Each of the models is integrated separately by the following algorithm. First, we solve numerically the task of the global model forecast for periods $t_1, t_2, t_3, \ldots, t_n$ with storing the values of the dependent variables in the coarse grid nodes that belong to the boundaries of the area of mesoscale models. According to the obtained values, by interpolation in space and time, the values of the dependent variables in the boundary nodes of the fine grid are calculated at time instances that are appropriate to time steps for the numerical solution of the mesoscale model. This method of mesoscale forecast for nested grids method was called "unilateral effect", and the boundary conditions are stated as follows:

$$\Re = \Re_b \, , \tag{5.2.3}$$

where $\Re_b$ is of meteorological variables on the boundary of solutions mesoscale tasks specified on the basis of numerical solution of the global model.

In the numerical implementation of global hydrodynamic models, the decision, as a rule, is obtained on a discrete set of points, the distance between them is much greater than that on the difference grids used in the numerical solution of the regional hydrodynamic models. Therefore, the use of solutions to global model as the boundary conditions for mesoscale problems is associated with the problem of interpolation. There is a great variety of methods of interpolation, e.g. interpolation formulas of Bessel, Lagrange, Newton, Steffensen, Stirling, Everett and others (Korn & Korn, 1968). Among the known effective methods of interpolation, one can call the methods of cubic splines and smooth filling of grid functions (Marchuk, 1973). We use the latter, as it has good approximating and algorithmic properties. Using the approach of smooth filling, we will extend it in the two-dimensional case.

Let

$$\aleph_{hs} = \left\{\left(x_{1i}, x_{2j}\right), 1 \le i \le N, 1 \le j \le M\right\}$$

is a fixed grid with steps $h=L_1 / (N - 1)$, $s = L_2 / (M - 1)$, in the nodes of which the values are given of some functions $\Re_{ij}$, $i = \left(\overline{1, N}\right)$, $j = \left(\overline{1, M}\right)$; L_1 and L_2 are linear dimensions of the solution of the problem in the direction x_1 and x_1 respectively; N and M are some fairly large natural numbers. Let us construct the interpolation function $\Im\left(x_1, x_2\right)$ of the smoothness class C^p $\left(p \ll N, M\right)$.

Suppose that the interpolation function $\Im\left(x_1, x_2\right)$ is defined in the region

$$\left[\left(x_{1i+1}, x_{2j+1}\right), \left(x_{1i+1}, x_{2j}\right), \left(x_{1i}, x_{2j+1}\right), \left(x_{1i}, x_{2j}\right)\right]$$

as $\wp_{ij}\left(x_1, x_2\right)$, $i = \left(\overline{1, N}\right)$, $j = \left(\overline{1, M}\right)$ where $\wp_{ij}\left(x_1, x_2\right)$ is a polynomial of degree p, satisfying the following conditions

$$\begin{aligned}
&D^k \wp_{ij}\left(x_1, x_2\right) = D^k \Im_{ij}\left(x_1, x_2\right), \\
&D^k \wp_{i+1j}\left(x_1, x_2\right) = D^k \Im_{i+1j}\left(x_1, x_2\right), \\
&D^k \wp_{ij+1}\left(x_1, x_2\right) = D^k \Im_{ij+1}\left(x_1, x_2\right), \\
&D^k \wp_{i+1j+1}\left(x_1, x_2\right) = D^k \Im_{i+1j+1}\left(x_1, x_2\right), \\
&k = \left(\overline{0, p}\right), i = \left(\overline{1, N}\right), \ j = \left(\overline{1, M}\right),
\end{aligned} \tag{5.2.4}$$

where D^k is the derivative with respect to x_1,x_2 of the order k, and $\Im_{ij}, \Im_{i+1j}, \Im_{ij+1}, \Im_{i+1j+1}$ are the Lagrange polynomials of the degree at most p,

$$\Im_{nm}\left(x_1, x_2\right) = \sum_{i=1}^{k}\sum_{j=1}^{m} \Re_{ij} \prod_{n=1}^{k}\prod_{l=1}^{m} \frac{\left(x_1 - x_{1i}\right)\left(x_2 - x_{2j}\right)}{\left(x_{1n} - x_{1i}\right)\left(x_{2l} - x_{2j}\right)},$$

taking the given values $\Re$ in the grid nodes

$$\left(x_{1i+1}, x_{2j+1}\right), \left(x_{1i+1}, x_{2j}\right), \left(x_{1i}, x_{2j+1}\right), \left(x_{1i}, x_{2j}\right),$$

respectively. Such polynomials $\wp_{ij}\left(x_1, x_2\right)$ exist, and the conditions (5.2.4) are uniquely defined (Marchuk (1973)).

It is sufficient to restrict ourselves to the case of $p=1$ for this model. Then the polynomial of the third degree $\wp_{ij}\left(x_1, x_2\right)$ in the region

$$\left[\left(x_{1i+1}, x_{2j+1}\right), \left(x_{1i+1}, x_{2j}\right), \left(x_{1i}, x_{2j+1}\right), \left(x_{1i}, x_{2j}\right)\right]$$

can be written as

$$\begin{aligned}
&\wp_{ij}\left(x_1,\ x_2\right) = C_{ij}^{(0)} + C_{ij}^{(1)}\left(x_1 - x_{1i}\right)^1 + \ldots + C_{ij}^{(3)}\left(x_1 - x_{1i}\right)^3 \\
&+\left(x_2 - x_{2j}\right)^1\left[C_{ij}^{(4)} + C_{ij}^{(5)}\left(x_1 - x_{1i}\right)^1 + \ldots + C_{ij}^{(7)}\left(x_1 - x_{1i}\right)^3\right] \\
&+\left(x_2 - x_{2j}\right)^2\left[C_{ij}^{(8)} + C_{ij}^{(9)}\left(x_1 - x_{1i}\right)^1 + \ldots + C_{ij}^{(11)}\left(x_1 - x_{1i}\right)^3\right] \\
&+\left(x_2 - x_{2j}\right)^3\left[C_{ij}^{(12)} + C_{ij}^{(13)}\left(x_1 - x_{1i}\right)^1 + \ldots + C_{ij}^{(15)}\left(x_1 - x_{1i}\right)^3\right]
\end{aligned} \tag{5.2.5}$$

with coefficients calculated by the formulas

$$\begin{aligned}
&C_{ij}^{(0)} = \mathfrak{R}_{ij},\ C_{ij}^{(1)} = \Lambda_1\mathfrak{R}_{ij},\ C_{ij}^{(2)} = -\Lambda_1^2\mathfrak{R}_{ij}, \\
&C_{ij}^{(3)} = h^{-1}\Lambda_1^2\mathfrak{R}_{ij},\ C_{ij}^{(4)} = \Lambda_2\mathfrak{R}_{ij},\ C_{ij}^{(5)} = \Lambda_1\Lambda_2\mathfrak{R}_{ij}, \\
&C_{ij}^{(6)} = -\Lambda_1^2\Lambda_2\mathfrak{R}_{ij}, \Lambda_{ij}^{(7)} = h^{-1}\Lambda_1^2\Lambda_2\mathfrak{R}_{ij}, C_{ij}^{(8)} = -\Lambda_2^2\mathfrak{R}_{ij}, \\
&C_{ij}^{(9)} = -\Lambda_1\Lambda_2^2\mathfrak{R}_{ij}, C_{ij}^{(10)} = -\Lambda_1^2\Lambda_2^2\mathfrak{R}_{ij}, C_{ij}^{(11)} = h^{-1}\Lambda_1^2\Lambda_2^2\mathfrak{R}_{ij}, \\
&C_{ij}^{(12)} = s^{-1}\Lambda_2^2\mathfrak{R}_{ij}, C_{ij}^{(13)} = s^{-1}\Lambda_1\Lambda_2^2\mathfrak{R}_{ij}, \\
&C_{ij}^{(14)} = -s^{-1}\Lambda_1^2\Lambda_2^2\mathfrak{R}_{ij}, C_{ij}^{(15)} = h^{-1}s^{-1}\Lambda_1^2\Lambda_2^2\mathfrak{R}_{ij}, i = \left(\overline{1, N}\right), j = \left(\overline{1, M}\right)
\end{aligned}$$

where the difference operators are defined by expressions

$$\begin{aligned}
&\Lambda_1\mathfrak{R}_{ij} = \left(\mathfrak{R}_{i+1j} - \mathfrak{R}_{ij}\right) / h, \Lambda_1^2\mathfrak{R}_{ij} = \left(\mathfrak{R}_{i+2j} - 2\mathfrak{R}_{i+1j} + \mathfrak{R}_{ij}\right) / h^2, \\
&\Lambda_2\mathfrak{R}_{ij} = \left(\mathfrak{R}_{ij+1} - \mathfrak{R}_{ij}\right) / s, \Lambda_2^2\mathfrak{R}_{ij} = \left(\mathfrak{R}_{ij+2} - 2\mathfrak{R}_{ij+1} + \mathfrak{R}_{ij}\right) / s^2, \\
&\Lambda_1\Lambda_2\mathfrak{R}_{ij} = \left(\mathfrak{R}_{i+1j+1} - \mathfrak{R}_{i+1j} - \mathfrak{R}_{ij+1} + \mathfrak{R}_{ij}\right) / \left(hs\right).
\end{aligned}$$

Through the above interpolation method, we can specify the boundary conditions on the boundaries of the regional model even in the case where the distances between the grid nodes of the regional and macro-scale models are not a multiple of one relative to the other.

BASICS OF MATHEMATICAL MODELING OF THE SPREAD OF POLLUTANTS IN THE ATMOSPHERE

The loss prevention problem from air pollution has both social and economic orientation. It has long been not a problem of individual countries, but and even continents, it has acquired an international character and has become common for almost all countries of the world. A representative example of this problem at the international level is radioactive contamination of air, water, and soil after Chornobyl nuclear accident in 1986, the effect of which has become significant not only for the territories of Ukraine, Belarus, and Russia but also other countries of Europe. Emissions of greenhouse and ozone-depleting gasses lead to global climate change and create the preconditions to an ecological catastrophe on Earth.

Reducing the negative impact of man-made air pollution can only be based on a set of measures such as the development of general schemes of the rational location of industrial enterprises and the development of plans for economic development; the use of efficient gas scrubbers and dust collectors; establishing of expedient industrial emissions. For the latter, research should contribute to developing methods for calculating the dispersion of gas and aerosol emissions from the chimneys of industrial and chemical plants, thermal and nuclear power plants during normal and emergency modes, taking into account the actual weather conditions, as well as methods for assessing the impact of harmful pollutants in the biosphere.

As noted in section 2.3, the transfer and diffusion of impurities at a certain distance from the source is determined by atmospheric motions on different scales (wind and turbulence). Horizontal distribution of impurities in the atmosphere after its release from the source is determined mainly by the wind field. It depends on the wind speed what is the distance by which the substance is transferred aerosol, and what the impurity concentration is in the plume and in the transfer flow. The wind speed and the roughness of the underlying surface determine the intensity of the dynamic turbulence. The direction of the spread of contamination depends on the direction of the wind, and the variability of the latter effects on the value of the gas-aerosol plume. In reality, variability in wind direction results (due to the influence of atmospheric vortices of all sizes) in the meandering shape of the plume. Scattering plume generally includes a dispersion of individual clubs and random walk of the center of mass relative to the mean wind.

The temperature stratification is of great importance. It determines the state of the atmosphere, affecting the intensity of the turbulence and thickness of the mixing layer; it influences on the nature of the impurity scattering in the vertical direction, and hence, exchange of contaminated air from the upper clean. The impurity is carried up by convection currents if the stratification is unstable. At the same time, there is vortexlike smoke over chimneys and a series of waves at considerable superadiabatic temperature gradients. When unstable stratification, the pollution concentration decreases uniformly with distance from the source, and the shape of its distribution is conic. Stable stratification is particularly favorable for air pollution as a plume impurities extend horizontally above the underlying surface in the direction of the wind. If the inversion layer is lower than the source of emission, the plume vortexlike goes up, otherwise, the plume emission goes down, contaminating the surrounding area.

As the atmosphere is a chaotic mixture of all sorts of organic and chemical substances, the physical and chemical changes of the impurity are connected with meteorological characteristics such as water vapor or droplets, air temperature, the intensity of solar radiation and the presence of other atmospheric pollutants. Also, processes of leaching of impurities precipitation, gravitational settling of suspended particles and the absorption of the underlying surface are weather dependent.

Consequently, the mathematical modeling of the spreading gas-aerosol impurities from the emission source in the real atmosphere is a difficult complex problem. It is a part of the more general problem of environmental protection and rational use of natural resources. Previously, atmospheric pollution studies were conducted with respect to local air pollution within the boundaries of the enterprise, urban district or town, but now they have developed research on the macro-scale effects of pollution spreading and their role in the ecological and climatic processes (Prusov, Grymov & Kopiiko, 1996; Prusov & Doroshenko, 2003).

The history of the theoretical modeling of pollutants in the atmosphere from industrial sources has almost five decades. During this period, one can note a few basic steps of evolution of the quality of the models:

- A simple model of inflow and outflow of the substance to assess the impurity concentration in the boundary layer of the atmosphere over a limited area; the impurity concentration is taken proportional to the intensity of its receipt, and inversely proportional to the mean wind speed and the thickness of the mixed layer.
- Diffusion transport model, integrated with the help of methods of mathematical physics equations (eigenfunction expansion, the sine and cosine transform, Fourier transform, Laplace transform, and others.).
- Diffusion transport model, integrated on the basis of job analysis profiles of wind speed and turbulence in the boundary layer of the atmosphere.
- Gaussian model of transfer and turbulent impurity scattering, which describes the binormal distribution law of its concentration.
- Diffusion transport model, integrated by numerical methods.

Fairly simple and, at the same time, physically reasonable models have been proposed in (Marchuk, 1982). They paid attention to the natural mechanisms that have a dominant importance in settling the gas-aerosol impurities from the atmospheric boundary layer (sedimentation of particulate impurities, horizontal transfer with a variable height profile of the wind, the cross scattering of impurities, the vertical turbulent diffusion with a sharp increase in the intensity of mixing in the middle of boundary layer, "dry deposition" to the underlying surface). The intensity of the contamination of the area from a linear source located across the direction of the wind speed at the height of the source is given as follows:

$$I = \frac{Q}{\sigma\sqrt{2\pi}}\frac{v\left(H\right)}{HL}\mathbf{\Omega}\left(s,\tau\right)\exp\left(\frac{y^2}{2\sigma}\right),$$

where s,τ are dimensionless similarity criteria; Q, H are, respectively, power and height of the source; L is the distance parameter; $v(H)$ is the speed at the height of the source; σ is dispersion of impurities due to the transverse component of the wind speed and the horizontal turbulent diffusion; $\mathbf{\Omega}$ is a function featuring the stationary and uniformity of the atmospheric boundary layer in the horizontal direction. The formula above was obtained under the condition that there was no flow of particles through its upper limit, and depositing on the underlying surface is determined only by the sedimentation of particles. Ω function is expressed analytically in terms of the gamma function and Gegenbauer polynomials (Bayin, 2006).

Simple models share a common approach based on the use of simple algebraic relations for calculating pollution characteristics. These algebraic curves were obtained on the basis of empirical generalization of a large number of experimental and field observations, or as a result of generalization of a large number of calculated data for the equation of turbulent diffusion in different types of stratification of the atmospheric boundary layer, or, eventually, as a result of solving individual problems with some simplifying assumptions. The advantage of these models is their simplicity and the possibility of preliminary calculations or approximate assessment of air pollution. However, the use of simplified models in some cases leads to considerable errors. The numerical results are sometimes even not qualitatively consistent with the actual concentrations of impurities.

Analyzing and forecasting the spread of contaminants from industrial sources at the present stage of development of computational tools is already impossible to study only under the hypothetical idea of atmospheric parameters. Any application-oriented research should be able to model and calculate the concentration of both inert, almost weightless, and heavy impurity in the actual meteorological conditions and its interaction with the inhomogeneous underlying surface. Obviously, this problem is dominated by the general problem of the numerical weather-ecology forecasting. Recently, some research institutions working in the field of mathematical modeling of processes of industrial emission are increasingly turning to more detailed models. Among them are those that are based on the solution of the initial-boundary value problems for the turbulent diffusion equations (Euler models) and the Cauchy problem for systems of ordinary differential equations (Lagrange trajectory models) together with solving the closed system of equations of the atmospheric boundary layer. The important place is given to the modeling of the process of turbulent exchange in the surface boundary layer of the atmosphere, which requires the use of prognostic meteorological information.

In this section, we consider a model of transfer and diffusion of the dust-gas mixture in the atmosphere, based on an equation that describes the processes, the domain of definition and properties of solutions. There are given setting problems for instantaneous and stationary point sources of emission, the problems of setting the initial and boundary conditions are discussed.

We use the equation of the diffusion approximation (3.2.28) in environmental problems

$$\rho_i \frac{\partial s_i}{\partial t} + \rho_i \mathbf{V} \nabla s_i = -\varrho \rho_i s_i + \nabla \left(\rho_i \mathrm{d} \nabla s_i \right) \tag{5.3.1}$$

where each i-th component (ingredient) is considered to be passive (coincidence of movements of the mixture and its individual components). The mixture velocity in the equation (5.3.1) can be determined from the equations of the dynamics of a homogeneous medium (5.2.3) with the values of physical constants corresponding to the mixture.

The equation of spreading impurities from the source of pollution (5.3.1) describes the essence of the process, but in some way idealizes the real processes that are much more complex in its physical content. From field observations, it follows that the impurity in the atmosphere is likely spreading and forms a very complex distribution of aerosols for away from the emission source. This is because even in calm weather the atmosphere is a turbulent environment in which the momentum is transferred by small-scale fluctuations and eddy formations, moving chaotically. The physics of fluctuation effects at this time is sufficiently well studied. However, these effects arise only statistically, and there is no possibility of their prediction or at least fixing in the real world. That is why; the mathematical description of the effects of fluctuation is still implemented by semi-empirical correlations, which are based on Reynolds number turbulence model (Marchuk, 1982; Paskonov, Polezhaev & Zhukov, 1984; Prusov & Doroshenko, 2006).

Let us designate the volume concentration of impurities as a whole by s, i.e., assume that $s = \sum_i \rho_i s_i$.

Taking into account the incompressibility equation, we can write the equation of transfer and diffusion of impurities (5.3.1) in divergence form

$$\frac{\partial s}{\partial t} + \nabla \cdot \left(\mathbf{V}\mathrm{s}\right) = -\varrho s + \nabla\left(\mathrm{d}\nabla \mathrm{s}\right) \tag{5.3.2}$$

We represent the function s as a sum of averaged $\overline{s}$ and fluctuation s' components. With regard to the latter, we integrate equation (5.3.2) with respect to t within the range $t \le \tau \le t + T$, where T is time interval at which the function $\overline{s}\left(t\right)$ changes little. As a result, we obtain an equation for the averaged component

$$\frac{\partial \overline{s}}{\partial t} + \nabla \cdot \left(\overline{\mathbf{V}}\overline{\mathrm{s}}\right) + \nabla \cdot \left(\overline{\mathbf{V}'\mathrm{s}'}\right) = -\overline{\varrho s} + \nabla\left(\mathrm{d}\nabla \overline{\mathrm{s}}\right), \tag{5.3.3}$$

which differs from the equation (5.3.2) by the presence of fluctuating moments $\nabla \cdot \left(\overline{\mathbf{V}'\mathrm{s}'}\right)$. It is assumed that this additional term on the left side of the equation (5.3.3) should provide modeling erosion of impurities by the flow of air masses, entraining particles of polluting substances.

Numerous studies have shown that the vector $\overline{\mathbf{V}'\mathrm{s}'}$ can be represented for atmospheric processes as a semi-empirical averaged field of the substance:

$$\overline{\mathbf{V}'\mathrm{s}'} = -\frac{\nu}{\xi}\nabla \overline{\mathrm{s}}, \tag{5.3.4a}$$

where $\nu=(\nu_1,\nu_2,\nu_3)$ is the turbulent coefficient of (virtual) viscosity, which, in contrast to the molecular viscosity coefficient ν_L, is not a physical constant and is determined experimentally; ξ is the dimensionless Schmidt criterion in a turbulent flow. The dash in the formula $\overline{s}$ means the average value determined as ordinary integral average

$$\overline{\mathrm{s}}\left(x_1, x_2, x_3, t\right) = \frac{1}{T}\int_{t-T/2}^{t+T/2} s\left(x_1, x_2, x_3, \tau\right) d\tau \tag{5.3.4b}$$

during the time interval T, called the averaging period. We assume that for each considered turbulent motion in the atmosphere there is a sufficiently large compared to the period of turbulent fluctuations, but small compared to the averaged turbulent motion time interval, the constant period T, which being used in time smoothing (5.3.4b), leads to the averaged value, that repeated smoothing does not change it, i.e., $\overline{\overline{\mathrm{s}}} = \overline{\mathrm{s}}$.

If the relations (5.3.4a) and (5.3.4b) are taken into account in (5.3.3), we obtain the diffusion approximation of impurity propagation in atmosphere

$$\frac{\partial s}{\partial t} + \nabla \cdot \left(\mathbf{V} \cdot \mathrm{s}\right) = -\varrho s + \nabla \left[\left(d + \frac{\nu}{\xi}\right)\nabla \mathrm{s}\right] \tag{5.3.5}$$

where, for the sake of convenience, a dash at the top of the averaged values is omitted. In most applications they neglect the molecular diffusion coefficient, assuming $d = 0$, as it is several orders of magnitude smaller in the real atmosphere than the coefficient of "turbulent diffusion" $d_T = \nu / \xi$. Therefore, we take as a fundamental equation of an impurity distribution in the atmosphere

$$\frac{\partial s}{\partial t} + \nabla \cdot \left(\mathbf{V} \cdot \mathrm{s}\right) = -\phi s + \nabla \left(\frac{\nu}{\xi}\nabla \mathrm{s}\right). \tag{5.3.6}$$

We investigate now the set of additional conditions to enable correct formulation and ensuring the uniqueness of the solution of the equation (5.3.6).

Correct Statement of Convection-Diffusion Problem

Let the domain of solving the problem is a cylinder G with full surface Q, consisting of the side surface S, and lower and upper S_b and S_h grounds, respectively. We introduce in G a rectangular Cartesian coordinate system $\boldsymbol{X} = (x_1, x_2, x_3)$ and designate the velocity by $\boldsymbol{V} = \left(v_1, v_2, v_3\right)$ and the turbulent viscosity by $\boldsymbol{V} = \left(v_1, v_2, v_3\right)$.

We set, for convenience, $\xi = 1$ in the equation (5.3.6)

$$\frac{\partial s}{\partial t} + \nabla \cdot \left(\mathbf{V}\mathrm{s}\right) = -\varrho s + \nabla \left(\nu \nabla \mathrm{s}\right). \tag{5.3.7}$$

and consider the incompressibility equation

$$\nabla \cdot \mathbf{V} = 0\,, \tag{5.3.8}$$

initial data

$$s = s_0, if\ t = 0\,, \tag{5.3.9}$$

and boundary conditions

$$\begin{aligned}
&s = s_S, if\ v_n < 0\ on\ S,\\
&\frac{\partial s}{\partial n} = 0, if\ v_n \geq 0\ on\ S,\\
&\frac{\partial s}{\partial n} = 0\ on\ S_h,\\
&\frac{\partial s}{\partial n} = \beta s\ on\ S_b.
\end{aligned} \tag{5.3.10}$$

In addition, we assume that

$$\begin{aligned}
&x_3 = 0, \text{ if } \boldsymbol{X} \in S_b\ ,\\
&x_3 = H, \text{ if } \boldsymbol{X} \in S_h\ ,\\
&\nu_3 = 0, \text{ if } x_3 = 0, x_3 = H.
\end{aligned} \tag{5.3.11}$$

In the given conditions (5.3.9) – (5.3.11), $\beta \geq 0$ is a function that characterizes the interaction of impurities with the underlying surface; s_0 and s_s are specified functions; v_n is the projection of the velocity vector on the external normal to the surface S; $\partial s / \partial n$ is the derivative in the direction of the external normal to the surfaces S, S_h or S_b; H is the upper boundary of the computational domain.

In accordance with the methodology proposed in (Marchuk, 1982), we multiply equation (5.3.7) by s and integrate the result over the time $0 \leq t \leq T$, and the space G:

$$\begin{aligned}
&\int_G \frac{s^2}{2} dG\Big|_{t=T} - \int_G \frac{s^2}{2} dG\Big|_{t=0} + \int_0^T dt \int_G \nabla \frac{\mathbf{V} s^2}{2} dG + \varrho \int_0^T dt \int_G s^2 dG\\
&= -\int_0^T dt \int_G \sum_{i=1}^{3} \nu_i \left(\frac{\partial s}{\partial x_i}\right)^2 dG + \int_0^T dt \left[\nu \int_S s \frac{\partial s}{\partial n} dS + \nu_3 \left(\int_{S_h} s \frac{\partial s}{\partial x_3} dS - \int_{S_b} s \frac{\partial s}{\partial x_3} dS\right)\right]
\end{aligned} \tag{5.3.12}$$

According to the Ostrogradskii-Gauss formula we have

$$\int_G \nabla \frac{\mathbf{V} s^2}{2} dG = \int_Q \frac{v_n s^2}{2} dQ\,. \tag{5.3.13}$$

Note that by virtue of (5.3.5) v_n turns to zero at $x_3 = 0$ and $x_3 = H$, so integration over Q in (5.3.13) can be replaced by integration over the side of the cylindrical surface S. Considering (5.3.9) - (5.3.11) and (5.3.13), we transform the relation (5.3.12) into the form:

$$\begin{aligned}
&\int_G \frac{s^2}{2} dG\Big|_{t=T} - \int_G \frac{s_0^2}{2} dG + \int_0^T dt \int_S \frac{v_n^+ s^2}{2} dS + \varrho \int_0^T dt \int_G s^2 dG\\
&= -\int_0^T dt \int_G \sum_{i=1}^{3} \nu_i \left(\frac{\partial s}{\partial x_i}\right)^2 dG - \nu_3 \int_0^T dt \int_{S_b} \beta s^2 dS + \nu \int_0^T dt \int_S s_S \frac{\partial s}{\partial n} dS - \int_0^T dt \int_S \frac{v_n^- s_S^2}{2} dS.
\end{aligned} \tag{5.3.14}$$

Here, $v_n = v_n^+ + v_n^-$; $v_n^+ = v_n$, if $v_n > 0$ and $v_n^- = 0$, if $v_n < 0$.

The identity (5.3.14) allows us to prove the uniqueness of the solution of problem (5.3.7) - (5.3.11). Indeed, suppose that equation (5.3.7), and the same initial (5.3.9) and boundary conditions (5.3.10) are satisfied the two different solutions of the problem. Let these solutions will be s_1 and s_2. For their difference $\psi = s_1 - s_2$, we obtain the equation

$$\frac{\partial \psi}{\partial t} + \nabla \cdot \left(\mathbf{V}\psi\right) = -\varrho\psi + \nabla\left(\nu\nabla\psi\right), \tag{5.3.15}$$

initial data

$$\psi = 0, if\ t = 0, \tag{5.3.16}$$

and boundary conditions

$$\begin{gathered} \psi = 0\ on\ S, if\ v_n < 0, \\ \frac{\partial \psi}{\partial n} = 0\ on\ S, if\ v_n \geq 0, \\ \frac{\partial \psi}{\partial n} = 0\ on\ S_h, \\ \frac{\partial \psi}{\partial n} = \beta\psi\ on\ S_b. \end{gathered} \tag{5.3.17}$$

For the formulated problem, the relation of the type of (5.3.14) takes the form

$$\int_G \frac{\psi^2}{2} dG \Bigg|_{t=T} + \int_0^T dt \int_S \frac{v_n \psi^2}{2} dS + \varrho \int_0^T dt \int_G \psi^2 dG + \int_0^T dt \int_G \sum_{i=1}^3 \nu_i \left(\frac{\partial \psi}{\partial x_i}\right)^2 dG + \nu_3 \int_0^T dt \int_{S_b} \beta \psi^2 dS = 0 \tag{5.3.18}$$

Since the quantities v_n, v, ϱ, β are non-negative, then this relation is zero iff $\psi = 0$; i.e., when $s_1 = s_2$. And this ensures the uniqueness of the solution.

It has been considered above the general statement of the problem (5.3.7) - (5.3.10), but in some cases, for stationary sources of emission, it is advisable to use a stationary approximation, which can be obtained on the basis of physically reasonable assumptions.

Suppose that in different periods of time $t \in \left[0, T\right]$ in the given region that is close to a stationary source of emissions, there are implemented certain thermodynamic atmospheric processes which on each of the time intervals $\Delta t_n = t_{n+1} - t_n$, $n = \left(\widehat{0, N-1}\right)$ can be regarded as finally established. After each such period, there is a reorganization of the atmospheric processes and a new steady state is set. The time of processes adjustment τ we assume a lot less than the time Δt_n, i.e. $\tau \ll \Delta t_n$. From this, one can assumed that the change of types of atmospheric processes happens instantly. Consider the non-stationary problem corresponding to the n-th type of steady thermodynamic atmospheric processes,

$$\begin{aligned}
&\frac{\partial s}{\partial t} + \nabla \cdot \left(\mathbf{V} s\right) = -\varrho s + \nabla\left(\nu \nabla s\right), t_n + \tau \le t \le t_{n+1}, \\
&s = s_0, if\ t = t_n \\
&s = s_S, if\ on\ S\ v_n < 0, \\
&\frac{\partial s}{\partial n} = 0, if\ on\ S\ v_n \ge 0, \\
&\frac{\partial s}{\partial n} = 0, on\ a\ S_h, \\
&\frac{\partial s}{\partial n} = \beta s, on\ a\ S_b.
\end{aligned} \tag{5.3.19}$$

where the vector function $\mathbf{V}$, v for established processes will be considered as independent on t. The uniqueness of the solution of (5.3.19), under appropriate assumptions on the smoothness of the $\mathbf{V}$, v, is proved in the same way as for the problem (5.3.7) – (5.3.10).

Integrating the equation (5.3.19) in the range $[t_n, t_{n+1}]$ and denoting

$$\overline{s} = \frac{1}{\Delta t_n} \int_{t_n}^{t_{n+1}} s dt ,$$

we obtain the equation

$$\nabla \cdot \left(\mathbf{V} \overline{s}\right) = -\varrho \overline{s} + \nabla\left(\nu \nabla \overline{s}\right),$$

for which the stationary problem is simply formulated:

$$\begin{aligned}
&\nabla \cdot \left(\mathbf{V} \overline{s}\right) = -\varrho \overline{s} + \nabla\left(\nu \nabla \overline{s}\right), \\
&\overline{s} = \overline{s}_S\ on\ a\ S\ on\ condition\ v_n < 0, \\
&\frac{\partial \overline{s}}{\partial n} = 0\ on\ a\ S\ on\ condition\ v_n \ge 0, \\
&\frac{\partial \overline{s}}{\partial n} = 0\ on\ a\ S_h, \\
&\frac{\partial \overline{s}}{\partial n} = \beta \overline{s}\ on\ a\ S_b.
\end{aligned} \tag{5.3.20}$$

It is easy to verify that the identity corresponding to (5.3.14) has the form

$$\int_S \frac{v_n^+ \overline{s}^2}{2} dS + \varrho \int_G \overline{s}^2 dG = -\int_G \sum_{i=1}^{3} \nu_i \left(\frac{\partial \overline{s}}{\partial x_i}\right)^2 dG - \nu_3 \int_{S_b} \beta \overline{s}^2 dS + \nu \int_S s_S \frac{\partial \overline{s}}{\partial n} dS - \int_S \frac{v_n^- s_S^2}{2} dS . \tag{5.3.21}$$

We can show that (5.3.20) has a unique solution with the method outlined above. It means that the average over the period $[t_n, t_{n+1}]$, the solution of the problem (5.3.20) coincides with the solution of the problem (5.3.7) – (5.3.10).

The Initial and Boundary Conditions for the Problem of Transport and Diffusion of Pollutants in the Atmosphere from a Variety of Sources of Emissions

Consider the equation (5.3.6) in the rectangular Cartesian coordinate system $\mathbf{X}=(x_1,x_2,x_3)$ of transfer and diffusion of a substance $s(x_1,x_2,x_3)$ of aerosol emitted by a point source

$$\frac{\partial s}{\partial t}+v_j\frac{\partial s}{\partial x_j}+\varrho s=Q_s+\frac{\partial}{\partial x_j}\left(\frac{\nu_j}{\xi}\frac{\partial s}{\partial x_j}\right), j=1,2,3, \tag{5.3.22}$$

which provides a summation over twice repeated indices. Equation (5.3.22) describes the migration of the aerosol concentration of the substance s, along with an elementary volume of air in the atmosphere at a velocity of $\mathbf{V}=(v_1,v_2,v_3)$ and its diffusion by atmospheric turbulence with the exchange coefficient $\nu=(\nu_1,\nu_2,\nu_3)$. Projection v_3 of the velocity can consist of two components $v_3= v_3 + v_g$, where v_g denotes the absolute value of the vertical velocity of the particles under the influence of gravitational force. Q_s characterizes the outflow ($Q_s < 0$) or inflow ($Q_s > 0$) of the aerosol substance s through the boundary surface of the area of solving the problem; ϱs is the internal source (sink) due to impurities transformations as a result of leaching, chemical reactions, etc.

The equation (5.3.22) is written for monodisperse impurities. The equation for polydispersed particles flow is determined by the superposition of streams of particles of different sizes. It does not contain terms of molecular transport d_L, because, at high Reynolds numbers typical of motions in the atmosphere, these terms are negligibly small everywhere except isothermal calm areas. The main characteristics in its solution are the trajectory of impurity clouds, the intensity of the expansion of its areas of mixing, and nature of changes in the axes clouds excessive concentrations.

The equation (5.3.22) is of the first order with respect to time t, and the second with respect to the space variables $\mathbf{X}=(x_1,x_2,x_3)$. Therefore, one initial condition for t and two boundary conditions for x_1,x_2,x_3, respectively, should be imposed.

When modeling processes of industrial atmospheric pollution from thermal and nuclear power plants operating at design conditions, emission source can be considered stationary. The coincidence of the solutions (5.3.7) - (5.3.10) and (5.3.20) allows us to go to the stationary equation

$$v_j\frac{\partial s}{\partial x_j}+\varrho s=Q_s+\frac{\partial}{\partial x_j}\left(\frac{\nu_j}{\xi}\frac{\partial s}{\partial x_j}\right), j=1,2,3, \tag{5.3.22a}$$

i.e., to take $\partial s / \partial t = 0$ and thereby to simplify the overall task.

The Initial Conditions in the Case of Elevated Stationary Sources

The main factors influencing the shape of the torch of impurity distribution in theoretical studies of the transport processes and the diffusion of harmful emissions from the chimneys of industrial enterprises can be divided into two groups:

1. **Aerodynamic:** That especially are displayed in the immediate vicinity of the source and characterized mainly by temperature and gas emission rate of suspended particles with impurities.
2. **Meteorological:** That determine the propagation of impurities, both before and after gasses lose their initial kinetic and thermal energy.

From experimental studies, field observations and analysis of the photographs, it follows that the contaminated area close to of the chimney is some divergent cone curved in the direction of the wind (Schötz, 1984). The bending radius of the cone is increased if the impurity is ejected at high speed and the temperature is considerably greater than the ambient temperature, what gives rise to the buoyancy force. The plume is raised high above the chimney in these conditions, and then being bent it stretches in the direction of the wind. In the closest area near the source, emissions are dissipated by means of the jet mixing. It is only at a certain distance from the mouth of the tube when the axial velocity in the jet equals the entraining velocity of the wind, the diffusion begins to dominate the scattering mechanism.

Increasing wind velocity causes a decrease in the lift height of the plume and a strong stretching it in the direction of the wind. The height of the plume rising also depends on the vertical temperature gradient as the unstable stratification contributes to the rise of the plume and the inversion delays it. Thus, the height of impurity emissions depends on the geometric dimensions of the source (the height of the tube and the diameter of the mouth), the state of impurity (its fractional composition, temperature, and emission rate) and the prevailing weather conditions (wind speed and direction, the vertical temperature gradient).

The problem associated with a description and calculation of the initial phase of penetration of round jet into the stream in the atmosphere is one of a very complex and poorly studied in the theory of free turbulent flows. Along with the great importance, this problem has a number of features that make it difficult the theoretical description of such jets. Among these features are a primarily curvilinear jet and the asymmetry of its basic parameters, due to the influence of smashing flow, buoyancy and stratification of the environment. At the moment, in fact, there is not a recognized model of the flow.

In earlier theoretical works (Sedov, 1976; Marchuk, 1982), in which the problem of this class has been examined, a simple schematic representation of the actual flow was used when the bent jet was seen as a quasi-solid body. The results of these studies have been limited to a determination of the trajectory of the jet as a simple empirical relationship in a very narrow range of speed, temperature, and the initial expiration corners. The authors of other works (Oke, 1978; Berlyand, 1975) tried to take into account the transfer of momentum due to turbulent mixing. These studies have had some progress and contributed to a better understanding of the mechanism of interaction and mixing of the jet with crossflow. The models under certain assumptions allowed to describe the velocity field, temperature and concentration of the impurities either in the problems of free turbulence and boundary layer. However, despite the apparent abundance of these models, most of them are very close to each other in the sense of the description of the physics phenomena, and the difference is mainly in the process of completion, simulation variants of unknown correlation terms and computational methods of solving the problem.

Critical analysis of existing models and methods of their implementation indicates that construction principles, assumptions, and limitations put in their base, create certain difficulties in solving this class of problems for the real atmosphere. This is due primarily to the incommensurability of geometric scale of the jet near the chimney and the entire region of increased impurity concentration in the atmosphere, caused by the emission.

Strictly speaking, the impurity emissions are carried from bulk sources that have certain geometrical dimensions. However, these dimensions are small compared to the distances at which created fields of concentration are investigated. Therefore, we can accept the premise, allowing to treat real sources as points. This assumption replaces the task of impurity scattering of a tube of height x_3^0, located at the point $\left(x_1^0, x_2^0\right)$, by the task of turbulent diffusion from a point source displaced from the axis of the tube at some effective distance $\Delta r^2 = \Delta x_1^2 + \Delta x_2^2$ and elevated above the tube to some effective height Δx_3. In this case, the law of conservation of mass implies that the average concentration of impurities at a distance from the mouth of the tube will be proportional to α^{-2} (α is an effective local radius of the cone). In this formulation, a point source is described by of δ-function and, if to denote the effective coordinates $x_1^{eff} = x_1^0 + \Delta x_1$, $x_2^{eff} = x_2^0 + \Delta x_2$, $x_3^{eff} = x_3^0 + \Delta x_3$, initial condition takes the form

$$\mathbf{V}s = M\delta\left(x_1 - x_1^{eff}\right)\delta\left(x_2 - x_2^{eff}\right)\delta\left(x_3 - x_3^{eff}\right) \tag{5.3.23}$$

where *M* is the substance emission from the source per unit of time.

Recall that for an arbitrary function $\varpi\left(\eta\right)$ δ-function, by definition, satisfies the relation

$$\int_{\eta_1}^{\eta_2} \varpi\left(\eta\right)\delta\left(\eta - \alpha\right)d\eta = \varpi\left(\alpha\right)$$

for values ϱ, falling within the interval $\left(\eta_1, \eta_2\right)$, and

$$\int_{\eta_1}^{\eta_2} \varpi\left(\eta\right)\delta\left(\eta - \alpha\right)d\eta = 0$$

for other values of α.

The simplest assessment of the effective displacement of the source coordinates is based on the results of some experimental studies (Csanady, 1961) for reasons of dimension, introduced the value

$$F_\tau = \frac{s}{T_s} V_s R_s^2 \left(T_s - T_\infty\right),$$

characterizing the effect of the rate of emission impurities V_s and overheating $T_s - T_\infty$ (the difference between the temperature of the air-gas mixture and the temperature of ambient air), and the length scale $M_\tau = F_\tau / V_\infty^3$. Based on the processing of the experimental data he has received the relationship

$$\frac{x_3^{eff} + \Delta x_3}{M_\tau} = \left(\frac{r^{eff} + \Delta r}{M_\tau}\right)^{2/3}, \tag{5.3.24a}$$

which is known as the "law of 2/3". In this relationship $r^2 = x_1^2 + x_2^2$, $\Delta r^2 = \Delta x_1^2 + \Delta x_2^2$ and the wind direction V_∞ identical to *r*.

In the work (Briggs, 1969), to determine the effective height Δx_3 as a function of Δr, offers the dependence

$$\Delta x_3 = 2F_\tau^{1/3} \Delta r^{2/3}. \tag{5.3.24b}$$

In the above semi-empirical results, it is believed that the initial portion of the plume above the chimney can be seen as extending up stream with an initial speed V_s in which there is a gradual attenuation of velocity *U* with height due to the expansion of the jet. The value Δx_3 is found as the level x_3 (measured from the source up), where the value of the vertical jet velocity *U* is small compared to the rate of razing flow and transport of impurities is approximately horizontal. However, if we assume that the initial rise Δx_3 coincides with the level, where the velocity *U* is small compared with the speed of emission V_s, or define Δx_3 from some relation between *U* and wind speed V_∞ (for example, $U(x_3) \approx \varepsilon V_\infty$ ($\varepsilon \ll 1$)), in both cases, the choice of the limit value is arbitrary.

In (Prusov & Doroshenko, 2006) for determining the effective displacement of coordinates of the source there is used extensive experimental material on research of depending the trajectory of central point of the active part of the jet blown into razing flow at an angle of 90° on the parameter values

$$\vartheta = \left(\frac{\rho_s}{\rho_\infty}\right)\left(\frac{V_s}{V_\infty}\right)^2,$$

where, as above, ρ_s and V_s are the density and velocity of the gas mixture; ρ_∞ and V_∞ are the density and velocity of the surrounding air.

Generalization obtained in (Prusov & Doroshenko, 2006) and shown in Fig. 2 experimental data allows obtaining a universal empirical formula for calculating the effective coordinate of the beginning of the passive portion of the plume $x_1^{eff} = x_1^0 + \Delta x_1$, $x_2^{eff} = x_2^0 + \Delta x_2$, $x_3^{eff} = x_3^0 + \Delta x_3$

$$\frac{\Delta x_3}{d_j} = 1.2 \frac{V_s}{V_\infty} \sqrt{\frac{\rho_s}{\rho_\infty}} \ln\left[\left(\frac{\Delta r}{d_j}\right)^{0.57} + 1\right], \tag{5.3.24c}$$

Figure 2. The trajectories of the jets, injected into razing air flow at various values of the parameter ϑ

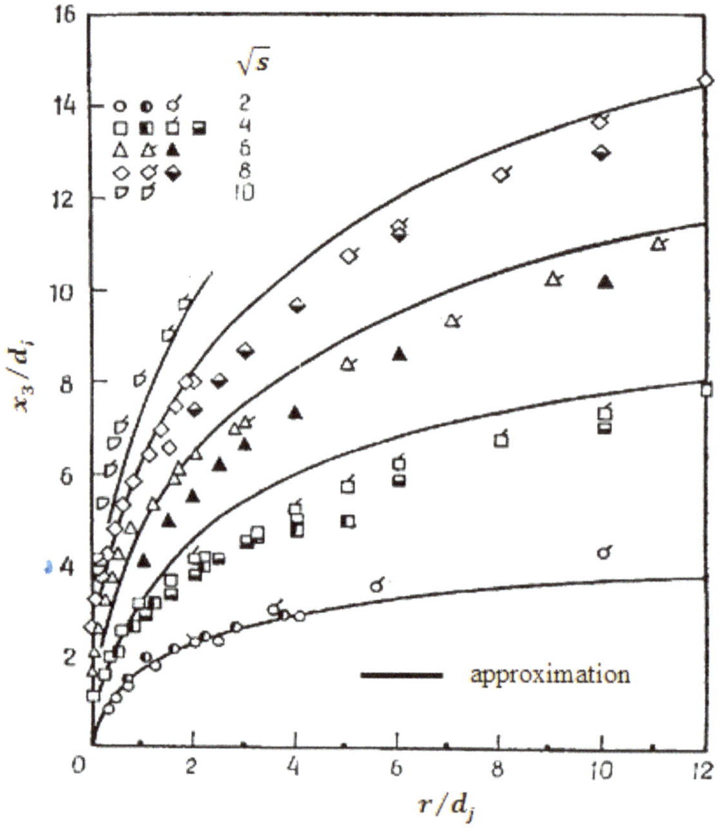

where d_j is the diameter of the mouth of the emission source and $r^2 = x_1^2 + x_2^2$. The effective height Δx_3 of the passive portion of the plume is determined from the condition $d\left(\Delta x_3\right) / dr \to 0$.

Thus, in simulating processes of concentration fields from stationary point sources, one needs to set the coordinates and the physical parameters of the original source (by the passport data of industrial enterprises) and then to calculate according to (5.3.24) the effective coordinates taken as initial data for passive plume area.

When a visual observation of the plume is used, and the effective height is defined as the height where the plume axis takes a horizontal position, there is still uncertainty in the choice of the degree of horizontality of the plume. Often, there is a gradual rise of the smoke plume at long distances from the source. Therefore, using empirical relationships derived from field observations can lead to distortion of the field of impurity concentrations, especially at relatively close distances from the source.

The Initial Conditions for an Explosion (Instantaneous Power)

Independent processing of international data on the height of clouds from more than 60 nuclear land, surface and air explosions (Prusov, Doroshenko (2006)) allows to obtain empirical relationship to calculate the height of the center of the cloud h_c after its stabilization

$$h_c = \frac{284M^{0.22}}{b_\tau^{1/3}} - \frac{0.36V_\infty}{b_\tau}, \tag{5.3.25}$$

where *M* is the charge capacity of the nuclear explosion in tons of TNT equivalent; $b_\tau = 0.01 - \partial\theta / \partial x_3$ is a parameter determining the state of the atmosphere. In the case of isothermal conditions, a stable stratification, or conditions close to neutral, the formula (5.3.25) takes the form

$$h_c = 1400M^{0.22} - 52. \tag{5.3.26}$$

The comparative analysis of the formulas (5.3.25) and (5.3.26) shows that the rise height of the explosion cloud insignificantly depends on the weather conditions. However, it cannot be used when $\partial\theta / \partial x_3 > 0.01$, i.e., under the conditions of unstable stratification, in which the rise due to buoyancy forces is theoretically unlimited. By the way, according to different authors, clouds rise height of the explosion at the condition of stable stratification is close to the height calculated using the formula (5.3.26). Since the coefficient of *M* is changed from the value of 700 for chemical explosives (BP) to 1600 for nuclear explosions with a capacity of *M* = 1 ÷ 100 kt of TNT, and the exponent has a value of 0.2 - 0.25.

An instant source of emissions is not necessarily regarded as a point. In the event of an accidental emission from the explosion type, the source can be described as a fungus, consisting of a cloud and a leg. Each of these entities has a different effect on the formation of the concentration field in the space. From the analysis of nuclear explosions, the cloud contains about 80 - 90% of impurities, the rest is in the leg. Based on internationally published information and general ideas about the rise of the cloud, it is possible to estimate the basic geometric characteristics of the original source after its stabilization from the following formulas (Prusov, Doroshenko (2006)):

- **The Cloud Diameter:** $D_0 = 1600M^{0.117}$.
- **The Vertical Size of the Cloud:** $H_0 = 1430M^{0.246}$. (5.3.27)
- **The Diameter of the Leg:** $D_h = 1420M^{0.184}$.

These formulas are valid for $M = 10 \div 10^5$ kt of TNT equivalent. They can be used to set the initial parameters provided accident emission sources of explosive character at industrial facilities, nuclear power plants, chemical explosions BP, i.e., when the original source is formed, which is a universal form.

Boundary Conditions

Formulated in general terms, the boundary conditions (5.3.10) ensuring uniqueness of the solution of equation (5.3.7), are universal for a wide range of transport problems and turbulent diffusion of impurities, except the conditions at the lower boundary.

The problem (5.3.7) - (5.3.10), in the case of a homogeneous fourth term in (5.3.10), models the transport and diffusion of impurities in the infinite space for the plume (the cloud) located at a great height from the underlying surface. The contact it with the underlying surface leads to dynamic interaction of air stream comprising impurities with a non-homogeneous underlying surface, and specific physical and chemical interaction of impurities with the underlying surface (adsorption, adhesion, saltation, gravitational loss - sedimentation, and chemical reactions). In general case, the lower boundary condition, written in the form of impurities balance equation at the level of the underlying surface, has the form

$$\nu_3 \frac{\partial s}{\partial x_3} = \beta s + \gamma \text{ on condition } x_3 = 0. \quad (5.3.28)$$

Here, the left side of the equation represents the vertical turbulent flow of impurities $Q = \nu_3 \partial s / \partial x_3$ with turbulent exchange coefficient for this impurity v_3. The second term on the right describes the conversion of impurities due to chemical reactions and the first term – the impurity interaction with the underlying surface. In the absence of impurity transformation due to chemical reactions ($\gamma = 0$), the condition $\beta \to \infty$ corresponds to the total absorption of impurities underlying surface, i.e.,

$$s = 0, if \ x_3 = 0, \quad (5.3.28a)$$

and the condition $\beta = 0$ corresponds to total reflection (adhesion) from the underlying surface impurities, i.e.,

$$\nu_3 \frac{\partial s}{\partial x_3} = 0, if \ x_3 = 0. \quad (5.3.28b)$$

Available empirical coefficients β and γ hinders practical implementation of the boundary condition (5.3.28). When setting the parameter β, one should take into account the type of the underlying surface (soil, vegetation, or water), the parameter of its roughness z_0 and a shape of roughness elements. Experimental studies have revealed the presence of minimum value of β in the region of impurity particle diameter of a few micrometers and its gradual smoothing with increasing roughness, as well as a significant increase in β with increasing z_0 for all impurities particle size. The value of β depends on the dynamic processes taking place in the near-wall zones. For very small particles of impurities, it is necessary to consider the Brownian diffusion, and for large – their inertia. An important role is played here by the effect of heat flow stratification. Unfortunately, it is very difficult to predict all of these conditions with sufficient accuracy. This again emphasizes the need for more in-depth experimental and theoretical study of the interaction with the underlying surface impurities. Today, we can recommend a simplified connection between the parameter β and dry deposition velocity of impurities v_g given the constant height of the vertical turbulent flow of impurity (Prusov, Doroshenko (2006)):

$$\beta = v_g \left[1 - \frac{v_g}{kV_\infty} \ln\left(\frac{x_3}{z_g} \right) \right]^{-1},$$

where $\ln z_g$ is the integration constant, k is Karman constant.

Research in the field of chemical transformations in the atmosphere and on the underlying surface is paid much less attention than research in physics. At the same time, the question of the scattering of chemical contaminants, except the problem setting the boundary conditions on the underlying surface, has an independent meaning. Thus, the sulfur dioxide SO_2 is the main impurity generated from the combustion of liquid fuel can be oxidized and turn into sulfuric anhydride SO_3, which reacts with water (steam) H_2O in the presence of catalysts, thus causing droplets of sulfuric acid H_2SO_4. This acid reacts with other impurities, which leads to sulfate particles. The droplets of sulfuric acid and sulfate particles are poisoning the water in lakes and rivers, located on the leeward side of the large industrial centers.

Nitrogen oxides, which are mainly the result of emissions from motor vehicles, due to the impact of solar radiation and in the presence of hydrocarbons, are converted, in the course of a series of complex chemical reactions, in a variety of secondary impurities, especially oxidants, causing irritation to the eyes and throat, destroying the plants. This again emphasizes the need to study the characteristics of the local industrial production, transportation, and climate. However, even knowing these features, it is difficult with sufficient accuracy to set the boundary conditions due to the complexity of the processes of chemical transformations on the underlying surface.

Polydisperse Mixture

Industrial emissions are almost always polydisperse. They also have a different particle density ρ_s which deposition rate depends on. Dispersed characteristics of impurities, emitted into the atmosphere after purification, depends on the efficiency of those devices and a number of other factors. Heavy particles are better captured by typical pollution control equipment, which operates on the principle of inertia. As a result, the increase in the proportion of dust decreases the maximum size of the particles, making a substantial contribution to its range.

It follows from field observations (Prusov & Doroshenko, 2006) that for a weightless impurity (v_g=0) land concentration s_b decreases with increasing height of the source while maintaining other conditions at all distances from the source. This important feature of atmospheric diffusion is not always performed in the propagation of heavy particles. A characteristic feature of the heavy impurities is associated with a partial loss of it near the source. It turns out that the dependence of the land concentration s_b on loss rate v_g at near and far distances r from the source is opposite. For small r, s_b increases with increasing concentration v_g and decreases at large r; and the more r, the more decrease. With increase source height, these features in changing r are preserved, but the distance, over which the maximum of land concentration s_b is achieved, increases dependently on v_g.

The polydispersity leads to offset positions of impurity concentration peaks for individual fractions of the wind direction and therefore contributes to reduce the total concentration maximum. It also turns out that the location of the land maximum of concentration in this case is substantially less dependent on the source height than in the case of monodisperse impurities, but still it is somewhat removed from the source if altitude increases.

In modeling of the transport processes and the diffusion of polydispersed particles, it is convenient to consider not separately the density and variety of particle sizes but the more stable their complex characteristics $P(v_g)$ that is a distribution function of particles weight with different sedimentation rates v_g. General recommendation is to split the typical spectrum $P(v_g)$ into equal intervals (sufficiently into ten intervals). For each interval, one must specify an average sedimentation rate v_{gi}. Then, the total concentration of s can be found from the formula

$$s = \sum_i s_i P\left(v_{gi}\right), \tag{5.3.29}$$

where the speed of the fall of the individual fractions is determined by the well-known Stokes formula for spherical particles

$$v_{gi} = \frac{2\rho_{gi} g}{9\rho_\infty \nu_\infty} \varrho^2 . \tag{5.3.30}$$

There ρ_{gi}, ϱ are the density and radius of i-th impurity particles, respectively; g is the acceleration due to gravity; ρ_∞, ν_∞ are density and the coefficient of molecular viscosity of air, respectively.

SUMMARY

This chapter analyzes the problems of modeling of atmospheric processes of regional and local scales where the assumption of hydrostatic equilibrium is still justified. A common vertical coordinate is a pressure as in this system the basic equations of the atmospheric circulation take rather a simple form, what makes a much easier comparison of synoptic weather maps for different heights. Some basic points are formulated that are common to many existing models of atmospheric circulation of various scales, despite the diversity of approaches to the formulation of initial and boundary conditions. The methods of the parameterization of sub-scale processes, methods of parameterization processes of horizontal and vertical turbulent exchange, dry and moist convection, large-scale condensation, and radiative heat fluxes are considered. Physical and mathematical foundations of the theory of atmospheric diffusion, laws, and principles of the spread dispersion modeling in the atmosphere of gasses, aerosols, and particulate matter are outlined.

The developed mathematics is then applied to modeling mesoscale and macroscale effects of the spread of contaminants from separate sources with huge emissions and permanent industrial sources under real weather conditions, considering the topography of the underlying surface and atmospheric stratification.

REFERENCES

Bayin, S. S. (2006). *Mathematical Methods in Science and Engineering*. Hoboken, NJ: Wiley-Interscience. doi:10.1002/0470047429

Berlyand, M. E. (1975). *Modern problems of atmospheric diffusion and air pollution*. Leningrad: Gidrometeoizdat. (in Russian)

Briggs, G. (1969). *Plume rise*. USAEC Critical Review Series, TID-25075, Clearinghouse for Federal Scientific and Technical Information Materials of IAEA Meeting.

Csanady, G. (1961). Some observations on smoke plumes. *International Journal of Air and Water Pollution*, *4*, 47–51. PMID:13718690

Korn, G. A., & Korn, T. M. (1968). *Mathematical handbook for scientists and engineer's definitions, theorems and formulas for reference and review*. New York: McGraw-Hill.

Marchuk, G. I. (1973). *Methods of Computational Mathematics*. Novosibirsk: Nauka. (in Russian)

Marchuk, G. I. (1982). *Mathematical modeling in environmental problem*. Moscow: Science. (in Russian)

Oke, T. R. (1978). *Boundary layer climates*. New York: John Wiley & Sons. doi:10.4324/9780203407219

Paskonov, V. M., Polezhaev, V. I., & Chudov, L. A. (1984). *Numerical modeling of heat and mass transfer*. Moscow: Science.

Prusov, V., & Doroshenko, A. (2003). Modeling and Forecasting Atmospheric Pollution over Region. *Annales Univ. Sci.*, *46*, 55–73.

Prusov, V. A., & Doroshenko, A. Yu. (2006). *Modeling Natural and Technogenic Atmospheric Processes*. Kyiv: Naukova Dumka. (in Ukrainian)

Prusov, V. A., Grymov, R. Y., & Kopiiko, O. V. (1996). The prediction of environmental pollution in the case of an accident at atomic power plants or in the field of chemical industry. In The problems of the Chernobyl exclusion zone-1996. Kyiv: Administartion of Chernobyl Exclusion Zone. (in Ukrainian)

Sedov, L. I. (1976). *Continuum Mechanics (vols. 1-2)*. Moscow: Nauka. (in Russian)

Chapter 6
Methods of Calculating Turbulent Processes in the Atmosphere

ABSTRACT

The hydrodynamics laws describe well the laminar motion of a continuous medium where the particle trajectories and streamlines have fully defined regular character. However, random deviations, or "disturbances", significantly change the nature of the initial motion, and cause a transition to a new or constant movement, or to some chaotic. This chapter studies forms of motion of a viscous fluid that is very common in the atmosphere and is called the turbulent motion.

BRIEF INTRODUCTION

Experts in the field of hydrodynamics know that the system of equations (3.2.29) – (3.2.35) well describes the laminar motion of a continuous medium, i.e., those cases where the particle trajectories, streamlines, velocity and pressure fields in these movements have a fully defined "regular" character. The example is Poiseuille movement of viscous fluid in a circular tube. It clearly confirms the compliance of the theoretically calculated characteristics (parabolic profile of velocity, the values of flow rate and resistance) and the experimental data. Available in a variety of natural conditions, mostly small in size, random deviations, or "disturbances", may either very little change the motion under consideration, indicating the constancy of motion with respect to small perturbations, or completely distort it, what is typical for the unsteady flow. In the latter case, small at first perturbations increase, significantly changing the nature of the initial motion, and cause a transition to a new or constant movement, or to some chaotic movement formed by liquid elemental masses (moles), irregularly moving and interacting with each other. This form of motion of a viscous fluid is very common in nature and is called the *turbulent motion*.

According to modern concepts, a turbulence arises from external disturbances and irregularities in the flow of a viscous fluid because of a chaotic variety of vortex tubes, thus appearing. These tubes are stretched in certain directions under the influence of deformations of the main flow and move in ran-

DOI: 10.4018/978-1-5225-2636-0.ch006

dom directions in conjunction with each other. In the developed turbulent flow, vortex lines constantly change direction and intertwist. Vortex lines near the wall are the most confused, they can curl up in harnesses and overlap. Thus, despite the presence of the dominant direction, turbulence is essentially a three-dimensional process.

Dimensions in vortices developed turbulent flow vary widely. In order of magnitude, the size of the largest eddies is of the size of the flow region, and the size of the smallest eddies is of the size of the field, across which can be carried out the transfer of momentum, heat and mass under the action of molecular viscosity. The smaller vortices are exposed to strain created by larger vortices; in this way, there is a transfer of energy between vortices. In the process of these deformations, the vorticity of smaller vortices increases and thus increases their energy due to the energy of larger vortex. Energy can be exchanged only eddies of comparable dimensions. Thus, there is a cascade of energy transfer process between adjacent turbulent flow vortices in size, and the energy is transmitted to all of smaller vortices and ultimately converted into internal energy of the thermal motion by the action of viscous forces on the smallest eddies.

Chaotic velocity field produced by elementary eddy tubes leads to intensive mixing and the moving fluid, and, as a consequence, the exchange of momentum, heat and mass in the transverse direction. The turbulent velocity field, pressure, temperature, concentration, and other hydrodynamic quantities have a very complex structure. The complexity of the structure of these fields is due to the extremely irregular and random nature of their change in space and time. Fields of flow variables in a turbulent flow, unsteady as always, are very dependent on the initial conditions, which are generally known far from complete. At present, there is no general and universal theory of the study of these fields.

The search of acceptable for meteorological practice forms the mathematical description of turbulent processes in the atmosphere, i.e., the turbulence models, occupies the minds of many outstanding mathematicians and meteorologists for over 120 years (starting with the classic work of (Reynolds, 1895)). This is due to the exceptional complexity of the turbulence as a physical phenomenon and the fact that it is this form of turbulent air movement that is most often realized in nature.

Direct Methods for the Numerical Simulation of Turbulent Flows

Experimental studies show that the space-time scales of turbulence are always several orders of magnitude superior to the space-time scale of molecular movement, and, according to modern concepts, the turbulent processes in the atmosphere obey classical Navier-Stokes and convective diffusion. Therefore, in recent years in dealing with the application of aerodynamics problems, there are being applied more and more approaches to the modeling of turbulence, based on primary principles of aerodynamics – *Direct Numerical Simulation (DNS)*. The interest to the DNS application in modeling hydrodynamics is due primarily to the fact that, at the distances of an order of magnitude of turbulent non-homogeneities, all the fields of related functions change smoothly, functions are differentiable, and hence the use of the Navier-Stokes differential equations is quite justified (Spalart, 2000; Moser, Kim, Mansour, 1999; Shur et al, 1996).

The basic assumption of the DNS is that the Navier-Stokes equations describe adequately not only laminar but turbulent processes also. According to this approach, the calculation of turbulent processes is carried out by means of direct (without any prior averaging) the numerical solution of the Navier-Stokes equations. Irrespective of the nature of the process (regardless of whether it is a two-dimensional or three-dimensional, static or nonstationary) three-dimensional unsteady Navier-Stokes equations must be used, since turbulence is essentially three-dimensional and time-dependent phenomenon. In addition, for a sufficiently accurate solution of all spatial and temporal scales of turbulence, DNS implies the need for three-dimensional non-stationary calculations on sufficiently fine grids even in cases where the purpose of the calculation is to determine the parameters of the averaging process only.

Solving the system of equations (3.2.29) – (3.2.35) is impossible without the use of numerical methods in which the flow field can be approximated by a finite number of calculation points. The main difficulty in this calculation of turbulent flow is that the turbulence scale is much smaller than the distances between the nodes used in the practice of computational grids. Although estimates of various authors differ, calculations show that for the numerical simulation of the turbulent vortex motion is necessary, at least 10 mesh nodes. The scale of the smallest eddies typically 1000 times smaller than the area of the flow along the solid surface. For typical flows, one may need 10^5 mesh nodes in the calculation of the area of the flow volume of 1 cm^3. For these reasons, direct calculation fields of turbulent flows on modern computing are practically impossible.

Currently, DNS is used only for calculating the hydrodynamic problems with relatively low numbers of Re and mainly in basic research, the purpose of which is to obtain detailed information on the structure and major patterns of turbulence. The rationale for this is estimation that the difference grid step should be of the order of the Kolmogorov scale $\eta_k = \left(\nu^3 / \varepsilon\right)^{1/4}$ (ν is the viscosity of the continuous medium, ε is the dissipation of turbulent kinetic energy) and, therefore, the computational cost is proportional to the number of nodes in each coordinate direction and the number of time steps, i.e., approximately Re^3.

Thus, although the turbulent motion of air obeys classical Navier-Stokes equations, computing capabilities, despite significant progress in its development, is still not sufficient for solving these equations for direct modeling the atmospheric circulation and weather forecast of practical interest. More recently, DNS has been used to solve fairly simple hydrodynamic problems (such as, for example, the flow of incompressible fluid between two parallel plates, Poiseuille flow in a flat channel and in cylindrical coordinates in ring-shaped tubes) (Betchov, 1975; Schumann, Patterson, 1978). The basic equations subject to the integration have been written in the differential form with constant parameters of a moving medium as follows

$$\nabla \cdot \mathbf{V} = 0,$$

$$\frac{d\boldsymbol{V}}{dt} = -\frac{1}{\rho}\nabla\, p + \nu\Delta\mathbf{V} + \mathbf{F}, \tag{6.1.1}$$

$$\frac{dT}{dt} = k\Delta T + \mathrm{Q}.$$

Here, $\mathbf{V} = \left(v_1, v_2, v_3\right)$ is wind velocity vector, *t* is time, *p* is pressure, ρ is density, ν is the kinematic viscosity, **F** is the force per unit mass, *T* is temperature, *k* is the thermal diffusivity, Q is the intensity of the heat sources per unit mass divided by the specific heat, $\left(\nabla \cdot\right)$ is the operator of divergence, ∇ is the gradient operator, Δ is Laplace operator.

Convective members

$$\left(\mathbf{V} \cdot \nabla\right) \mathfrak{H} = v_1 \frac{\partial \mathfrak{H}}{\partial x_1} + v_2 \frac{\partial \mathfrak{H}}{\partial x_2} + v_3 \frac{\partial \mathfrak{H}}{\partial x_3}, \ \mathfrak{H} = \left(\mathbf{V}, T\right)$$

have been written in the so-called "conservative form"

$$\left(\mathbf{V} \cdot \nabla\right) \mathfrak{H} = \frac{\partial \left(v_1 \mathfrak{H}\right)}{\partial x_1} + \frac{\partial \left(v_2 \mathfrak{H}\right)}{\partial x_2} + \frac{\partial \left(v_3 \mathfrak{H}\right)}{\partial x_3},$$

whose advantage is that the finite difference approximation of equations allows their partial integration.

Even such simple hydrodynamic problems for the DNS required effective algorithms for the numerical solution of equations. Their role was usually performed by cost-effective finite-difference (Williams, 1969) and spectral methods (Orszag, Patterson, 1972). These methods needed calculations of computer time from several hours to several days (Cooley, Tukey, 1965). Even the most optimistic estimations show that such situation will remain at least until the second half, and even before the end, of the XXI century. In this regard, the key issue in dynamic meteorology remains still open – how to find an acceptable compromise between the physical model adequacy and acceptable for practical use level of its complexity.

LARGE EDDY SIMULATION METHOD

The prospective method, known as the Large Eddy Simulation (LES), makes attempts to simulate large-scale turbulent motion, based on the numerical solution of "filtered" Navier-Stokes equations describing this large-scale three-dimensional unsteady motion (Ferziger, 1996; Garnier, Adams, Sagaut, 2009; Piomelli et al, 1990). The main role of small-scale vortices is to obtain the kinetic energy from large-scale eddies and transform it in the heat due to the viscous dissipation. Herewith, semi-empirical modeling techniques are used for approximation of turbulence effects, the scale of which is smaller than the difference grid cells. The article (Andren et al., 1994) contains a comparison of four different LES-models with different numerical schemes and different subgrid parameterization. It shows that the main characteristics of the neutral atmospheric boundary layer are reproduced even at coarse resolution and weakly depend on the choice of the numerical scheme and the method of parameterization of subgrid turbulence. Recently, calculations were made of the neutral atmospheric boundary layer with a higher resolution, and the use of more modern closure schemes and numerical schemes of a higher order of accuracy (Esau, 2004; Esau, 2003; Zilitinkevich, Esau, 2002; Zilitinkevich, Esau, 2003; Zilitinkevich, Esau, 2005; Glazunov, 2006). Such calculations seem to be very promising for atmospheric circulation models, but their cost is too great at the moment in order to be an instrument of operational practices in meteorological forecasting centers (HMC).

Thus, despite the rapid progress in the development of computer technology, its advanced capabilities are still insufficient for solving the Navier-Stokes equations DNS and LES methods for modeling of atmospheric processes of practical interest. This however does not mean that the hydrodynamic equations in general cannot be used in the simulation of atmospheric turbulence.

Methods of Statistical and Spectral Analysis

In practice, detailed knowledge of the fields of the hydrodynamic quantities in most cases is not required, as one can be interested only in averaged (statistical) characteristics of turbulent fields. Random velocities in the turbulent flow field produced by the elementary eddy tubes within the cascade energy transfer from larger eddies to smaller eddies cannot be described by mathematical deterministic relations, since each modeling of this phenomenon will play only one of the possible results. Therefore, the analysis of turbulence involves a statistical description, which does not allow to predict the outcome of only one way, but makes it possible to determine the probability of occurrence of this result. The statistical models, being used for describing physical processes in the developed turbulent flow, assume that such a movement is generated by a randomly varying set of vortex whose size and angular rotation speed vary widely both in space and in time. Despite significant irregularities, one can determine statistically distinguishable average of hydrodynamic parameters of the state of a turbulent flow. Therefore, to describe physical processes in the developed turbulent motion, one can apply the methods of statistical and spectral analysis (Frost, Mould, 1980). Statistical analysis allows describing the energy transfer between vortices of different scales or various sizes, and the spectral analysis responds the exchange of kinetic energy corresponding to the vortices of different sizes or different pulsation frequencies in the turbulent motion. However, the limits of applicability of these methods are not yet fully known and must be analyzed in each individual case.

Thus, in spite of its more than a century, mathematical modeling of turbulent flows, remains the relevant area of exclusively classical and computational fluid dynamics.

Methods Based on Reynolds Stress Equations of the Second Order

Despite the rapid growth of computer performance and the significant progress made in the development of efficient numerical methods and algorithms that solve the problems of modeling of the atmosphere and the weather forecast, calculation of turbulent processes, which is an integral part of these problems, it is today one of the most difficult problems of dynamic meteorology. Reliable prediction of turbulent characteristics of processes in the atmosphere is still the exception rather than the rule, due to the exceptional physical complexity of turbulence, in particular, its stochastic nature, essentially three-dimensional non-stationary nature and a wide range of spatial and temporal scales.

Due to the fact that the individual implementation of hydrodynamic fields obeys the equations of hydrodynamics, it is clear that the statistical characteristics of these fields also must be subordinated to them. The latter allows applying the methods of averaging to atmospheric processes and to receive processes of "laminar type", i.e. slow and smooth, which can be investigated by conventional methods. A wide range of scales between the microscopic and macroscopic motions allows justifying the suitability of the averaging procedure. It is known that almost all of the real atmospheric turbulence has a continuous spectrum. The averaging procedure assumes in advance that the turbulent flow can be divided into slow and fast (pulsating) movement. In this context, the question of averaging and rules for calculating

Figure 1. Communication between $\mathcal{F}, \bar{\mathcal{F}}$ and $\mathcal{F}'$ at a) steady flow; b) unsteady flow

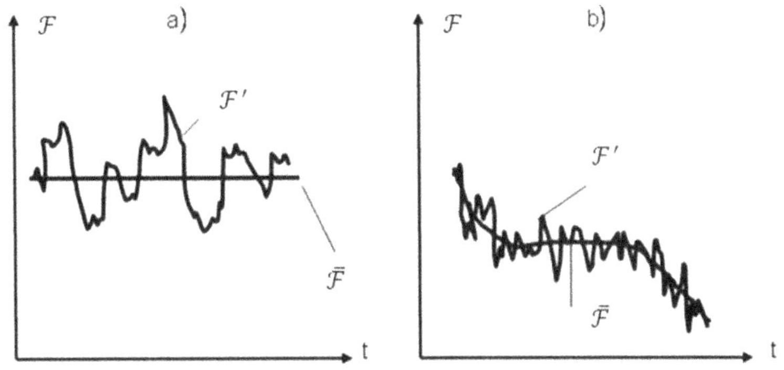

average values is a matter of the theory of turbulence. In practice, the averaging is performed either for a finite period of time or in some region of space, using a probability-theoretic approach that treats the field of hydrodynamic quantities in turbulent flow as a random field. Herewith, the rules of averaging applied to the system of atmospheric circulation equations provide a simple equation for the mean values.

Techniques that allow establishing links between the averaged characteristics of the fields of hydrodynamic quantities in the case of an incompressible isothermal fluid have been first given by O. Reynolds in (Reynolds, 1895). So far, these techniques are fairly described in the literature (Byzova, Ivanov, Garger, 1989; Kollman, 1984; Prusov, Romaniuk, 1998; Frost, Mould, 1980; Galperin et al., 1988). Therefore, we only briefly recall their essence. As it is known, O. Reynolds proposed to represent the values of all hydrodynamic quantities in a turbulent flow as the sum of averaged and pulsation components and to study only average values, relatively smoothly changing in space and in time. To determine the average values of a quantity he proposed to apply the usual averaging over a certain time interval. Note that in addition to the above method of averaging there are possible others, in particular, averaging over a certain area in a given time or averaged for a large number of fields, which vary both from one point to another and from one instant to another (averaging over the ensemble). Without discussing here the advantages and disadvantages of different ways of averaging, we note only that usually used conventional instrumentation allows measurement of the average values of the time, so the time averaging appears the most appropriate.

Following O. Reynolds, we represent the instantaneous value of each of the unknown variables in the system of equations (3.2.29) – (3.2.35) or any combination thereof as the sum of averaged $\bar{\mathcal{F}}$ and pulsating $\mathcal{F}'$ components (see Figure 1).

By averaging, we mean time averaging

$$\bar{\mathcal{F}} = \frac{1}{\vartheta}\int_{t_0}^{t_0+\vartheta} \mathcal{F} dt ,$$

and the averaging interval ϑ is assumed to be large compared with the characteristic period of the pulsation of the field and significantly small compared with the period averaged field. The very form of Reynolds averaged equations of hydrodynamics and heat and mass transfer is very weakly dependent on the extent of using filters or model grid spacing. Hence Reynolds averaged equations are usually given in the publication without any indication of the scale of averaging performed, and sometimes even without mentioning that this homogenization has occurred. As the meteorological observations show (Prusov, Doroshenko, 2006), there is a failure in the field perturbation with a period of about 1 hour in the spectrum of fluctuations of meteorological variables in the real atmosphere. Therefore, it seems reasonable 1-hour averaging. The results of this averaging are equivalent to the data given on the smoothed mesoscale and other synoptic charts.

By applying the basic rules of averaging

$$\overline{\mathcal{F}'} = 0,\ \overline{\mathcal{F} + \mathfrak{H}} = \bar{\mathcal{F}} + \bar{\mathfrak{H}},\ \overline{\mathcal{F} \cdot \mathfrak{H}} = \bar{\mathcal{F}} \cdot \bar{\mathfrak{H}},\ \overline{\frac{\partial \mathcal{F}}{\partial r}} = \frac{\partial \bar{\mathcal{F}}}{\partial r},$$

$$\overline{\mathcal{F} \cdot \mathfrak{H}} = \overline{\left(\bar{\mathcal{F}} + \mathcal{F}'\right)\left(\bar{\mathfrak{H}} + \mathfrak{H}'\right)} = \bar{\mathcal{F}} \cdot \bar{\mathfrak{H}} + \overline{\mathcal{F}' \cdot \mathfrak{H}'}$$

the system of Navier-Stokes equations we obtain the well-known equation of Reynolds.

We will call the models that provide closing the Reynolds equations as Reynolds strain models (RSM). The pioneering work (Karman, 1921), there is proposed to close the Reynolds equations system by sequentially calculating the unknown strain tensor of ever-increasing order. The equations for the strain tensor of the second order contain the third order strain tensor, the equations for the strain tensor of the third order contain the strain tensor of the fourth order, etc. The result is an endless chain of interrelated equations. It is assumed that the closure of the transfer equations at the level of equations for the correlation moments of the higher order should lead to a more general RSM, as the equations for higher order moments contain a large amount of information regarding the dynamics of the fluctuation characteristics of turbulent flows. Provided that this information is not lost due to the failure simulation of members for the closure of the system, the considered approach should become more versatile with increased order closure. However, the continuation of the procedure of derivation of the exact equations for emerging unknown correlation moments leads to an ever-growing number of unknown variables, what leads to a significant increase of the empirical information required for the closure. Therefore, the chain of equations is usually interrupted by neglecting higher-order tensors. The models of RSM have no connection of turbulent strains with an average strain rate, but they contain a very large number of equations and constants. So, apparently, these models are more likely to become universal "adequate" models of turbulence if success could be achieved in solving the time-averaged Navier-Stokes equations. However, the RSM still require to make assumptions in the modeling members included in the equations that cannot currently any way be accurately measured or found by an exact solution. Until now, RSM did not pass experimental testing for many types of processes in the atmosphere and therefore at most are used as a tool in the study of turbulence, rather than as a means to tackle the operational challenges in meteorological services. RSM are still in the development stage, and it will take time before they will be refined and tested so that they will be used in circulation models of the atmosphere.

The literature often classifies models of turbulent closures according to the number of additional differential equations in partial derivatives that must be addressed for achieving the model parameters. This number can vary from zero in the case of the simplest algebraic models to 12 in the case of the most complex RSM. According to this terminology, the Reynolds strains are calculated in a model of a closure of the first order as functions of only the averaged velocity and the geometry of the problem. The second order closure model uses a solution of the model equation for the transfer of one or more characteristics of turbulence. Currently, the most widely used application are the transport equations closure techniques at the level of correlation equations for moments of the second order (Kollman, 1984). A huge role in the development of this approach to the study of turbulence played a hypothesis (Kolmogorov, 1941; Rotta, 1951), which allowed to close the system of equations involving ratios for the second moments of pulsations. Today these hypotheses specified and summarized a number of researchers (Kollman, 1984); Frost, Mould, 1980).

It is important in the mesoscale modeling of atmospheric processes that the influence of density stratification on the structure of movements detected in the field measurements in the atmospheric boundary layer can be described by a direct extension of Rotta hypothesis (Rotta, 1951) and missing dimensionless constants can be set on the laboratory data for neutral conditions. The models in which the system of equations is closed with the assistance of the correlation moments of the second order, allow to reproduce the stabilizing or destabilizing effects in the boundary layer of the atmosphere that arise due to orographic heterogeneity of the underlying surface.

Considered models of closure of turbulent transport equations are not used for numerical prediction of atmospheric processes of the planetary scale, because of the lack of capacity of modern computers, so we limit ourselves by the system of Reynolds strains balance equations obtained in (Donaldson, 1973) presented in projections on the axis of Cartesian coordinate system.

$$
\begin{aligned}
&\frac{\partial \overline{v_i' v_j'}}{\partial t} + V_k \frac{\partial \overline{v_i' v_j'}}{\partial x_k} = -\overline{v_i' v_k'} \frac{\partial V_j}{\partial x_k} - \overline{v_j' v_k'} \frac{\partial V_i}{\partial x_k} + \frac{1}{\overline{\theta}} \left(g_i \overline{v_j' \theta'} + g_j \overline{v_i' \theta'} \right) - 2 \left(\varepsilon_{ikl} \Omega_k \overline{v_i' v_j'} + \varepsilon_{jlk} \Omega_l \overline{v_k' v_i'} \right) \\
&- \frac{\partial \overline{v_k' v_i' v_j'}}{\partial x_k} - \frac{1}{\rho} \left(\overline{v_i' \frac{\partial p'}{\partial x_j}} + \overline{v_j' \frac{\partial p'}{\partial x_i}} \right) + v_L \left(\frac{\partial^2 \overline{v_i' v_j'}}{\partial x_k \partial x_k} - 2 \overline{\frac{\partial v_i'}{\partial x_k} \frac{\partial v_j'}{\partial x_k}} \right),
\end{aligned}
$$

$$
\begin{aligned}
&\frac{\partial \overline{v_i' \mathfrak{H}'}}{\partial t} + V_j \frac{\partial \overline{v_i' \mathfrak{H}'}}{\partial x_j} = -\overline{v_i' v_j'} \frac{\partial \mathfrak{H}}{\partial x_k} - \overline{v_j' \mathfrak{H}'} \frac{\partial V_i}{\partial x_j} + g_i \frac{\overline{\mathfrak{H}'^2}}{\mathfrak{H}} \\
&- 2\varepsilon_{ikl} \boldsymbol{\Omega}_j \overline{v_k' \mathfrak{H}'} - \frac{\partial \overline{v_i' v_j' \mathfrak{H}'}}{\partial x_j} - \overline{\frac{\mathfrak{H}'}{\rho} \frac{\partial p'}{\partial x_i}} - \nu_L \overline{\frac{\mathfrak{H}' \partial^2 v_i'}{\partial x_j \partial x_j}} - \overline{\frac{d_l \partial^2 \mathfrak{H}'}{\partial x_j \partial x_j}},
\end{aligned}
\qquad (6.4.1)
$$

$$
\frac{\partial \overline{\mathfrak{H}^2}}{\partial t} + V_j \frac{\partial \overline{\mathfrak{H}^2}}{\partial x_j} = -2\overline{v_j' \mathfrak{H}'} \frac{\partial \mathfrak{H}^2}{\partial x_j} - \frac{\partial \overline{v_j' \mathfrak{H}'^2}}{\partial x_j} + d_L \left(\frac{\partial^2 \overline{\mathfrak{H}'^2}}{\partial x_j \partial x_j} - 2 \overline{\frac{\partial \mathfrak{H}'}{\partial x_j} \frac{\partial \mathfrak{H}'}{\partial x_j}} \right),
$$

where, for the sake of simplicity of notation, summation is over repeated indices (i, j, k = 1,2,3). There uppercase characters indicate the average values and lower case indicate pulsation. The first equation contains three members of:

- Transportation:

$$\frac{\partial \overline{v_k' v_i' v_j'}}{\partial x_k}. \tag{6.4.2a}$$

- Correlation (containing pressure gradient and velocity):

$$\frac{1}{\rho}\left(\overline{v_i' \frac{\partial p'}{\partial x_j} + v_j' \frac{\partial p'}{\partial x_i}}\right). \tag{6.4.2b}$$

- Dissipation:

$$\nu_L \left(\frac{\partial^2 \overline{v_i' v_j'}}{\partial x_k \partial x_k} - 2 \overline{\frac{\partial v_i'}{\partial x_k} \frac{\partial v_j'}{\partial x_k}} \right). \tag{6.4.2c}$$

Similar terms are included in the following two equations (6.4.1). To close the system of equations (6.4.1) one must obtain model presentations.

Here we confine ourselves to the simplest possible second-order closure scheme, in which the weighting coefficients are assumed to be constant. In more complex models, these factors are considered as a function of dimensionless parameters, their values must be determined from experiments, what reduces the practical value of these schemes. As mentioned above, the expression (6.4.2c) describes the effect of viscous dissipation on the structure of Reynolds strains, that is, it defines the basic mechanism of the turbulent kinetic energy scattering. A hypothesis belongs to Kolmogorov that small-scale turbulence is isotropic, even if the larger scale perturbations not have this property. According to this hypothesis, given the possibility of anisotropy, dissipation processes can be written

$$\nu_L \overline{\frac{\partial v_i'}{\partial x_k} \frac{\partial v_j'}{\partial x_k}} = C_1 \delta_{ij} \frac{k^3}{\Lambda} + \nu_L \overline{v_i' v_j'} \frac{C_2}{\Lambda^2}, \tag{6.4.3}$$

where $k = \overline{v_i' v_i'} / 2$ is the turbulent kinetic energy density, Λ is the length scale of the process of turbulent exchange, C_1 and C_2 are empirical constants, δ_{ij} is the symbol of Kronecker: $\delta_{ij} = 1$ for *i=j* and $\delta_{ij} = 0$ for *i≠j*.

The expression in (6.4.2b) describes the redistribution of turbulent energy by correlating the pressure fluctuations with a strain rate. Rewrite it in the form

$$\frac{1}{\rho}\left(\overline{v_i' \frac{\partial p'}{\partial x_j}} + \overline{v_j' \frac{\partial p'}{\partial x_i}}\right) = \frac{1}{\rho}\left[\overline{\frac{\partial v_i' p'}{\partial x_j}} + \frac{\partial v_j' p'}{\partial x_i} - \overline{p' \left(\frac{\partial v_i'}{\partial x_j} + \frac{\partial v_j'}{\partial x_i} \right)}\right]. \tag{6.4.4}$$

Following Rotta, the term in parentheses on the right side can be represented as

$$-\overline{\frac{p'}{\rho}\left(\frac{\partial v_i'}{\partial x_j}+\frac{\partial v_j'}{\partial x_i}\right)}=\frac{k}{\Lambda}\left(\overline{v_i'v_j'}-\delta_{ij}\frac{k^2}{3}\right),$$

and the first two terms can be combined with the first expression (6.4.2a). The simplest model of the third moments rate fluctuations formula can be used:

$$\frac{\partial\overline{v_k'v_i'v_j'}}{\partial x_k}+\frac{1}{\rho}\left(\frac{\partial\overline{v_i'p'}}{\partial x_j}+\frac{\partial\overline{v_j'p'}}{\partial x_i}\right)=C_3\frac{\partial\overline{k\Lambda}}{\partial x_k}\frac{\partial\overline{v_i'v_j'}}{\partial x_k},\tag{6.4.5}$$

where C_3 is an empirical constant.

For closing (6.4.3) - (6.4.5), one must specify how to determine the scale of turbulence. There are several approaches to solving this problem. The scale Λ can be defined as empirical functions, based on the features of the flow geometry, or be found as a solution of a semi-empirical model of a differential equation. Each of these approaches has its advantages and disadvantages. In the simulation of atmospheric processes, properties of the averaged flow are little sensitive to moderate changes in the macroscale length. This allows using simple closure models. Some examples of such models and values of included therein empirical constants are given in (Frost, Mould, 1980).

The system of equations (6.4.1) - (6.4.5) can be solved only by means of numerical methods on a computer. Since the second order models contain a large number of computed variables their use increases the cost of solving the hydrodynamic problem as a whole. Consequently, it is hardly advisable to use these models in the cases to which simple models are applicable, if only the second order model do lead to a significant improvement results. Very often, the latter is not the case.

Methods Based on the Concept of Turbulent Viscosity

By applying the basic rules of averaging the system of equations (3.2.29) – (3.2.35) and suggesting the possibility of permutation of operations of averaging and differentiating by the coordinates and time, we obtain a system of equations for the mean motion:

$$\frac{d\mathcal{R}}{dt}\equiv\frac{\partial\mathcal{R}}{\partial t}+\left(\mathbf{V}\cdot\nabla\right)\mathcal{R}=\nabla\cdot\mathbf{V},$$

$$\frac{d\boldsymbol{V}}{dt}=-e^{-\mathcal{R}}\nabla\, p+\mathbf{g}+2\mathbf{V}\times\boldsymbol{\Omega}+\nabla\cdot\left(\vec{\tau}_n-\overline{\mathrm{v}'\mathrm{v}'}\right),$$

$$\frac{d\theta}{dt}=\nabla\cdot\left(k_T\nabla\theta-\overline{\mathrm{v}'\theta'}-F_R\right)-L\left(Q_H-S_k+S_u\right),$$

$$\frac{dq}{dt}=\nabla\cdot\left(k_d\nabla\mathrm{q}-\overline{\mathrm{v}'q'}\right)+Q_H-S_k+S_u,\tag{6.5.1}$$

$$\frac{d\delta}{dt} = \nabla \cdot \left(k_d \nabla \delta - \overline{\mathrm{v}'\delta'}\right) - Q_H + S_u + \frac{\partial F_g}{\partial x_3},$$

$$\frac{ds}{dt} = \nabla \cdot \left(k_d \nabla \mathrm{s} - -\overline{\mathrm{v}'s'}\right) - S_s - J_s,$$

$$p = e^{-\mathcal{R}} RT,$$

where averaging symbols are omitted for simplicity.

Note that in the derivation of these equations, the terms containing pulsations of density ρ, viscosity ν, and diffusion coefficient k_d, are neglected in comparison with the members, containing their averaged values. In the equations (6.5.1), the values of $-\overline{\mathrm{v}'\mathrm{v}'}$ have sense of tensor components of additional strains, and $\overline{\mathrm{v}'\mathfrak{H}'}$ have sense of additional transfer arising from the presence of turbulent velocity fluctuations and the scalar quantity $\mathfrak{H} = \left(\theta, q, \delta, s\right)$. They appear as a result of the averaging advective-convective terms in the system of equations (6.5.11.2.28) - (6.5.11.2.34). The tensor $-\overline{\mathrm{v}'\mathrm{v}'}$, at equal components $\overline{v_1'v_2'} = \overline{v_2'v_1'}$, $\overline{v_1'v_3'} = \overline{v_3'v_1'}$ and $\overline{v_3'v_2'} = \overline{v_2'v_3'}$, has the symmetry property that allows presenting it in the form of a matrix

$$\begin{matrix} -\overline{v_1'v_1'} & -\overline{v_1'v_2'} & -\overline{v_1'v_3'} \\ -\overline{v_2'v_1'} & -\overline{v_2'v_2'} & -\overline{v_2'v_3'} \\ -\overline{v_3'v_1'} & -\overline{v_3'v_2'} & -\overline{v_3'v_3'} \end{matrix}, \tag{6.5.2}$$

and turbulent flow of scalar quantity $\mathfrak{H}$ is a vector with components

$$-\overline{\mathrm{v}'\mathfrak{H}'} = \left(-\overline{v_1'\mathfrak{H}'}, -\overline{v_2'\mathfrak{H}'}, -\overline{v_3'\mathfrak{H}'}\right). \tag{6.5.3}$$

Thus, for a viscous heterogeneous medium, a unclosed system of equations for the mean motion is obtained from the closed system of equations (3.2.29) – (3.2.35). The hypothesis of O. Reynolds has generated a huge number of works associated with the closure of a system of equations for the turbulent motions (6.5.2). A very detailed presentation and description of fundamental models of turbulence are given in (Rodi, 1978).

In (Boussinesq, 1877), the author has speculated that the apparent turbulent shear strains can be associated with an average strain rate through apparent (effective) scalar turbulent (or "eddy") viscosity ν. This gives for the tensor (6.5.2) of Reynolds strains the general form

$$-\overline{\mathrm{v}'\mathrm{v}'} = \nu \left(\frac{\partial \overline{v_i}}{\partial x_j} + \frac{\partial \overline{v_j}}{\partial x_i}\right) - \frac{2}{3}\delta_{ij}\left(\nu \frac{\partial \overline{v_k}}{\partial x_k} + k\right), \tag{6.5.4}$$

where ν is the coefficient of turbulent viscosity, δ_{ij} is the symbol of Kronecker. Unlike the molecular viscosity coefficient, the eddy viscosity coefficient ν does not characterize the physical properties of the medium, but the statistical properties of pulsating movement. Therefore, the coefficient ν is a function of space and time. It is important to emphasize that the turbulent viscosity is much greater than the molecular viscosity at the distance from the solid surface.

By analogy with the kinetic theory of gasses, one can expect the turbulent viscosity to be written with sufficient accuracy as

$$\nu = U \cdot \Lambda , \tag{6.5.5}$$

where U and Λ are the characteristic scales of turbulence speed and length respectively.

We will refer models that use the Boussinesq hypothesis to the class of *models of turbulent viscosity* (*MTV*). Most models used to date in engineering calculations fall into this group. Experimental observations show that the turbulent viscosity hypothesis is suitable for many hydrodynamic problems. There are exceptions, however, when, firstly, MTV lead to incorrect results and, secondly, for these cases, there is no physical basis fairness for MTV.

The simplest, based on the concept of turbulent viscosity, are the first-order model, as they include in the equations the moments of the first order (medium speed, temperature, concentration, etc.), and the closure of their constituent moments of the second order with the help of the moments of the first order. The first order models have been used for several decades in hydrodynamics applications. A number of satisfactory results were produced with their help, despite the fact that they are based in principle on misconceptions. This is because the constant parameters of the models were selected on the base of experimental data, i.e., turbulent viscosity was chosen so that the average characteristics of the profiles of some of the standard currents were predicted to be right.

In other words, there is created a new medium with different physical properties, that has, however, the same characteristics of resistance, heat and mass transfer, as the turbulence under appropriate outer conditions.

Following the hypothesis of Boussinesq (6.5.4) and Reynolds analogy

$$-\overline{v_j' \mathfrak{H}'} = d \frac{\partial \mathfrak{H}}{\partial x_j} \left(j = 1,2,3 \right), \tag{6.5.6}$$

we assume that the relationship between the coefficient of turbulent viscosity ν and the coefficients of turbulent heat and mass transfer d is realized by the turbulent Prandtl number Pr and Schmidt number Sc. Designating $\zeta = \left(Pr, Sc \right)$, we can write this relationship in the form $d = \nu / \zeta$. Although the turbulent transfer coefficients are generally coordinates and time functions, the dimensionless combination of Pr and Sc are usually values, changing little, allowing in many cases of practical importance to assume them permanent. The numerical values of Pr and Sc are currently uncertain. A few experimental data on the values of these numbers is very contradictory, however, we can assume that in the flow of boundary layer near the solid surface Pr and Sc are close to 1.

The system of equations (6.5.4) - (6.5.6) by itself does not introduce turbulence model, and only defines the structure of a model; the main challenge now is to define the function ν. There are developed various models of turbulence for the concept of turbulent viscosity ν. These models have varying degrees of complexity, ranging from simple, where ν is expressed through the local characteristics of the averaged velocity fields, temperature, and concentration, to quite complex models, including the transport equation for the characteristic velocity and linear scale of turbulent flows.

Models of Turbulence of Prandtl and Karman

The simplest models describing the distribution of ν are built on the hypothesis proposed by Prandtl in the 20-s of XX century, according to which the turbulent viscosity coefficient is proportional to the scale of the speed and scale of turbulence,

$$\nu = \ell^2 \left| \frac{\partial v_1}{\partial x_3} \right|, \tag{6.5.7a}$$

where v_1 is the velocity component in the direction of the main flow, ℓ is a characteristic length scale (mixing path). The product of $\ell \partial v_1 / \partial x_3$ can be interpreted as a characteristic turbulent velocity U (see (6.5.5)).

For cases where the air flow direction does not coincide with the direction of the axis of x_1, Prandtl formula is usually written in the form

$$\nu = \ell^2 \left[\left(\frac{\partial v_1}{\partial x_3} \right)^2 + \left(\frac{\partial v_2}{\partial x_3} \right)^2 \right]^{1/2}. \tag{6.5.7b}$$

In this formula, eddy viscosity is regarded as a scalar and gives a qualitatively correct results, especially of wall currents, as there is experimental evidence that the outer layer of turbulent viscosity should be regarded as a tensor (i.e., a quantity depending on the direction of deformation) to ensure better agreement with the experimental results.

Laboratory and field experiments in the boundary layer observations indicate that the scale of turbulence may be fairly well approximated either by using the Prandtl model

$$\ell = \chi x_3 \text{ on condition } x_3 \leq \delta,$$

$$\ell = \chi \delta \text{ on condition } x_3 > \delta, \tag{6.5.8a}$$

or by using the Karman model

$$\ell = \chi_1 \left| \frac{\partial v_1 / \partial x_3}{\partial^2 v_1 / \partial x_3^2} \right|, \tag{6.5.8b}$$

where χ, χ_1 are constant, δ is the height of the boundary layer.

Selecting one of these two schemes of closure is to some extent a matter of taste, but in some cases, one of them may be more convenient to use than the other.

The assumption of an isotropic coefficient of turbulent viscosity (6.5.7) in models of turbulence Prandtl and Karman is a strong simplification that causes limited their usefulness in the interpretation of complex flows, in particular, those for which the action of mass forces is a predominant direction.

There is a model below in which the vector of turbulent viscosity coefficient is taken proportional to absolute value of the vorticity vector and therefore has different components in different directions.

We assume that the coefficient of turbulent viscosity ν at the interface between the two layers inside the air flow is completely determined by the distribution of average velocity $\bar{\mathbf{V}}$ near the boundary of these layers, i.e. by the set of rotations of the field $\mathrm{rot}\bar{\mathbf{V}}$, $\mathrm{rot}\left(\mathrm{rot}\bar{\mathbf{V}}\right)$, $\mathrm{rot}\left[\mathrm{rot}\left(\mathrm{rot}\bar{\mathbf{V}}\right)\right]$, …. The very speed $\bar{\mathbf{V}}$ cannot enter into this set. This is explained by that, associating an elementary volume of air moving along the boundary layers with constant velocity $\bar{\mathbf{V}}$ and progressively moving coordinate system implies the classical principle of Galileo. It says that all dynamic processes with respect to inertial reference system occur equally, regardless of the system $\bar{\mathbf{V}}$ forward velocity.

Let the coefficient of turbulent viscosity ν is a function of the values of rotating field $\mathrm{rot}\,\mathbf{V}$ of viscous medium (the vortex velocity)

$$\mathrm{rot}\mathbf{V} = \boldsymbol{i}\left(\frac{\partial v_3}{\partial x_2} - \frac{\partial v_2}{\partial x_3}\right) + \boldsymbol{j}\left(\frac{\partial v_1}{\partial x_3} - \frac{\partial v_3}{\partial x_1}\right) + \boldsymbol{k}\left(\frac{\partial v_2}{\partial x_1} - \frac{\partial v_1}{\partial x_2}\right)$$

and rotation of the vortex field $\mathrm{rot}\ \mathrm{rot}\mathbf{V}$.

We use the theory of dimensions and assume that the turbulent viscosity ν can be represented by one-term dependence

$$\nu \sim \left(\mathrm{rot}\mathbf{V}\right)^{\alpha}\left(\mathrm{rot}\ \mathrm{rot}\mathbf{V}\right)^{\beta}. \tag{6.5.9}$$

or in the physical dimensions of the system

$$m^2 \cdot c^{-1} = \left(m \cdot c^{-1} \cdot m^{-1}\right)^{\alpha}\left(m \cdot c^{-1} \cdot m^{-2}\right)^{\beta} = c^{-\alpha} \cdot c^{-\beta} \cdot m^{-\beta}.$$

Equating the exponents at the dimensions *c* and *m* in the left and right sides of the last expression, we obtain a system of equations:

$$\alpha + \beta = 1,\ \beta = -2,$$

which has a unique solution

$$\alpha = 3,\ \beta = -2.$$

Consequently, the relationship (6.5.9) leads to the formula:

$$\nu = \chi \left(\mathrm{rot}\mathbf{V}\right)^3 \left(\mathrm{rot}\ \mathrm{rot}\mathbf{V}\right)^{-2}, \tag{6.5.10}$$

where χ is a proportionality coefficient, some dimensionless constant.

According to the formula of vector analysis

$$\mathrm{rot}\ \mathrm{rot}\mathbf{V} = \nabla\left(\nabla \cdot \mathbf{V}\right) - \Delta\mathbf{V}\,. \tag{6.5.11a}$$

By definition of the scalar product of two vectors, the scalar square of the vector is equal to the square of its module

$$\left(\mathrm{rot}\mathbf{V}\right)^2 = \left|\mathrm{rot}\mathbf{V}\right|^2 = \left(\frac{\partial v_3}{\partial x_2} - \frac{\partial v_2}{\partial x_3}\right)^2 + \left(\frac{\partial v_1}{\partial x_3} - \frac{\partial v_3}{\partial x_1}\right)^2 + \left(\frac{\partial v_2}{\partial x_1} - \frac{\partial v_1}{\partial x_2}\right)^2. \tag{6.5.12}$$

Given (6.5.11a) and (6.5.12) the formula (6.5.10) can be written as

$$\nu = \chi \frac{\left|\mathrm{rot}\mathbf{V}\right|^2}{\left[\nabla\left(\nabla \cdot \mathbf{V}\right) - \Delta\mathbf{V}\right]^2} \mathrm{rot}\mathbf{V}\,.$$

Wishing to emphasize the sign of ν, we rewrite this formula in the form of:

$$\nu = \chi \frac{\left|\mathrm{rot}\mathbf{V}\right|^2}{\left[\nabla\left(\nabla \cdot \mathbf{V}\right) - \Delta\mathbf{V}\right]^2} \sqrt{\left(\mathrm{rot}\mathbf{V}\right)^2}\,. \tag{6.5.13}$$

When using the proposed model (6.5.13), one should be aware of the incompleteness of information obtained with its help. The limitation of this type of models is in the local equilibrium of the simulated turbulence. Local equilibrium is due to the fact that at each point of space there is a balance of generation and dissipation of turbulent energy, which is not affected by the development of the previous process. Thus, the model is not applicable in cases when the background plays an essential role in the process. It is conceptually very simple, very easy to use and, as will be shown in Section 8.1.2, does not cause computational difficulties. Of course, it is possible to refuse it only in cases where the need for alternative approaches is obvious.

If the medium is incompressible

$$\nabla \cdot \mathbf{V} = \frac{\partial v_1}{\partial x_1} + \frac{\partial v_2}{\partial x_2} + \frac{\partial v_3}{\partial x_3} = 0 \tag{6.5.14}$$

then, as it follows from (6.5.11a), we have

$$\text{rot rot}\mathbf{V} = -\Delta\mathbf{V}, \tag{6.5.11b}$$

and the formula (6.5.13) takes the form

$$\nu = \chi\left(\frac{|\text{rot}\mathbf{V}|}{\Delta\mathbf{V}}\right)^2 \sqrt{|\text{rot}\mathbf{V}|^2}. \tag{6.5.15}$$

It is interesting that if we restrict ourselves to the case of motion of a homogeneous incompressible flow along the surface of the solid and superpose the axis x_1 and direction of the flow rate, then at the condition

$$v_2 = 0, \frac{\partial v_1}{\partial x_1} = \frac{\partial v_2}{\partial x_1} = \frac{\partial v_3}{\partial x_1} = 0, \frac{\partial v_1}{\partial x_2} = \frac{\partial v_2}{\partial x_2} = \frac{\partial v_3}{\partial x_2} = 0$$

The incompressibility equation (6.5.14) leads to the identity $\partial v_3 / \partial x_3 = 0$, and the equation (6.5.15) is simplified and takes the form of formula Karman

$$\nu_1 = \chi\frac{\left(\partial v_1 / \partial x_3\right)^2}{\left(\partial^2 v_1 / \partial x_3^2\right)^2}\left|\frac{\partial v_1}{\partial x_3}\right|,$$

Or, for the characteristic length scale (the mixing),

$$\ell = \chi\frac{\left(\partial v_1 / \partial x_3\right)^2}{\left(\partial^2 v_1 / \partial x_3^2\right)^2}. \tag{6.5.16}$$

Since it is usually assumed that the empirical constant χ should be a universal dimensionless number being the same for all the turbulent flows, the coincidence of the formulas (6.5.16) and Karman (6.5.8b) implies that $\chi = 0.4$.

In stratified boundary layers, the effect of mass forces arising from buoyancy or curvature of the streamlines, the distribution of scale of turbulence can be accounted for by the formula (Thom (1975))

$$\ell = \ell_0\left(1 - \varsigma_1 \text{Ri}\right) \text{ if Ri} > 0,$$

$$\ell = \ell_0\left(1 - \varsigma_2 \text{Ri}\right)^{-1/4} \text{ if Ri} < 0, \tag{6.5.17}$$

where ℓ_0 is a scale of turbulence on the relations (6.5.8) and (6.5.16) at Ri = 0; $\varsigma_1 \approx 7$, $\varsigma_2 \approx 14$ are empirical constants; Ri is the Richardson number, defined by the expression

$$\mathrm{Ri} = -\frac{g}{\rho}\frac{\partial \rho / \partial x_3}{\left(\partial V_g / \partial x_3\right)^2}.$$

Models with the Turbulent Kinetic Energy Equation

Turbulence is generated in the boundary layer mainly near the underlying surface, and due to the turbulent diffusion, it is transferred to the upper edge of the boundary layer. Models (6.5.7) - (6.5.9) neglect this transfer and, in addition, require zeros of turbulent viscosity everywhere velocity gradient is zero. These circumstances reveal the main drawbacks of the model (6.5.7) - (6.5.9), and therefore Prandtl suggested a significant generalization of the theory of turbulence calculation by introducing the concept of kinetic energy density of turbulent pulsating movement

$$k = \frac{\rho}{2}\sum_{i=1}^{3}\overline{v_i'^2}$$

as the velocity pulsation characteristic scale.

Without dwelling on the exact derivation of the transport equation for *k* from the Navier-Stokes equations, we give the model representation in the form in which it is widely used in the mesoscale models of the environment (Kollman, 1984)

$$\frac{dk}{dt} = \frac{\partial}{\partial x_i}\left(\frac{\nu}{\zeta_k}\frac{\partial k}{\partial x_i}\right) + \nu\left(\frac{\partial v_i}{\partial x_j} + \frac{\partial v_j}{\partial x_i}\right)\frac{\partial v_i}{\partial x_j} - C_d\frac{k^{3/2}}{L} + \frac{g}{T}\frac{\nu}{\zeta_T}\left(\frac{\partial T}{\partial x_3} + \gamma_a\right), \left(i, j = 1,2,3\right). \qquad (6.5.18)$$

We use here summing over the same indices commonly used for the analysis of tensor quantities in Cartesian coordinates. The first term on the right side of the equation describes the diffusion transfer due to pulsations of velocity and pressure. The second term describes a generation of energy, caused by the interaction of the Reynolds strains and mean velocity gradients. The third term describes the viscous dissipation of energy into heat and, finally, the fourth term describes the transition of the turbulent kinetic energy into potential one and vice versa due to buoyancy forces. Since the molecular viscosity of the medium is a small value, the main contribution to the dissipation is provided by small vortices with large gradients of wind speed. When the generated turbulent energy is large, an equilibrium condition is rapidly reached and the left-hand side of the equation (6.5.18) can be neglected $\left(\partial k / \partial t = 0\right)$. Under these conditions, generation of the dynamic and convective turbulence at the bottom of the boundary layer is compensated approximately by a dissipation process. However, relatively large vortices are generated, and dissipation occurs at the smallest eddies. This implies the continuous energy transfer in the entire spectrum from large to small vortices. In the upper part of the boundary layer, the values of the terms describing the generation of energy, are small. The kinetic energy balance is achieved here due to the converting into the heat of momentum of the turbulent flow of the lower layers of the liquid and the rate of dissipation.

The relationship between the characteristic scale of speed pulsation k and eddy viscosity is established by the Kolmogorov-Prandtl relationship

$$\nu = C_k L\sqrt{k}\,, \tag{6.5.19}$$

which, like the dissipative term of the equation (6.5.18), comprises a linear scale of L, to be determined for closing the model of turbulence. In the model (6.5.18), (6.5.19), the turbulent viscosity (diffusion) ν is assumed to be isotropic, that obviously narrows a set of solving applications with this model. For the near-wall boundary layers, a value of the scale L can be set using the simple empirical models (6.5.8). It is assumed that both C_d and C_k are universal constants whose values for all trends can be determined in advance (Kollman, 1984).

The model (6.5.18), (6.5.19) allows taking into account the convective and diffusive transfer in the background process. Therefore, these models are preferable the models (6.5.7), (6.5.13) in cases where these factors play an important role. At the same time, the range of applicability of the model (6.5.18) - (6.5.19) is limited mainly to simple unseparated boundary layers, as for more complex separated flows. it is difficult to obtain the empirical distribution of characteristic linear scale L. Consequently, they are not applicable to very common types of atmospheric motions: mountain-valley wind, breeze, foehn, bora, etc. In this regard, there were developed a model in which the linear scale is determined from the transfer equation.

Linear scale L characterizes the size of the big energy vortices. Stretching vortex tubes associated with an energy cascade reduces this size and the dissipation destroys small vortices what leads to an increase in the average size of vortices. The balance of these processes can be expressed by the model transport equation of the scale L, which sets its distribution in both time and space.

A Model with Two Transport Equations

The most commonly used option for the Prandtl-Kolmogorov model is a model in which the dissipation rate of turbulent kinetic energy is used as the dependent variable in an additional transport equation

$$\varepsilon = C_d k^{3/2} L\,. \tag{6.5.20}$$

The exact equation for ε can be obtained from the Navier-Stokes equations for the pulsation of the vortex. Omitting an analysis of model assumptions about members to describe diffusion, generation and dissipation, we give here the speed transfer equation of dissipation of the turbulent kinetic energy in the form

$$\frac{d\varepsilon}{dt}=\frac{\partial}{\partial x_i}\left(\frac{\nu}{\zeta_\varepsilon}\frac{\partial\varepsilon}{\partial x_i}\right)+\frac{\varepsilon}{k}C_1\nu\left(\frac{\partial v_i}{\partial x_j}+\frac{\partial v_j}{\partial x_i}\right)\frac{\partial v_i}{\partial x_j}+\frac{\varepsilon}{k}\left[C_2\frac{g}{T}\frac{\nu}{\zeta_T}\left(\frac{\partial T}{\partial x_3}+\gamma_a\right)-C_3\varepsilon\right],\left(i,j=1,2,3\right). \tag{6.5.21}$$

where ζ_ε and C_1, C_2, C_3 are empirical constants; $\gamma_a = 0.0098$ is the dry adiabatic temperature gradient. Turbulent viscosity is given by

$$\nu=\frac{C_4 k^2}{\varepsilon}. \tag{6.5.22}$$

As constant values in the equations (6.5.18) - (6.5.22) are usually chosen values shown in Table 1.

Semi-empirical model (6.5.18) - (6.5.22), known in the scientific community as a $\left(k-\varepsilon\right)$-turbulence model is one of the most tested models of the boundary layer. As shows the practice of its application, it gives satisfactory results with the same values of the empirical constants (Table 6.1) in complex cases of recirculating three-dimensional flows. This indicates that the $\left(k-\varepsilon\right)$-model is sufficiently versatile and more general than the classic approach using local dependencies for turbulent viscosity. However, the use of $\left(k-\varepsilon\right)$-turbulence model in dynamic meteorology dedicated simulation of atmospheric circulation rises quite complex problems. The reason is that in the integration of the differential equations forming the basis of $\left(k-\varepsilon\right)$-model, there is a need to formulate the initial and boundary conditions for the variables *k* and ε e along with the usual requirements for the variables of averaged motion (meteorological values). This is often not a trivial task, especially when setting the initial conditions for *k* and ε because meteorological variables of averaged motion are not very accurate.

For a long time, the need for formulation of nontrivial boundary conditions on the underlying surface has arisen from the complexity of the technical implementation of the conditions of adhesion on computers with limited resources, on the one hand, and the inapplicability of the standard $\left(k-\varepsilon\right)$-model, on the other hand. Obviously, the rate of dissipation ε vanishes on the wall, while the energy of the pulsations on the wall *k* is also zero. This means that the fractional value of k^2/ε on the underlying surface tends to infinity.

These above problems remain with the other models, which are a combination of turbulent kinetic energy density *k* and length scale *l* in a form $f=k^\alpha l^\beta$. This product is then multiplied by a specially selected non-dimensional function, the value of which depends on the stability (Richardson number). Thus, knowing physically reasonable turbulent length scale *l*, which is a physical characteristic of an

Table 1. The empirical constants

The empirical constants	C_d	C_k	ζ_ε	C_1	C_2	C_3	C_4
Their values	0,3	0,3	1,13	1,43	0,29	1,92	0,09

amount of energy of atmospheric turbulent eddies, is important to calculate the turbulent transfer coefficients. However, in some models, such as (Blackadar, 1962), to calculate l, a simple formula is used for the local diagnostic neutral stratification of the flat underlying surface. The formula gives the asymptotic behavior $l \sim \chi x_3$ near the underlying surface and $l \sim l_0$ on the upper edge of the boundary layer. The size l_0 in different models varies from 200 to 500 m. If the lower asymptotic behavior performed quite well and is supported by numerous experimental data, there are problems in the top of the boundary layer. Thus, if the atmospheric boundary layer in the upper part comprises a temperature inversion layer, that is a resistant layer, then it should attenuate the turbulent eddies, and thus, the value of l should aspire to the minimum value. However, the calculation formula at the top of the boundary layer gives maximum values for l.

In addition, in the equation for turbulent kinetic energy k, the mixing length l is included in the denominator in the term describing the scattering of the kinetic energy of turbulence. This leads to the fact that l increases and dissipation decreases in a stable layer, increasing the turbulent kinetic energy k. Although, from a physical point of view, there should be the opposite in a stable layer: l should decrease and the dissipation should increase, thereby reducing the value of k.

In areas of active convection, using simple local diagnostics for the mixing formula l also leads to a nonphysical result. Indeed, in the process of convection heated air particles can rise from the boundary layer to a considerable height, that depends on k at a starting level, and the stratification at different levels, located above (below) the original level. However, the calculation formula determines only the local dependence on the height l. These and other considerations indicate that there are problems when using the length scale l in a model of atmospheric turbulence.

In addition, the model (6.5.18) - (6.5.22) contains a larger number of calculated variables and is more complex than the model (6.5.7), (6.5.13). Naturally, its use increases the cost of the solution. Consequently, the choice of a particular model of turbulence depends on the nature of the problem being solved. It is hardly appropriate to use complex models in cases coming within the scope of simple models, and when it is not expected to significantly improve the results.

"SELF-DIFFUSION" MODELS

There is a method that greatly simplifies the problem of turbulence closure of a model where the change in the eddy viscosity ν_t is given by a separate differential equation describing transfer of this quantity. In scientific publications (Prusov, Romaniuk, 1998; Prusov, Sologub, 2011; Nee, Kovasznay, 1969; Blackadar, 1962; Spalart, Allmaras, 1992; Gulyaev et al., 1993; Durbin, Mansour, Yang, 1994), some modifications of this turbulence model are proposed in which the link between the turbulent viscosity and the parameters of the mean flow are given directly by the non-stationary nonlinear equation for the turbulent viscosity ν_t

$$\frac{\partial \nu_t}{\partial t} + \mathrm{v}_j \frac{\partial \nu_t}{\partial x_j} = D_{\nu_t} + G_{\nu_t} - \varepsilon_{\nu_t}, j = 1,2,3. \qquad (6.5.23)$$

Here and further below, for simplicity, the dashes of averaging the wind speed vector and its components are omitted.

Following the work of the authors (Prusov, Romaniuk, 1998; Prusov, Sologub, 2011), we will focus on the construction of empirical relationships for terms on the right side of the equation (6.5.23): D_{ν_t} – diffusion of turbulent viscosity ν_t; G_{ν_t} – generation of ν_t due to wind shear and buoyancy; ε_{ν_t} – attenuation of ν_t due to dissipation.

The Diffusion of the Turbulent Viscosity

Let $\mathfrak{H}$ is some scalar substance external to given point. Recall the known laws of the heat conduction by Fourier and of diffusion by Fick, which argue that the instant amount of scalar substance $\mathfrak{H}$ passing through a unit area is proportional to the gradient of the averaged substance $\mathfrak{H}$. Then, the growth rate of $U_{\mathfrak{H}}$ taken per unit volume of $\mathfrak{H}$ may be determined from the balance equation

$$\int_Q U_{\mathfrak{H}} dQ = \int_{\Im} \lambda \frac{\partial \mathfrak{H}}{\partial n} d\Im = \int_Q \operatorname{div}\left(\lambda \operatorname{grad}\mathfrak{H}\right) dQ,$$

composed for an arbitrary amount of Q, bounded by a closed surface $\Im$, which contains the given point. By virtue of the arbitrariness of the volume Q we will have

$$U_{\mathfrak{H}} = \operatorname{div}\left(\lambda \operatorname{grad}\mathfrak{H}\right).$$

If as the scalar substance $\mathfrak{H}$ to take turbulent viscosity ν_t, and as the proportionality factor to keep in mind the overall kinematic viscosity $\lambda = \nu + \nu_t$, then the member D_{ν_t}, which describes the diffusion of turbulent viscosity ν_t in equation (2) will take the form

$$D_{\nu_t} = \operatorname{div}\left[\left(\nu + \nu_t\right)\operatorname{grad}\nu_t\right] = \sum_{j=1}^{3} \frac{\partial}{\partial x_j}\left[\left(\nu + \nu_t\right)\frac{\partial \nu_t}{\partial x_j}\right]. \qquad (6.5.24)$$

Since the turbulent viscosity is not a fundamental physical quantity and it is not required for it to satisfy the laws of conservation, the diffusion turbulent viscosity ν_t will be presented as:

$$D_{\nu_t} = c_1 \sum_{j=1}^{3} \frac{\partial}{\partial x_j}\left[\left(\nu + \nu_t\right)\frac{\partial \nu_t}{\partial x_j}\right], \qquad (6.5.24a)$$

or

$$D_{\nu_t} = c_1 \sum_{j=1}^{3} \left\{ \frac{\partial}{\partial x_j}\left[\left(\nu + \nu_t\right)\frac{\partial \nu_t}{\partial x_j}\right] + c_2 \frac{\partial \nu_t}{\partial x_j}\frac{\partial \nu_t}{\partial x_j} \right\}. \qquad (6.5.24b)$$

Calibration of constants performed on the tasks of the boundary layer and free shear flows, has led to the following results: $c_1 = 2/3$, $c_2 = 0.622$.

The Rate of Generation of Eddy Viscosity

Experimental studies show that the maximum energy of turbulence in the boundary layer of the atmosphere is located in the area of the larger vortices. The physical explanation of this phenomenon lies in the fact that smaller vortices degenerate faster than big ones, as the greater velocity gradients are created in smaller vortices, that is, the tangent strains are prevailing. The concept of turbulent viscosity suggests that the transfer of momentum, heat and mass is similar to the transfer due to molecular motion. Turbulent viscosity is postulated in such a way that the average profiles of the main characteristics of the standard currents would be predicted near the right values. In other words, we refer to some new air, having different physical properties, but with the same characteristics of resistance, heat and mass transfer, that the actual air has, and which follows the motion of the turbulent regime at respective ambient conditions. Being subjected to fair criticism as physically ill-founded, the concept, however, is widely used because it allows getting quite acceptable results in meteorological practice. Consequently, at the Reynolds numbers typical of dynamic meteorology, we formulate generation G_{ν_t} of turbulent viscosity ν_t as follows:

$$G_{\nu_t} = c_2 \frac{\nu_t}{k} \wp_{in}, \tag{6.5.25}$$

where $\wp_{in}$ is a member of the generation of turbulent kinetic energy k, which is determined by the product of the Reynolds strains and average velocity gradients (characterizes the transfer of energy from the averaged flow to the vortices), c_2 is an empirical constant.

According to modern concepts, a turbulence arises from external disturbances and irregularities in the flow of viscous fluid and is due to a chaotic variety of vortex tubes to appear. These tubes are stretched in certain directions under the influence of deformations of the main flow and move in random directions in conjunction with each other. Despite the presence of the dominant direction, the turbulence is preferably a three-dimensional process.

Dimensions of vortices in a developed turbulent flow vary widely. The largest eddies are of an order of magnitude of the height of the atmospheric boundary layer, and the size of the smallest eddies has the size of the field, across which the transfer of momentum, heat and mass under the action of molecular viscosity can be carried out. Because the smaller vortices are exposed to strains created by larger vortices, there is a transfer of energy between them. In the process of deformation, the vorticity of smaller vortices increases and thus increases their energy due to the energy of larger vortices. The intensity of the turbulent diffusion is changing quite quickly, depending on the size of diffusing vortices. The strongest increase the intensity of the turbulent diffusion is in the vicinity of the underlying surface; it quickly leads to a change in the basic flow rate above it. This area of influence extends over the underlying surface, and a vertical velocity profile obtains the logarithmic form classical to turbulent flow at neutral stratification. Chaotic velocity field produced by the eddy elementary tubes leads to intensive mixing and the moving medium and, as a consequence, the exchange of momentum, heat and mass in the transverse direction. So, the generalized Newton's law to mean velocity $\bar{V}$ must lie at the base of a rheological equation for the member $\wp_{in}$ that describes the power generation, due to wind shear, of the turbulent viscosity ν_t.

The formula of the generalized Newton's law for air is:

$$P = 2\nu_t \dot{S} - \left(\frac{p}{\rho} + \frac{2}{3}\nu_t \mathrm{div}\mathbf{V}\right)\varepsilon,$$

Or, in the matrix form

$$\begin{pmatrix} p_{11} & p_{21} & p_{31} \\ p_{12} & p_{22} & p_{32} \\ p_{13} & p_{23} & p_{33} \end{pmatrix} = 2\nu_t \begin{pmatrix} \frac{\partial v_1}{\partial x_1} & \frac{1}{2}\left(\frac{\partial v_1}{\partial x_2} + \frac{\partial v_2}{\partial x_1}\right) & \frac{1}{2}\left(\frac{\partial v_1}{\partial x_3} + \frac{\partial v_3}{\partial x_1}\right) \\ \frac{1}{2}\left(\frac{\partial v_1}{\partial x_2} + \frac{\partial v_2}{\partial x_1}\right) & \frac{\partial v_2}{\partial x_2} & \frac{1}{2}\left(\frac{\partial v_2}{\partial x_3} + \frac{\partial v_3}{\partial x_2}\right) \\ \frac{1}{2}\left(\frac{\partial v_1}{\partial x_3} + \frac{\partial v_3}{\partial x_1}\right) & \frac{1}{2}\left(\frac{\partial v_2}{\partial x_3} + \frac{\partial v_3}{\partial x_2}\right) & \frac{\partial v_3}{\partial x_3} \end{pmatrix}.$$

$$-\left(\frac{p}{\rho} + \frac{2}{3}\nu_t \mathrm{div}\mathbf{V}\right)\begin{pmatrix} 1 & 0 & 0 \\ 0 & 1 & 0 \\ 0 & 0 & 1 \end{pmatrix}$$

The density distribution of the power of the internal forces is given by:

$$\begin{aligned} \wp_{in} &= \mathrm{div}\left(P\boldsymbol{V}\right) - \boldsymbol{V}\cdot\mathrm{divP} \\ &= P\cdot\mathrm{div}\left(\boldsymbol{V}\right) + \boldsymbol{V}\cdot\mathrm{div}\left(P\right) - \boldsymbol{V}\cdot\mathrm{divP} = P\cdot\mathrm{div}\left(\boldsymbol{V}\right)\end{aligned}.$$

Expanding the differential tensor (dyad) of velocity field $\mathrm{div}\left(\boldsymbol{V}\right)$ to the symmetric $\dot{S}$ and antisymmetric $\dot{A}$ parts and noting that for the antisymmetric tensor $\dot{A}$ the equality $P\cdot\dot{A} = 0$, we will find

$$\wp_{in} = P\cdot\dot{S} = 2\nu_t \dot{S}^2 + \frac{p}{\rho}\mathrm{div}\mathbf{V} - \frac{2}{3}\nu_t\left(\mathrm{div}\mathbf{V}\right)^2,$$

where the second term on the right-hand side expresses 1 sec. work of pressure, and the third term expresses the work of internal friction forces in air compression.

Thus, the generation or the generation of turbulent kinetic energy *k* is

$$\wp_{in} = 2\nu_t\dot{S}^2 - \frac{2}{3}\nu_t\left(\mathrm{div}\mathbf{V}\right)^2$$
$$= \nu_t\left\{2\left[\left(\frac{\partial v_1}{\partial x_1}\right)^2 + \left(\frac{\partial v_2}{\partial x_2}\right)^2 + \left(\frac{\partial v_3}{\partial x_3}\right)^2\right] + \left(\frac{\partial v_2}{\partial x_3} + \frac{\partial v_3}{\partial x_2}\right)^2 + \left(\frac{\partial v_3}{\partial x_1} + \frac{\partial v_1}{\partial x_3}\right)^2 + \left(\frac{\partial v_1}{\partial x_2} + \frac{\partial \mathrm{v}_2}{\partial x_1}\right)^2\right\}^2 .$$
$$-\frac{2}{3}\nu_t\left(\frac{\partial v_1}{\partial x_1} + \frac{\partial v_2}{\partial x_2} + \frac{\partial v_3}{\partial x_3}\right)^2$$

It is useful to give similar operations result in tensor notation. Recall that the tensor notation implies summation over the repeated indices (in this case, for the "*i*" and "*j*"). The result is:

$$\wp_{in} = \nu_t\left[\left(\frac{\partial v_i}{\partial x_j} + \frac{\partial v_j}{\partial x_i}\right)\frac{\partial v_i}{\partial x_j} - \frac{2}{3}\left(\frac{\partial v_i}{\partial x_i}\right)^2\right],\ i,j = 1,2,3 . \tag{6.5.26}$$

In stratified atmospheric boundary layers, the effect of mass forces on ν_t behavior arising from buoyancy or curvature of the streamlines can be accounted for using the relationship:

$$G_{T\nu_t} = c_3\frac{\nu_t}{k}\wp_T , \tag{6.5.27}$$

where $\wp_T$ is the generation of turbulent kinetic energy *k* which is determined by the Archimedes force

$$\wp_T = \frac{g_j}{\theta}\frac{\nu_t}{\sigma_T}\frac{\partial\theta}{\partial x_j} . \tag{6.5.28}$$

Calibration of constants c_2 and c_3 performed on the task of decay of turbulence in a uniform shear field, has led to the following result: $c_2 = c_3 = 0.05 \div 0.08$. Sometimes the constants c_2 and c_3 reach the values of 0.1.

Attenuation Turbulent Viscosity Due to Dissipation

The maximum energy of turbulence is usually in the region located between the areas of generation and dissipation. The physical explanation for this shift is that the turbulence energy is taken from the main flow and concentrated mainly in the large eddies. Energy transfer in the region of small-scale vortices occurs as a result of the process of crushing of large eddies into smaller eddies or stretching (diameter reduction) of the vortex tube. As a result, energy is dissipated in the small vortices under the influence of viscosity. It is physically justified that dissipation occurs mainly in small eddies, since smaller vortices are of greater velocity gradients and, hence, more significant tangent strains.

At high Reynolds numbers, number of vortices, in which dissipation of kinetic energy occurs, is significantly smaller number of vortices which receive energy directly from the main flow, and it can be assumed that the statistical characteristics of the small-scale vortices do not depend on the geometry of the main flow. Since the scale of the largest eddies does not affect the process of dissipation, the rate of their destruction is not dependent on the molecular viscosity ν. Thus, the rate of dissipation of turbulent kinetic energy always has a negative sign, and has the greatest impact on small vortices.

Consequently, at Reynolds numbers typical of dynamic meteorology, we can formulate the attenuation ε_{ν_t} of turbulent viscosity ν_t due to dissipation as follows:

$$\varepsilon_{\nu_t} = c_4 \frac{\nu_t}{k} \wp_{dis}, \tag{6.5.29}$$

where $\wp_{dis}$ characterizes the rate of conversion of kinetic energy into heat as small vortices are deformed under the influence of the viscous strains; c_4 is an empirical constant.

The rate of conversion of kinetic energy into heat is

$$\wp_{dis} = \nu_t \Omega_i \cdot \Omega_i, \tag{6.5.30}$$

where the vector of vorticity Ω_i is defined as the rotor of the vector of the average speed $\bar{\mathbf{V}}$

$$\Omega_i = e_{ijk} \frac{\partial \mathrm{v}_k}{\partial x_j}.$$

Here, as previously, the operations are shown in tensor notation. Recall that the tensor notation implies summation over the repeated indices (in this case, for the "i", "j", and "k"), e_{ijk} is an antisymmetric unity tensor of third rank, whose components change their sign when any two indices permuted; $e_{123} = 1$ and it is assumed that repeated indices do not explicitly specify summation from 1 to 3.

Energy dissipation $\wp_{dis}$ as the sum of the squares is a non-negative value that expresses the irreversibility of the transition kinetic viscous fluid flow energy into heat.

In the form of (6.5.29), the dissipative term $\wp_{dis}$ is aimed to a logarithmic region of the boundary layer. It decreases in the outer region of the boundary layer by reducing $\bar{\mathbf{V}}$ velocity gradients and causes suppression of the turbulence when to approach the underlying surface where the gradients are maximal.

Turbulence Model

Placing a common factor out of the brackets in the equations (6.5.25), (5.6.27) and (6.5.29) and substituting these equations together with (6.5.24), (5.6.26), (5.6.28) and (6.5 .30) into the equation (6.5.23), we obtain the turbulence model in the form:

$$\frac{\partial \nu_t}{\partial t}+v_j\frac{\partial \nu_t}{\partial x_j}=c_1\sum_{j=1}^{3}\frac{\partial}{\partial x_j}\left[\left(\nu+\nu_t\right)\frac{\partial \nu_t}{\partial x_j}\right]+\frac{\nu_t}{k}\left\{\begin{aligned}&c_2\nu_t\left[\left(\frac{\partial v_i}{\partial x_j}+\frac{\partial v_j}{\partial x_i}\right)\frac{\partial v_i}{\partial x_j}-\frac{2}{3}\left(\frac{\partial v_i}{\partial x_i}\right)^2\right]\\&+c_3\frac{g_j}{\theta}\frac{\nu_t}{\sigma_T}\frac{\partial \theta}{\partial x_j}-c_4\nu_t\Omega_i\cdot\Omega_i\end{aligned}\right\},\ j=1,2,3. \tag{6.5.31}$$

In the publications (Bradshaw, Ferriss, 1971; Townsend, 1956) based on the experimental results for the boundary layers and mixing layers, there is formulated the assumption that the ratio of the Reynolds strain $\tau_{ij}=-\rho\overline{v_i'v_j'}$ to turbulent fluctuations of energy k is a constant that is defined as

$$\frac{\tau_{ij}}{k}=\sqrt{C_\mu}\,. \tag{6.5.32}$$

Proportionality (6.5.32) of Reynolds strains $\tau_{ij}=-\rho\overline{v_i'v_j'}$ and energy of turbulent fluctuations k means that a change of the ratio τ_{ij}/k is much less pronounced than the change in k, what actually takes place in different currents. The exception, as shown by the experimental studies, are the regions near the axis of symmetrically-axis currents (jets, tracks, flows in channels, etc.).

The system of equations (6.2.4) and (5.6.32), assuming $\sqrt{C_\mu}=1/3$, allows finding the formula for k in the form of:

$$k=\frac{3\nu_T}{1+2\delta_{ij}}\left(\frac{\partial \mathrm{v}_i}{\partial x_j}+\frac{\partial \mathrm{v}_j}{\partial x_i}\right),\ i,j=1,2,3. \tag{6.5.33}$$

where $\delta_{ij}=1$ for $i=j$ and $\delta_{ij}=0$ for $i\neq j$.

Substituting (6.5.33) to (6.5.31), we obtain a model of turbulence "of self-diffusion" in its final form:

$$\begin{aligned}&\frac{\partial \nu_t}{\partial t}+\mathrm{v}_j\frac{\partial \nu_t}{\partial x_j}=c_1\sum_{j=1}^{3}\frac{\partial}{\partial x_j}\left[\left(\nu+\nu_t\right)\frac{\partial \nu_t}{\partial x_j}\right]\\&+\frac{\nu_t}{\dfrac{3}{1+2\delta_{ij}\left(\dfrac{\partial v_i}{\partial x_j}+\dfrac{\partial v_j}{\partial x_i}\right)}}\left\{c_2\left[\left(\frac{\partial v_i}{\partial x_j}+\frac{\partial v_j}{\partial x_i}\right)\frac{\partial v_i}{\partial x_j}-\frac{2}{3}\left(\frac{\partial v_i}{\partial x_i}\right)^2\right]+c_3\frac{g_j}{\theta\sigma_T}\frac{\partial \theta}{\partial x_j}-c_4\Omega_i\cdot\Omega_i\right\},\ j=1,2,3.\end{aligned} \tag{6.5.34}$$

Calibration constants c_4 and σ_T has led to the following result: $c_4=\sigma_T=1.0$.

To illustrate the accuracy of the "self-diffusion" turbulence model (6.5.34) the results of calculation of the main characteristics of the boundary layer (dimensionless velocity profiles $\overline{\mathrm{v}}=v_1/v_{1\infty}$ and dimensionless turbulent viscosity $\overline{\nu}=0.002\nu_t/\nu$) are compared in Figure 2 to experimental results

Figure 2. Comparison of the results of calculations of the turbulent boundary layer on a plate with the experimental data

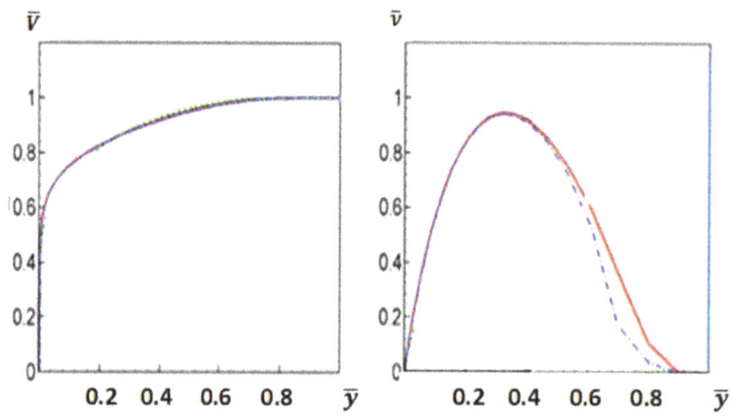

(Klebanoff, 1955). As one can see from the figure, the solution on the basis of (6.5.34) is almost identical with the experimental data.

Currently, the most complete version of the existing models of turbulence with a single equation (6.5.23) is the model described in (Spalart, Allmaras, 1992), in which instead of using eddy viscosity ν_t, the relative variable $\tilde{\nu} = \nu_t / f_{v1}$ is used, where

$$f_{v1} = \bar{\tilde{\nu}}^{-3} / \left(\bar{\tilde{\nu}}^{-3} + C_{v1}^3\right), \bar{\tilde{\nu}} = \tilde{\nu} / \nu, C_{v1} = 0.71.$$

Diffusion of turbulent viscosity $\tilde{\nu}$ and is given by

$$D_{\tilde{\nu}} = \frac{1}{\sigma}\frac{\partial}{\partial x_j}\left[\left(\nu + \tilde{\nu}\right)\frac{\partial \tilde{\nu}}{\partial x_j}\right], \ \sigma = \frac{2}{3};$$

or

$$D_{\tilde{\nu}} = \frac{1}{\sigma}\left\{\frac{\partial}{\partial x_j}\left[\left(\nu + \tilde{\nu}\right)\frac{\partial \tilde{\nu}}{\partial x_j}\right] + c_{b2}\frac{\partial \tilde{\nu}}{\partial x_j}\frac{\partial \tilde{\nu}}{\partial x_j}\right\}, \ c_{b2} = 0.622,$$

because turbulent viscosity is not a fundamental physical quantity, and it is not required to satisfy the laws of conservation for it.

In the term describing the generation of $\tilde{\nu}$, the speed derivative is replaced by the value of the invariant tensor $G_{\nu_t} = c_{b1}\tilde{S}\tilde{\nu}$,, where

$$\tilde{S} = S + \tilde{\nu} f_{v2} / \left(\chi^2 d^2\right),$$

$$S = \sqrt{2\Omega_{ij}\Omega_{ij}},$$

$$\Omega_{ij} = \left(\partial v_i / \partial x_j - \partial v_j / \partial x_i\right) / 2,$$

$\chi = 0.41$, d is a distance from the solid surface, $f_{v2} = 1 - \bar{\tilde{\nu}} / \left(1 + \bar{\tilde{\nu}} f_{v1}\right)$.

The dissipative term ε_{ν_t} in the right side of (5.6.32) is defined by the expression

$$\varepsilon_{\nu_t} = \left(c_{w1} f_w - \frac{c_{b1}}{\chi^2} f_{t2}\right)\left(\frac{\tilde{\nu}}{d_w}\right)^2,$$

in which d_w is a height of the first calculated level above the underlying surface; k is the Karman constant;

$$f_w = \xi \cdot \left(\frac{1 + c_{w3}^6}{\xi^6 + c_{w3}^6}\right)^{1/6}, \quad \xi = \varepsilon\left[1 + c_{w2}\left(\varepsilon^5 - 1\right)\right], \quad \varepsilon = \frac{\tilde{\nu}}{\tilde{S}\chi^2 d_w^2}.$$

The function f_{t2}, providing suppression in the boundary layer of so-called "spontaneous" or "numerical" transition from laminar to turbulent motion is given by

$$f_{t2} = c_{t3} \cdot \exp\left(-c_{t4}\chi^2\right).$$

The most complete model (Spalart, Allmaras, 1992) contains the term $f_{t1}\Delta U^2$ in the right-hand side of (6.5.32) that is designed to initiate the laminar-turbulent transition in predetermined point, and is calculated using the formula:

$$f_{t1} = c_{t1}\xi_t \cdot \exp\left[-c_{t2}\frac{\varsigma_t^2}{\Delta U^2}\left(d_w^2 + \xi_t d_t^2\right)\right],$$

$$\xi_t = \min\left(0.1;\ \frac{\Delta U}{\varpi_t} l_t\right), \quad \varpi_t = \omega_{W,trip},$$

$$d_t^2 = \left(x - x_{trip}\right)^2 + \left(y - y_{trip}\right)^2 + \left(z - z_{trip}\right)^2,$$

$$\Delta U = \left|V - V_{trip}\right|.$$

Here, the subscript "trip" refers to the values determined at the point on the transition line which is located at a minimum distance from the considered point of the flow, $l_t = \left(\Delta x^2 + \Delta y^2\right)^{1/2}$ is a cell diagonal of the grid at the streamlined surface at this point, and the value ΔU is the module of the velocity vector difference at the considered point and the transition point (in the case of a fixed underlying surface, $\Delta U = \left|V\right|$.

The empirical constants of the model are:

$$\sigma = 2/3;\ \chi = 0,41;\ c_{b1} = 0.1355;\ c_{b2} = 0.622;$$

$$c_{w1} = \frac{c_{b1}}{\chi^2} + \frac{1 + c_{b2}}{\sigma};\ c_{w2} = 0.3;\ c_{w3} = 2;$$

$$c_{t1} = 1;\ c_{t2} = 2;\ c_{t3} = 1.2;\ c_{t4} = 0.5.$$

The boundary conditions for the equation (6.5.32) are defined as follows. On the solid underlying surface, the modified turbulent viscosity, as well as genuine one, must vanish $\tilde{\nu} = 0$. On the input side of the border areas of the computational domain, there must be given the condition of the first kind for $\tilde{\nu}$, and on output sides, the value of $\tilde{\nu}$ is extrapolated to the boundary of the internal points of the domain.

When calculating the air movements under the inversion layer, the particular choice of the value of $\tilde{\nu}$, defined at the outer edge of the inversion layer, is as follows. If the point (line) of the transition from laminar to turbulent motion is known from the preliminary calculation of the vertical temperature profile, then $\tilde{\nu}$ at the outer edge of the inversion layer of the computational domain is to be equal to a small value (about 10^{-3} of the molecular viscosity), the specific choice of which does not play a role. In this case due to the presence in the equation (6.5.32) "trip term", the transition from laminar to turbulent flow occurs in the small neighborhood of the given point (line) of transition.

In those cases, when the position of the point (line) of the transition is not known beforehand, but an inversion layer is still present so that the portion of laminar boundary layer on the underlying surface can be neglected, it is possible to use the so-called "fully turbulent" approach to the description of the transition. In this approach, at the outer edge of the inversion layer of the computational domain, a high enough level of $\tilde{\nu}$ ($\tilde{\nu} = \left(1 \div 5\right)\nu$) is established. The equation (6.5.32) provides a fast (for several grid steps), the transition to a developed turbulent flow in the boundary layer.

Finally, we note that for the "diagnostics" of the state of the boundary layer, it is helpful to use so-called turbulence index, the value of which in the developed turbulent boundary layer is close to 1. The turbulence index is calculated by the formula of (Spalart, Allmaras, 1992)

$$i = \frac{1}{\chi v^*}\frac{\partial \tilde{\nu}}{\partial n},$$

where the "friction velocity" $v^* = \sqrt{\tau_w / \rho}$ in the boundary layer on the wall in the general case of three-dimensional flow is calculated as $v^* = \sqrt{\nu\omega}$ (ω is a vorticity unit at a given point on the underlying surface).

When modeling a barrage gate (vortex with horizontal axis) and a tornado (vortex with a vertical axis), the curvature of the current flow and the rotation of the lines have a significant effect on the turbulence characteristics and can lead both to its significant intensification and to suppress. Within the framework of this model, it is necessary to introduce specific amendments for their accounting. The most successful of these amendments is, apparently, the amendment of Spalart-Shur (Spalart, Shur, 1997; Shur M. et al., 2000; Spalart, 2000), that takes into account the common nature of the curvature of the effects of the current lines and rotation and is applicable in general to any linear model.

In the framework of the model (6.5.32) the generation member G_{ν_t} is modified by multiplying by an empirical function

$$f_{r1} = \frac{2r^*\left(1 + c_{r1}\right)}{1 + r^*}\left[1 - c_{r3}\text{arctg}\left(c_{r2}\tilde{r}\right)\right] - c_{r1},$$

which depends on two parameters r^* and $\tilde{r}$, which are, respectively, measures of the curvature of the lines of current and rotation. The expressions for these parameters are as follows:

$$r^* = \left|\frac{S}{\omega}\right|, \ \tilde{r} = \frac{2\omega_{ik}S_{jk}\left(DS / Dt\right)_{ij}}{D^4}.$$

Here, $\left(DS / Dt\right)_{ij}$ denotes the corresponding component of the total time derivative of the strain rate tensor, and the value of D is given by:

$$D^2 = \frac{1}{2}\left(S^2 + \omega^2\right),$$

where

$$S = \left(2S_{ij}S_{ij}\right)^{1/2}, \ S_{ij} = \frac{1}{2}\left(\frac{\partial u_i}{\partial x_j} + \frac{\partial u_j}{\partial x_i}\right),$$

$$\omega = \left(2\omega_{ij}\omega_{ij}\right)^{1/2}, \ \omega_{ij} = \frac{1}{2}\left(\frac{\partial u_i}{\partial x_j} - \frac{\partial u_j}{\partial x_i}\right).$$

Three new constants in the last expressions have the following meanings:

$$c_{r1} = 1, \ c_{r2} = 12, \ c_{r3} = 1.$$

THE COEFFICIENT OF HORIZONTAL TURBULENT DIFFUSION

All turbulence models described above are based on Reynolds equations of motion and heat and mass transfer for the averaged quantities. Essentially, Reynolds equations themselves are a model of turbulent flows, as some of the characteristic properties of the flow field can be omitted when using the averaging procedure. For example, it is known that the basis of planetary processes in the atmosphere is a large-scale orderly movement. Therefore, macroscale hydrodynamic models of the atmosphere motion usually use an averaging procedure on three levels by dividing the flow field in the averaged (basic), large-scale turbulent motion, and processes of small length scales. Small-scale turbulent motion is regarded as the transfer of the main course and large vortices isotropic turbulence, i.e., it is using one of the described above the turbulence model. A large-scale turbulent motion is modeled by simple deterministic models describing the behavior of large eddies in a turbulent flow, and the probability at this level is accounted by postulating the chance of occurrence of large vortices in time and their location in space.

At the macroscale modeling of currents in the atmosphere, the closure of averaged over the ensemble Navier-Stokes equations do not provide the information applicable to the averages of horizontal. This causes a problem of numerical implementation of the model, since, mainly because of the limited capacity of existing computer technology, the distance between the grid points horizontally are unacceptably high in comparison with the turbulence scales. In addition, the macro-scale models feature hydrostatic approximation, which eliminates the possibility of direct determination of the characteristics of small-scale turbulence. As shown by evaluation studies, customarily used horizontal turbulent diffusion coefficients are several orders of magnitude more of those coefficients calculated by a model of isotropic turbulence. That is why in modern hydrodynamic models of macroscale circulation, the rates of change of momentum, heat and humidity caused by turbulent diffusion, are presented separately for the horizontal and vertical directions. Horizontal diffusion is introduced to describe clearly non-calculated directly horizontal advection processes of sub-grid scale and blanking grid step scale noise.

In the work (Smagorinsky, 1963), the tensor components of additional strains $-v_1'^2, -v_1'v_2', -v_2'^2$ are parametrized by expressions

$$-v_1'^2 = -v_2'^2 = k_g D_c,\ -v_1'v_2' = k_g D_d, \tag{6.5.35}$$

where k_g is a coefficient of horizontal turbulent diffusion, D_c and D_d are strain rates of the averaged flow having the following form:

- In Cartesian coordinate system:

$$D_c = \frac{\partial v_1}{\partial x_1} - \frac{\partial v_2}{\partial x_2},\ D_d = \frac{\partial v_1}{\partial x_2} + \frac{\partial v_2}{\partial x_1}; \tag{6.5.36}$$

- In spherical coordinates:

$$D_c = \frac{1}{r\cos\varphi}\frac{\partial v_1}{\partial \lambda} - \frac{\cos\varphi}{r}\frac{\partial}{\partial \varphi}\left(\frac{v_2}{\cos\varphi}\right),$$

$$D_d = \frac{1}{r\cos\varphi}\frac{\partial v_2}{\partial \lambda} + \frac{\cos\varphi}{r}\frac{\partial}{\partial \varphi}\left(\frac{v_1}{\cos\varphi}\right). \tag{6.5.37}$$

The coefficient of horizontal turbulent diffusion of (5.6.23) can be chosen as

$$k_g = \left(\frac{\chi}{\sqrt{2}}\Delta s\right)^2 |D|,\ |D| = \sqrt{D_c^2 + D_d^2}, \tag{6.5.38}$$

where $\chi = 0.4$ is Karman constant, $\Delta s = \sqrt{\Delta x_1^2 + \Delta x_2^2}$ or $\Delta s = \sqrt{\Delta\lambda^2 + \Delta\varphi^2}$ (depending on the coordinate system) are the parameters of the finite grid.

SUMMARY

This chapter studies the motion of a viscous fluid, very common in the atmosphere, that is called the *turbulent motion*. A turbulence arises from external disturbances and irregularities in the flow of a viscous fluid because of a chaotic variety of vortex tubes. These tubes are stretched in certain directions under the influence of deformations of the main flow and move in random directions in conjunction with each other. Vortex lines constantly change direction, intertwist and are the most confused near the wall. Thus, despite a dominant direction, turbulence is essentially a three-dimensional process.

Chaotic velocity field produced by elementary eddy tubes leads to intensive mixing and the moving air masses and, therefore, the exchange of momentum, heat, and mass in the transverse direction. According to modern concepts, the turbulent processes in the atmosphere obey classical Navier-Stokes and convective diffusion. The chapter describes several methods of modeling turbulent processes in the atmosphere, including direct methods for the numerical simulation of turbulent flows, large eddy simulation method, methods of statistical and spectral analysis, methods based on Reynolds stress equations of the second order, and methods based on the concept of turbulent viscosity.

REFERENCES

Andren, A., Brown, A. R., Graf, J., Mason, P. J., & Moeng, C.-H. (1994). Large-eddy simulation of a neutrally stratified boundary layer: A comparison of four computer codes. *Quarterly Journal of the Royal Meteorological Society*, *120*(520), 1457–1484. doi:10.1002/qj.49712052003

Betchov, R. (1975). Numerical simulation of isotropic turbulence. *Physics of Fluids*, *18*(10), 1230–1236. doi:10.1063/1.861006

Blackadar, A. K. (1962). The vertical distribution of wind and turbulence exchange in a neutral atmosphere. *Journal of Geophysical Research*, *67*(8), 3095–3102. doi:10.1029/JZ067i008p03095

Boussinesq, J. (1877). Essai sur la théorie des eaux courantes. *Mémoires présentés par divers savants à l'Académie des Sciences*, *23*(1), 1-680.

Bradshaw, P., & Ferriss, D. H. (1971). Calculation of boundary layer development using the turbulent energy equation: Compressible flow on adiabatic walls. *Journal of Fluid Mechanics*, *46*(1), 83–110. doi:10.1017/S0022112071000417

Byzova, I. L., Ivanov, V. N., & Garger, E. K. (1989). *The turbulence in the boundary layer of the atmosphere*. Leningrad: Gidrometeoizdat. (in Russian)

Computational Fluid Dynamics. (1980). Washington, DC.: Hemisphere Publishing Corp.

Cooley, J. W., & Tukey, J. W. (1965). An algorithm for the machine calculation of complex Fourier series. *Mathematics of Computation*, *19*(90), 297–301. doi:10.1090/S0025-5718-1965-0178586-1

Donaldson, C. (1973) *Atmospheric turbulence and the dispersal of atmospheric pollutants*. Boston: Science Press.

Durbin, P. A., Mansour, N. N., & Yang, Z. (1994). Eddy viscosity transport model for turbulent flow. *Physics of Fluids*, *6*(2), 1007–1015. doi:10.1063/1.868334

Esau, I. (2003). The Coriolis effect on coherent structures in planetary boundary layers. *Journal of Turbulence*, *4*(1), 1–19.

Esau, I. (2004). Simulation of Ekman boundary layers by large eddy model with dynamic mixed subfilter closure. *Environmental Fluid Mechanics*, *4*(3), 273–303. doi:10.1023/B:EFMC.0000024236.38450.8d

Ferziger, J. H. (1996) Recent Advances in Large Eddy Simulation. In Engineering Turbulence Modelling and Experiments (vol. 3, pp. 163-175). Amsterdam: Elsevier Science. doi:10.1016/B978-0-444-82463-9.50022-8

Frost, E.D.W., & Mould, T. (1980). *Turbulence. Principles and Applications*. Moscow: Mir. (in Russian)

Galperin, B., Kantha, L., Hassid, S., & Rosati, A. (1988). A quasi-equilibrium turbulent energy model for geophysical flows. *Journal of the Atmospheric Sciences*, *45*(1), 55–62. doi:10.1175/1520-0469(1988)045<0055:AQETEM>2.0.CO;2

Garnier, E., Adams, N., & Sagaut, P. (2009). *Large eddy simulation for compressible flows*. Groningen: Springer Netherlands. doi:10.1007/978-90-481-2819-8

Glazunov, A.V. (2006). Modeling neutral stratified turbulent air flow over a rough surface. *Math. Russian Academy of Sciences. The physics of the atmosphere and ocean, 42*(3), 307-325.

Gulyaev, A. N., Kozlov, V. E., & Secundov, A. N. (1993). To creating a universal one-parameter model for the turbulent viscosity. [in Russian]. *Fluid Dynamics*, *4*, 69–81.

Karman, Th. (1921). Uber laminare und turbuiente Reibund. *Zeitschr. f. And Math. und Mech.*, *1*, 233-252.

Klebanoff, P. S. (1955). *Characteristics of turbulence in a boundary layer with zero pressure gradient.* Tech. Rep. 1247. Washington, DC: NACA.

Kolmogorov, A. N. (1941). The local structure of turbulence in an incompressible fluid at very high Reynolds numbers. In Doklady Academii Nauk SSSR (vol. 30). Moskow: AN SSSR. (in Russian)

Moser, R., Kim, J., & Mansour, N. N. (1999). Direct Numerical Simulation of Turbulent Channel Flow up to Reτ = 590. *Physics of Fluids*, *11*(4), 943–945. doi:10.1063/1.869966

Nee, V. W., & Kovasznay, L. S. G. (1969). Simple phenomenological theory of turbulent, shear flows. *Physics of Fluids*, *12*(3), 473–484. doi:10.1063/1.1692510

Orszag, S. A., & Patterson, G. S. (1972). Numerical simulation of three-dimension homogeneous isotropic turbulence. *Physical Review Letters*, *28*(2), 76–79. doi:10.1103/PhysRevLett.28.76

Piomelli, U., Zang, T. A., Speziale, C. G., & Hussaini, M. Y. (1990). On the large-eddy simulation of transitional wall-bounded flows. *Physics of Fluids. A, Fluid Dynamics*, *2*(2), 257–265. doi:10.1063/1.857774

Prusov, V. A., & Doroshenko, A. Yu. (2006). *Modeling Natural and Technogenic Atmospheric Processes*. Kyiv: Naukova Dumka. (in Ukrainian)

Prusov, V. A., & Romaniuk, A. P. (1998). The mathematical model of turbulence for stratified media. *Tech. Rep. of Ukrainian Research Hydro-Meteorology Institute*, *246*, 35–45. (in Ukrainian)

Prusov, V. A., & Sologub, T. A. (2011). Model of vertical column horizontally homogeneous atmosphere. *Tech. Rep. of Ukrainian Research Hydro-Meteorology Institute*, *259*, 61–73. (in Ukrainian)

Reynolds, O. (1895). On the dynamical theory of incompressible viscous fluids and the determination of the criterion. *Phil.Trans. Roy. Soc. A*, *186*(0), 123–184. doi:10.1098/rsta.1895.0004

Rodi, W. (1978). *Turbulence models and their application in hydraulics - a state of the art review. Report No. SFB 80/T/127*. University of Karlsruhe.

Rotta, J. (1951). Statistische Theorie nichthomogener Turbulenz. *Zeitschrift fur Physik*, *129*(6), 547–572. doi:10.1007/BF01330059

Schumann, U., & Patterson, G. S. (1978). Numerical study of the return of axisymmetric turbulence to isotropy. *Journal of Fluid Mechanics*, *88*, 685–709. doi:10.1017/S0022112078002347

Shur, M. L., Spalart, P. R., Strelets, M. Kh., & Travin, A. K. (1996) Navier–Stokes simulation of shedding turbulent flow past a circular cylinder and a cylinder with a backward splitter plate. In G.A. Desideri, C. Hirsch, P. Le Tallec, M. Pandolfi, & J. Periaux (Eds.), *Third ECCOMAS CFD Conference*, Paris, September, 1996. Chichester, UK: John Wiley & Sons.

Shur, M. L., Strelets, M. K., Travin, A. K., & Spalart, P. R. (2000). Turbulence modeling in rotating and curved channels: Assessment of the Spalart-Shur correction term. *AIAA Journal*, *38*(5), 784–792. doi:10.2514/2.1058

Smagorinsky, J. (1963). General circulation experiments with the primitive equations.1: The basic experiment. *Monthly Weather Review*, *91*(3), 99–164. doi:10.1175/1520-0493(1963)091<0099:GCE WTP>2.3.CO;2

Spalart, P. R. (2000). Strategies for turbulence modeling and simulations. *International Journal of Heat and Fluid Flow*, *21*(3), 252–263. doi:10.1016/S0142-727X(00)00007-2

Spalart, P. R., & Allmaras, S. R. (1992). A one-equation turbulence model for aerodynamic flows. *AIAA conference Paper 92-0439*. Reno, NV: AIAA.

Spalart, P. R., & Shur, M. L. (1997). On the sensitization of simple turbulence models to rotation and curvature. *Aerospace Science and Technology*, *1*(5), 297–302. doi:10.1016/S1270-9638(97)90051-1

Townsend, A. A. (1956). *The structure of turbulent shear flow*. Cambridge, UK: Cambridge University Press.

Williams, G. P. (1969). Numerical integration of the three-dimensional Navier-Stokes equations for incompressible flow. *Journal of Fluid Mechanics*, *37*(04), 727–750. doi:10.1017/S002211206900084X

Zilitinkevich, S., & Esau, I. (2002). On integral measures of the neutral barotropic planetary boundary layer. *Boundary-Layer Meteorology*, *104*(3), 371–379. doi:10.1023/A:1016540808958

Zilitinkevich, S., & Esau, I. (2003). The effect of baroclinicity on the depth of neutral and stable planetary boundary layers. *Quarterly Journal of the Royal Meteorological Society*, *129*(595), 3339–3356. doi:10.1256/qj.02.94

Zilitinkevich, S. S., & Esau, I. N. (2005). Resistance and heat transfer laws for stable and neutral planetary boundary layers: Old theory, advanced and re-evaluated. *Quarterly Journal of the Royal Meteorological Society*, *131*(609), 1863–1892. doi:10.1256/qj.04.143

Chapter 7
Parameterization of Subgrid-Scale Processes

ABSTRACT

In Chapters 4 and 5, we considered a system hydrodynamics equations and boundary conditions that constitute the mathematical basis of the circulation models of the atmosphere of a scale. They contain terms describing sources (sinks) of mass and energy involved in the phase transformation of atmospheric moisture and radiative processes in the system atmosphere – the Earth. Direct inclusion of these microscale and mesoscale processes in the atmospheric circulation model is inappropriate as: 1) this leads to an increase in the total number of grid points in the decision area, and 2) not all these processes can be described by precise differential equations.

PARAMETERIZATION OF THE INTERACTION OF THE ATMOSPHERE WITH THE UNDERLYING SURFACE

The interaction of the atmosphere with the underlying surface occurs through the physical processes of the boundary layer. Regarding the atmosphere, the latter is a "sink" of the momentum and the "source" of heat and moisture. At the same time, atmospheric processes themselves significantly affect the structure of the boundary layer. However, the models take into account its properties only with a certain degree of approximation as possibilities of description of the boundary layer in the atmospheric circulation models are limited. These options depend on the time and spatial scales of the simulated atmospheric processes and phenomena.

In the simulation of the general circulation of the atmosphere and long-range weather forecasting, it is important to consider the dissipation of the average amount of motion in the boundary layer, as well as heat and moisture flow from the planetary boundary layer. At the same time, for the calculation of dynamic processes in mesoscale systems, the boundary layer associated with the topography and thermal heterogeneity of the underlying horizontal surface, and the spatial variability of surface turbulence play especially an important role. Therefore, the creation of numerical models of atmospheric circulation of various types of different resolution in both vertical and horizontal makes demands on adequate

DOI: 10.4018/978-1-5225-2636-0.ch007

description of the boundary layer. Unfortunately, there is no well-founded answer to the question, what are details necessary to consider the boundary layer in the models, and what is a sensitivity of different scale to a method for its disclosure.

The main way to account the interaction of the atmosphere with the underlying surface in the given numerical model is implemented based on the method of parameterization of the planetary boundary layer (PBL) of the atmosphere.

Small-scale turbulence may be the most significant in layers with significant gradients of wind speed, temperature, and humidity, i.e., in the atmospheric boundary layer. The height *h* in the case of PBL neutral stratification can be approximately estimated by equating the orders of magnitude of the forces of turbulent viscosity and the Coriolis force,

$$O\left[\frac{\partial}{\partial x_3}\left(\nu\frac{\partial v_1}{\partial x_3}\right)\right]=O\left[2\mathbf{\Omega} v_2\sin\varphi\right]=O\left[lv_2\right].$$

It follows that $h\approx\sqrt{\nu/l}$. More detailed assessment based on the definition of *h* as a size of the layer in which the direction of the actual and geostrophic wind is different, lead to $h\approx 1$km.

Designate the vertical turbulent flows of momentum of the horizontal components of the wind speed (v_1,v_2) by F_λ and F_φ, and turbulent flow of heat and moisture by F_θ and F_q respectively, i.e.,

$$F_\lambda=\left[\frac{\partial}{\partial x_3}\left(\nu\frac{\partial v_1}{\partial x_3}\right)\right], F_\varphi=\left[\frac{\partial}{\partial x_3}\left(\nu\frac{\partial v_2}{\partial x_3}\right)\right],$$

$$F_\theta=\left[\frac{\partial}{\partial x_3}\left(\frac{\nu}{\xi_\theta}\frac{\partial \theta}{\partial x_3}\right)\right],_q=\left[\frac{\partial}{\partial x_3}\left(\frac{\nu}{\xi_q}\frac{\partial q}{\partial x_3}\right)\right].$$

We confine ourselves to the PBL. We assume that $\xi_\theta=\xi_q=1$, and replace the values $\nu\partial v_1/\partial x_3$, $\nu\partial v_2/\partial x_3$, $\nu\partial v_1/\partial x_3$, $\nu\partial\theta/\partial x_3$, $\nu\partial q/\partial x_3$ by finite differences with emphasis on their values when $x_3=0$ and $x_3=h$. The result is:

$$F_\lambda=\frac{1}{h}\left[\left(\nu\frac{\partial v_1}{\partial x_3}\right)_h-\left(\nu\frac{\partial v_1}{\partial x_3}\right)_0\right], F_\varphi=\frac{1}{h}\left[\left(\nu\frac{\partial v_2}{\partial x_3}\right)_h-\left(\nu\frac{\partial v_2}{\partial x_3}\right)_0\right],$$

$$F_\theta=\frac{1}{h}\left[\left(\nu\frac{\partial \theta}{\partial x_3}\right)_h-\left(\nu\frac{\partial \theta}{\partial x_3}\right)_0\right], F_q=\frac{1}{h}\left[\left(\nu\frac{\partial q}{\partial x_3}\right)_h-\left(\nu\frac{\partial q}{\partial x_3}\right)_0\right].$$

Assuming that at $x_3=h$, i.e. at the upper limit of PBL, vertical flows of momentum, heat and moisture are zero, we find:

$$F_{\lambda} = -\frac{1}{h}\left(\nu\frac{\partial v_1}{\partial x_3}\right)_0 = -\frac{\tau_{01}}{\rho h}, F_{\varphi} = -\frac{1}{h}\left(\nu\frac{\partial v_2}{\partial x_3}\right)_0 = -\frac{\tau_{02}}{\rho h},$$

$$F_{\theta} = -\frac{c_p}{h}\left(\nu\frac{\partial \theta}{\partial x_3}\right)_0 = \frac{H_0}{\rho h}, F_q = -\frac{1}{h}\left(\nu\frac{\partial q}{\partial x_3}\right)_0 = \frac{Q_0}{\rho h}. \quad (7.1.1)$$

Thus, to determine the quantities $F_{\lambda}, F_{\varphi}, F_{\theta}, F_q$ in PBL, it is sufficient to determine the turbulent vertical streams $\tau_{01}/(\rho h)$, $\tau_{02}/(\rho h)$, $H_0/(\rho h)$, and $Q_0/(\rho h)$ at the earth's surface. For these flows, following relations of parameters are widespread:

$$\tau_{01} = -\rho C|U|v_1\big|_{x_3=h_1}, \tau_{02} = -\rho C|U|v_2\big|_{x_3=h_1},$$

$$H_0 = -\rho C_h^d|U|\big|_{x_3=h_1}\left(\theta_{h_1} - \theta_0\right), Q_0 = -\rho C_q^d|U|f\big|_{x_3=h_1}, \quad (7.1.2)$$

where $U = \left(v_1^2 + v_2^2\right)^{1/2}$;*f* is the relative humidity; C, C_h^d and C_q^d are dimensionless coefficients gleaned empirically. The index h_1 indicates that the corresponding values are calculated at the upper sub-layer of permanent flows (in modern operational weather prediction models, $h_1 \in [50, 70]$ m). The sub-layer of permanent flows is often called in the literature the surface boundary layer. The values of the index "0" refers to the Earth's surface $x_3 = z_0$ with roughness parameter z_0, ensuring the fulfillment of the condition $U\big|_{x_3=z_0} = 0$.

At neutral stratification in the surface layer, there is a logarithmic wind speed profile

$$U(x_3) = \frac{\mathrm{v}^*}{\chi}\ln\frac{x_3}{z_0},$$

where $\mathrm{v}^* = \sqrt{\tau_0/\rho}$ is dynamic speed, χ = 0.4 is the Karman constant.

In fact, the logarithmic profile can be applied beginning from a height equal to several z_0, and below of this level, the air is exposed to local pressure forces occurring near the individual elements of surface roughness. For large values of z_0 (tall grass, forest), they introduce displacement height *d*, i.e., the level above which it can be assumed the wind velocity being of logarithmic profile

$$U(x_3) = \frac{\mathrm{v}^*}{\chi}\ln\frac{x_3 - d}{z_0}.$$

When stratification is different from neutral, according to the similarity theory of A. Monin and A. Obukhov (Obukhov, 1988) for the surface boundary layer, they introduce the dimensionless height

$\zeta = x_3 / L^*$, called stratification parameter, where $L^* = \mathrm{v}^* / \left(\chi^2 \beta T^*\right)$ is the length scale; $T^* = -H_0 / \left(\rho \chi c_p \mathrm{v}^*\right)$, $Q^* = -Q_0 / \left(\rho \chi \mathrm{v}^*\right)$ are the scales, respectively, of temperature and humidity; β is a constant.

It is assumed that the dimensionless wind velocity U / v^*, temperature θ / T^*, and humidity of air q / Q^* are universal features of the single parameter ζ, called the parameter of static stability. According to the known values of U, θ and q at the level z_0, values $U\left(x_3\right), \theta\left(x_3\right)$ and $q\left(x_3\right)$ in the surface boundary layer can be found from the relations:

$$U\left(x_3\right) = \frac{\mathrm{v}^*}{\chi}\left[\phi_U\left(\zeta\right) - \phi_U\left(\zeta_0\right)\right],$$

$$\theta\left(x_3\right) = \theta\left(z_0\right) T^*\left[\phi_\theta\left(\zeta\right) - \phi_\theta\left(\zeta_0\right)\right],$$

$$q\left(x_3\right) = q\left(z_0\right) Q^*\left[\phi_q\left(\zeta\right) - \phi_q\left(\zeta_0\right)\right], \tag{7.1.3}$$

where the dimensionless variables ζ and ζ_0 are appropriate to the height of x_3 and z_0; ϕ_U, ϕ_θ, and ϕ_q are the universal function for wind speed, temperature, and humidity, respectively. For these functions, based on the processing of instrumental observation data, there is obtained a variety of approximation, while it is assumed that $\phi_U = \phi_\theta = \phi_q$. The functions that ϕ_U and ϕ_θ are frequently used in the form:

$$\phi_U = \begin{cases} 1 + \beta_U \zeta, & \zeta > 0, \\ \left(1 - 15\zeta\right)^{-1/4}, & \zeta < 0, \end{cases}$$

$$\phi_\theta = \begin{cases} 0.74 + \beta_\theta \zeta, & \zeta > 0, \\ \left(1 - 9\zeta\right)^{-1/2}, & \zeta < 0, \end{cases} \tag{7.1.4}$$

The values of constants β_U and β_θ in various models of the atmosphere are taken in the range from 4 to 10; it is often used a value of about 5.

THERMAL CONDITIONS OF THE UNDERLYING SURFACE

A heat mode of the Earth surface is influenced by the result of heat flow to the surface which depends on the characteristics of the atmosphere, soil, and the underlying surface. Solar radiation arriving on the Earth surface is partially absorbed in the thin layer of soil and partially reflected into the atmosphere. In addition to solar radiation, the Earth's surface receives heat in the form of infrared radiation from the atmosphere and lose heat by the thermal radiation. Along with radiation, the heat balance of the underlying surface is influenced by the turbulent heat flows and moisture (heat expenditure for evaporation) on the boundary of the atmosphere-land, as well as the transfer of heat in the soil. The basic equation for calculating the temperature of the soil is the heat equation:

$$\frac{\partial T_{so}}{\partial x_3} = \frac{1}{\rho c}\frac{\partial}{\partial x_3}\left(\lambda \frac{\partial T_{so}}{\partial x_3}\right). \tag{7.1.5}$$

where T_{so} is the temperature of the upper layer of the soil; ρ is the density of soil; c is its specific heat; λ is a coefficient of thermal conductivity of the soil.

As the upper boundary condition (at the upper limit of the soil), there are taken radiation flows and turbulent flows of explicit and latent heat. The bottom boundary condition for the equation of heat conduction in the soil is given as climatic temperature in the lower layer; this temperature is assumed to be constant in time, but different in different nodes of the grid.

To solve the equation (7.1.5) used levels

$$x_{3k} = 0.01 \cdot 3^{k-1}, k = 1,2,\ldots,8,$$

where k is a number of layers in the soil. At the upper boundary of the soil, heat flow $\lambda \partial T_{so} / \partial x_3$ is calculated as the sum of the radiation balance and turbulent flows of explicit and latent heat. Then the equation (7.1.5) for the upper layer of soil can be written:

$$\left(\frac{\partial T_{so}}{\partial x_3}\right)_{k=1} = \frac{1}{\rho c \cdot x_3}\left[\frac{\left(T_{so}\right)_{k=2} - \left(T_{so}\right)_{k=1}}{\left(x_3\right)_{k=2} - \left(x_3\right)_{k=1}} + G_{sfc}\right], \tag{7.1.6}$$

$$G_{sfc} = H_{sfc} + F_{qsfc} + Q_{rad}. \tag{7.1.7}$$

Here, H_{sfc} is the explicit heat flow; $F_{q\,sfc}$ is the latent heat flow; Q_{rad} is the radiation balance of the soil surface.

Under the influence of the turbulent heat exchange with the atmosphere, the soil loses (receives) the amount of sensible heat equal to

$$H_{sfc} = \rho c_p C_h^d \left|V_h\right| \left(T_a - T_{sfc}\right), \tag{7.1.8}$$

where ρ is air density; C_h^d is a coefficient of turbulent heat exchange; $\left|V_h\right|$ is the module of wind speed; $\left(T_a - T_{sfc}\right)$ is a difference between the air temperature in the lower estimated level of the model of the atmosphere and surface temperature; c_p is the heat capacity of air at constant pressure.

The flow of latent heat (heat expenditure for evaporation) is defined as the total evaporation:

$$F_{qsfc} = \rho C_q^d \left|V_h\right| \left(q - q_{sfc}\right), \tag{7.1.9}$$

where C_q^d is a coefficient of turbulent water exchange; $\left(q - q_{sfc}\right)$ is a difference between the humidity on the lower estimated level of the model of the atmosphere and humidity on the surface.

The incoming part of the radiation balance of the surface includes absorbed direct and diffused solar (shortwave) radiation plus a heat flow from the atmosphere, and the expenditure part is the thermal radiation of the Earth's surface. The method of calculating the radiation balance is based on the solution of radiative transfer equation in the two-stream approach, considering the effects of scattering, absorption and emission of radiation by atmospheric gasses, clouds and aerosols (Ritter, Geleyn, 1992). The basic system of equations for the radiation flow in describing the transfer of solar and longwave radiation in scattering and absorbing the atmosphere by means of the two-stream approach is as follows:

$$\frac{dF^{\uparrow}}{d\tau} = \alpha_1 F^{\uparrow}\left(\tau\right) - \alpha_2 F^{\downarrow}\left(\tau\right) - \alpha_3 J,$$

$$\frac{dF^{\downarrow}}{d\tau} = \alpha_2 F^{\uparrow}\left(\tau\right) - \alpha_1 F^{\downarrow}\left(\tau\right) + \alpha_4 J, \tag{7.1.10}$$

$$\frac{dS}{d\tau} = \frac{S}{\mu_0}.$$

Here $F^{\uparrow}$ and $F^{\downarrow}$ are diffuse radiation flows directed, respectively, upwards and downwards; *S* is the flow of direct sunlight; coefficients $\alpha_1 \div \alpha_4$ are functions of the optical properties of the atmosphere layers, namely: of optical thickness τ, single scattering albedo, the probability of survival of the quantum ω, and the elongation parameter of scattering function; μ_0 is sinus of elevation of the Sun. For solar radiation $J = S / \mu_0$, for longwave radiation $J = \pi B$, where *B* is blackbody radiation. Solving the system of equations (7.1.10) recorded for each layer of the atmosphere is reduced to solving a system of algebraic equations to compute the unknown streams at the boundaries of layers in a multilayer atmosphere depending on the factors characterizing each of these layers. Knowing the flow of solar and longwave radiation, it is possible to obtain values of the radiation balance.

PARAMETERIZATION OF LARGE-SCALE CONDENSATION, CLOUDS, AND PRECIPITATION

Modeling atmospheric cloudiness is important due to the fact that it essentially determines the radiation regime and atmospheric precipitation. In the published works devoted to the modeling of the general circulation of the atmosphere, empirical relations like Smagorinsky graphs (Smagorinsky, 1953), or

Table 1. Laboratory experiments on temperature prediction error

T °C	16	15	11	10	6	5	1	0	-4	-5	-9	-10	-14	-15
Δq_s g/m³	0.81		0.61		0.46		0.35		0.25		0.19		0.13	

relations close thereto are used to calculate the amount of cloud cover. They establish a linear correlation between the amount of cloud cover *n* and relative humidity *f* in the form $n_k = \alpha_k f - \beta_k$, where:

- For low tier clouds $\alpha_1 = 0.33, \beta_1 = 2.0$.
- For middle tier clouds $\alpha_2 = 2.0, \beta_2 = 0.7$.
- For top tier clouds $\alpha_3 = 1.73, \beta_1 = 0.43$.

In general, the establishment of such links between the amount of cloudiness inside the cell of the numerical integration and average values of meteorological variables in this cell is a solution to some inverse problem, the correctness of which is not obvious.

As it is known (Mazin, Shmeter, 1978; Matveev, 1981; Mason, 1961; Shmeter, 1987), mists and clouds in the atmosphere are formed by the condensation and sublimation of water vapor when cooling the humid air to a temperature below the dew point. The reason for cooling the humid air can be expansion or pressure reduction, or mixing with a portion of cold air, or, eventually, evaporation of the water contained therein. In real conditions, temperature decrease is the result of the mixing of air masses with different thermic and hygrometric properties at upward movement, advection and turbulent mixing. When the portions of moist air mix with different temperatures, temperature equalization occur with following condensation (sublimation) of excess water vapor in the warmer air. Appearing drops (crystals) of water spread throughout the volume. Since the mixing temperature is above the cold air increases, then there arises saturation deficit and the rest of drops (crystals) forms a cloud or mist.

In cases where the conditions in the cloud (such as high water saturation, intense turbulent mixing, the presence of ice crystals, and others.) contribute to the multiple increase of cloud particles, the particles fall out of the clouds and reach the Earth's surface in the form of precipitation, overcoming the effect of upward flows and evaporation.

Thus, from its origin as a primary cloud of small particles to the final fall from the cloud in the form of liquid or solid precipitation, increasing the particle is affected by many interrelated micro- and macro-scale phenomena and thermodynamic processes. Existing cloud parameterization schemes (Akimov, 2001; Jonson, 1993; Sundqvist, 1993), typically, do not account for the fact that in different temperature intervals (at the same pressure), lowering the temperature by 1°C in a saturated air condenses different specific conductivity quantity $\Delta\hat{\delta} = \Delta q_s$ (see the Table 1).

Furthermore, laboratory experiments listed in Table 1 show that the temperature prediction error at 1°C and humidity at 0.2 g/m³ in the saturated air could lead to errors, commensurate with an average moisture value in the cloud (0.1 ÷ 0.2 g/m³).

The complexity and inadequacy of interrelated micro- and macroscale thermodynamic processes greatly complicates the problem of mathematical modeling formation of mists, clouds, and precipitation. Obviously, some simplifying assumptions are required.

When modeling the processes leading of the formation of mists, clouds, and precipitation, we will proceed from macrophysical features of their formation. We will ignore the microphysical properties, structure, processes, merging and evaporation of cloud particles. We will create the mathematical model using both theoretical and empirical dependencies. Just assume that the actual processes are still strictly described by the relevant equations of the state, regardless of their infrequency and the presence of the effects of accidental circumstances associated with the presence in the atmosphere of supersaturated vapor and overcooled water.

If the atmosphere is saturated with water vapor then the terms of stability of dry atmosphere take place only for moist air which moves to a higher temperature area (goes down), because an amount of moisture, which the examined volume can contain, as a rule, increases at heating (lowering). Vice versa, in the air which moves to a lower temperature area (raises) the amount of moisture, which can be held in a volume, diminishes, what results in the processes of condensation (crystallizations) and heat release. As a consequence, the floatation of an air volume increases in comparison to that which would be observed in the absence of condensation (crystallizations). The condensation, which results in cloudiness and precipitation refers to be the non-deterministic process and therefore is described as prognostic and circulation patterns in the atmosphere models by means of parameterization.

To derive the parameterization equations, we represent the temperature θ, specific humidity q and the specific conductivity of δ as a sum:

$$\theta = \tilde{\theta} + \hat{\theta}, q = \tilde{q} + \hat{q}, \delta = \tilde{\delta} + \hat{\delta},$$

where $\tilde{\theta}$, $\tilde{q}$ and $\tilde{\delta}$ are variables whose values are changed due to convection-diffusion processes and which are described by following homogeneous equations:

$$\frac{d\tilde{\theta}}{dt} = \nabla \cdot \left(k_T \nabla \tilde{\theta} - F_{rad}\right), \quad (7.2.1')$$

$$\frac{d\tilde{q}}{dt} = \nabla \cdot \left(k_d \nabla \tilde{q}\right), \quad (7.2.2')$$

$$\frac{d\delta}{dt} = \nabla \cdot \left(k_d \nabla \delta\right). \quad (7.2.3')$$

$\hat{\theta}$, $\hat{q}$ and $\hat{\delta}$ are variables whose values are changed solely due to moisture phase transitions of the first kind and are described by the equations:

$$\frac{\partial \hat{\theta}}{\partial t} = \left(\frac{p_{00}}{p}\right)^{R/C_p} \frac{L}{C_p} Q_H, \quad (7.2.1'')$$

$$\frac{\partial \hat{q}}{\partial t} = Q_H, \quad (7.2.2'')$$

$$\frac{\partial \hat{\delta}}{\partial t} = -Q_H \, . \tag{7.2.3''}$$

Changing the amount of saturated water vapor per unit of volume per time unit (the rate of condensation) Q_H we define follows:

$$Q_H = \begin{cases} 0, & if\ q < q_s; \\ -\dfrac{\partial q_s}{\partial t} & otherwise. \end{cases}$$

where q_s is the specific humidity at saturation.

For the equilibrium of the system "water-water vapor or ice-water vapor" it is necessary, that the vapor pressure in the air would correspond to saturation. Dependence of pressure of the saturated vapor *E* on temperature *T* in differential form can be given based on thermodynamics laws and described by equation of Clausius-Clapeyron

$$\frac{dE}{dT} = \frac{LE}{R_w T^2}, \tag{7.2.5}$$

where $R_w = 461.50$ J kg^{-1} kcal^{-1}·is the universal gas constant for water vapor.

For determination of pressure of the saturated vapor over water we will perform integration of (7.2.5) from the temperature $T_0 = 273.15\,\mathrm{K}$ and corresponding value of pressure of the saturated vapor $E_0 = 6.1078$ of Mb to some values *T* and *E*. Accepting the first approach approximation as $L = const$, we will get

$$\ln\frac{E}{E_0} = -\frac{L}{1.6R}\left(\frac{1}{T} - \frac{1}{T_0}\right) = \frac{L}{1.6RT_0}\frac{T - T_0}{T}, \tag{7.2.6a}$$

or

$$E = E_0 \exp\left(C_w \frac{T - T_0}{T}\right), \tag{7.2.6b}$$

where $C_w = L / \left(1.6RT_0\right)$.

When determining the dependence of resilience of the saturated steam on temperature over ice E_i it is necessary to take the heat of sublimation $L_s = L + L_m$ instead of the heat of evaporation *L*, where L_m=335.008 J·kg^{-1}, is a heat of melting of ice. Then like (7.2.6a) we will get:

$$\ln\frac{E_i}{E_0} = -\frac{L+L_m}{1.6R}\left(\frac{1}{T}-\frac{1}{T_0}\right) = \frac{L+L_m}{1.6RT_0}\frac{T-T_0}{T}, \tag{7.2.7a}$$

or performing the same transformation as above, we will find

$$E_i = E_0 \exp\left(C_i \frac{T-T_0}{T}\right), \tag{7.2.7b}$$

where

$$C_i = \frac{L+L_m}{1{,}6RT_0}.$$

Now we will find the difference between the saturated vapor resilience over overcooled water E and the saturated vapor resilience over ice E_i at the same temperature. This difference, in view of (7.2.6*b*) and (7.2.7*b*), can be written as:

$$\Delta E = E_0\left[\exp\left(C_w \frac{T-T_0}{T}\right) - \exp\left(C_i \frac{T-T_0}{T}\right)\right].$$

To find extreme of the function $\Delta E(T)$ we hold differentiation left and right sides of the resulting equation with respect to T and equate ∂ (ΔE) /∂T to zero:

$$\frac{T_0 E_0}{T^2}\left[C_w \exp\left(C_w \frac{T-T_0}{T}\right) - C_i \exp\left(C_i \frac{T-T_0}{T}\right)\right] = 0.$$

Factor $T_0 E_0 / T^2$ is not equal to zero, so the expression in square brackets is equal to zero. Hence we obtain:

$$T = T_0\left[1 - \frac{\ln(C_w / C_i)}{C_i - C_w}\right]^{-1}. \tag{7.2.8}$$

The result is that when $C_W = 19.80$ and $C_i = 22.46$ maximum value of the difference $\Delta E(T)$ is about 0.269 Mb and corresponds to the point with the temperature $T \approx 260.4$K or $t \approx 12.6$°C. Subsequently, the value of this temperature we will take as the critical temperature below of which crystallization replaces the condensation process.

We will assume that the change of the temperature of some elementary volume of moist air, in which steam is reached saturation state q_s and condensation (or crystallization) takes place, occurs without the exchange of heat between the dedicated air mass and the environment. Assume also that the condensation (or crystallization) products are still left inside the considered volume element of air. Then its total moisture content does not change but only its ratio in the gas and the condensed water vapor phases. In other words, we will consider only the case of the moist-adiabatic process in which the temperature change associated with the change in humidity $dq_s\left(p,T\right)$ in saturated air, according to the first law of thermodynamics, is described by the equation

$$c_p d\hat{T} - dp / \rho + Ldq_s = 0$$

which can be converted to

$$\frac{c_p \hat{T}}{\hat{\theta}} d\hat{\theta} + Ldq_s = 0 .$$

From this observation we obtain

$$d\hat{\theta} = -\frac{L}{c_{ps}}\left(\frac{p_{00}}{p}\right)^{R_s/c_{ps}} dq_s , \tag{7.2.9}$$

where $c_{ps} = 4R_s$ is the specific heat of water vapor at constant pressure.

Here, as in the system of equations for the mathematical model of the atmospheric circulation, L is the latent heat of vaporization, which is calculated by the formula (Gill, 1982)

$$L = \left(2500.8 - 2.3t\right) \times 10^3 J \cdot kg^{-1} . \tag{7.2.10a}$$

At temperature being t<-12.6°C, when the ice is formed instead of water, q_s is replaced by the corresponding value for ice q_i, and instead of the latent heat of vaporization we introduce the latent heat of sublimation:

$$L = 2.839 \times 10^6 - 3.6\left(t + 35\right)^2 J \cdot kg^{-1} . \tag{7.2.10b}$$

In general, when the value of q_s is changing, the value of coefficient

$$A = -\frac{L}{c_{ps}}\left(\frac{p_{00}}{p}\right)^{R_s/c_{ps}}$$

in the equation (7.2.9) is also changing. But for a fixed value q_s coefficient A is a number that does not depend on dq_s. It is known from the calculus that if $A \neq 0$, that is if $d\hat{\theta} \neq 0$, then the differentiability of $\hat{\theta}$ at q_s means that the main part of the increase of the function $\Delta\hat{\theta}$ at q_s is a linear function with respect to $\Delta\hat{q}$. The increase and the differential are equivalent infinitesimals when $\Delta\hat{q} \to 0$. Thus, the differential equation (7.2.9) can be written as

$$\Delta\hat{\theta} = -\frac{L}{c_{ps}}\left(\frac{p_{00}}{p}\right)^{R_s/c_{ps}} \Delta q_s + \alpha\left(\Delta q_s\right). \tag{7.2.11}$$

where $\alpha\left(\Delta q_s\right) = 0\left(\Delta q_s\right)$ when $\Delta q_s \to 0$.

The equation (7.2.11) allows determining the amount of change in temperature if we know the amount of change of humidity $\Delta q_s = q - q_s$, which occurs as a result of phase transitions in the wet air when water vapor is saturated to q_s.

Many experimental research efforts carried out in laboratory and in vivo indicate, that to condensate water vapor in the atmosphere the vapor resilience in the air e is needed to be greater than the vapor resilience E over the surface being formed by new phase particles, and that there should be fine particles in the air that could serve as nuclei of condensation. And droplets formed at the nuclei can grow to form cloud droplets only with oversaturation, i.e. when the relative air humidity $f = e / E$ will be greater than 1. For the crystallization process in the presence of ice crystals oversaturation must be determined with respect to the surface of the ice. Since the vapor resilience over the ice is less than over the water, then at low negative temperatures oversaturation can occur at a relative humidity in the air less than 1 with respect to the flat surface of the overcooled water.

We introduce the so-called critical relative humidity f_{or} above which there will always be the condensation of water vapor in the atmosphere. We associate f_{or} with a ratio of saturated vapor resilience over ice E_i and the same quantity over overcooled water E at the same temperature:

$$f_{cr} = E_i\left(T\right) / E\left(T\right). \tag{7.2.12}$$

The values of E and E_i calculated by formulas (7.2.6*b*) and (7.2.7*b*) do not coincide with the experimental data. On the basis of the experimental data presented in the Smithsonian meteorological tables, there have been offered empirical formulas (Gill, 1982). According to these tables saturation vapor resilience $E(T)$ – the pressure of saturated water vapor – over a flat surface water at temperatures in the range of ± 40 °C can be calculated by the formula

$$E\left(T\right) = \gamma \exp\left[\left(0.08006T - 20.047\right) / \left(0.00412T - 0.12476\right)\right]. \tag{7.2.13}$$

The value of the factor γ in (7.2.13) considered to atmospheric conditions is within 1 ÷ 1.006, and can be found using the following equation:

$$\gamma = 1 + p\left[4.5 + 0.0006\left(T - T_0\right)^2\right] \cdot 10^{-6}. \tag{7.2.14}$$

Saturated vapor tension $E_i(T)$ above the flat surface of ice at temperatures ranging from 0 to -40 °C satisfies the condition

$$E_i\left(T\right) = E\left(T\right)\exp\left(0.00422T - 1.15206\right). \tag{7.2.15}$$

Values calculated by formulae (7.2.13) and (7.2.15) coincide with values from Smithsonian meteorological tables N°94 and N°96 accordingly (Gill, 1982) providing absolute error ~ 0,0002 mbar within the temperature range of -40 °C ÷ +40 °C. Substituting equalities (7.2.13) and (7.2.15) into (19) we obtain:

$$f_{cr} = E_i\left(T\right) / E\left(T\right) = \exp\left(0.00422T - 1.15206\right). \tag{7.2.16}$$

It can be seen from the resulting dependence that the relative humidity f_{or} significantly increases with temperature. For example, at the temperature t=0°C relative humidity f_{or}=1 and at t=-30°C it drops to values f_{or}=0.88. In numerical simulation of clouds, we will take that saturation, i.e. condensation (at t≥-12.6°C) or crystallization (at t<-12.6°C) of water vapor occurs when the relative humidity is greater than or equal to the value f_{or} calculated using the formula (7.2.16).

In this connection let us return to the question of calculating the amount of change in the specific humidity Δq_s, which is part of the equation (7.2.11), for moist air in which water vapor is reached saturation. We will carry out following chain of elementary transformations:

$$\Delta q_s = q - q_s = \xi\left[\frac{e}{p-\left(1-\xi\right)e} - \frac{E}{p-\left(1-\xi\right)E}\right] = \xi\frac{E}{p-\left(1-\xi\right)E}\left[\frac{e}{E}\frac{p-\left(1-\xi\right)E}{p-\left(1-\xi\right)e} - 1\right],$$
$$\approx \xi\frac{E}{p-\left(1-\xi\right)E}\left(f - f_{cr}\right) \tag{7.2.17}$$

where $\xi = R/ = 0.62197$ and take into account that indeed there is always $p>>e$ and $p>>E$.

When calculating the temperature change by the formula (7.2.11) and changes in specific humidity by the formula (7.2.17) occurring as a result of phase transitions in the humid air in which water vapor reached a state of saturation, it is necessary to take into account that a change in temperature leads to a change in the conditions of saturation. Hence, these calculations should be performed using the method of successive approximations (iterations).

From the system of equations (7.2.1'') – (7.2.3'') and the equality $Q_H = -\partial q_s / \partial t$ it follows:

$$\frac{\partial\hat{\theta}}{\partial t} = -\left(\frac{p_{00}}{p}\right)^{R/C_p}\frac{L}{C_p}\frac{\partial q_s}{\partial t},$$

$$\frac{\partial \hat{q}}{\partial t} = -\frac{\partial q_s}{\partial t},$$

$$\frac{\partial \hat{\delta}}{\partial t} = \frac{\partial q_s}{\partial t}.$$

Integrating this system of equations in the interval $t = \left[0, t_{pr}\right]$, we find that the temperature, specific humidity and specific conductivity will change during the time t_{pr}, respectively, by an amount of:

$$\Delta\hat{\theta} = -\left(\frac{p_{00}}{p}\right)^{R/C_p} \frac{L}{C_p} \Delta q_s, \Delta\hat{q} = -\Delta q_s, \Delta\hat{\delta} = \Delta q_s. \quad (7.2.18)$$

Stratus Clouds

The boundaries and extent of a cloud layer are determined on the basis of the analysis of the vertical distribution of moisture. The layer in the atmosphere is considered to be cloudy when the condition $f \geq f_{or}$ holds, where f is a value of relative humidity in the layer, f_{or} is the critical relative humidity, defined by the formula (7.2.16).

After determining the spatial extent and boundaries of the cloud layer, one calculates values of water content in this layer δ (kg/m^3), using the relations (Matveev, 1981; Matveev, 2000):

$$\hat{\delta} = \begin{cases} q - q_s, & if\ \hat{\delta} \geq 0.1q, \\ 0.27\dfrac{T}{p}\exp\left[17.86\left(1 - \dfrac{258.0}{T}\right)\right], & if\ \hat{\delta} < 0.1q. \end{cases} \quad (7.2.19)$$

Convective Clouds

Defining the boundaries and extent of cloud layer of convective type, as well as the quantities of water content in this layer are produced by other relations. The criterion for the presence of convective instability in the atmosphere is the inequality $\gamma > \gamma_a$, where γ (K/m) is the temperature gradient in the surface layer of the atmosphere, $\gamma_a = 0.0098$ (K/m) is the dry adiabatic gradient.

Condensation level (the lower limit of the convective clouds) h_c is defined as the level where the following equalities hold:

$$q\big|_{x_3=h_c} = q_s\left(T^*\right), T^*\big|_{x_3=h_c} = \tau\big|_{x_3=h_c},$$

where T^* is a temperature of rising elemental volume of air, τ is the dew point temperature.

For the level of condensation, Taylor series to the first order of magnitude gives the relation

$$T_0^* + h_c \frac{dT^*}{dx_3} = \tau_0 + h_c \frac{d\tau}{dx_3}.$$

We find from this the height of the condensation level

$$h_c = \frac{T_0^* - \tau_0}{\dfrac{d\tau}{dx_3} - \dfrac{dT^*}{dx_3}} = \frac{T_0^* - \tau_0}{\dfrac{d\tau}{dx_3} + \gamma_a}. \tag{7.2.20}$$

To determine $d\tau / dx_3$ we use the fact that the maximum of the vapor elasticity at the temperature of the dew point τ must be equal to the actual vapor elasticity of an e, i.e. $E = e$, and specific humidity $q = 0.622E / p$ must remain constant at $x_3 \leq h_c$. Taking logarithms and differentiating the expression for q by x_3 and taking into account the equation of statics, we get

$$\frac{d\tau}{dx_3} = -\frac{g}{RT} \cdot \frac{1}{\dfrac{1}{E}\dfrac{dE}{d\tau}}. \tag{7.2.21}$$

From the equation of Clausius-Clapeyron, we have

$$\frac{1}{E}\frac{dE}{d\tau} = \frac{L}{R_s \tau^2} = \frac{L}{1.6R\tau^2}.$$

Substituting this expression in (7.2.21), we obtain

$$\frac{d\tau}{dx_3} = -1.6 \frac{g\tau^2}{LT}.$$

Consequently, the height h_c of the condensation level is

$$h_c = \frac{T_0^* - \tau_0}{\gamma_a - 1.6 \dfrac{g\tau^2}{LT}}.$$

Water content, resulting from the rise in the convective layer of the thickness h_c is calculated according to the following formula:

$$\hat{\delta} = q_0 - q\big|_{x_3=h_c} - a \cdot h_c \, , \tag{7.2.22}$$

where $a = 10^{-5} \div 10^{-3}$ (m^{-1}) is an index of inclusion.

We do not dwell on the theory of formation and loss of liquid or crystal cloud particles, which can be found, for example, in (Matveev, 1981). Just note that the comprehensive and physically robust mathematical models of liquid and solid precipitation does not exist due to the complexity and inadequacy of processes. Therefore, below we have to limit ourselves to the construction of a semi-empirical model of precipitation.

We assume that, regardless of the origin of clouds and cloud systems (front, convective, orographic, and others.), the following parameters should be determining in a mathematical model of precipitation:

- Power (geometric dimensions) of a cloud.
- Specific humidity q.
- The speed of the vertical upward movement v_3.
- The temperature θ_{lbc} at the cloud base.
- The temperature θ_{ubc} at the cloud top.

Particularly, in constructing a mathematical model of precipitation, we will also assume that there are known the vertical dimension h_{cl} and the rate of change of the specific water content $\partial\hat{\delta} / \partial t$. Expected precipitation I dropped from a single column of clouds during the period of time t_{pr} is estimated by the formula:

$$I = \int_0^{t_{pr}} \int_0^{h_{cl}} \frac{\partial \hat{\delta}}{\partial t} dx_3 dt \, . \tag{7.2.23}$$

Integrating (7.2.23) with respect to t in the interval $t = \left[0, t_{pr}\right]$, we obtain for $\Delta\hat{\delta} = \Delta q_s$:

$$I = \int_0^{h_{cl}} \Delta q_s dx_3 \, . \tag{7.2.24}$$

Experimental studies in (Matveev, 1976) showed that the clouds with the water content of 1 g/m^3 and a height of 6000 meters in 1 hour provide so much sediment that their amount exceeds the supply of cloud moisture by 4.9 times. To account this feature in the formula (7.2.21), they introduce the coefficient C, the value of which is determined experimentally. The final formula for the calculation of the intensity of precipitation in the node (i,j) takes the form

$$I_{ij} = C \int_0^{h_{cl}} \left(\Delta q_s\right)_{ij} dx_3 \, . \tag{7.2.24a}$$

and it may be implemented using numerical integration.

Formulas (7.2.24) and (7.2.24*a*) do not take into account the individual variation of the specific humidity in the horizontal movement of air. Typically, the effect of this change on the intensity of rainfall is little (Matveev, 1981). Exceptions are cases when a cold period the intense advection of warm air, followed by a transformational change in temperature, i.e. when calculating the drizzling rain.

Formulas (7.2.24) and (7.2.24a) also do not include cloud microphysics and suggest that the condensation is immediate and completely falls as precipitation. Obviously, this does not correspond to the natural processes, where it is known that from existing clouds, there may be no precipitation at all.

In the calculation of precipitation intensity by the method described in (Akimov, 2001) it is assumed that initially, the cloud layer is a colloid-stable, and sedimentation rate is determined by the change in cloud distribution function of particles in the layer over the time interval Δt.

A measure colloidal instability of the cloud layer which characterizes the intensity of precipitation is determined by the parameter β_1 of the distribution function formed within the time interval Δt. It is calculated by the formula:

$$\beta_1 = \beta_0 + \left[(\alpha+2)\frac{\bar{E}\hat{\delta}}{4\rho_w} k_1 \beta_0^2 + \frac{16}{3}\frac{\hat{\delta}}{\rho_w}\beta_0 \sqrt{\frac{2\varepsilon}{15\pi\nu}} + P(1-P)\frac{\rho D}{2\rho_{cr}(1+\alpha)\beta_0}\Delta q_s \right]\Delta t, \quad (7.2.25)$$

where β_0 is the initial value of the parameter of the distribution function, which characterizes the average size of the cloud layer of particles prior to the formation of precipitation, $\alpha = 2$ is a parameter, $\bar{E} = 0.15$ is the value of the average coefficient of capture of cloud droplets, $\varepsilon = 14.0 \cdot 10^{-4}$ m^2/s^3 is the characteristic value of the turbulence energy for stratus-form clouds, ν is the kinematic viscosity of the air, ρ_w is the density of water, ρ is air density, $D = 0.22 \cdot 10^{-4}$ m^2/sec is the diffusion coefficient of water vapor, Δq_s is the difference between the values of saturation humidity of the water and the ice, *P* is the share of ice crystals in the cloud layer, given as a function of temperature, according to (Sundqvist, 1993).

Note that in the case of applying the method for the layer of clouds at a negative temperature, conductivity value $\hat{\delta}$ is a specific amount of cloud water of two phases (water and ice), the relationship between which is determined by the quantity *P*.

The value β_0 of the parameter of distribution is calculated from the mean radius of cloud particles $\bar{r}$ (Mazin, Hrgiana, 1989) by the relation:

$$\beta_0 = \frac{\bar{r}}{1+\alpha}, \quad (7.2.26)$$

and the mean size of cloud particles $\bar{r}$ is calculated from the values of the water content using the empirical relation (Jonson, 1993):

$$\bar{r} = a \cdot \delta + b, \quad (7.2.27)$$

where a = 11.0 and b = 4.0 are constants.

Thus, formulas (7.2.19), (7.2.22) and (7.2.27) compute the conductivity values and average particle radius of cloud points for the initial formation of a layer of sediment.

The critical water content δ_{cr}, characterizing the share of cloudy water remaining in the layer after the precipitation, calculated on the basis of the parameter values β_1, obtained by the formula (7.2.25), in accordance with (Akimov, 2001):

$$\delta_{cr}\left(t+\Delta t\right)=\hat{\delta}\left(t_0\right)\cdot I_{coag}\left(\beta_1,r_{cr}\right), \tag{7.2.28}$$

where

$$I_{coag}\left(\beta_1,r_{cr}\right)=1-\frac{1}{m!}\exp\left(\frac{r_{cr}}{\beta_1}\right)\sum_{i=0}^{m}\frac{m!}{\left(m-1\right)!}\left(\frac{r_{cr}}{\beta_1}\right)^{m-1},\ m=3+\alpha,$$

r_{or} = 50 mkm is the minimum value of the size of cloud particles, in which it can overcome the vertical movement at the lower boundary of the clouds and fall as a drop of rain.

Precipitation intensity I (mm/h), falling from the clouds layer thickness ΔH (m), in accordance with (Akimov, 2001), may be calculated from the difference between the water content of clouds and a critical water content, formed during time interval Δt:

$$I=\frac{3.6\cdot 10^6}{\rho_w}\cdot\frac{\hat{\delta}\left(t_0\right)-\delta_{cr}\left(t_0+\Delta t\right)}{\Delta t}\cdot H. \tag{7.2.29}$$

In this way, based on the increment of the distribution parameter determined according to (7.2.25), the value of the critical water content is calculated from the formula (7.2.28), and the value of the intensity of precipitation for each cloud layer is found. The calculated values of the intensity of precipitation can be changed as a result of strengthening the intensity of rainfall on the way to the underlying surface in the lower cloud layers and evaporation of precipitation in an under-cloud layer according to the formulas set out in (Akimov, 2001). On the way to the underlying surface, there also may be changes in the phase composition of precipitation particles.

THE PARAMETERIZATION OF CONVECTION PROCESSES

Convection in the atmosphere is a very complex physical process, so there is no an accurate mathematical model for it. Typically, simplified models are used. The characteristics of convection and its manifestations (extraction and absorption of heat, cloudiness, precipitation and others.) are described in these models by relatively simple equations and relations, expressed in terms of the parameters of large-scale atmospheric processes (averaged values of temperature, humidity and wind speed). As an example, we present a methodology of parameterization of convection, which is based on the hypothesis of convective adjustment proposed in (Manabe, Smagorinsky & Strickler, 1965) and slightly modified by other authors.

Let the moist-adiabatic instability occurred in the region between the two levels, $x_3 = z_k$ (lower level) and $x_3 = z_{k+1}$ (upper level). Static energy inside of the field equals

$$E = E^* = c_p T + g z_{k+1} + L q_s .$$

As unsaturated air is supplied through the lower level, the necessary condition for the existence of the moist-adiabatic instability is given by $E^*_{k+1} \le E_k$, or

$$c_p T_{k+1} + g z_{k+1} + L q_s \left(T_{k+1}, p_{k+1}\right) \le c_p T_k + g z_k + L q_k .$$

Let give the last relation to the form

$$\frac{T_k - T_{k+1}}{z_{k+1} - z_k} \ge \frac{g}{c_p} - \frac{L}{c_p} \cdot \frac{q(k) - q_s(k+1)}{z_{k+1} - z_k} = \frac{g}{c_p} - \frac{L}{c_p} \cdot \frac{q_s(k) - q_s(k+1)}{z_{k+1} - z_k} + \frac{L}{c_p} \cdot \frac{(1-r) q_s(k)}{z_{k+1} - z_k}, \tag{7.3.1}$$

where $r = q(k) / q_s(k)$.

It follows from (7.3.1) that

$$\frac{T_k - T_{k+1}}{z_{k+1} - z_k} \approx -\frac{\partial T}{\partial z} = \gamma, \frac{g}{c_p} = \gamma_a, -\frac{L}{c_p} \cdot \frac{q_s(k) - q_s(k+1)}{z_{k+1} - z_k} = \gamma_{ma},$$

where γ, γ_a and γ_{ma} are, respectively, the actual, the dry adiabatic, and moist-adiabatic vertical temperature gradients. Therefore, the condition of occurrence of moist convection can be rewritten as

$$\gamma \le \gamma_{ma} + \frac{L}{c_p} \cdot \frac{(1-r) q_s(k)}{z_{k+1} - z_k}. \tag{7.3.2}$$

We introduce the concept of critical relative humidity based on the condition that the temperature gradient cannot be greater than dry adiabatic, i.e.

$$\gamma = \gamma_{ma} + \frac{L}{c_p} \cdot \frac{(1-r_{cr}) q_s(k)}{z_{k+1} - z_k}.$$

From this, it follows

$$r_{cr} = 1 - \frac{\gamma_a - \gamma_{ma}}{\dfrac{L}{c_p} \dfrac{q_s(k)}{z_{k+1} - z_k}}. \tag{7.3.3}$$

When using (7.3.3), condition (7.3.2) takes the form

$$\gamma \le \gamma_a \frac{1-r}{1-r_{cr}} + \gamma_{ma} \frac{r-r_{cr}}{1-r_{cr}} \equiv \gamma_{cr} \, . \tag{7.3.4}$$

Changes in temperature and mass fraction of moisture in the convective layer are determined based on the conditions of the immutability of moist static energy in the process of convection in the layer

$$\int_{p_k}^{p_{k+1}} \left(c_p \delta T + L \delta r \right) dp = 0 \, , \tag{7.3.5}$$

setting the moist-adiabatic gradient in which the equivalent-potential temperature does not change with the vertical movements

$$\frac{\partial}{\partial p} \left[\theta^* \left(T + \delta T, r + \delta r, p \right) \right] = 0 \, , \tag{7.3.6}$$

and equality of the new humidity value to its value at saturation at the new value of temperature

$$r + \delta r = r_{\max} \left(T + \delta T \right) . \tag{7.3.7}$$

Under the conditions (7.3.5)–(7.3.7), δT and δr are increment values T and r, $\theta^* = \theta \exp\left(L r_{\max} / c_p T \right)$ is the equivalent-potential temperature.

To determine the values δT and δr between levels p_k and p_{k+1}, we introduce intermediate levels $p_{k+i/n}$ $\left(i = 1, 2, \ldots, n-1 \right)$. Then, the integral (7.3.5) and derivative (7.3.6) are approximated by finite differences of a given order of accuracy based on levels $p_{k+i/n}$ and we write (7.3.7) for each of these levels.

Thus, we obtain a system of linear algebraic equations for determining the unknown values δT_i and δr_i imposed on all levels. Solving this system and summing the found increments of δT_i and δr_i with baseline values T_i and r_i we obtain their new values. The process is repeated using these new values. Using criteria (7.3.4) one searches for new values p_k and p_{k+1}, draw up again a system of linear algebraic equations, etc. The described computational process is completed when the condition (7.3.4) holds at all levels of the vertical line.

The parameterization algorithm is like the algorithm of dry convection parameterization of moist convection given above provided $\gamma_{cr} = \gamma_a$.

RADIATION PARAMETERIZATION

The actual picture of radiative processes in the real atmosphere is very complex, depends on a large number of scattering and absorbing substances, the spectral absorption properties of radiation and scattering of gasses and fine air components, cloud structure and aerosol optical and geometrical characteristics of the underlying surface. The quantity of total radiation coming to the surface of all the above factors depends on the angle of incidence of sunlight and illumination duration. The horizontal surfaces and slopes, differently oriented with respect to the Sun and having a different steepness, receive different amounts of total radiation. The most striking differences are observed between the differently oriented slopes, especially in the number of incoming direct radiation. Although scattered radiation falls on all slopes more evenly, still uneven incoming direct radiation significantly affect the value of the total radiation.

Today, there is no mathematical model that adequately describes the whole picture of the multi parameter stochastic radiative processes and their interaction with physical processes in the atmosphere.

Existing simplified models and methods for calculating the components of the radiation fluxes and radiation balance can be divided into two groups: based on the theoretical calculation and relations, and derived from empirical formulas (Ritter, Geleyn, 1992; Geleyn, 1977; Zdunkowski, Weltch & Korb, 1980).

Theoretical methods include two possible approaches to describing the process of heat transfer by radiation: the *phenomenological and statistical.*

The phenomenological approach particular features of the environment and radiation process, as well as their interaction, are not considered. Instead, there is established and used some fairly general relations simulating a particular process, and the physical properties of the medium in which the process is playing, recorded in these relation set by some factors determined empirically. Classic examples of the phenomenological approach are relations (the laws) of Newton, as well as Fourier and Fick, describing the process of friction, heat transfer and thermal conductivity of the impurity diffusion. The properties of the medium in these relations are recorded using the coefficients of viscosity, thermal conductivity and diffusion, respectively.

In principle, such a consideration is preserved also in the description of the process of heat transfer by radiation with the aid of Maxwell's equations written for various physical media. But the electromagnetic processes of thermal radiation are random, so Maxwell's equations used to be regarded as some stochastic equations. It should be also noted that the system of Maxwell's equations describing the laws of behavior of a thermal electromagnetic field in the space filled with the medium (substance) is an incomplete system (from a mathematical point of view). This system of equations must be supplemented by certain relations that take into account the specific properties of the medium, the conditions on the emitting and absorbing surfaces, etc., that, of course, are not derived from the main system.

The objective of the statistical description of the radiation heat transfer process is obtaining a macroscopic or integral characteristics of the system by the given microscopic properties, i.e. the properties of its constituent particles. In this representation, an electromagnetic field is considered as a collection of particles (photons or light quanta), and the radiation heat transfer process – as a result of the interaction of these particles with the particles of the substance (a substance interacting with photons). This interaction is usually described using the Boltzmann equation written with respect to the specific frequency f of the radiation intensity.

$$\frac{\partial f}{\partial t}+\frac{\boldsymbol{P}}{m}\frac{\partial f}{\partial \boldsymbol{X}}+\boldsymbol{F}\frac{\partial f}{\partial \boldsymbol{P}}=\left.\frac{\partial f}{\partial t}\right|_{coll},$$

where $\boldsymbol{X}$ is a coordinate, $\boldsymbol{P}$ is momentum, $\boldsymbol{F}(\boldsymbol{X},t)$ is a force field acting on the particles in the air, m is the particle mass. Its right-hand side contains the source-sink terms that describe the processes of absorption, emission and scattering, which are functions of temperature. Consequently, for the Boltzmann equation, it is necessary to know the temperature distribution in the region under consideration, because only under this condition, the right part of this equation becomes determined. Since the temperature distribution is generally not known a priori, and in particular, is determined by the process of radiation transfer, for a general solution of the heat exchange problem, Boltzmann equation is to be supplemented by an equation with respect to temperature.

Mutual influence and interdependency of processes of radiative transfer and heat transfer causes a great difficulty to solve this problem by theoretical methods. This greatly restricts the use of the phenomenological and statistical methods.

The basis of the theoretical calculation methods developed and used by individual researchers, is formed by some general guidelines, not always clearly articulated, but revealed in the compilation and critical analysis of the various methods. Development of these basic principles allows representing actinometry not as a summary of empirical methods, based on purely statistical comparisons, but as a certain system of functional relationships, physically well-founded. This approach helps to assess the existing calculation methods, to establish their accuracy and to determine the conditions under which the method can give the best results. At the same time, it facilitates the development of new, not previously applied methods and improvement of existing ones, which are especially necessary due to the increased requirements for the accuracy of the calculations. Many of the methods of calculation, which give satisfactory results in the calculation of climatic parameters, are very rough approximations and of little use to calculate radiation characteristics in the prediction models (Black, 1956; Fu, Rich, 2002; Rich et al., 1994; Briegleb, 1992). Identification of opportunities and ways of more accurate calculation methods can only be based on the study of physical laws that determine the intensity and the amount of each type of radiation.

There most often arises the need to calculate the short-wave characteristics. This is explained by the fact that shortwave balance in the equation of radiative balance, which is subject to the greatest changes in time, has the predominant importance. And since the main factors affected the weather are shortwave radiation fluxes, the methods of calculating the characteristics of short-wave radiation mode are the most developed (Houghton, 1954; Monteith, 1962; Davies, Schertzer & Nunez, 1975). In these models, the short-wave radiation distribution ϱ_{sh}, coming on top of the atmosphere, are calculated according to the laws of celestial mechanics as a purely geometric problem. Next, given coefficients of absorption, scattering and reflection of the different components of the atmosphere (water vapor, gasses, dust, cloud particles), it is determined the part of ϱ_{sh}, reached the Earth's surface. Particular attention is paid to the physical justification of recommended methods and the establishment of general laws binding characteristics of the radiation mode with each other and with the impact of meteorological factors on them.

Much more complicated is the calculation of long-wave radiation fluxes. Methods of direct measurement are still far from perfect. Applied for their measurement, devices have different designs systematic errors of various origins, that are yet insufficiently investigated. In such a situation there is no confidence in the accuracy and comparability of the data, and systematization, compilation, and evaluation of methods of calculation of the long-wave flows is difficult. Among the works in which were proposed models for the calculation of long-wave radiation from the Earth's surface, it may be noted (Yamamoto, 1952; Elsasser, Culbertson, 1960; Zhang, Stamnes & Bowling, 2001; Prata, 1996). In these works, complicated absorption and emission spectra of water vapor, carbon dioxide, and ozone are replaced by integral characteristics of absorption and emission as the temperature and humidity distribution functions by the beam path (Ramanathan & Inamdar, 2006). Vertical profiles of temperature and humidity are set on predictive data.

Currently, the above model, due to its bulkiness and uncertainties, are used mainly in basic research to obtain detailed information about the structure and the basic laws of radiation. The possibilities of computer technology, despite significant progress in its development, is still not sufficient to meet these challenges in modeling the atmospheric circulation and weather forecast of practical interest.

Other, less rigorous approach is to find empirical functional relationships between the sought radiation fluxes and forecast meteorological variables. The empirical constants (predictors), included in these relations are determined by regression analysis. The obtained formulas typically are not difficult and does not require lengthy calculation. The accuracy of the results is about the same as the accuracy of the calculation of phenomenological and statistical methods. In most cases, the required data on radiation mode are obtained using approximate calculation methods. Application of these methods is possible due to the fact that basic characteristics of the radiation mode are related among themselves and with the most important meteorological characteristics. Moreover, the reliability of the calculation is provided by the fact that these relations are of physical nature. That is why established quantitative relationships are stable and can be generalized to a variety of weather and climate conditions.

The existence of such dependencies allows calculating most of the individual characteristics of the radiation mode or their combination on the meteorological data and a minimal amount of solar radiation data. If necessary, calculation can be made only on meteorological data at the cost of some decrease in the accuracy of results. The main attention is paid to the methods of calculating the integrated features that are most important for solving the problems of atmospheric physics. As for the spectral radiation characteristics, despite their importance for the physics of the atmosphere, the development of methods of their calculation is difficult because of the lack of reliable baseline data.

The theoretical basis of empirical parameterization methods of radiation is radiant energy transfer equation for stationary radiation field (Brunt, 1932; Idso, Jackson, 1969; Monteith, 1962; Monteith, 1973; Swinbank, 1963; Choudhury, Idso, 1985).

To simplify the problem of calculating the radiation heat influx $\varrho = \nabla \cdot F_R$ to a certain volume of air (see. equation 3.2.31), we assume that the atmosphere is a plane-parallel and horizontally uniform. Then the operator $\nabla \cdot \left(*\right)$ degenerates into a derivative related to the vertical coordinate x_3. If we think of the radiation in the form of plane-parallel flows of downward longwave $\varrho_{l\downarrow}^i$ and rising $\varrho_{l\uparrow}^i$ radiation and short-wave radiation downward ϱ_{sh}^i, where the superscript marked values relating to some i-th spectral interval, the equation for these streams can be written as:

$$\frac{\partial \varrho_{l\downarrow}^i}{\partial x_3} = \alpha_i \rho_{st} \left(\varrho_{l\downarrow}^i - B\right),$$

$$\frac{\partial \varrho_{l\uparrow}^i}{\partial x_3} = \alpha_i \rho_{st} \left(B - \varrho_{l\uparrow}^i\right),$$

$$\frac{\partial \varrho_{sh}^i}{\partial x_3} = \frac{\beta_i \rho_{st}}{\cos \vartheta_0} \varrho_{sh}^i, \tag{7.4.1}$$

where α_i and β_i are integral absorption coefficients for long-wave and short-wave radiation fluxes, respectively, related to the i-th spectral range; ρ_{st} is the vapor density, ϑ_0 is the zenith distance of the Sun. Boundary conditions are taken as follows:

$$\varrho_{sh}^i = \chi_i \varrho_{sh}^0 \cos\vartheta_0 \text{, if } x_3 \to \infty \text{, } \varrho_{l\downarrow}^i = 0 \text{,}$$

$$\varrho_{l\downarrow}^i = \eta_i B\left(T_0^*\right) \text{, if } x_3 = 0, \tag{7.4.2}$$

where χ_i and η_i are parts of the long-wave and short-wave radiation in this spectral range; ϱ_{sh}^0 is the solar constant; T_0^* is the surface temperature; $B = \sigma T^4$. The solution of equations (7.4.1) with boundary conditions (7.4.2) can be written as

$$\begin{aligned}
&\varrho_{l\downarrow}^i\left(x_3\right) = \alpha_i \int_{x_3}^{\infty} \left[B\left(T\right)\rho_{st} \exp\left(-\alpha_i \int_{x_3}^{\varkappa} \rho_{st} d\varkappa\right)\right] d\varkappa \text{ ,}\\
&\varrho_{l\uparrow}^i\left(x_3\right) = \alpha_i \int_{x_3}^{\infty} \left[B\left(T\right)\rho_{st} \exp\left(-\alpha_i \int_{x_3}^{\varkappa} \rho_{st} d\varkappa\right)\right] d\varkappa + \eta_i B\left(T_0^*\right) \exp\left(-\alpha_i \int_0^{x_3} \rho_{st} dx_3\right), \\
&\varrho_{sh}^i = \chi_i \varrho_{sh}^0 \cos\vartheta_0 \exp\left(-\frac{\beta_i}{\cos\vartheta_0} \int_{x_3}^{\infty} \rho_{st} dx_3\right),
\end{aligned} \tag{7.4.3}$$

where $\varkappa$ is the variable of integration.

The radiation balance at the height x_3 determined by summing over all the spectral range:

$$\varrho = \sum_i \left(\varrho_{sh}^i - \varrho_{l\uparrow}^i + \varrho_{l\downarrow}^i\right). \tag{7.4.4}$$

When applying the integrated transmission functions to complete radiation and heat flow per unit volume $\varrho_{\breve{v}}$, following expressions are used:

$$\begin{aligned}
&\varrho_{l\downarrow}^i\left(\omega\right) = -\int_{\omega}^{\omega_\infty} B\left(T\right) dP\left(\omega' - \omega\right),\\
&\varrho_{l\uparrow}^i\left(\omega\right) = -\int_0^{\omega} B\left(T\right) dP\left(\omega - \omega'\right) + \delta B\left(T_0^*\right) P\left(\omega\right) - \left(1-\delta\right) \int_0^{\omega_\infty} B\left(T\right) dP\left(\omega + \omega'\right), \\
&\varrho_{sh}\left(\omega\right) = \left(1 - \bar{A}\right) \varrho_{sh}^0 \cos\vartheta_0 P_{sh}\left(\frac{\omega_\infty - \omega'}{\cos\vartheta_0}\right),\\
&\varrho_{\breve{v}} = \frac{\partial \varrho}{\partial x_3} = \frac{\partial}{\partial x_3}\left(\varrho_{sh}^i - \varrho_{l\uparrow}^i + \varrho_{l\downarrow}^i\right).
\end{aligned} \tag{7.4.5}$$

In these relations,

$$\omega = \int_0^{x_3} \rho_{st} dx_3 ,$$

where ρ_{st} is still water vapor content, and the content of other absorbing and emitting materials (carbon dioxide and ozone) is taken into account by the by the coefficients in the expression for the transmission function $P(\xi)$.

Under the integral transmission function $P(\xi)$, one refers the total radiation flux ratio $\varrho(x_3)$, incoming in the layer of thickness Δx_3 and containing the mass ξ of absorbing and emitting substances, in the full radiation flux $\varrho(x_3 + \Delta x_3)$, coming out from this layer:

$$P(\xi) = \frac{\varrho(x_3)}{(x_3 + \Delta x_3)} .$$

Typically, the transmission function $P(\xi)$ are defined on the basis of specific experiments and calculations. There are several formulas (Oke (1978)), allowing to calculate $P(\xi)$ for a long-wavelength radiation

$$P(\xi) = 0.471 \exp\left(-0.696\sqrt{\xi}\right) + 0.529 \exp\left(-8.94\sqrt{\xi}\right), \tag{7.4.6a}$$

$$P(\xi) = 0.515 \exp\left(-0.8\sqrt{\xi}\right) + 0.485 \exp\left(-2.0\sqrt{\xi}\right), \tag{7.4.6b}$$

and for short-wave radiation

$$P_{sh}(\xi) = 1 - 0.09\xi^{0.303} . \tag{7.4.7}$$

The parameterization of radiative heat transfer on the basis of integral transmission functions can be generalized to the case of a cloudy atmosphere. The easiest way to do it for unlimited horizontal cloud layers. Assuming that within the cloud effective radiation is $\left(\varrho_{l\downarrow}^{i} - \varrho_{l\uparrow}^{i}\right)$ and short-wave radiation flux ϱ_{sh} is equal to zero, we can integrate the equation of radiative transfer separately for the over-cloud and under-cloud parts of the atmosphere, and for cloudless layers. We can then calculate the radiation balance for each level and the influx of heat to individual layers. The influx of heat, when the number of clouds is *n* (score), is described by the relation

$$\varrho_{\breve{o}}(n) = \frac{n}{10}\varrho_{\breve{o}}(10) + \left(1 - \frac{n}{10}\right)\varrho_{\breve{o}}(0), \tag{7.4.8}$$

where $\varrho_{\breve{o}}(10)$ and $\varrho_{\breve{o}}(0)$ are the influx of heat for overcast and clear sky, respectively.

SUMMARY

The system hydrodynamics equations and boundary conditions considered in previous chapters constitute the mathematical basis of the circulation models of the atmosphere of a scale. They contain terms describing sources (sinks) of mass and energy involved in the phase transformation of atmospheric moisture and radiative processes in the system Atmosphere – the Earth. However, direct inclusion of these microscale and mesoscale processes in the atmospheric circulation model is inappropriate as 1) this would lead to an unrealistic increase in the total number of grid points in the decision area, and 2) contemporary mathematical models of the atmosphere cannot adequately reproduce the whole picture of the turbulent multi parameter stochastic, phase and radiative processes.

Therefore, the processes of interaction of the atmosphere with the underlying surface should be referred to as sub grid scale, and creating a semi-empirical (parametric) model of subgrid-scale processes is needed.

In this chapter, we apply a variety of approaches and methods considered in previous chapters for the parameterization of turbulent exchange with the underlying surface, cloud and precipitation formation, convection, and radiation.

REFERENCES

Akimov, I. V. (2004). Precipitation calculation method based on parameterization of distribution function evolution and its performance in global spectral atmospheric model [in Russian]. *Journal of Hydrology (Amsterdam)*, *288*(1-2), 105–120. doi:10.1016/j.jhydrol.2003.11.012

Black, J. N. (1956). The distribution of solar radiation over the earth's surface. *Arch. Met. Geoph. Biokl. Ser. B.*, *7*, 165–189.

Briegleb, B. P. (1992). Delta-Eddington approximation for solar radiation in the NCAR Community climate model. *Journal of Geophysical Research*, *97*(D7), 7603–7612. doi:10.1029/92JD00291

Brunt, D. (1932). Notes on radiation in the atmosphere. *Quarterly Journal of the Royal Meteorological Society*, *58*(247), 389–420. doi:10.1002/qj.49705824704

Choudhury, B. I., & Idso, S. B. (1985). An empirical model for stomatal resistance of field - grown wheat. *Agricultural and Forest Meteorology*, *36*(1), 65–82. doi:10.1016/0168-1923(85)90066-8

Davies, J. A., Schertzer, W., & Nunez, M. (1975). Estimating global solar radiation. *Boundary-Layer Meteorology*, *9*(1), 33–52. doi:10.1007/BF00232252

Elsasser, W. M., & Culbertson, M. F. (1960). *Atmospheric radiation tables*. Boston: Amer. Meteorol. Soc. doi:10.1007/978-1-940033-48-8

Fu, P., & Rich, P. M. (2002). A Geometric Solar Radiation Model with Applications in Agriculture and Forestry. *Computers and Electronics in Agriculture*, *37*(1-3), 25–35. doi:10.1016/S0168-1699(02)00115-1

Geleyn, J.-F. (1977) Parameterization of radiative transfer at ECMWF. *Proc. ECMWF 1977 seminar on the parameterization of the physical processes in the free atmosphere*, 273–345.

Gill, A. E. (1982). *Atmosphere-Ocean Dynamics*. London: Academic Press.

Houghton, H. G. (1954). On the annual heat balance of the Northern Hemisphere. *Journal of Meteorology, 11*(1), 1–9. doi:10.1175/1520-0469(1954)011<0001:OTAHBO>2.0.CO;2

Idso, S., & Jackson, R. D. (1969). Thermal radiation from the atmosphere. *Journal of Geophysical Research, 74*(23), 5397–5403. doi:10.1029/JC074i023p05397

Johnson, D. W. (1993) Parameterization of the cloud topped boundary layer: Aircraft measurements. In *Proceedings of the workshop held at ECMWF on parameterization of the cloud topped boundary layer* (pp. 77-117). Reading, UK: European Centre for Medium-Range for Weather Forecasts.

Manabe, S., Smagorinsky, J., & Strickler, R. F. (1965). Simulated climatology of a general circulation model with a hydrologic cycle. *Monthly Weather Review, 93*(12), 769–798. doi:10.1175/1520-0493(1965)093<0769:SCOAGC>2.3.CO;2

Mason, W. J. (1961). *The physics of clouds*. Leningrad: Gidrometeoizdat. (in Russian)

Matveev, L. T. (1976). *Course of general meteorology. Atmospheric Physics*. Leningrad: Gidrometeoizdat. (in Russian)

Matveev, L. T. (1981). *Dynamics of clouds*. Leningrad: Gidrometeoizdat. (in Russian)

Matveev, L. T. (2000). *Atmospheric Physics. Saint-Petersburg*. Gidrometeoizdat. (in Russian)

Mazin, I. P., & Hrgiana, A. H. (1989). *Clouds and cloudy atmosphere*. Leningrad: Gidrometeoizdat. (in Russian)

Mazin, I. P., & Shmeter, S. M. (1978). *Clouds, structure and physical education*. Leningrad: Gidrometeoizdat. (in Russian)

Monteith, J. I. (1962). Attenuation of solar radiation: A climatological study. *Quarterly Journal of the Royal Meteorological Society, 88*(378), 508–521. doi:10.1002/qj.49708837812

Monteith, J. L. (1973). *Principles of environmental physics*. London: Edward Arnold.

Obukhov, A. M. (1988). *Turbulence and dynamic atmosphere*. Leningrad: Gidrometeoizdat. (in Russian)

Prata, A. J. (1996). A new long-wave formula for estimating downward clear-sky radiation at the surface. *Quarterly Journal of the Royal Meteorological Society, 112*(533), 1127–1151. doi:10.1002/qj.49712253306

Ramanathan, V., & Inamdar, A. (2006). The radiative forcing due to clouds and water vapor. In J. T. Kiehl & V. Ramanathan (Eds.), *Frontiers of Climate Modeling* (pp. 119–151). New York: Cambridge University Press. doi:10.1017/CBO9780511535857.006

Rich, P. M., Dubayah, R., Hetrick, W. A., & Saving, S. C. (1994). Using Viewshed Models to Calculate Intercepted Solar Radiation: Applications in Ecology. *American Society for Photogrammetry and Remote Sensing Technical Papers., 4*, 524–529.

Ritter, B., & Geleyn, J. F. (1992). A comprehensive radiation scheme for numerical weather prediction models with potential applications in climate simulations. *Monthly Weather Review, 120*(2), 303–325. doi:10.1175/1520-0493(1992)120<0303:ACRSFN>2.0.CO;2

Shmeter, S. M. (1987). *Thermodynamics and physics of convective clouds*. Leningrad: Gidrometeoizdat.

Smagorinsky, J. (1953). On the dynamical prediction of large-scale condensation by numerical methods. *Qart. J. Roy. Met. Soc., 79*(341).

Sundstrom, A. (1986). Boundary conditions for limited-aria integration of the viscous forecast equations. *Beitrage zur Physik der Atmosphare*, *50*, 218–227.

Swinbank, W. C. (1963). Long-wave radiation from clear skies. *Quarterly Journal of the Royal Meteorological Society*, *89*(381), 339–348. doi:10.1002/qj.49708938105

Yamamoto, G. (1952). *On a radiation hart. Science Reports. Ser. 5, Geophys, 4*. Tohoku Univ.

Zdunkowski, W. G., Weltch, R. M., & Korb, G. J. (1980). An investigation of the structure of typical two-stream methods for the calculation of solar fluxes and heating rates in clouds. *Contributions to Atmospheric Physics*, *53*, 147–166.

Zhang, T., Stamnes, K., & Bowling, S. A. (2001). Impact of the Atmospheric Thickness on the Atmospheric Downwelling Longwave Radiation and Snowmelt under Clear-Sky Conditions in the Arctic and Subarctic. *Journal of Climate*, *14*(5), 920–939. doi:10.1175/1520-0442(2001)014<0920:IOTATO>2.0.CO;2

Chapter 8
Numerical Methods of Modeling Atmospheric Circulation and Its Physical Processes

ABSTRACT

The last decades are characterized by significant progress in the development and operational use of modern numerical hydrodynamic methods of the Earth's weather and climate. This was made possible primarily due to modern understanding of the laws governing the basic physical and thermodynamic processes in the atmosphere and the emergence of more advanced mathematical models and effective methods of their implementation. In this chapter, we develop new numerical techniques used to solve the non-stationary problem of general circulation of the atmosphere with a prehistory and the problem of planetary weather forecast.

A BRIEF OVERVIEW

The existing prediction models are classified according to the type of meteorological processes, specifics of their mathematical realization of the forecast period, as well as the size of the territory for which they are formulated. Until 1980 the operational weather services of the developed countries have focused on models of global forecasting of synoptic processes, despite some evidence that the quality of such forecasts remained almost constant. The increased interest in global models is explained, first of all, by political reasons and that they allow significant simplifications of the model equations and the use of effective methods of their numerical solution.

To implement a global model with a local high resolution with limited computing resources requires the use of efficient numerical methods. We will measure efficiency as the time that a single processor takes for the integration of model equations for one hour of the model time at a predetermined level of atmospheric circulation reproduction error. Of course, it is desirable that the methods used have an internal parallelism permitting efficient implementation on modern parallel computers.

DOI: 10.4018/978-1-5225-2636-0.ch008

The accuracy and efficiency of the existing numerical methods depend on the dimension of the problem. Choosing the best solution method is a difficult task as the purpose (accuracy) and means (cost) are contradictive to each other. Despite the constant improvement of computers, problems of accuracy and efficiency of numerical methods for solving problems of the circulation of the atmosphere remain valid today.

Currently, there are widely known various modifications of approaches to the numerical solution of the Navier-Stokes equations that are the basis of hydrodynamic models of atmospheric circulation. They can be divided into the following groups:

- **Direct Methods:** These methods (mainly Galerkin methods and their modifications (Fletcher, 1984; Morton & Parrot, 1980; Giraldo, 2000; Moorthi, Higgins, 1993) are accurate, efficient and cost-effective in required computer memory. However, their convergence is largely influenced by the choice of test functions. That is why direct methods are found in the use of convection problems with free or periodic boundaries, for which analytical solution of the linear problem is known.
- **Semi-Analytic Methods:** This approach combines such approximate methods as the method of integral relations, method of the straights, the method of the truncated series, etc., in which the solution of a system of differential equations in partial derivatives is reduced to solving a system of ordinary differential equations. The use of these methods is limited by difficulties in solving a system of ordinary differential equations, since their properties are usually radically different from those of the corresponding partial differential equations (e.g., stiffness of equations when applying direct methods). In studies of recent years, a number of very useful properties and promising results are received for the use of spectroscopic methods in numerical weather prediction based on hemispherical or global baroclinic hydrodynamic models (Boer et al., 1984; Carpenter et al., 1990; Ducrocq, Bougeault, 1995; Geleyn, Hollingsworth, 1979; Hortal, Simmons, 1991; Jakob-Chien, Hack, Williamson, 1995; Taylor, Tribbia, Iskandarani, 1997; Thuburn, Yong, 2000). Its essence consists in an approximation of some meteorological quantities at a time by a finite segment of the series expansion for linearly independent analytic functions defined in the whole domain of integration, and the analytic representation of spatial derivatives included in the equation. At present, the spectral models are used for operational numerical weather prediction in Australia, Canada and developed in other countries.
- **Finite Element Method:** (Argyris, Mareczek, Scharpf, 1970; Barrett, Morton, 1981; Griffiths, Mitchel, 1979; Mitchel, Griffiths, 1980; Cullen, 1973; Lele, 1992; Steger, Warming, 1981). The main advantage of this method is that the ruling equations are solved in a Cartesian coordinate system with the use of computational grids associated with the boundaries geometry of the field solutions and using an approximation based on variation principles. The main disadvantage of the method of finite elements is their cost inefficiency.
- **Finite Difference Methods:** These methods are the most versatile and widely applicable. The first attempts to use finite difference schemes for solving hydrodynamic problems relate yet to the 1930s. The advent of computers contributed to the further rapid development of finite difference methods for solving Navier-Stokes equations and the development of efficient algorithms for the solution of difference equations approximating the differential equations of elliptic, parabolic

and hyperbolic type (Mesinger, Arakava, 1976; Van Leer, 1974; Courtier, Geleyn, 1988; Girard, Delage, 1990). All these types of numerical methods encountered in various simplified versions defining equations of fluid dynamics. However, the system of equations can also have a mixed type.

- **Semi-Lagrangian Method:** In this method, there is no formally the property of norm conservation of values carried in the process of advection that is theoretically necessary for the integration model for long (several decades or more) time. Practice shows that this does not affect significantly the quality of the solution using a model with a resolution of about 1 degree and more for a medium-term weather forecasting. Currently, the semi-Lagrangian approach is widely used to model the short-term and medium-term weather forecasting (Bermejo, Staniforth, 1992; Gospodinov et al., 2002; Gravel, Staniforth, Côté, 1993; Hortal, 2002; Li, Bates, 1996; Naughton, Bourke, 1996; Ritchie, Tanguay, 1996; Williamson, Olson, 1998). There is also the successful experience of the application of semi-Lagrangian model for climate simulations (Bates et al., 1995; Bates, Chen, 1996; Williamson, Olson, 1998). Herewith, it is applied and restoring atmospheric moisture weight at each time step. A disadvantage should be noted that the semi-Lagrangian method is not applicable to tasks that require strict conservatism, for example, problems with discontinuous solutions (shock waves).

Until recently, the finite difference and spectral methods are the most widely used in the implementation of the tasks of atmospheric circulation.

The advantages of the finite difference methods are their locality and a linear scalability in the number of arithmetic operations on one coordinate with increasing resolution. However, the finite-difference model of the atmosphere on a regular latitude-longitude grid, in addition to the introduced phase error, has still one drawback. Due to the convergence of meridians near the poles, this grid has a large heterogeneity resolution in longitude and latitude. This deficiency leads to more restrictions on the Courant number in Euler models and problems in the use of parallel iterative algorithms.

The spectral method has no phase error and nonlinear instability. Other advantages of the method are the spectral uniformity of resolution on the field, trivial solutions of elliptic equations on the sphere, the possibility of using grids with a decreasing number of nodes in longitude while approaching the poles. The main drawback of the spectral method is the cubic increase in the number of operations with increasing resolution. In addition, in the case of a large number of harmonics on the sphere, there is a trouble of calculating sustainable basis functions – associated Legendre polynomials. Because of commonly used triangular Fourier series truncation, when implemented on parallel processors, a very difficult task is to balance the load of all processors.

Despite a number of positive qualities of spectral methods in the numerical integration of equations of hydrodynamics and the heat/mass transfer in problems of macroscale weather forecasting, we opted for the finite-difference method because:

- It is the most fully developed the basic theoretical concepts (approximation, convergence, and stability).
- It refers to the most versatile methods.
- It allows, by geometric splitting, to reduce a numerical solution of the multidimensional problem to a time sequence of one-dimensional tasks.
- It is possible for the development of parallel computing algorithms for multiprocessor computers.

NUMERICAL METHOD FOR SOLVING THE NON-STATIONARY PROBLEM WITH A PREHISTORY

This section is devoted to theoretical and practical aspects of the numerical solution of the problem of the first-order ordinary differential equation

$$\frac{dy}{dt} = f\left(y,t\right), \tag{8.1.1}$$

$$y\Big|_{t=t_0} = y_0, \tag{8.1.2}$$

where $y_0 \in \mathbb{R}, f \in C\left(\mathbb{R} \to \mathbb{R}\right)$, and the function f satisfies the Lipschitz condition.

The exact solution of so set the original problem exists and is unique.

In the practice of the numerical solution of the problem (8.1.1), (8.1.2) for the first-order differential equations, there are used, as a rule, one-step methods (Babenko, 1986; Bahvalov, Zhidkov, Kobelkov, 1987; Berezin, Zhidkov, 1962; Gorbunov, 1973; Krylov, Bobkov, Monastyrnyi, 1977; Milne, 1955; Ortega, Poole, 1986; Hall, Watt, 1979; Hamming, 1973; Gear, 1971; Henrici, 1962; Shampine, Gordon, 1975). One-step methods suggest using information about the problem to be solved only on the time interval of the one step length $\tau = t_l - t_{l-1}, l = \widehat{1,N}$. Such methods, having a number of convenient features for computing, suffer from one significant drawback – low accuracy. To provide the accuracy $O\left(\tau^p\right)$ of the methods, when p>1, such information at each stage of the process should be, in general, obtained again, what determines the complexity of most of the relevant rules of computing. Rejecting the one-step requirements, computational methods can be constructed so that a part of the received information on several adjacent steps of the computational process can be re-used. Such methods, using information on the problem to be solved on the time interval length of more than one step, will be the main object of our study in this section.

Auxiliary Definitions and Results

Our main goal is creation and analysis of basic structural properties of an efficient finite-difference l-steps one-stage scheme for solving the problem (8.1.1) - (8.1.2). We define the temporal difference grid, which will be used for numerical solution of the problem (8.1.1) - (8.1.2). For the sake of simplicity, we will assume below $t_0 = 0$, what does not limit the generality.

Definition 8.1.1: A finite set of nodes $t_n \in \left[0,T\right], n = \widehat{0,N}$, is called the (computational) grid on [0,T] if

$$t_0 = 0, t_n > t_{n-1}, n = \widehat{1,N}. \tag{8.1.3}$$

The values $\tau_n = t_n - t_{n-1} > 0, n = \widehat{1,N}$ are called grid steps, the value $\gamma = t_{n-1} / t_n$, $n = \widehat{1,N}$ is called the ratio of grid steps.

So, equivalently, the grid $t_n \in [0,T], n = \widehat{0,N}$ can be given by the set $\omega_\tau = \{\tau_n, n = \widehat{1,N}\}$ where $\tau_n = t_n - t_{n-1} > 0, n = \widehat{1,N}$.

Definition 8.1.2: We will call a grid $\bar{\omega}_\tau = \{\tau = T / N\}$ as regular if

$$t_n = n\tau, n = \widehat{1,N}, \; \gamma = 1;$$

otherwise, a grid $\omega_\tau = \{\tau_n, n = \widehat{1,N}\}$ is called irregular if

$$t_n \neq t_{n-1}, \; n = \widehat{1,N}, \; \gamma \neq 1. \tag{8.1.4}$$

The structure of the existing common *l*-step methods is very complicated (Bahvalov, Zhidkov, Kobelkov, 1987; Zalëtkin, 1986; Hall, Watt, 1976; Stetter, 1973). So we use here only a special class of single-stage *l*-step methods, in which the step procedure consists in pure drawing a linear combination of values y_i and f_i, $i = \widehat{n, n+l}$. No repeated substituting values of *f* (which constitute the essence of multi-stage methods) is not allowed.

In the *l*-step one-stage scheme for solving the problem (8.1.1), (8.1.2), the most widely used is the step procedure of the form

$$\sum_{i=0}^{l} a_i y_{n+i} - \tau_l \sum_{i=0}^{l} b_i f_{n+i} = 0, \; n = \widehat{0,N}. \tag{8.1.5}$$

Modifier τ_l is allocated to the second sum to make coefficients b_i and a_i to be dimensionless. The matrix

$$M = \begin{pmatrix} a_0 \dots a_N \\ b_0 \dots b_N \end{pmatrix}$$

of the size 2×(*N*+1) is called *generating matrix* of the scheme (8.1.5). If b_l=0, the computational methods of the form (8.1.5) is usually called *explicit,* otherwise it is called *implicit.*

Note that the formula (8.1.5) must be accurate in the case where *y*(*t*) is a constant $\left(\partial y / \partial t = f \equiv 0\right)$. This yields:

$$a_l \neq 0, \sum_{i=0}^{l} a_i = 0. \tag{8.1.6}$$

The linear l-step procedure (8.1.5) and its generating matrix, by performing replacement $q^i = y_i$ and $r^i = f_i$, $i = \widehat{0,l}$, can be associated with the "characteristic" polynomials

$$\Phi_a(q) = \sum_{i=0}^{l} a q^i, \; \Phi_b(r) = \sum_{i=0}^{l} b_i r^i. \tag{8.1.7}$$

If c is a number, then the number

$$\Phi_a(c) = \sum_{i=0}^{l} a_i c^i,$$

obtained by replacing the variable q by the value c and the subsequent executing all these operations, is called the value of the polynomial $\Phi_a(q)$ at q=c. If $\Phi_a(c) = 0$, then c is called the root of the polynomial $\Phi_a(q)$ (or of the equation $\Phi_a(q) = 0$).

Statement 8.1.3: A number c is the root of the polynomial $\Phi_a(q)$ if $\Phi_a(c)$ is equal to zero.

As the number one is a root of the characteristic polynomial Φ_a, it cannot be a zero of the characteristic polynomial Φ_b. Therefore, in every class of procedures, there is a unique linear l-step scheme and the single corresponding generating matrix satisfying the condition

$$\sum_{i=0}^{l} b_i = 1. \tag{8.1.8}$$

The following finite-difference scheme (8.1.5), subject to the conditions (8.1.6) and (8.1.8) is called *normalized.*

Coefficients a_i and b_i can be selected in various ways so that (8.1.5) will approximate $\partial y / \partial t - f = 0$. In the books (Bahvalov, Zhidkov, Kobelkov, 1987; Zalëtkin, 1986; Hall, Watt, 1976; Stetter, 1973), there are shown the methods of construction and investigation of approximation errors, stability, and convergence of some special classes of efficient linear l-step schemes. The systematization of those theoretical results shows that the basic requirements of the stability of single-stage multistep schemes relate mainly to the coefficients a_i of the characteristic polynomial Φ_a and formulated the following definitions.

Statement 8.1.4: The characteristic polynomial Φ_a satisfies the root criterion if its roots belong to the unit of circle area, and it has no multiple zeros on the unit circle.

Definition 8.1.5: The homogeneous linear l-step scheme is called strongly stable if its characteristic polynomial Φ_a satisfies the root criterion.

The simplest and at the same time the most common schemes of the form (8.1.5) can be obtained based on quadrature formulas and the method of undetermined coefficients. However, not all the schemes produced by such methods are stable. For example, explicit scheme

$$y_{n+l} + 4y_{n+l-1} - 5y_{n+l-2} - \frac{\tau}{6}\left(4f_{n+l-1} + 2f_{n+l-2}\right) = 0,$$

satisfying the normalization conditions (8.1.6), (8.1.8) is not stable and cannot converge because of its characteristic polynomial

$$\Phi_a\left(q\right) = q^2 + 4q - 5 = \left(q-1\right)\left(q+5\right)$$

has a root that goes beyond the unit interval.

The following method is chosen as a basis for constructing a family of multistep single-stage high-order schemes. We use the special relation between the coefficients of the characteristic polynomial Φ_a, according to which real coefficients of linear l-step scheme a priori satisfy the criterion of the root in the narrow sense, i.e., characteristic polynomial Φ_a has *only real roots inside the unit interval* of the real axis, and *there are no multiple zeros* at the end of the interval.

First, we consider some of the basic properties of divisibility of polynomials, which will be used in the following. In general, when the polynomial with real coefficients

$$\eta\left(q\right) = a_l q^l + a_{l-1} q^{l-1} + \ldots + a_1 q + a_0 = \sum_{i=0}^{l} a_i q^i \tag{8.1.9}$$

is divided by a linear polynomial

$$\varphi\left(q\right) = q - c \tag{8.1.10}$$

the resulting remainder will be some number ε (polynomial of the degree of zero). To find the remainder not performing division is possible by the following theorem.

Theorem 8.1.1: The remainder of the division of the polynomial $\eta\left(q\right)$ by the linear polynomial $\varphi\left(q\right)$ is the value of the polynomial $\eta\left(c\right)$ when $q = c$.

Proof: From elementary algebra, it is known that two polynomials with real coefficients $\eta\left(q\right)$ and $\varphi\left(q\right) = q - c$ can be found such polynomials $\psi\left(q\right)$ and ε, that

$$\eta\left(q\right) = \left(q-c\right)\psi\left(q\right) + \varepsilon. \tag{8.1.11}$$

It is also known that the polynomials $\psi(q)$ and ε, satisfying this condition, are determined uniquely.

Taking the values of both sides of the equation (8.1.11) at $q=c$, we obtain $\eta(c)=(c-c)\psi(c)+\varepsilon=\varepsilon$, what proves the theorem.

Corollary 8.1.1: When $c=1$ and the condition (8.1.6) holds, the remainder ε is zero.

Indeed, substituting $q=1$ in (8.1.9), we get

$$\eta(1)=a_l+a_{l-1}+\ldots+a_1+a_0=\sum_{i=0}^{l}a_i=0.$$

This implies the following theorem.

Theorem 8.1.2: The polynomial $\varphi(q)$ is the divisor of the polynomial $\eta(q)$ iff there exists a polynomial $\psi(q)$, satisfying the equation

$$\eta(q)=\varphi(q)\psi(q). \tag{8.1.12}$$

The polynomial $\psi(q)$, satisfying this condition, is determined uniquely.

Proof: Suppose there is another polynomial $\bar{\psi}(q)$, satisfying the equality

$$\eta(q)=\varphi(q)\bar{\psi}(q). \tag{8.1.13}$$

Equating the right sides of equations (8.1.12) and (8.1.13), we obtain:

$$\varphi(q)\left\{\psi(q)-\bar{\psi}(q)\right\}=0.$$

Since in this expression $\varphi(q)\neq 0$, it should be $\psi(q)-\bar{\psi}(q)=0$; and so $\psi(q)=\bar{\psi}(q)$, as required. Obviously, the degree $\varphi(q)$ is not greater than the degree $\eta(q)$.

Theorem 8.1.3: The number $c=d/a$, $a\neq 0$ is the root of the polynomial $\eta(q)$, iff $\eta(q)$ is divided by a linear polynomial aq-d.

Proof: Indeed, let

$$\eta(q) = (aq - d)\psi(q) = a(q - c)\psi(q).$$

Taking the value of both parts of this equation at $q=c$, we get:

$$\eta(c) = a(c - c)\psi(c) = 0,$$

what proves the theorem. Consequently, finding the roots of a polynomial $\eta(q)$ is equivalent to finding its linear divisors.

The Fundamental Theorem of Algebra

Every polynomial with real coefficients l-th degree

$$\eta(q) = a_l q^l + a_{l-1} q^{l-1} + \ldots + a_1 q + a_0 = \sum_{i=0}^{l} a_i q^i \tag{8.1.14}$$

has exactly l+1 real roots, provided that each root is counted as many times as its multiplicity, has the following decomposition:

$$\eta(q) = (a_0 q - 1)(a_1 q - 1)\ldots(a_l q - 1). \tag{8.1.15}$$

The expansion (8.1.15) is unique for the polynomial of (8.1.14) up to the order of the factors.

Proof: The theorems proved above allow us to assert the existence of the root $q_0 = a_0^{-1}$ for the $\eta(q)$.

Therefore, polynomial $\eta(q)$ has an expansion

$$\eta(q) = (q - q_0)\psi(q).$$

The coefficients of the polynomial $\psi(q)$ are real again, and therefore $\psi(q)$ has a root $q_1 = a_1^{-1}$, from where we have

$$\eta(q) = (q - q_0)(q - q_1)\xi(q).$$

Continuing in this way, we arrive at a finite number of steps to the decomposition of (8.1.15).

To prove the uniqueness of the expansion (8.1.15), we assume that there is also a decomposition

$$\eta(q) = (q - \zeta_0)(q - \zeta_1)\ldots(q - \zeta_l). \tag{8.1.16}$$

From (8.1.15) and (8.1.16), it implies that

$$\left(q-q_0\right)\left(q-q_1\right)\ldots\left(q-q_l\right)=\left(q-\zeta_0\right)\left(q-\zeta_1\right)\ldots\left(q-\zeta_l\right). \tag{8.1.17}$$

If the root q_l of the multiplicity s=1 was different from all ζ_j, $j = 0,1,\ldots,l$, then substituting q_l in (8.1.17) instead of unknown q, we would get a zero on the left and a number other than zero, on the right, i.e., a contradiction. Thus, any root q_j of the multiplicity s=1 is equal to a root ζ_j and back.

Suppose further that the expansion contains s>1 roots q_0 and among the roots ζ_j, $j = 0,1,\ldots,l$, m roots are equal to the root q_0. We have to prove that $s=m$. If we assume, for example, that $s>m$, then dividing both sides of (8.1.17) by a factor $(q-q_0)^m$, we come to the equation, the left part of which still contains the factor $q-q_0$, but the right side does not contain it. Above it was shown that this leads to a contradiction. Thus, the uniqueness of the decomposition (8.1.15) for the polynomial (8.1.17) is proved.

Although the fundamental theorem of algebra and the majority of her previous results are true for complex coefficients a_j(j=0,1,…,l), we shall consider them real everywhere, focusing on our applications.

CREATING A FAMILY OF STRONGLY RESISTANT LINEAR *l*-STEP FINITE DIFFERENCE SCHEMES ON A UNIFORM GRID

A numerical implementation of dynamic models of atmospheric circulation on a computer, as always, puts forward very stringent performance requirements of computational methods (Young, 1968):

- The method must be identical to the explicit one, to provide quick calculations to determine the problem values at each time step.
- The method should have favorable properties of strong stability to exclude the excessive propagation of error for the problems with decreasing components of the solution.
- The method should provide the ability to control the length of time step on the entire range of integration of equations, and the step change should not complicate the implementation of the algorithm of the method.

Despite the mutual contradiction of these requirements, it is possible to construct a family of very simple linear l-step one-stage methods that satisfy all three requirements.

Based on the methodology required to build a family of one-stage multistep methods, we begin on a uniform grid $\bar{\omega}_\tau$. Then we extend this result to the grid ω_τ with a variable step. We will get some of the consequences of the theory of complex numbers that are related to polynomials with real coefficients. Essentially, these implications make the fundamental theorem extremely important what will be discussed below.

Let have a polynomial with real coefficients

$$\eta\left(q\right)=a_l q^l+a_{l-1}q^{l-1}+\ldots+a_1 q+a_0$$

That has a complex root α, i.e.

$$a_l\alpha^l + a_{l-1}\alpha^{l-1} + \ldots + a_1\alpha + a_0 = 0.$$

This equality is not violated if all numbers in it are replaced by the conjugate ones. But all the coefficients $a_i\left(i = \widehat{0,l}\right)$, and the number 0 in the polynomial $\eta\left(q\right)$ to the right remain real and unchanged at the replacement held. So we arrive at the equality:

$$\eta\left(\bar{\alpha}\right) = a_l\bar{\alpha}^l + a_{l-1}\bar{\alpha}^{l-1} + \ldots + a_1\bar{\alpha} + a_0 = 0.$$

Thus, we can formulate the following

Definition 8.1.6: If a complex number α is the root of the polynomial $\eta\left(q\right)$ with real coefficients, then the conjugate number $\bar{\alpha}$ is also root of this polynomial.

The polynomial $\eta\left(q\right)$ is dividable, therefore, by the quadratic trinomial

$$\psi\left(q\right) = \left(q - \alpha\right)\left(q - \bar{\alpha}\right) = q^2 - \left(\alpha + \bar{\alpha}\right)q + \alpha\bar{\alpha}, \tag{8.1.18}$$

whose coefficients are known from the theory of complex numbers to be real. Using this, we can prove that the roots α and $\bar{\alpha}$ of the polynomial $\eta\left(q\right)$ are of the same multiplicity.

Suppose, on the contrary, that the polynomial $\eta\left(q\right)$ has the roots, respectively, of multiplicity k_1 and k_1 and let, for example, $k_1 > k_2$. Then $\eta\left(q\right)$ is dividable by k_2-th degree of the polynomial $\psi\left(q\right)$, $\eta\left(q\right) = \psi^{k_2}\left(q\right)\xi\left(q\right)$.

The polynomial $\xi\left(q\right)$, as the quotient of two polynomials with real coefficients, also has real coefficients, but in contradiction to the established above, it has α as its $\left(k_1 - k_2\right)$-fold root, while the number $\bar{\alpha}$ is not its root. It follows that $k_1 = k_2$. Thus, we arrive at the well-known the fact from algebra that the complex roots of any polynomial with real coefficients are pairwise conjugated. From this and the above proved the uniqueness of expansions (8.1.15), we obtain the following final result.

Corollary 8.1.2: Every polynomial $\Phi_a\left(q\right)$ with real coefficients is represented in a unique way (up to the order of the factors) as a product of its leading coefficient a_l and several polynomials with real coefficients of linear type $q - \alpha$, corresponding to its real roots, and of square type (8.1.18), the respective pairs of conjugate complex roots.

The above result suggests the following.

Theorem 8.1.3: For strongly stable homogeneous linear l-step finite-difference scheme (8.1.5) with real coefficients a_i and b_i $\left(i=\widehat{0,l}\right)$, obeying the normalization conditions (8.1.6) and (8.1.8), the characteristic polynomial $\Phi_a\left(q\right)$ has only real roots that satisfy the strong root criteria (when, except for simple one $q_0 = 1$, the rest of roots q_i $\left(i=\widehat{1,l}\right)$ are located inside the open unit interval), and is defined in a unique way:

$$\Phi_a\left(q\right)=\left(q-1\right)\prod_{i=1}^{l}\left(2q-1\right)^i. \tag{8.1.19}$$

Proof: Let us first consider the polynomial (8.1.7) with such the coefficients $\overline{a}_i$, $\left(i=\widehat{1,l}\right)$ that all the roots $\overline{q}_i$ of the polynomial

$$\overline{\Phi}_a\left(q\right)=\sum_{i=1}^{l}\overline{a}_i q^i$$

lie inside the open unit interval. Next, consider the polynomial (8.1.15) which has only a simple root $q_0 = 1$ and define the connection between the coefficients a_i and $\overline{a}_i$ $\left(i=\widehat{1,l}\right)$ by the equality

$$\Phi_a\left(q\right)=\left(q-q_0\right)\overline{\Phi}_a\left(q\right). \tag{8.1.20}$$

We expressed the coefficients of the polynomial $\overline{\Phi}_a\left(q\right)$ in terms of its roots with the help of Viet formulas:

$$\overline{a}_0=\left(-1\right)^{l-1},$$

$$\overline{a}_1=\left(-1\right)^{l-2}\left(\alpha_1+\alpha_2+\ldots+\alpha_l\right), \tag{8.1.21}$$

$$\overline{a}_{l-1}=\left(\alpha_1\alpha_2\ldots\alpha_{l-1}+\alpha_1\alpha_2\ldots\alpha_{l-2}\alpha_l+\ldots+\alpha_2\alpha_3\ldots\alpha_l\right),$$

$$\overline{a}_l=\alpha_1\alpha_2\ldots\alpha_l.$$

Then the coefficients of the polynomial $\Phi_a\left(q\right)$ and the equality (8.1.20) take the form:

$$a_0 = (-1)^l,$$

$$a_1 = (-1)^{l-1}(\alpha_1 + \alpha_2 + \ldots + \alpha_l + 1),$$

$$a_2 = (-1)^{l-2}(\alpha_1 + \alpha_2 + \ldots + \alpha_l + \alpha_1\alpha_2 + +\alpha_1\alpha_3 + \ldots + \alpha_{l-1}\alpha_l), \quad (8.1.22)$$

$$a_{l-1} = -(\alpha_1\alpha_2 \ldots \alpha_l + \alpha_1\alpha_2 \ldots \alpha_{l-1} + \alpha_1\alpha_2 \ldots \alpha_{l-2}\alpha_l + \ldots + \alpha_2\alpha_3 \ldots \alpha_l),$$

$$a_l = \alpha_1\alpha_2 \ldots \alpha_l .$$

Expand the function $y(t)$ to the Taylor series at the points l,l-2,…,0 by its value at the point l-1

$$y_{n+l} = y_{n+l-1} + \frac{\tau}{1!} f_{n+l-1} + \frac{\tau^2}{2!}\left(\frac{\partial f}{\partial t}\right)_{n+l-1} + \ldots + \frac{\tau^k}{k!}\left(\frac{\partial^{k-1} f}{\partial t^{k-1}}\right)_{n+l-1},$$

$$y_{n+l-2} = y_{n+l-1} - \frac{\tau}{1!} f_{n+l-1} + \frac{\tau^2}{2!}\left(\frac{\partial f}{\partial t}\right)_{n+l-1} + \ldots + (-1)^k \frac{\tau^k}{k!}\left(\frac{\partial^{k-1} f}{\partial t^{k-1}}\right)_{n+l-1}, \quad (8.1.23)$$

$$y_{n+l-3} = y_{n+l-1} - 2\frac{\tau}{1!} f_{n+l-1} + 4\frac{\tau^2}{2!}\left(\frac{\partial f}{\partial t}\right)_{n+l-1} + \ldots + (-1)^k 2^k \frac{\tau^k}{k!}\left(\frac{\partial^{k-1} f}{\partial t^{k-1}}\right)_{n+l-1},$$

$$y_n = y_{n+l-1} - (l-1)\frac{\tau}{1!} f_{n+l-1} + (l-1)^2 \frac{\tau^2}{2!}\left(\frac{\partial f}{\partial t}\right)_{n+l-1} + \ldots + (-1)^k (l-1)^k \frac{\tau^k}{k!}\left(\frac{\partial^{k-1} f}{\partial t^{k-1}}\right)_{n+l-1}.$$

Multiplying the left and right side of the first equality (8.1.23) by the coefficient a_l determined by the system (8.1.22), of the second equation (8.1.23) – by the factor a_{l-2}, of the third – by a_{l-3}, etc., and then adding the resulting expressions term by term, we arrive at the equality

$$\sum_{i=0}^{l} a_i y_{n+i} = \tau \sum_{i=1}^{k-1} \frac{c_i}{(i+1)!}\left(\frac{\partial^i f}{\partial t^i}\right)_{n+l-1} \quad (8.1.24)$$

where

$$c_1 = a_l - a_{l-2} - 2a_{l-3} - 3a_{l-4} - \ldots,$$

$$c_2 = a_l + a_{l-2} + 4a_{l-3} + 9a_{l-4} - \ldots,$$

$$c_3 = a_l - a_{l-2} - 8a_{l-3} - 27a_{l-4} - \ldots,$$

$$c_4 = a_l + a_{l-2} + 16a_{l-3} + 81a_{l-4} - \ldots,$$

$$c_l = a_l + (-1)^l \left(a_{l-2} + 2^l a_{l-3} + 3^l a_{l-4}\right).$$

If the right side of the differential equation (8.1.1) is continuous and bounded together with its derivatives of n-th order, the approximation (8.1.24) determines a high accuracy due to the low coefficient of the remainder, especially with decreasing the step τ. If the right-hand side has not listed derivatives, the ultimate order of accuracy of the scheme cannot be realized. Then, as a rule, there are used the schemes of lower order accuracy corresponding to the order of the derivative.

Missing the consideration of the simplest two-step scheme of first and second order as enough investigated so far, we construct here a family of the third (in the case of explicit scheme) and the fourth (in the case of an implicit scheme) order of accuracy schemes and analyze the basic ideas of the methods. As a source of expression for characteristic polynomial $\Phi_a(q)$, take the quadratic trinomial

$$\Phi_a(q) = (q-1)(a_1 q - 1) = a_1 q^2 - (1 + a_1)q + 1, \tag{8.1.25}$$

one of real roots of which is the root $q_0 = 1$. In accordance with the result of the section 8.2, the second root of the polynomial (8.1.25) is also real.

We give the coefficient a_1 so that the characteristic polynomial (8.1.25) corresponds to a linear procedure (8.1.5), which approximates the equation $\partial y / \partial t - f(t, y) = 0$. As an initial expression, we take the equation (8.1.24) for $l = 2$

$$\begin{aligned} a_1 y_{n+l} - (a_1 + 1) y_{n+l-1} + y_{n+l-2} &= (a_1 - 1)\tau f_{n+l-1} + (a_1 + 1)\frac{\tau^2}{2!}\left(\frac{\partial f}{\partial t}\right)_{n+l-1} \\ &+ (a_1 - 1)\frac{\tau^3}{3!}\left(\frac{\partial^2 f}{\partial t^2}\right)_{n+l-1} + (a_1 + 1)\frac{\tau^4}{4!}\left(\frac{\partial^3 f}{\partial t^3}\right)_{n+l-1} + \ldots \end{aligned} \tag{8.1.26a}$$

and supplement it with the Taylor series for the function as follows:

$$0 = -\tau f_{n+l-2} + \tau f_{n+l-1} - 2\frac{\tau^2}{2!}\left(\frac{\partial f}{\partial t}\right)_{n+l-1} + 3\frac{\tau^3}{3!}\left(\frac{\partial^2 f}{\partial t^2}\right)_{n+l-1} - 4\frac{\tau^4}{4!}\left(\frac{\partial^3 f}{\partial t^3}\right)_{n+l-1} + \ldots \tag{8.1.26b}$$

Multiply (8.1.26b) by the factor $(a_1+1)/2$ and add the obtained expression to (8.1.26a). The first two terms on the right side of the number (8.1.26b) contain the coefficients with opposite signs, equal modulo τ. Therefore, their algebraic sum, multiplied by the same factor $(a_1+1)/2$ yields zero. Therefore, to satisfy the condition (8.1.8) the coefficient a_1-1 at f_{n+l-1} must be equal to 1, so a_1=2. The resulting value of a_1 leads to the following formula for the explicit difference scheme:

$$y_{n+l} = y_{n+l-1} + \frac{1}{2}\left[y_{n+l-1} - y_{n+l-2} + \frac{\tau}{2}\left(5f_{n+l-1} - 3f_{n+l-2}\right)\right] + O\left(\tau^3\right). \quad (8.1.27)$$

When a_1=2 the roots of the characteristic equation for the scheme (8.1.27) are real and different and are subject to dependencies q_0=1 and $0<q_1<1$. Consequently, in accordance to the definition 8.1.5 the homogeneous scheme (8.1.27) is one of the strongly stable schemes.

To construct the implicit scheme, we supplement the systems (8.1.26a) and (8.1.26b) by Taylor series

$$0 = -\tau f_{n+l} + \tau f_{n+l-1} + 2\frac{\tau^2}{2!}\left(\frac{\partial f}{\partial t}\right)_{n+l-1} + 3\frac{\tau^3}{3!}\left(\frac{\partial^2 f}{\partial t^2}\right)_{n+l-1} - 4\frac{\tau^4}{4!}\left(\frac{\partial^3 f}{\partial t^3}\right)_{n+l-1} + \ldots \quad (8.1.28)$$

Now multiply (8.1.26b) by some factor A and (8.1.28) – by some factor B and add obtained expressions to (8.1.26a). Determine the values of the factors A and B so that the coefficients $\left(\partial f / \partial t\right)_{n+l-1}$ and $\left(\partial^2 f / \partial t^2\right)_{n+l-1}$ vanish. Hence, we obtain a system of algebraic equations

$$2A - 2B = -\left(a_1 + 1\right),$$

$$3A + 3B = -\left(a_1 - 1\right). \quad (8.1.29)$$

As, like in (8.1.26b), the first two terms on the right hand side of (8.1.28) with equal coefficients have opposite signs, to satisfy the condition (8.1.8) it is necessary the coefficient $(a_1$-1) at f_{n+l-1} on the right side of (8.1.26a) is equal to 1. This condition is still satisfied by the equality a_1=2. Then, from the solution of (8.1.29), we obtain A=-11/12, B=7/12 and the following formula for the implicit difference scheme:

$$y_{n+l} = y_{n+l-1} + \frac{1}{2}\left[y_{n+l-1} - y_{n+l-2} + \frac{\tau}{12}\left(11f_{n+l} + 8f_{n+l-1} - 7f_{n+l-2}\right)\right] + O\left(\tau^4\right). \quad (8.1.30)$$

Construct another family of finite difference schemes for l=3. Give the characteristic polynomial $\Phi_a\left(q\right)$ in the form of a cubic polynomial

$$\Phi_a\left(q\right) = \left(q-1\right)\left(a_1 q - 1\right)\left(a_2 q - 1\right) = a_1 a_2 q^3 - \left(a_1 + a_2 + a_1 a_2\right) q^2 + \left(1 + a_1 + a_2\right) q - 1, \quad (8.1.31)$$

the roots of which are different reals q_0=1, $q_1 = 1/a_1 = 1/2$. In accordance with the corollary of 8.1.2 the third root of (8.1.31) is also real.

Represent the equation (8.1.24) at l=3 and additional Taylor series for the function f in the following form:

$$a_1 a_2 y_{n+l} - \left(a_1 + a_2 + a_1 a_2\right) y_{n+l-1} + \left(1 + a_1 + a_2\right) y_{n+l-2} - y_{n+l-3}$$
$$= \left(a_1 a_2 - a_1 - a_2 + 1\right) \tau f_{n+l-1} + \left(a_1 a_2 + a_1 + a_2 - 3\right) \frac{\tau^2}{2!} \left(\frac{\partial f}{\partial t}\right)_{n+l-1} , \tag{8.1.32a}$$
$$+ \left(a_1 a_2 - a_1 - a_2 + 7\right) \frac{\tau^3}{3!} \left(\frac{\partial^2 f}{\partial t^2}\right)_{n+l-1} + \left(a_1 a_2 + a_1 + a_2 - 15\right) \frac{\tau^4}{4!} \left(\frac{\partial^3 f}{\partial t^3}\right)_{n+l-1}$$

$$0 = -\tau f_{n+l-2} + \tau f_{n+l-1} - 2 \frac{\tau^2}{2!} \left(\frac{\partial f}{\partial t}\right)_{n+l-1} + 3 \frac{\tau^3}{3!} \left(\frac{\partial^2 f}{\partial t^2}\right)_{n+l-1} - 4 \frac{\tau^4}{4!} \left(\frac{\partial^3 f}{\partial t^3}\right)_{n+l-1} + \ldots \tag{8.1.32b}$$

$$0 = -\tau f_{n+l-3} + \tau f_{n+l-1} - 4 \frac{\tau^2}{2!} \left(\frac{\partial f}{\partial t}\right)_{n+l-1} + 12 \frac{\tau^3}{3!} \left(\frac{\partial^2 f}{\partial t^2}\right)_{n+l-1} - 32 \frac{\tau^4}{4!} \left(\frac{\partial^3 f}{\partial t^3}\right)_{n+l-1} + \ldots \tag{8.1.32c}$$

Add (8.1.32a) and the equations (8.1.32b) and (8.1.32c) multiplied by some coefficients A and B, respectively. Determine the values of the factors A and B so that the coefficients $\left(\partial f / \partial t\right)_{n+l-1}$ and $\left(\partial^2 f / \partial t^2\right)_{n+l-1}$ vanish. As a result, we obtain a system of equations:

$$2A + 4B = -\left(a_1 a_2 + a_1 + a_2 - 3\right),$$

$$3A + 12B = -\left(a_1 a_2 - a_1 - a_2 + 7\right). \tag{8.1.33}$$

Note that the first two terms in the right side of (8.1.32b) and the right side of (8.1.32c) remain equal and have opposite signs. Therefore, to satisfy the condition (8.1.8), the coefficient at f_{n+l-1} on the right side of (8.1.32a) should be equal to 1:

$$\left(a_1 a_2 - a_1 - a_2 + 1\right) = 1. \tag{8.1.34}$$

The equation (1.8.34) is valid if $a_1 = a_2 = 2$, as required. At the same time, from the solution of (8.1.33), we obtain A=88/12, B=-29/12 and the following formula for the explicit difference scheme:

$$y_{n+l} = y_{n+l-1} + \frac{1}{4}\left[4y_{n+l-1} - 5y_{n+l-2} + y_{n+l-3} + \frac{\tau}{12}\left(71f_{n+l-1} - 88f_{n+l-2} + 29f_{n+l-3}\right)\right] + O\left(\tau^3\right). \tag{8.1.35}$$

The roots of the characteristic equation for the scheme (8.1.35) are different reals and subject to dependencies $q_0=1$, $0<q_1<1$ and $0<q_2<1$. Consequently, in accordance with the definition 8.1.5, the homogeneous scheme (8.1.35) is strongly stable.

To construct the implicit scheme, supplement the system (1.8.32) by the Taylor series

$$0 = -\tau f_{n+l} + \tau f_{n+l-1} + 2\frac{\tau^2}{2!}\left(\frac{\partial f}{\partial t}\right)_{n+l-1} + 3\frac{\tau^3}{3!}\left(\frac{\partial^2 f}{\partial t^2}\right)_{n+l-1} + 4\frac{\tau^4}{4!}\left(\frac{\partial^3 f}{\partial t^3}\right)_{n+l-1} + 5\frac{\tau^5}{5!}\left(\frac{\partial^4 f}{\partial t^4}\right)_{n+l-1} + \ldots \tag{8.1.36}$$

Multiply (1.8.36) by some factor A and add the result obtained to the equations (8.1.32) and (8.1.33), multiplied by factors B and C, respectively. Determine the values of factors A, B and C so that the coefficients $\left(\partial f / \partial t\right)_{n+l-1}$, $\left(\partial^2 f / \partial t^2\right)_{n+l-1}$ and $\left(\partial^3 f / \partial t^3\right)_{n+l-1}$ vanish. This gives a system of equations:

$$2A - 2B - 4C = -\left(a_1 a_2 + a_1 + a_2 - 3\right),$$

$$3A + 3B + 12C = -\left(a_1 a_2 - a_1 - a_2 + 7\right),$$

$$4A - 4B - 32C = -\left(a_1 a_2 + a_1 + a_2 - 15\right). \tag{8.1.37}$$

It should be noted once again that the first two terms on the right sides of the series (8.1.32a), (8.1.32b) and (8.1.32c) have opposite signs, so the algebraic sum of equal by modulo coefficients multiplied by the same number in each of these rows is zero. Therefore, to satisfy the condition (8.1.8) it is necessary the only non-zero coefficient $\left(a_1 a_2 - a_1 - a_2 + 1\right)$ at f_{n+l-1} on the right-hand side of (8.1.31) is equal to 1, i.e. we again arrive at (8.1.34) and equality $a_1 = a_2 = 2$. By solving the system (8.1.37), we obtain A=-41/24, B=53/24, C=-17/24 and the following formula for the implicit difference scheme:

$$y_{n+l} = y_{n+l-1} + \frac{1}{4}\left[4y_{n+l-1} - 5y_{n+l-2} + y_{n+l-3} + \frac{\tau}{24}\begin{pmatrix}41f_{n+l} + 19f_{n+l-1} \\ -53f_{n+l-2} + 17f_{n+l-3}\end{pmatrix}\right] + O\left(\tau^5\right). \tag{8.1.38}$$

Thus, we have shown that a family of strongly stable linear l-step schemes (8.1.35) and (8.1.38) for l=3, regardless of the order of local accuracy p, determines the characteristic polynomial $\Phi_a\left(q\right)$ with the same values of a_1 and a_2, defined by (8.1.34). It remains to prove that these values of a_1 and a_2 are zeros of the polynomial $\Phi_a\left(q\right)$ of the strongly stable l-step scheme at another value of l.

For this purpose, we construct the characteristic polynomial $\Phi_a(q)$ of the fourth degree

$$\begin{aligned}\Phi_{\blacklozenge}(q) &= (q-1)(a_1q-1)(a_2q-1)(a_3q-1)\\ &= a_1a_2a_3q^4 - (a_1a_2 + a_1a_3 + a_2a_3 + a_1a_2a_3)q^3\\ &+(a_1 + a_2 + a_3 + a_1a_2 + a_1a_3 + a_2a_3)q^2 - (a_1 + a_2 + a_3 + 1)q + 1\end{aligned}, \tag{8.1.39}$$

the roots of which are different reals $q_0=1$, $q_1 = 1/a_1 = 1/2$ and $q_2 = 1/a_2 = 1/2$. In accordance with the corollary 8.1.2 the fourth root of (8.1.39) will also be real.

The factor a_3 choose so that the characteristic polynomial (8.1.39) corresponds to a linear procedure (8.1.24) which approximates the equation $\partial y/\partial t - f(t,y) = 0$. To do this, multiply both sides of the first series of the system (8.1.23) by the coefficient of q^4 of the characteristic polynomial (8.1.39), both parts of the second series of the system (8.1.23) is multiplied by the coefficient at q^2 in the same polynomial, both parts of the third series of the system (8.1.23) is multiplied by the coefficient at q in (8.1.39), and add these three series to the fourth series of the system (8.1.23). As the result, we will have:

$$\begin{aligned}&\alpha_1\alpha_2\alpha_3 y_{n+l} - (\alpha_1\alpha_2\alpha_3 + \alpha_1\alpha_2 + \alpha_1\alpha_3 + \alpha_2\alpha_3)y_{n+l-1}\\ &+(\alpha_1\alpha_2 + \alpha_1\alpha_3 + \alpha_2\alpha_3 + \alpha_1 + \alpha_2 + \alpha_3)y_{n+l-2}(a_1 + a_2 + a_3 + 1)y_{n+l-3} - y_{n+l-4}\\ &= (a_1a_2a_3 - a_1a_2 - a_1a_3 - a_2a_3 + a_1 + a_2 + a_3)\tau f_{n+l-1}\\ &+\left[a_1a_2a_3 + a_1a_2 + a_1a_3 + a_2a_3 - 3(a_1 + a_2 + a_3) + 5\right]\frac{\tau^2}{2!}\left(\frac{\partial f}{\partial t}\right)_{n+l-1}\\ &+\left[a_1a_2a_3 - a_1a_2 - a_1a_3 - a_2a_3 + 7(a_1 + a_2 + a_3) - 19\right]\frac{\tau^3}{3!}\left(\frac{\partial^2 f}{\partial t^2}\right)_{n+l-1}\\ &+\left[a_1a_2a_3 + a_1a_2 + a_1a_3 + a_2a_3 - 15(a_1 + a_2 + a_3) + 65\right]\frac{\tau^4}{4!}\left(\frac{\partial^3 f}{\partial t^3}\right)_{n+l-1} + \ldots\end{aligned} \tag{8.1.40}$$

To satisfy condition (8.1.8) one must make the coefficient at τf_{n+l-1} on the right side of (8.1.40) equal to one, i.e. to put

$$a_1a_2a_3 - a_1a_2 - a_1a_3 - a_2a_3 + a_1 + a_2 + a_3 - 1 = 1.$$

This equation is valid if

$$a_1 = a_2 = a_3 = 2, \tag{8.1.41}$$

what is required, and uniform scheme (8.1.40) appears to be strongly stable.

Increase the order of the scheme (8.1.40) by deleting derivatives $\left(\partial f / \partial t\right)_{n+l-1}$, $\left(\partial^2 f / \partial t^2\right)_{n+l-1}$ and $\left(\partial^3 f / \partial t^3\right)_{n+l-1}$ of the right-hand side, with the aid of series:

$$0 = -\tau f_{n+l} + \tau f_{n+l-1} + 2\frac{\tau^2}{2!}\left(\frac{\partial f}{\partial t}\right)_{n+l-1} + 3\frac{\tau^3}{3!}\left(\frac{\partial^2 f}{\partial t^2}\right)_{n+l-1} + 4\frac{\tau^4}{4!}\left(\frac{\partial^3 f}{\partial t^3}\right)_{n+l-1} + 5\frac{\tau^5}{5!}\left(\frac{\partial^4 f}{\partial t^4}\right)_{n+l-1} + \ldots,$$

$$0 = -\tau f_{n+l-2} + \tau f_{n+l-1} - 2\frac{\tau^2}{2!}\left(\frac{\partial f}{\partial t}\right)_{n+l-1} + 3\frac{\tau^3}{3!}\left(\frac{\partial^2 f}{\partial t^2}\right)_{n+l-1} - 4\frac{\tau^4}{4!}\left(\frac{\partial^3 f}{\partial t^3}\right)_{n+l-1} + 5\frac{\tau^5}{5!}\left(\frac{\partial^4 f}{\partial t^4}\right)_{n+l-1} + \ldots,$$

$$0 = -\tau f_{n+l-3} + \tau f_{n+l-1} - 4\frac{\tau^2}{2!}\left(\frac{\partial f}{\partial t}\right)_{n+l-1} + 12\frac{\tau^3}{3!}\left(\frac{\partial^2 f}{\partial t^2}\right)_{n+l-1},$$
$$-32\frac{\tau^4}{4!}\left(\frac{\partial^3 f}{\partial t^3}\right)_{n+l-1} + 80\frac{\tau^5}{5!}\left(\frac{\partial^4 f}{\partial t^4}\right)_{n+l-1} + \ldots$$

$$0 = -\tau f_{n+l-4} + \tau f_{n+l-1} - 6\frac{\tau^2}{2!}\left(\frac{\partial f}{\partial t}\right)_{n+l-1} + 27\frac{\tau^3}{3!}\left(\frac{\partial^2 f}{\partial t^2}\right)_{n+l-1},$$
$$-108\frac{\tau^4}{4!}\left(\frac{\partial^3 f}{\partial t^3}\right)_{n+l-1} + 405\frac{\tau^5}{5!}\left(\frac{\partial^4 f}{\partial t^4}\right)_{n+l-1} + \ldots$$

As a result, we obtain the explicit scheme in the form of

$$y_{n+l} = y_{n+l-1} + \frac{1}{8}\left[12y_{n+l-1} - 18y_{n+l-2} + 7y_{n+l-3} - y_{n+l-4} + \frac{\tau}{24}\begin{pmatrix}325f_{n+l-1} - 617f_{n+l-2}\\ +415f_{n+l-3} - 99f_{n+l-4}\end{pmatrix}\right] + O\left(\tau^5\right). \tag{8.1.42}$$

and the nervous system in the form

$$y_{n+l} = y_{n+l-1} + \frac{1}{8}\begin{bmatrix}12y_{n+l-1} - 18y_{n+l-2} + 7y_{n+l-3} - y_{n+l-4}\\ +\frac{\tau}{720}\left(2321f_{n+l} + 466f_{n+l-1} - 4584f_{n+l-2} + 3166f_{n+l-3} - 649f_{n+l-4}\right)\end{bmatrix} + O\left(\tau^6\right). \tag{8.1.43}$$

Continuing in this way and increasing the value of l in the normalized scheme (8.1.24), we can see that the characteristic polynomial with real coefficients $\Phi_a\left(q\right)$ has the form (8.1.19), as required.

CREATING A FAMILY OF STRONGLY RESISTANT LINEAR *l*-STEP FINITE DIFFERENCE SCHEMES ON IRREGULAR GRID

We assume without loss of generality, that the solution of the problem (8.1.5) is implemented at the initial numerical calculation step of the difference grid ω_τ, i.e. when n=0. In this case, many of the technical details of the proves become much easier due to the fact that the dependence of the functions from (8.1.5) on n disappears and the dependence on l becomes trivial:

$$\sum_{i=0}^{l} a_i y_{n+i} - \tau_l \sum_{i=0}^{l} b_i f_{n+i} = 0. \tag{8.1.5a}$$

Theorem 8.1.4: For every strongly stable homogenous one-step linear multistep one-stage finite-difference scheme (8.1.5a) with real coefficients a_i and b_i $\left(i = \widehat{0,l}\right)$ obeying the normalization condition (8.1.6) and (8.1.8), the characteristic polynomial $\Phi_a\left(q\right)$ is unique, has only real roots which satisfy a strong root criterion, and is defined as:

$$\Phi_a\left(q\right) = \left(q - q_0\right)\prod_{i=1}^{l}\left(s_i q - 1\right)^i, \tag{8.1.44}$$

wherein the sequence of real numbers $s_i\left(i = \widehat{1,l}\right)$:

1. Is common for linear l-step procedures (8.1.5a) (l=k and l=k+1) for a given grid ω_τ with $i = \widehat{0,k}$;
2. Has the values of its elements on an irregular grid ω_τ within the range $1 < s_i \leq 2\left(i = \widehat{1,l}\right)$ depending solely on the ratio of grid steps.

Proof: Replace the Taylor series expansion (8.1.23) of the function $y(t)$ on a regular grid $\bar{\omega}_\tau$ at points l,l-2,…,0 on its value at the point l-1 by Taylor series on irregular grid ω_τ

$$y_l = y_{l-1} + \frac{\tau_l}{1!} f_{l-1} + \frac{\tau_l^2}{2!}\left(\frac{\partial f}{\partial t}\right)_{l-1} + \ldots + \frac{\tau_l^p}{p!}\left(\frac{\partial^{p-1} f}{\partial t^{p-1}}\right)_{l-1},$$

$$y_{l-2} = y_{l-1} - \frac{\tau_{l-1}}{1!} f_{l-1} + \frac{\tau_{l-1}^2}{2!}\left(\frac{\partial f}{\partial t}\right)_{l-1} + \ldots + \left(-1\right)^p \frac{\tau_{l-1}^p}{p!}\left(\frac{\partial^{p-1} f}{\partial t^{p-1}}\right)_{l-1}, \tag{8.1.45}$$

$$y_{l-3} = y_{l-1} - 2\frac{\tau_{l-1} + \tau_{l-2}}{1!} f_{l-1} + \frac{\left(\tau_{l-1} + \tau_{l-2}\right)^2}{2!}\left(\frac{\partial f}{\partial t}\right)_{l-1} + \ldots + \left(-1\right)^p \frac{\left(\tau_{l-1} + \tau_{l-2}\right)^p}{p!}\left(\frac{\partial^{p-1} f}{\partial t^{p-1}}\right)_{l-1},$$

$$y_0 = y_{l-1} - 2\frac{\tau_{l-1} + \tau_{l-2} + \ldots + \tau_0}{1!} f_{l-1} + \frac{\left(\tau_{l-1} + \tau_{l-2} + \ldots + \tau_0\right)^2}{2!}\left(\frac{\partial f}{\partial t}\right)_{l-1}.$$
$$+ \ldots + \left(-1\right)^p \frac{\left(\tau_{l-1} + \tau_{l-2} + \ldots + \tau_0\right)^p}{p!}\left(\frac{\partial^{p-1} f}{\partial t^{p-1}}\right)_{l-1}$$

Multiply consequently the left and right side of the first equality (8.1.45) by the coefficient a_l determined by the system (8.1.22), the second equation (1.8.45) – by the factor a_{l-2}, the third equation – by a_{l-3}, etc., and add the resulting expressions term by term. We arrive at the equality

$$\sum_{i=0}^{l} a_i y_i = \tau_l \sum_{i=0}^{p-1} \frac{c_i}{\left(i+1\right)!}\left(\frac{\partial^i f}{\partial t^i}\right)_{l-1}, \tag{8.1.46}$$

where

$$c_0 = a_l - \frac{\tau_{l-1}}{\tau_l} a_{l-2} + \frac{\tau_{l-1} + \tau_{l-2}}{\tau_l} a_{l-3} - \frac{\tau_{l-1} + \tau_{l-2} + \tau_{l-3}}{\tau_l} a_{l-4} + \ldots;$$

$$c_1 = \tau_l \left[a_l + \left(\frac{\tau_{l-1}}{\tau_l}\right)^2 a_{l-2} + \left(\frac{\tau_{l-1} + \tau_{l-2}}{\tau_l}\right)^2 a_{l-3} + \left(\frac{\tau_{l-1} + \tau_{l-2} + \tau_{l-3}}{\tau_l}\right)^2 a_{l-4} + \ldots\right];$$

$$c_2 = \tau_l^2 \left[a_l - \left(\frac{\tau_{l-1}}{\tau_l}\right)^3 a_{l-2} - \left(\frac{\tau_{l-1} + \tau_{l-2}}{\tau_l}\right)^3 a_{l-3} - \left(\frac{\tau_{l-1} + \tau_{l-2} + \tau_{l-3}}{\tau_l}\right)^3 a_{l-4} - \ldots\right];$$

$$c_3 = \tau_l^3 \left[a_l + \left(\frac{\tau_{l-1}}{\tau_l}\right)^4 a_{l-2} + \left(\frac{\tau_{l-1} + \tau_{l-2}}{\tau_l}\right)^4 a_{l-3} + \left(\frac{\tau_{l-1} + \tau_{l-2} + \tau_{l-3}}{\tau_l}\right)^4 a_{l-4} + \ldots\right]; \tag{8.1.47}$$

$$c_2 = \tau_l^p \left[\begin{array}{l} a_l + \left(-1\right)^{p+1}\left(\frac{\tau_{l-1}}{\tau_l}\right)^{p+1} a_{l-2} + \left(-1\right)^{p+1}\left(\frac{\tau_{l-1} + \tau_{l-2}}{\tau_l}\right)^{p+1} a_{l-3} \\ + \left(-1\right)^{p+1}\left(\frac{\tau_{l-1} + \tau_{l-2} + \tau_{l-3}}{\tau_l}\right)^{p+1} a_{l-4} + \ldots \end{array}\right].$$

Neglecting in the right side of (8.1.46) all the terms containing derivatives of the function $f(t)$, we get the scheme the order p=1

$$\sum_{i=0}^{l} a_i y_i = \tau_l c_0 f_{l-1} + O\left[\bar{\tau}^2\right], \tag{8.1.48}$$

where $\bar{\tau} = \max_{i=0,l} \tau_i$ is the maximum grid step of ω_τ.

To satisfy the condition (8.1.1) it is necessary to equate the factor c_0 at f_{l-1} to one, i.e. to put

$$a_l - \frac{\tau_{l-1}}{\tau_l} a_{l-2} + \frac{\tau_{l-1} + \tau_{l-2}}{\tau_l} a_{l-3} - \frac{\tau_{l-1} + \tau_{l-2} + \tau_{l-3}}{\tau_l} a_{l-4} + \ldots + a_0 \left(\sum_{i=0}^{l} \tau_{l-i} \right) / \tau_l = 1 . \tag{8.1.49}$$

To increase the order of local accuracy of the explicit l-step scheme (8.1.48), one can supplement the equality (8.1.46) by Taylor series with respect to the function f_{l-1}

$$0 = -f_l + f_{l-1} + \frac{\tau_l}{1!}\left(\frac{\partial f}{\partial t}\right)_{l-1} + \frac{\tau_l^2}{2!}\left(\frac{\partial^2 f}{\partial t^2}\right)_{l-1} + \ldots + \frac{\tau_l^p}{p!}\left(\frac{\partial^p f}{\partial t^p}\right)_{l-1},$$

$$0 = -f_{l-2} + f_{l-1} - \frac{\tau_{l-1}}{1!}\left(\frac{\partial f}{\partial t}\right)_{l-1} + \frac{\tau_{l-1}^2}{2!}\left(\frac{\partial^2 f}{\partial t^2}\right)_{l-1} + \ldots + (-1)^p \frac{\tau_{l-1}^p}{p!}\left(\frac{\partial^p f}{\partial t^p}\right)_{l-1},$$

$$0 = -f_{l-3} + f_{l-1} - \frac{\tau_{l-1} + \tau_{l-2}}{1!}\left(\frac{\partial f}{\partial t}\right)_{l-1} + \frac{\left(\tau_{l-1} + \tau_{l-2}\right)^2}{2!}\left(\frac{\partial^2 f}{\partial t^2}\right)_{l-1} + \ldots + (-1)^p \frac{\left(\tau_{l-1} + \tau_{l-2}\right)^p}{p!}\left(\frac{\partial^p f}{\partial t^p}\right)_{l-1}, \tag{8.1.50}$$

$$0 = -f_0 + f_{l-1} - \frac{1}{1!}\left(\sum_{i=1}^{l-1} \tau_{l-i}\right)\left(\frac{\partial f}{\partial t}\right)_{l-1} + \frac{1}{2!}\left(\sum_{i=1}^{l-1} \tau_{l-i}\right)^2\left(\frac{\partial^2 f}{\partial t^2}\right)_{l-1} + \ldots + (-1)^p \frac{1}{p!}\left(\sum_{i=1}^{l-1} \tau_{l-i}\right)^p\left(\frac{\partial^p f}{\partial t^p}\right)_{l-1}$$

and exclude the derivatives of the function f of order $k = \left(\widehat{1, p}\right)$ from the system of equations (8.1.46) and (8.1.50).

It is necessary to pay attention here that the first two terms on the right sides of each of the series (8.1.50) have coefficients equal in modulus and with opposite signs. Therefore, in the process above of exclusion the derivatives of the function f k-th order from the system of equations (8.1.46), (8.1.50), algebraic sum of the coefficients at the function f in (8.1.50) multiplied by the same arbitrary factor will be equal to zero and, therefore, condition (8.1.1) will be provided exclusively by (8.1.49).

Equation (8.1.49) contains l+1 of unknown $a_i\left(i = \widehat{0, l}\right)$. At first glance, there is no single solution since there is only one equation for the definition of l+1 of unknowns. However, it is not. To prove the existence of a unique solution is possible, applying the method of induction. We start the proof with the l-step scheme at l=2. As a source of expression for characteristic polynomial $\Phi_a(q)$ (8.1.7) we take the quadratic trinomial

$$\Phi_a(q) = (q-1)(a_1 q - 1) = a_1 q^2 - (1+a_1)q + 1, \tag{8.1.51}$$

one of the roots of which is real q_0=1. In accordance with the corollary of the statement 8.1.2, the second root of the polynomial (8.1.51) will also be real.

We give the coefficient a_1 so that the characteristic polynomial (8.1.51) corresponds to a linear procedure (8.1.5a), approximating equation $\partial y / \partial t - f(t,y) = 0$. To do this, multiply both sides of the first series (8.1.45) by a_1 and add to the second one of (8.1.45). The result is following:

$$\begin{aligned} a_1 y_l - (1+a_1) y_{l-1} + y_{l-2} &= \left(a_1 - \frac{\tau_{l-1}}{\tau_l}\right)\frac{\tau_l}{1!} f_{l-1} + \left[a_1 + \left(\frac{\tau_{l-1}}{\tau_l}\right)^2\right]\frac{\tau_l^2}{2!}\left(\frac{\partial f}{\partial t}\right)_{l-1} \\ &+ \left[a_1 - \left(\frac{\tau_{l-1}}{\tau_l}\right)^3\right]\frac{\tau_l^3}{3!}\left(\frac{\partial^2 f}{\partial t^2}\right)_{l-1} + \left[a_1 + \left(\frac{\tau_{l-1}}{\tau_l}\right)^4\right]\frac{\tau_l^4}{4!}\left(\frac{\partial^3 f}{\partial t^3}\right)_{l-1} + \ldots \end{aligned} \tag{8.1.52}$$

Neglecting in the right side of (8.1.52) all terms containing derivatives of the function $f(t)$, we get the scheme of the order $p = 1$.

$$a_1 y_l - (1+a_1) y_{l-1} + y_{l-2} = \left(a_1 - \frac{\tau_{l-1}}{\tau_l}\right)\tau_l f_{l-1} + O\left[\bar{\tau}^2\right]. \tag{8.1.52a}$$

To satisfy condition (8.1.8), the coefficient at $\tau_l f_{l-1}$ must be equated to one, i.e. $a_1 - \tau_{l-1} / \tau_l = 1$, from where we have

$$a_1 = 1 + \tau_{l-1} / \tau_l = 1 + \gamma, \tag{8.1.53}$$

For $\tau_{l-1} > 0$ and $\tau_l > 0$ for any γ we have $1<a_1\leq 2$, therefore, the characteristic polynomial of the family of three-point schemes (8.1.52) has two distinct real roots q_0=1 and $0<q_1<1$. In accordance with the definition 8.1.5, *the family of regular schemes (8.1.52) is strongly stable.*

To increase the of order accuracy of explicit local l-step scheme for l=2 is possible by adding the equation (8.1.52) to the second Taylor series of (8.1.50) and excluding the first derivative $(\partial f / \partial t)_{l-1}$ from these equations. Multiply both sides of the second Taylor series of (8.1.50) by a factor

$$\frac{1}{2}\frac{\tau_l}{\tau_{l-1}}\left[a_1 + \left(\frac{\tau_{l-1}}{\tau_l}\right)^2\right]\tau_l$$

and add the resulting expression to (8.1.52). The result is:

$$a_1 y_l - \left(1 + a_1\right) y_{l-1} + y_{l-2} = \frac{\tau_l}{2}\left(f_l - f_{l-1}\right)\left[a_1 + \left(\frac{\tau_{l-1}}{\tau_l}\right)^2\right] + \left(a_1 - \frac{\tau_{l-1}}{\tau_l}\right)\tau_l f_{l-1} + O\left[\bar{\tau}^3\right]. \quad (8.1.52b)$$

Since the functions f_{l-2} and f_{l-1} in parentheses in the first term of the right-hand side of this equation have opposite signs, the algebraic sum of multipliers $-\tau_l\left[a_1 + \left(\tau_{l-1} / \tau_l\right)^2\right] / 2$ and $+\tau_l\left[a_1 + \left(\tau_{l-1} / \tau_l\right)^2\right] / 2$ is equal to zero. Therefore, to satisfy the condition (8.1.18), the factor $a_1 - \tau_{l-1} / \tau_l$ at f_{l-1} of the second term on the right-hand side (8.1.52b) should identically be equal to one. This condition is satisfied by the equation (8.1.53), as stated in the theorem. Substituting the value a_1 of (8.1.53) in (8.1.52b), we obtain the final formula for the explicit difference scheme of order p=2.

For the construction of l-step implicit scheme of the third order for l=2, we supplement the equation (8.1.52) by the first and second Taylor series for the function f in (8.1.50). We multiply the first row by the uncertain yet factor A, the second row – by the uncertain factor B and add the obtained expression to (8.1.52). Define the values of factors A and B so that the sum of the coefficients at $\left(\partial f / \partial t\right)_{l-1}$ and $\left(\partial^2 f / \partial t^2\right)_{l-1}$ vanish. Hence we obtain the system of algebraic equations:

$$\tau_l A - \tau_{l-1} B + \frac{\tau_l^2}{2!}\left[a_1 + \left(\frac{\tau_{l-1}}{\tau_l}\right)^2\right] = 0,$$

$$\frac{\tau_l^2}{2} A + \frac{\tau_{l-1}^2}{2} B + \frac{\tau_l^2}{3!}\left[a_1 - \left(\frac{\tau_{l-1}}{\tau_l}\right)^3\right] = 0,$$

which has the solution of the form

$$A = -\frac{1}{6}\frac{\tau_l^2}{\tau_l + \tau_{l-1}}\left[a_1\left(2 + 3\frac{\tau_{l-1}}{\tau_l}\right) + \left(\frac{\tau_{l-1}}{\tau_l}\right)^3\right], \quad (8.1.54)$$

$$B = \frac{1}{6}\frac{\tau_l^2}{\tau_l + \tau_{l-1}}\left\{a_1\left[3 - 2\left(\frac{\tau_{l-1}}{\tau_l}\right)^2\right] + 5\left(\frac{\tau_{l-1}}{\tau_l}\right)^2\right\}.$$

Before demonstrating the result of the addition of (8.1.52) and the first two of the series of (8.1.50), multiplied by the obtained values of A and B, respectively, we note once again that the first two terms on the right sides of these rows with equal coefficients have opposite signs; therefore, the algebraic sum of the coefficients equal in magnitude, multiplied by the same number in each of these rows is zero. So, to satisfy the condition (8.1.8) the only non-zero coefficient $a_1 - \tau_{l-1} / \tau_l$ at f_{l-1} in the right-hand side of (1.8.52) is needed identically equal to one, what is consistent with (8.1.53) and is stated in the theorem.

Combining the system of equations consisting of equations (8.1.52) and the first two of the series (8.1.50) in light of the values of the unknowns A and B, we obtain the following formula for the highly stable implicit difference schemes of order p=3:

$$a_1 y_l - \left(1 + a_1\right) y_{l-1} + y_{l-2} = -A f_l + \left[\left(a_1 - \frac{\tau_{l-1}}{\tau_l}\right)\tau_l + A + B\right] f_{l-1} - B f_{l-2} + O\left[\bar{\tau}^4\right]. \qquad (8.1.52c)$$

Thus, we have shown that a family of highly stable linear l-step schemes (8.1.52) at l=2, regardless of the order of local accuracy p, determines the characteristic polynomial (8.1.51) with the same value of a_1 determined by the equation (8.1.53). It remains to prove that the roots of the polynomial Φ_a at (8.1.44) of strongly stable homogeneous l-step scheme will remain valid for other values of l. o do this, we consider the family of finite difference schemes for $l = 3$. We construct The characteristic polynomial (8.1.44) with the given values of the real and distinct roots q_0=1 and $q_1 = 1 / a_1 = 1 + \tau_{l-1} / \tau_l$

$$\begin{aligned} \Phi_a(q) &= (q-1)(a_1 q - 1)(a_2 q - 1) \\ &= a_1 a_2 q^3 - (a_1 + a_2 + a_1 a_2) q^2 + (a_1 + a_2 + 1) q - 1, \\ &= \beta_1 q^3 - \beta_2 q^2 + \beta_3 q - 1 \end{aligned} \qquad (8.1.55)$$

where designation holds

$$\beta_1 = a_1 a_2,$$

$$\beta_2 = a_1 + a_2 + a_1 a_2,$$

$$\beta_3 = 1 + a_1 + a_2.$$

In accordance with the corollary 8.2, the third root (8.1.55) will also be real. We choose the factor a_2 so that the characteristic polynomial (8.1.55) corresponds to a linear procedure (8.1.15a), approximating the equation $\partial y / \partial t - f(t, y) = 0$. To do this, multiply both sides of the first row of the system (8.1.45) by a factor at q^3 in the characteristic polynomial (8.1.55), multiply both parts of the second row of (8.1.45) by a factor of at q in the same polynomial and the two add to the third series of the system (8.1.45). The result is:

$$\beta_1 y_l - \beta_2 y_{l-1} + \beta_3 y_{l-2} - y_{l-3} = \left(\beta_1 - \beta_3 s_1 + s_{1,2}\right)\tau_l f_{l-1}$$
$$+\frac{1}{2!}\tau_l\left(\beta_1 + \beta_3 s_1^2 - s_{1,2}^2\right)\tau_l\left(\frac{\partial f}{\partial t}\right)_{l-1} + \frac{1}{3!}\tau_l^2\left(\beta_1 - \beta_3 s_1^3 + s_{1,2}^3\right)\tau_l\left(\frac{\partial^2 f}{\partial t^2}\right)_{l-1}, \tag{8.1.56}$$
$$+\frac{1}{4!}\tau_l^3\left(\beta_1 + \beta_3 s_1^4 - s_{1,2}^4\right)\tau_l\left(\frac{\partial^3 f}{\partial t^3}\right)_{l-1} + \frac{1}{5!}\tau_l^4\left(\beta_1 - \beta_3 s_1^5 + s_{1,2}^5\right)\tau_l\left(\frac{\partial^4 f}{\partial t^4}\right)_{l-1} + O\left[\bar{\tau}^6\right]$$

where

$$s_1 = \frac{\tau_{l-1}}{\tau_l},\ s_{1,2} = \frac{\tau_{l-1} + \tau_{l-2}}{\tau_l}.$$

We first consider the explicit schemes of this family of schemes. Neglecting all terms containing derivatives of the function $f(t)$ in the right side of (8.1.56), we get the scheme of the order p=1

$$\beta_1 y_l - \beta_2 y_{l-1} + \beta_3 y_{l-2} - y_{l-3} = \left(\beta_1 - \beta_3 s_1 + s_{1,2}\right)\tau_l f_{l-1} + O\left[\bar{\tau}^2\right], \tag{8.1.56a}$$

To satisfy condition (8.18), the factor at $\tau_l f_{l-1}$ is necessary to equate to one:

$$a_1 a_2 - \left(a_1 + a_2 + 1\right)\frac{\tau_{l-1}}{\tau_l} + \frac{\tau_{l-2} + \tau_{l-1}}{\tau_l} = 1. \tag{8.1.57}$$

when $a_1 = 1 + \tau_{l-1} / \tau_l$, it follows from (8.1.57) that

$$a_2 = 1 + \frac{\tau_{l-1}}{\tau_l} a_1 - \frac{\tau_{l-2}}{\tau_l} = 1 + \frac{\tau_{l-1}}{\tau_l}\left(1 + \frac{\tau_{l-1}}{\tau_l}\right) - \frac{\tau_{l-2}}{\tau_l}. \tag{8.1.58}$$

From this, using the steps relation, we have:

$$a_2 = 1 + \frac{\tau_{l-1}}{\tau_l}\left(1 + \frac{\tau_{l-1}}{\tau_l}\right) - \frac{\tau_{l-2}}{\tau_l} = 1 + \frac{\tau_{l-1}}{\tau_l}\left(1 + \frac{\tau_{l-1}}{\tau_l}\right) - \frac{\tau_{l-2}}{\tau_{l-1}}\frac{\tau_{l-1}}{\tau_l} = 1 + \gamma\left(1 + \gamma\right) - \gamma^2 = 1 + \gamma,$$

i.e. for all γ we have $1 < a_2 \le 2$. Thus, the roots of the characteristic equation for the family of schemes (8.1.56) are real and distinct and are subject to dependencies: q_0=1, $0 < q_1 < 1$ and $0 < q_2 < 1$. Consequently, in accordance with the definition of 8.5, the family of four-point homogeneous schemes (8.1.56) is highly stable.

To increase the order of local accuracy of the explicit l-step scheme for l=3, one can supplement of (8.1.56) by the second Taylor series for the function f from (8.1.50). Delete the first derivative $\left(\partial f / \partial t\right)_{l-1}$ from these equations, multiplying both sides of the second series (8.1.50) by a factor

$$\frac{1}{2}\frac{\tau_l}{\tau_{l-1}}\left[a_{l-2}a_{l-1}+\left(a_{l-2}+a_{l-1}+1\right)\left(\frac{\tau_{l-1}}{\tau_l}\right)^2-\left(\frac{\tau_{l-2}+\tau_{l-1}}{\tau_l}\right)^2\right]$$

and add the obtained expression to (1.8.56). As a result, we obtain the following formula for the explicit difference scheme of the order p=2:

$$\begin{aligned}&\beta_1 y_l-\beta_2 y_{l-1}+\beta_3 y_{l-2}-y_{l-3}=\left(\beta_1-\beta_3 s_1+s_{1,2}\right)\tau_l f_{l-1}\\&+\frac{\tau_l}{2}\left\{\left[\left(2+\frac{\tau_l}{\tau_{l-1}}\right)\beta_1-\beta_3 s_1+\left(1-\frac{\tau_{l-2}}{\tau_{l-1}}\right)s_{1,2}\right]f_{l-1}-\frac{\tau_l}{\tau_{l-1}}\left(\beta_1+\beta_3 s_1^2-s_{1,2}^2\right)f_{l-2}\right\}+O\left[\overline{\tau}^3\right].\end{aligned}\tag{8.1.56b}$$

Now construct an explicit highly stable homogeneous l-step scheme of the third-order local accuracy for l=3 by expanding the equation (8.1.56) with the second and third members of Taylor series for the function f from (8.1.50). Multiply the second member of (8.1.50) by an indefinite factor A, the third member – by an indefinite factor B and add the obtained expression to the equation (8.1.56) so that the sum of the coefficients at $\left(\partial f/\partial t\right)_{l-1}$ and $\left(\partial^2 f/\partial t^2\right)_{l-1}$ vanish. Hence we obtain a system of algebraic equations:

$$-s_1 A-s_{1,2}B+\frac{1}{2}\left(\beta_1+s_1^2\beta_3-s_{1,2}^2\right)=0,$$

$$s_1^2 A+s_{1,2}^2 B+\frac{1}{3}\left(\beta_1-s_1^3\beta_3+s_{1,2}^3\right)=0,$$

the solution of which has the form:

$$A=\frac{1}{6}\frac{\tau_l^3}{\tau_{l-1}\tau_{l-2}}\left[\left(2+3s_{1,2}\right)\beta_1+s_1^2\left(s_{1,2}+2\frac{\tau_{l-2}}{\tau_l}\right)\beta_3-s_{1,2}^3\right],$$

$$B=-\frac{1}{6}\frac{\tau_l^3}{\tau_{l-2}\left(\tau_{l-1}+\tau_{l-2}\right)}\left[\left(2+3s_{1,2}\right)\beta_1+s_1^3\beta_3+\frac{2\tau_{l-2}-\tau_{l-1}}{\tau_l}s_{1,2}^2\right].\tag{8.1.59}$$

The system of equations consisting of the equation (8.1.56), the second and third members of the (8.1.50) in light of the values of the unknowns A and B, and the equations (8.1.55), (8.1.44), allows obtaining the following formula for a homogeneous strongly stable explicit difference schemes of the order p=3:

$$\beta_1 y_l-\beta_2 y_{l-1}+\beta_3 y_{l-2}-y_{l-3}=\left[A+B+\tau_l\left(\beta_1-\beta_3 s_1+s_{1,2}\right)\right]f_{l-1}-Af_{l-2}-Bf_{l-3}+O\left[\tau^4\right].\tag{8.1.56c}$$

To construct of l-step implicit scheme of the fourth-order at l=3, supplement the equation (8.1.56), by the first, second and third member of Taylor series for the function f from (8.1.50). We multiply the first member of (8.1.50) by yet an indefinite factor A, the second – by an indefinite factor B, the third – by an indefinite factor C and add the obtained expression to the equation (8.1.56) so that when the sum of the coefficients at $\left(\partial f / \partial t\right)_{l-1}$, $\left(\partial^2 f / \partial t^2\right)_{l-1}$ and $\left(\partial^3 f / \partial t^3\right)_{l-1}$ vanishes. Hence we obtain a system of algebraic equations:

$$A - s_1 B - s_{1,2} C + \frac{\tau_l}{2}\left(\beta_1 + s_1^2 \beta_3 - s_{1,2}^2\right) = 0,$$

$$A + s_1^2 B + s_{1,2}^2 C + \frac{\tau_l}{3}\left(\beta_1 - s_1^3 \beta_3 + s_{1,2}^3\right) = 0,$$

$$A - s_1^3 B - s_{1,2}^3 C + \frac{\tau_l}{4}\left(\beta_1 + s_1^3 \beta_3 - s_{1,2}^3\right) = 0,$$

the solution of which has the form:

$$A = -\frac{\tau_l^3}{12\left(\tau_l + \tau_{l-1}\right)\left(\tau_l + \tau_{l-1} + \tau_{l-2}\right)}\left\{\begin{aligned}&\left[1 + s_1 + \left(2 + 3s_1\right)\times\left(1 + 2s_{1,2}\right)\right]\beta_1\\ &+ s_1^3\left(s_{1,2} + \frac{\tau_{l-2}}{\tau_l}\right)\beta_1 - s_{1,2}^3\frac{\tau_{l-1} - \tau_{l-2}}{\tau_l}\end{aligned}\right\};$$

$$B = \frac{\tau_l^3}{12\tau_{l-2}\tau_{l-1}\left(\tau_l + \tau_{l-1}\right)}\left\{\left(1 + 2\frac{\tau_{l-1} - \tau_{l-2}}{\tau_l}\right)\beta_1 + s_1^3\left[2 + s_1 + 2\frac{\tau_{l-2}}{\tau_{l-1}}\left(3 + 2s_1\right)\right]\beta_3 - s_{1,2}^3\left(2 + s_{1,2}\right)\right\};$$

$$C = -\frac{\tau_l^4}{12\tau_{l-2}\left(\tau_{l-1} + \tau_{l-2}\right)\left(\tau_l + \tau_{l-1} + \tau_{l-2}\right)}\left[\begin{aligned}&\left(1 + 2s_1\right)\beta_1 + \left(2 + \frac{\tau_{l-1}}{\tau_l}\right)s_1^3\beta_3\\ &+ \left(\frac{3\tau_{l-2} - \tau_{l-1}}{\tau_l} + 2\frac{2\tau_{l-2} - \tau_{l-1}}{\tau_{l-2} + \tau_{l-1}}\right)s_{1,2}^3\end{aligned}\right]. \qquad (8.1.60)$$

The system of equations consisting of the equation (8.1.56), the first, second and third members of the (8.1.50) in light of the values of the unknowns A, B and C, and the equations (8.1.55), (1.8.44), allows getting the following formula for a homogeneous strongly stable implicit difference scheme of the order p=4:

$$\begin{aligned}&\beta_1 y_l - \beta_2 y_{l-1} + \beta_3 y_{l-2} - y_{l-3}\\ &= -Af_l + \left(A + B + C + \beta_1 - s_1\beta_3 + s_{1,2}\right)\tau_l f_{l-1} - Bf_{l-2} - Cf_{l-3} + O\left[\overline{\tau}^5\right]\end{aligned}. \qquad (8.1.56d)$$

The above process of building a highly stable linear l-step schemes, in the framework of the proof of Theorem 8.1.4, shows that within each family of similar schemes, i.e. at a particular value of l, the coefficients of the characteristic polynomial Φ_a are only functions of difference grid steps and, regardless of the local order of the scheme, remain real.

To complete the proof of Theorem 8.1.4 one need to show that the roots of the characteristic polynomial Φ_a remain real for multi-step scheme for l=4. We construct for this the characteristic polynomial Φ_a with given values of real roots q_0=1, $q_1 = 1/a_1$ and $q_2 = 1/a_2$

$$\begin{gathered}\Phi_a(q) = (q-1)(a_1q-1)(a_2q-1)(a_3q-1) = a_1a_2a_3q^4 - (a_1a_2 + a_1a_3 + a_2a_3 + a_1a_2a_3)q^3 \\ +(a_1 + a_2 + a_3 + a_1a_2 + a_1a_3 + a_2a_3)q^2 - (a_1 + a_2 + a_3 + 1)q + 1 = \beta_1q^4 - \beta_2q^3 + \beta_3q^2 - \beta_4q + 1,\end{gathered} \tag{8.1.61}$$

where

$a_1 = 1 + \tau_{l-1}/\tau_l$ and $a_2 = 1 + a_1\tau_{l-1}/\tau_l - \tau_{l-2}/\tau_l$

and following designations are used:

$\beta_1 = a_1a_2a_3$,

$\beta_2 = a_1a_2 + a_1a_3 + a_2a_3 + a_1a_2a_3$,

$\beta_3 = a_1 + a_2 + a_3 + a_1a_2 + a_1a_3 + a_2a_3$,

$\beta_4 = 1 + a_1 + a_2 + a_3$,

The factor a_3 is chosen so that the characteristic polynomial (8.1.61) corresponds to a linear procedure (8.1.15a), approximating equation $\partial y/\partial t - f(t,y) = 0$. To do this, multiply both sides of the first members of the system (8.1.45) by the coefficient at q^4 in the characteristic polynomial (8.1.61), both parts of the second row of the system (8.1.45) are multiplied by the coefficient at q^2 in the same polynomial, both parts of the third row of the system (8.1.45) are multiplied by the coefficient at q in (8.1.61), and the three members are added to the fourth member of the system (8.1.45). The result is:

$$\begin{gathered}\beta_1y_l - \beta_2y_{l-1} + \beta_3y_{l-2} - \beta_4y_{l-3} + y_{l-4} = (\beta_1 - s_1\beta_3 + s_{1,2}\beta_4 - s_{1,3})\tau_lf_{l-1} \\ +\frac{1}{2!}\tau_l(\beta_1 + s_1^2\beta_3 - s_{1,2}^2\beta_4 + s_{1,3}^2)\tau_l\left(\frac{\partial f}{\partial t}\right)_{l-1} + \frac{1}{3!}\tau_l^2(\beta_1 - s_1^3\beta_3 + s_{1,2}^3\beta_4 - s_{1,3}^3)\tau_l\left(\frac{\partial^2 f}{\partial t^2}\right)_{l-1}, \\ +\frac{1}{4!}\tau_l^3(\beta_1 + s_1^4\beta_3 - s_{1,2}^4\beta_4 + s_{1,3}^4)\tau_l\left(\frac{\partial^3 f}{\partial t^3}\right)_{l-1} + \frac{1}{5!}\tau_l^4(\beta_1 - s_1^5\beta_3 + s_{1,2}^5\beta_4 - s_{1,3}^5)\tau_l\left(\frac{\partial^4 f}{\partial t^4}\right)_{l-1} + O[\overline{\tau}^6]\end{gathered} \tag{8.1.62}$$

where

$$s_{1,3} = \frac{\tau_{l-3} + \tau_{l-2} + \tau_{l-1}}{\tau_l}.$$

To satisfy condition (8.1.18), there must be the coefficient at $\tau_l f_{l-1}$ equal to one:

$$\begin{aligned} & a_1 a_2 a_3 - \left(a_1 + a_2 + a_3 + a_1 a_2 + a_1 a_3 + a_2 a_3\right)\frac{\tau_{l-1}}{\tau_l} \\ & + \left(a_1 + a_2 + a_3 + 1\right)\frac{\tau_{l-2} + \tau_{l-1}}{\tau_l} - \frac{\tau_{l-3} + \tau_{l-2} + \tau_{l-1}}{\tau_l}. \\ & = \beta_1 - s_1\beta_3 + s_{1,2}\beta_4 - s_{1,3} = 1 \end{aligned} \qquad (8.1.63)$$

It is followed from (8.1.63):

$$a_3 = \frac{1 + \frac{\tau_{l-3}}{\tau_l} + \left[a_1 a_2 - \left(a_1 + a_2\right)\frac{\tau_{l-2}}{\tau_{l-1}}\right]\frac{\tau_{l-1}}{\tau_l}}{a_1 a_2 - \left(a_1 + a_2\right)\frac{\tau_{l-2}}{\tau_{l-1}} + \frac{\tau_{l-2}}{\tau_l}}, \qquad (8.1.64)$$

using the relationship of steps

$$a_1 = 1 + \frac{\tau_{l-1}}{\tau_l} = 1 + \gamma,$$

$$a_2 = 1 + \frac{\tau_{l-1}}{\tau_l}\left(1 + \frac{\tau_{l-1}}{\tau_l}\right) - \frac{\tau_{l-2}}{\tau_l} = 1 + \gamma,$$

we have

$$a_3 = \frac{1 + \frac{\tau_{l-3}}{\tau_{l-2}}\frac{\tau_{l-2}}{\tau_{l-1}}\frac{\tau_{l-1}}{\tau_l} + \left[a_1 a_2 - \left(a_1 + a_2\right)\frac{\tau_{l-2}}{\tau_{l-1}}\right]\frac{\tau_{l-1}}{\tau_l}}{a_1 a_2 - \left(a_1 + a_2\right)\frac{\tau_{l-2}}{\tau_{l-1}} + \frac{\tau_{l-2}}{\tau_{l-1}}\frac{\tau_{l-1}}{\tau_l}} = \frac{1 + \gamma^3 + \left[\left(1+\gamma\right)^2 - 2\left(1+\gamma\right)\gamma\right]\gamma}{\left(1+\gamma\right)^2 - 2\left(1+\gamma\right)\gamma - \gamma^2} = 1 + \gamma,$$

i.e., we have $1 < a_3 \le 2$ for all γ. Thus, the roots of the characteristic polynomial for the family of schemes (8.1.62) are real and different, and are subject to and dependencies: q_0=1, $0 < q_1 < 1$, $0 < q_2 < 1$

and $0 < q_3 < 1$. Consequently, in accordance with the definition of the family 8.1.5, the homogeneous four-point scheme (8.1.56) is a highly stable one.

As will be shown below with reference to the circulation problems of the atmosphere, to obtain the difference representation Λ of first and second order derivatives included in the differential operator *D* of equations of hydrodynamics, and heat, and mass transfer, they will be approximated by fourth-order difference operators. So, it is necessary to build a strongly stable explicit difference scheme of the order *p*=4 for ordinary differential equation $\partial y / \partial t - f(t, y) = 0$, that is consistent with the order of approximation of the differential operator *D* in the equations of atmospheric circulation problems. At the same time, no doubt, it will be useful to write down a scheme for an arbitrary *n*-th point of difference grid.

Supplement the equation (8.1.62) by the second, third and fourth members of Taylor series for the function *f* from (8.1.50). Multiply the second member of (8.1.50) by yet an indefinite factor *A*, the third member – by an indefinite factor *B*, the fourth member – by an indefinite factor *C* and add the obtained expression to the equation (8.1.62). Determine the values of factors *A*, *B* and *C* so that when the sum of the coefficients at $(\partial f / \partial t)_{n+l-1}$, $(\partial^2 f / \partial t^2)_{n+l-1}$ and $(\partial^3 f / \partial t^3)_{n+l-1}$ vanishes. Hence we obtain a system of algebraic equations:

$$s_1 A + s_{1,2} B + s_{1,3} C = \frac{\tau_{n+l}}{2} \vartheta_1,$$

$$s_1^2 A + s_{1,2}^2 B + s_{1,3}^2 C = -\frac{\tau_{n+l}}{3} \vartheta_2, \qquad (8.1.65)$$

$$s_1^3 A + s_{1,2}^3 B + s_{1,3}^3 C = \frac{\tau_{n+l}}{4} \vartheta_3,$$

where the following notation takes place:

$$s_1 = \frac{\tau_{n+l-1}}{\tau_{n+l}},\ s_{1,2} = \frac{\tau_{n+l-1} + \tau_{n+l-2}}{\tau_{n+l}},\ s_{1,3} = \frac{\tau_{n+l-1} + \tau_{n+l-2} + \tau_{n+l-3}}{\tau_{n+l}},$$

$$\vartheta_1 = \beta_1 + s_1^2 \beta_3 - s_{1,2}^2 \beta_4 + s_{1,3}^2,$$

$$\vartheta_2 = \beta_1 - s_1^3 \beta_3 + s_{1,2}^3 \beta_4 - s_{1,3}^3,$$

$$\vartheta_3 = \beta_1 + s_1^4 \beta_3 - s_{1,2}^4 \beta_4 + s_{1,3}^4.$$

If in addition, we introduce the notation:

$$s_2 = \frac{\tau_{n+l-2}}{\tau_{n+l}},\ s_3 = \frac{\tau_{n+l-3}}{\tau_{n+l}},\ s_{2,3} = \frac{\tau_{n+l-2} + \tau_{n+l-3}}{\tau_{n+l}},$$

then the solution of the system (8.1.65) takes the form:

$$A = \frac{\tau_{n+l}}{12 s_1 s_2 s_{2,3}} \left[6 s_{1,2} s_{1,3} \vartheta_1 + 4\left(s_{1,2} + s_{1,3}\right) \vartheta_2 + 3\vartheta_3 \right],$$

$$B = -\frac{\tau_{n+l}}{12 s_2 s_3 s_{1,2}} \left[6 s_1 s_{1,3} \vartheta_1 + 4\left(s_1 + s_{1,3}\right) \vartheta_2 + 3\vartheta_3 \right],$$

$$C = \frac{\tau_{n+l}}{12 s_3 s_{2,3} s_{1,3}} \left[6 s_1 s_{1,2} \vartheta_1 + 4\left(s_1 + s_{1,2}\right) \vartheta_2 + 3\vartheta_3 \right].$$

The system of equations consisting of the equation (8.1.62), the second, third and fourth members of the (8.1.50) in light of the values of *A*, *B* and *C*, and equalities (8.1.61), (1.8.64), allows getting the following formula for the strongly stable explicit difference scheme of the order p=4:

$$\begin{aligned} &\beta_1 y_{n+l} - \beta_2 y_{n+l-1} + \beta_3 y_{n+l-2} - \beta_4 y_{n+l-3} + y_{n+l-4} \\ &= \left[A + B + C + \tau_{n+l}\left(\beta_1 - s_1 \beta_3 + \beta_4 s_{1,2} - s_{1,3} \right) \right] f_{n+l-1} \cdot \\ &- A f_{n+l-2} - B f_{n+l-3} - C f_{n+l-4} + O\left[\overline{\tau}^5 \right] \end{aligned} \qquad (8.1.66)$$

Thus, Theorem 8.4, proved above, allows building a set of single-stage *l*-step explicit and implicit numerical methods for solving the problem (8.1.1) (8.1.2). However, the question arises – whether the value of *l* should be chosen arbitrarily large if the right side of the differential equation (8.1.1), (8.1.2) is continuous and bounded together with its derivatives of *l*-th order? The increase in the value of *l* increases the cost of the solution, especially if applied to weather problems simulation of atmospheric circulation and weather forecast, as will be discussed below. So, the choice of a particular value of *l* depends on the nature of the problem being solved. It is hardly appropriate to use large values of *l* in those cases, in which techniques with reduced values of *l* are applicable and/or when significant improving the results is not expected.

Note in conclusion of this subsection that the constructed implicit scheme (8.1.30), (1.8.38) and (8.1.43), that includes an additional approximate value $f\left(t_{n+l}, y_{n+l}\right)$, do not give explicit expressions for finding y_{n+l}, and represent only the equation for this variable. In this case, one can use the two-stage algorithm: first determine y_{n+l} by the explicit scheme where the order accuracy is equal to $O\left[\tau^{l-1}\right]$, and then by calculating $f\left(t_{n+l}, y_{n+l}\right)$, determine y_{n+l} by the implicit scheme of the order $O\left[\tau^{l}\right]$. One can also find the value y_{n+l} by the implicit scheme using any method of successive approximations. The iteration processes essentially complicate the calculation y_{n+l}, what is a significant lack of implicit methods. This is especially the case when $f\left(t_{n+l}, y_{n+l}\right)$ is a value of difference expressions for convection and diffusion terms of nonlinear differential equations of hydrodynamics since the non-linearity of equations does not allow to ensure the convergence of the corresponding iterative process a priori. Therefore, below, we will not consider the multi-stage implicit scheme as independent constructions.

BRINGING INITIAL-BOUNDARY VALUE PROBLEMS FOR PARTIAL DIFFERENTIAL EQUATIONS TO THE CAUCHY PROBLEM FOR ORDINARY DIFFERENTIAL EQUATIONS

Let the circulation of the atmosphere in an area $\wp\left(r\right)$ is defined by the vector $\Re = \left(\mathbf{V}, p, \theta, \rho, q, \delta, s\right)$ of discrete values of analysis and forecast obtained from the model (4.1.5) - (4.1.7) or (4.1.14) - (4.1.16). To distinguish the time grid from the spatial grid, the value relating to the specific points of time will be marked by superscript indices and the values specified in the nodes of the spatial mesh will be marked by subscript indices. The state vector $\Re\left(r,t\right)$ in the area $\wp\left(r\right)$ is determined by a vector of discrete values $\Re\left(r,t^n\right) = \Re^n\left(r\right)$ in predetermined time instants $t=t^n$ (n=0,1,2,…) with the step of $\tau = t^n - t^{n-1}$.

To determine the state of the atmosphere in the area of $\wp\left(r\right)$ at $t > t^n$, it is required to solve the problem which has the vector view as:

$$\frac{d\Re\left(r,t\right)}{dt} = D\Re\left(r,t\right),\ t > t^n, n = 1,2,\ldots, \forall r \in \wp\left(r\right), \tag{8.1.67}$$

$$\Re\left(r,t^n\right) = \Re^n\left(r\right), n = 0,1,\ldots,$$

where, taking for certainty the system of equations (4.1.14) - (4.1.16), $D\Re\left(r,t\right) =$

- For the velocity component u:

$$-\frac{u}{r\cos\varphi}\frac{\partial u}{\partial\lambda} - \frac{v}{r}\frac{\partial u}{\partial\varphi} - \dot{\sigma}\frac{\partial u}{\partial\sigma} - \frac{u\dot{\sigma}}{r} + \frac{uv}{r}\text{tg}\varphi - \frac{1}{r\cos\varphi}\frac{\partial\Phi}{\partial\lambda} + \frac{1}{r\cos\varphi}\frac{\sigma}{p_B}\frac{\partial\Phi}{\partial\sigma}\frac{\partial p_B}{\partial\lambda} - 2\Omega\left(\text{w}\cos\varphi - v\sin\varphi\right) + F_u.$$

- For the velocity component v:

$$-\frac{u}{r\cos\varphi}\frac{\partial v}{\partial\lambda} - \frac{v}{r}\frac{\partial v}{\partial\varphi} - \dot{\sigma}\frac{\partial v}{\partial\sigma} - \frac{v\dot{\sigma}}{r} - \frac{u^2}{r}\text{tg}\varphi - \frac{1}{r}\frac{\partial\Phi}{\partial\varphi} + \frac{1}{r}\frac{\sigma}{p_B}\frac{\partial\Phi}{\partial\sigma}\frac{\partial p_B}{\partial\varphi} - 2\Omega u\sin\varphi + F_v.$$

- For the continuity equation:

$$-\frac{u}{r\cos\varphi}\frac{\partial p_B}{\partial\lambda}-\frac{v}{r}\frac{\partial p_B}{\partial\varphi}-\dot{\sigma}\frac{\partial p_B}{\partial\sigma}-p_B\left(\frac{1}{r\cos\varphi}\frac{\partial u}{\partial\lambda}+\frac{1}{r}\frac{\partial v}{\partial\varphi}+\frac{\partial\dot{\sigma}}{\partial\sigma}-\frac{v}{r}\mathrm{tg}\varphi\right).$$

- For scalar values $\mathfrak{H}=\left(\theta,q,\delta,s\right)$:

$$-\frac{u}{r\cos\varphi}\frac{\partial\mathfrak{H}}{\partial\lambda}-\frac{v}{r}\frac{\partial\mathfrak{H}}{\partial\varphi}-\left(\dot{\sigma}-w_g\right)\frac{\partial\mathfrak{H}}{\partial\sigma}-\phi\mathfrak{H}+f+F_{\mathfrak{H}}.$$

Let independent continuous variables λ,φ,σ of the coordinate system $r=\left(\lambda,\varphi,\sigma\right)$ in the area $\wp\left(r\right)$ take the values: $\alpha_1\leq\lambda\leq\alpha_2$, $\beta_1\leq\varphi\leq\beta_2$ and $\varsigma_1\leq\sigma\leq\varsigma_2$. The spatial grid ϖ_h in this area is formed by dividing the segments $\left[\alpha_1,\alpha_2\right]$, $\left[\beta_1,\beta_2\right]$ and $\left[\varsigma_1,\varsigma_2\right]$ and [ς_1, ς_2] into the set of *I*-1 elements of $\Delta\lambda_i$, *J*-1 elements of $\Delta\varphi_j$ and *K*-1 elements of $\Delta\sigma_k$ respectively. The value of the vector $\Re\left(\lambda_i,\varphi_j,\sigma_k,t\right)$ in the nodes of the grid ϖ_h

$$\lambda_i=\alpha_1+\sum_{\zeta=1}^{i-1}\Delta\lambda_\zeta,1\leq i\leq I,$$

$$\varphi_i=\beta_1+\sum_{\zeta=1}^{i-1}\Delta\varphi_\zeta,1\leq j\leq J,$$

$$\sigma_i=\varsigma_1+\sum_{\zeta=1}^{i-1}\Delta\sigma_\zeta,1\leq k\leq K$$

at time instances $t=t^n$ $\left(n=0,1,2,\ldots\right)$ will be denoted by $\Re^n_{i,j,k}$.

By approximating the operations of differentiation of the first and second order of the $D\Re\left(r,t\right)$ by the spatial variables λ,φ,σ using formulas (8.1.70) and (1.8.71) below, we thereby approximate equation partial differential-difference equations that represent a large number of system of ordinary differential equations

$$\frac{\partial\Re_{i,j,k}}{\partial t}=\Lambda\Re_{i,j,k},1\leq i\leq I,1\leq j\leq J,1\leq k\leq K,$$

where Λ is the difference representation of the differential operator *D*.

To determine the state of the atmosphere at the node $r_{i,j,k}$ in the field of grid ϖ_h of the area $\wp(r)$ for $t > 0$, we shall solve the problem

$$\frac{\partial \Re_{i,j,k}}{\partial t} = \Lambda \Re_{i,j,k}, t > 0, \tag{8.1.68}$$

$$\Re_{i,j,k}\Big|_{t=t^n} = \Re^n_{i,j,k}, \forall r_{i,j,k} \in \varpi_h, n = 1, 2, \ldots. \tag{8.1.69}$$

where $\Lambda \Re_{i,j,k}$ is the numerical value of the right side of the equation (8.1.67) at $r_{i,j,k}$ of the grid ϖ_h in the area $\wp(r)$.

Obviously, if we introduce the notation $y = \Re_{i,j,k}$ and $f(y,t) = \Lambda \Re^n_{i,j,k}$ for each particular value of i, j, k ($1 \le i \le I$, $1 \le j \le J$, $1 \le k \le K$) the Cauchy problem (8.1.1), (8.1.2) and the problem (8.1.68), (1.8.69) are equivalent, and therefore the problem (8.1.68), (1.8.69) can be solved using the single-stage l-step numerical methods proposed in section 8.1.1 for ordinary differential equations of the first order.

Thus, if one assigns the differential operator D in the equation (1.8.67) to a grid operator Λ satisfying the basic requirements of the approximation:

- The accuracy of a given order.
- Convergence ΛR → DR at $h \to 0$, where $h = (\Delta\lambda_i, \Delta\varphi_j, \Delta\sigma_k)$ is the spatial grid step.
- Portability (perturbation imposed on the function $\Re$ is carried by convection only in the direction of velocity).

and solve the problem (1.8.68), (1.8.69) at $r_{i,j,k}$ in of the grid ϖ_h of the area $\wp(r)$ for all specific values of indices $i \in (\overline{1,I})$), $j \in (\overline{1,J})$ and $k \in (\overline{1,K})$, then we obtain the solution of the problem (1.8.67).

Fourth-Order Approximation of Partial Derivatives in the Right Side of the Atmosphere Model Equations

It is believed that convective processes described by the operator

$$\frac{d}{dt} = \frac{\partial}{\partial} + \frac{u}{r\cos\varphi}\frac{\partial}{\partial\lambda} + \frac{v}{r}\frac{\partial}{\partial\varphi} + \dot{\sigma}\frac{\partial}{\partial\sigma},$$

included in almost all the equations of the atmospheric model (4.1.14) - (4.1.16), cause the greatest number of problems in the application of numerical methods. Estimates show that the approximation of horizontal convective terms are the main sources of errors (almost 40% of the total errors) in short-term numerical prediction (Young, 1968).

The spatial three-point finite-difference approximation of the convection equation generates the error of false representation, description of the phase velocity and computing dispersion, as well as the effects of nonlinear instability. Using a second order symmetric difference scheme leads exactly to two effects. First, the convection velocity is less the accurate convection velocity. The consequence of this error is a general slowdown in the convection process. Second, the convection velocity varies with wave number. Such parasitic dispersion, in particular, is quite significant for very short wavelengths. If to transfer this to the system is a superposition of waves, the parasitic dispersion will cause deformation of the system. This applies primarily mesoscale systems such as nascent fronts and cyclones, orographic and thermal perturbations et al. that are examples of the superposition of a set of waves. For this reason, in numerical forecasting, such system, if present in the initial field, begins to deform rapidly, taking a smoother shape. Because these features are very important fields in mesoscale processes, the effect of dispersion computing deserves good attention.

Obviously, the schemes with off-center spatial differences that prevent the spread of disturbances in the opposite direction to the physical convection are preferable schemes, in which the spatial derivative is approximated by central differences. However, a sufficiently strong smoothing effect and only the first order of accuracy of non-central difference operators significantly reduce these benefits compared with difference operators with second-order accuracy (Kalitkin, 1978; Samarskii, 1982).

The above analysis concerned the approximation of the spatial linear convection. But atmospheric circulation models have to deal with nonlinear equations of hydrodynamics, and their non-linearity is caused by the members describing namely convection. Nonlinear equations of hydrodynamics cause the appearance of nonlinear instability and "explosion" of the numerical solution if it is not prevented. The main difficulties, in particular, the errors of false representation, descriptions of the phase velocity, as well as computing dispersion, are solved in the most advanced meteorological centers by providing a model of atmospheric circulation equations in conservative form and by the application of difference schemes of high order of accuracy for the approximation of the convective terms.

In the following, an effective method of high order of accuracy is presented for the approximation of partial derivatives of first and second order, which are an integral part of the convection-diffusion terms of circulation models of the atmosphere of the system (4.1.14) - (4.1.16) (Prusov, Doroshenko, 2006; Prusov, Doroshenko, 2015; Prusov, Doroshenko, 2003; Prusov, Doroshenko, Tyrchak, 2005; Prusov, Doroshenko, 2005). When approximating these derivatives, we use the finite difference schemes, which have a uniform grid of the fourth order of accuracy.

Let η is one of the coordinate system axes $(\lambda, \varphi, \sigma)$ and discrete values $f(\eta_i) = f_i$ of the finction $f(\eta) = \Re^n(r)$ are defined at the nodes $< \eta_1 < \eta_2 < \ldots < \eta_{N-1} < b$, forming (for generality, uneven) the grid ϖ_h with steps $h_i = \eta_i - \eta_{i-1}$ $(i = 0, 1, \ldots, N-1)$.

Define approximate values ψ_i of the partial derivative of the first order $(\partial f / \partial \eta)_i$ and ξ_i of the partial derivative of the second order $(\partial^2 f / \partial \eta^2)_i$ at all nodal points $i \in (\overline{1, N})$ by the discrete values of the function f_i, $i \in (\overline{1, N})$.

Now expand functions $f(\eta)$ and $\psi(\eta)$ in a Taylor series at the points $i+1$, $i-1$ by their values at the point *i*

$$f_{i+1} = f_i + \frac{h_i}{1!}\psi_i + \frac{h_i^2}{2!}\left(\frac{\partial \psi}{\partial \eta}\right)_i + \ldots + \frac{h_i^p}{p!}\left(\frac{\partial^p \psi}{\partial \eta^p}\right)_i,$$

$$f_{i-1} = f_i - \frac{h_{i-1}}{1!}\psi_i + \frac{h_{i-1}^2}{2!}\left(\frac{\partial\psi}{\partial\eta}\right)_i + \ldots + (-1)^p \frac{h_{i-1}^p}{p!}\left(\frac{\partial^p\psi}{\partial\eta^p}\right)_i,$$

$$\psi_{i+1} = \psi_i + \frac{h_i}{1!}\left(\frac{\partial\psi}{\partial\eta}\right)_i + \frac{h_i^2}{2!}\left(\frac{\partial^2\psi}{\partial\eta^2}\right)_i + \ldots + \frac{h_i^p}{p!}\left(\frac{\partial^p\psi}{\partial\eta^p}\right)_i,$$

$$\psi_{i-1} = \psi_i - \frac{h_{i-1}}{1!}\left(\frac{\partial\psi}{\partial\eta}\right)_i + \frac{h_{i-1}^2}{2!}\left(\frac{\partial^2\psi}{\partial\eta^2}\right)_i + \ldots + (-1)^p \frac{h_{i-1}^p}{p!}\left(\frac{\partial^p\psi}{\partial\eta^p}\right)_i.$$

In the first two terms, it is considered equality $\psi_i = (\partial f / \partial\eta)_i$.

After the necessary transformations add up algebraically these expressions so that the coefficients at $(\partial\psi / \partial\eta)_i$ and $(\partial^2\psi / \partial\eta^2)_i$ vanish. The result is:

$$\frac{h_{i-1}}{h_i}\left[\frac{h_{i-1}}{h_i}\left(1 - \frac{h_{i-1}}{h_i}\right) + 1\right]\xi_{i+1} + \left(1 + \frac{h_{i-1}}{h_i}\right)\left[\frac{h_{i-1}}{h_i}\left(3 + \frac{h_{i-1}}{h_i}\right) + 1\right]\xi_i + \left[\frac{h_{i-1}}{h_i}\left(1 + \frac{h_{i-1}}{h_i}\right) - 1\right]\xi_{i-1}$$
$$= \frac{12}{h_i h_{i-1}}\left[\frac{h_{i-1}}{h_i} f_{i+1} - \left(1 + \frac{h_{i-1}}{h_i}\right) f_i + f_{i-1}\right] + \frac{h_i^2 h_{i-1}}{360}\left[1 - \left(\frac{h_{i-1}}{h_i}\right)^2\right]\left\{5\frac{h_{i-1}}{h_i} + 2\left[1 - \left(\frac{h_{i-1}}{h_i}\right)^2\right]\right\}\frac{\partial^5 f}{\partial\eta^5}, \quad (8.1.71)$$

It is obvious that the relations (8.1.70) and (1.8.71) are of the third-order approximation at $h_i \neq h_{i-1}$ and the fourth order at $h_i = h_{i-1}$.

We are interested in derivatives $\psi_i = (\partial f / \partial\eta)_i$, $\xi_i = (\partial^2 f / \partial\eta^2)_i$ that occurred in (1.8.70), (1.8.71)) implicitly. They can be calculated from the given values f_i, $i \in \left(\overline{1, N}\right)$ by solving systems of algebraic equations with sweep method for three diagonal matrices.

Methods for Solving Difference Equations System with a Three-Diagonal Matrix (Sweep Method)

The equation (8.1.70) and (8.1.71) in general form of can be written as:

$$\alpha_i U_{i+1} + \beta_i U_i + \gamma_i U_{i-1} = \omega_i, \quad (8.1.72)$$

where

$$U = \begin{Bmatrix} \psi \\ \xi \end{Bmatrix}, \; \alpha_i = \begin{Bmatrix} 1 \\ \frac{h_{i-1}}{h_i}\left[\frac{h_{i-1}}{h_i}\left(1 - \frac{h_{i-1}}{h_i}\right) + 1\right] \end{Bmatrix},$$

$$\beta_i = \left\{ \frac{2\left(1 + \dfrac{h_i}{h_{i-1}}\right)}{\left(1 + \dfrac{h_{i-1}}{h_i}\right)\left[\dfrac{h_{i-1}}{h_i}\left(3 + \dfrac{h_{i-1}}{h_i}\right) + 1\right]} \right\}, \ \gamma_i = \left\{ \frac{\dfrac{h_i}{h_{i-1}}}{\left[\dfrac{h_{i-1}}{h_i}\left(1 + \dfrac{h_{i-1}}{h_i}\right) - 1\right]} \right\},$$

$$\omega_i = \left\{ \frac{\dfrac{3}{h_i}\left\{ f_{i+1} - \left[1 - \left(\dfrac{h_i}{h_{i-1}}\right)^2\right] f_i - \left(\dfrac{h_i}{h_{i-1}}\right)^2 f_{i-1} \right\}}{\dfrac{12}{h_i h_{i-1}}\left[\dfrac{h_{i-1}}{h_i} f_{i+1} - \left(1 + \dfrac{h_{i-1}}{h_i}\right) f_i + f_{i-1}\right]} \right\}.$$

Since $\alpha_i > 0, \beta_i > 0, \gamma_i > 0$ and $\beta_i > \alpha_i + \gamma_i$, the value of the rounding error in the implementation of the sweep method by recurrence relations

$$U_i = E_i U_{i+1} + D_i, \ i = N-1, N-2, \ldots, 0, \tag{8.1.73a}$$

$$E_i = -\frac{\gamma_i}{\beta_i + \alpha_i E_{i-1}}, \ i = 0, 1, \ldots, N-1, \tag{8.1.73b}$$

$$D_i = -\frac{\omega_i - \alpha_i D_{i-1}}{\beta_i + \alpha_i E_{i-1}}, \ i = 0, 1, \ldots, N-1 \tag{8.1.73c}$$

will not grow (Godunov, Ryaben'kii, 1973).

To solve the equations (8.1.72) it is necessary to put the boundary conditions for U. We apply the boundary conditions of two types:

- If $U = \psi = \partial f / \partial \eta$, then the boundary conditions of the 1st kind are in the form

$$\psi_0 = \frac{1}{1 + c_1 - \chi\left(c_2 + c_3\right)}\left[\frac{f_1 - f_0}{h_0} + c_1 \frac{f_3 - f_0}{h_0 + h_1 + h_2} - \chi\left(c_2 \frac{f_2 - f_0}{h_0 + h_1} + c_3 \frac{f_4 - f_0}{h_0 + h_1 + h_2 + h_3}\right)\right], \tag{8.1.74a}$$

where

$$c_1 = \frac{h_0 h_1 \left(h_1 + h_2 + h_3\right)}{h_2 h_3 \left(h_0 + h_1 + h_2\right)},\ c_2 = \frac{h_1 + h_2 + h_3}{h_2 \left(h_0 + h_1\right)},$$

$$c_3 = \frac{h_1}{h_3 \left(h_0 + h_1 + h_2 + h_3\right)},\ \chi = \frac{h_0 \left(h_1 + h_2\right)}{h_2 + h_3},$$

if conditions are placed on the left border, and

$$\psi_N = \frac{1}{1 + c_1 - \chi\left(c_2 + c_3\right)} \left[\begin{array}{l} \dfrac{f_N - f_{N-1}}{h_{N-1}} + c_1 \dfrac{f_N - f_{N-3}}{h_{N-1} + h_{N-2} + h_{N-3}} \\ -\chi\left(c_2 \dfrac{f_N - f_{N-1}}{h_{N-1} + h_{N-2}} + c_3 \dfrac{f_N - f_{N-4}}{h_{N-1} + h_{N-2} + h_{N-3} + h_{N-4}}\right) \end{array} \right], \tag{8.1.74b}$$

where

$$c_1 = \frac{h_{N-1} h_{N-2} \left(h_{N-2} + h_{N-3} + h_{N-4}\right)}{h_{N-3} h_{N-4} \left(h_{N-1} + h_{N-2} + h_{N-3}\right)},\ c_2 = \frac{h_{N-2} + h_{N-3} + h_{N-4}}{h_{N-3} \left(h_{N-1} + h_{N-2}\right)},$$

$$c_3 = \frac{h_{N-2}}{h_{N-4} \left(h_{N-1} + h_{N-2} + h_{N-3} + h_{N-4}\right)},\ \chi = \frac{h_{N-1} \left(h_{N-2} + h_{N-3}\right)}{h_{N-3} + h_{N-4}}.$$

if the conditions are right border.

- If $U = \xi = \partial^2 f / \partial \eta^2$, then the boundary conditions of the 2nd kind are in the form

$$\left.\frac{\partial \xi}{\partial \eta}\right|_{\eta=a} = \phi_L, \tag{8.1.75a}$$

if conditions are placed on the left border, and

$$\left.\frac{\partial \xi}{\partial \eta}\right|_{\eta=b} = \phi_R, \tag{8.1.75b}$$

if the conditions are right border.

Obviously, the order of approximation of the boundary conditions (8.1.75a) and (8.1.75b) should be coordinated with the order of approximation (8.1.71). For the boundary conditions at the left margin expanded in Taylor series of the function $f(\eta)$, $\psi(\eta)$ and $\xi(\eta)$ at the point i=1 by their values at a point i=0

$$f_1 = f_0 + \frac{h_0}{1!}\psi_0 + \frac{h_0^2}{2!}\xi_0 + \frac{h_0^3}{3!}\left(\frac{\partial\xi}{\partial\eta}\right)_0 \ldots + \frac{h_0^p}{p!}\left(\frac{\partial^p\xi}{\partial\eta^p}\right)_0,$$

$$\psi_1 = \psi_0 + \frac{h_0}{1!}\xi_0 + \frac{h_0^2}{2!}\left(\frac{\partial\xi}{\partial\eta}\right)_0 + \ldots + \frac{h_0^p}{p!}\left(\frac{\partial^p\xi}{\partial\eta^p}\right)_0,$$

$$\xi_1 = \xi_0 + \frac{h_0}{1!}\left(\frac{\partial\xi}{\partial\eta}\right)_0 + \frac{h_0^2}{2!}\left(\frac{\partial^2\xi}{\partial\eta^2}\right)_0 + \ldots + \frac{h_0^p}{p!}\left(\frac{\partial^p\xi}{\partial\eta^p}\right)_0.$$

After the necessary transformations, add up algebraically these expressions so that the coefficients at $\left(\partial\xi / \partial\eta\right)_0$ and $\left(\partial^2\xi / \partial\eta^2\right)_0$ vanish. The result is:

$$\xi_1 - \xi_0 = \frac{6}{h_0}\left[\psi_1 + \psi_0 - \frac{2}{h_0}\left(f_1 - f_0\right)\right] + O\left[h_0^4\right]. \tag{8.1.76a}$$

Similarly, we obtain the boundary condition on the right edge, expanding function $f(\eta)$, $\psi(\eta)$ and $\xi(\eta)$ into the Taylor series at the point i=N-1 by their values at a point i=N

$$f_{N-1} = f_N - \frac{h_{N-1}}{1!}\psi_N + \frac{h_{N-1}^2}{2!}\xi_N - \frac{h_{N-1}^3}{3!}\left(\frac{\partial\xi}{\partial\eta}\right)_N + \ldots + \left(-1\right)^p \frac{h_{N-1}^p}{p!}\left(\frac{\partial^p\xi}{\partial\eta^p}\right)_N,$$

$$\psi_{N-1} = \psi_N - \frac{h_{N-1}}{1!}\xi_N + \frac{h_{N-1}^2}{2!}\left(\frac{\partial\xi}{\partial\eta}\right)_N + \ldots + \left(-1\right)^p \frac{h_{N-1}^p}{p!}\left(\frac{\partial^p\xi}{\partial\eta^p}\right)_N,$$

$$\xi_{N-1} = \xi_N - \frac{h_{N-1}}{1!}\left(\frac{\partial\xi}{\partial\eta}\right)_N + \frac{h_{N-1}^2}{2!}\left(\frac{\partial^2\xi}{\partial\eta^2}\right)_N + \ldots + \left(-1\right)^p \frac{h_{N-1}^p}{p!}\left(\frac{\partial^p\xi}{\partial\eta^p}\right)_N.$$

Adding these expressions so that the coefficients at $\left(\partial\xi / \partial\eta\right)_N$ and $\left(\partial^2\xi / \partial\eta^2\right)_N$ vanish, we get:

$$\xi_N - \xi_{N-1} = \frac{6}{h_{N-1}}\left[\psi_{N-1} + \psi_N - \frac{2}{h_{N-1}}\left(f_N - f_{N-1}\right)\right] + O\left[h_{N-1}^4\right]. \tag{8.1.76b}$$

The proposed approximation of boundary conditions allows the system of equations for ψ and ξ to save the fourth order of accuracy and the three diagonal matrix view up to the decision region boundaries.

The boundary conditions at the left edge allow determining the value E_0 and D_0 and the boundary conditions on the right border enable determining the value of U_N for the problem (8.1.73).

Thus, the equation (8.1.73) gives a double recursive procedure for solving three diagonal system of equations (8.1.72). Using equations (8.1.73b) and (8.1.73c), the grid is passed up the values of *i*, from the point *i* to the point *i*=*N*-1, while all values E_i and D_i on the grid are determined sequentially (forward sweep). By the values found for E_i and D_i, U_i values are determined consistently by the equation (8.1.73a), passing down the grid from the value *i*=*N*-1 to *i*=1 (reverse sweep).

It should be noted the main advantage of the proposed method of approximation of derivatives included in the differential equations of atmospheric circulation models comparing to the standard three-point approximation. Since the solution of algebraic equations (8.1.70) and (1.8.71) at all points *i* depends on the values of other points, it depends on f_i globally, not locally, as it is in the schemes of three-point difference approximations of the first and second orders.

Evaluation of the Effectiveness of the Method on the Test Problems

Let's start with the evaluation of discussed above explicit multistep one-stage (MS1S) method by the example of solving one of the problems of the Cauchy test, which is described by the equation

$$\frac{dy}{dt} + y\cos t = e^{-\sin t}, t > 0, t \in \left[0, 2\pi\right], \tag{8.1.77}$$

with the initial condition $y\left(0\right) = 0$, that has the exact solution $y\left(t\right) = te^{-\sin t}$.

This test problem is chosen to demonstrate the behavior of the solution of the difference scheme (8.1.66) on the class of problems admitting drop down and growing solutions as solutions of this type can sometimes arise in the meteorological problems. This task allows tracing the dependence of the error from the growing solutions.

The problem (8.1.77) was solved numerically using an explicit method MS1S and very popular method of Runge-Kutta on grids with steps $h = 2\pi / N$ when N = 25,50 and 100. The actual methods' error was evaluated by the maximum relative error within the integration interval. Qualitative assessment of the methods is shown in Figure 1), and quantitative – is shown in the Table 1. As the numerical results show (Figure 1 and Table 1), solving a class of problems (8.1.77) by MS1S explicit method is stable, accurate and, above all, more efficient than the method of Runge-Kutta, as it requires four times less than the computed value of the right-hand side of the differential equation.

The second example is taken from (Artemiev, Demidov, 1975; Skvortsov, 2007) and refers to the decision of the rigid non-linear problem, which is described by the equations

$$\frac{du}{dt} = v - \zeta \cdot u \cdot \left(u^2 + v^2 - 1\right),$$

Figure 1. The results of the numerical solution for the problem (8.1.77): ■ stands for MS1S; ● stands for Runge-Kutta method; ▬ stands for the exact solution; a) a grid with the steps $h = 2\pi / N$ at N=25; b) a grid with N=50; c) a grid with N=100

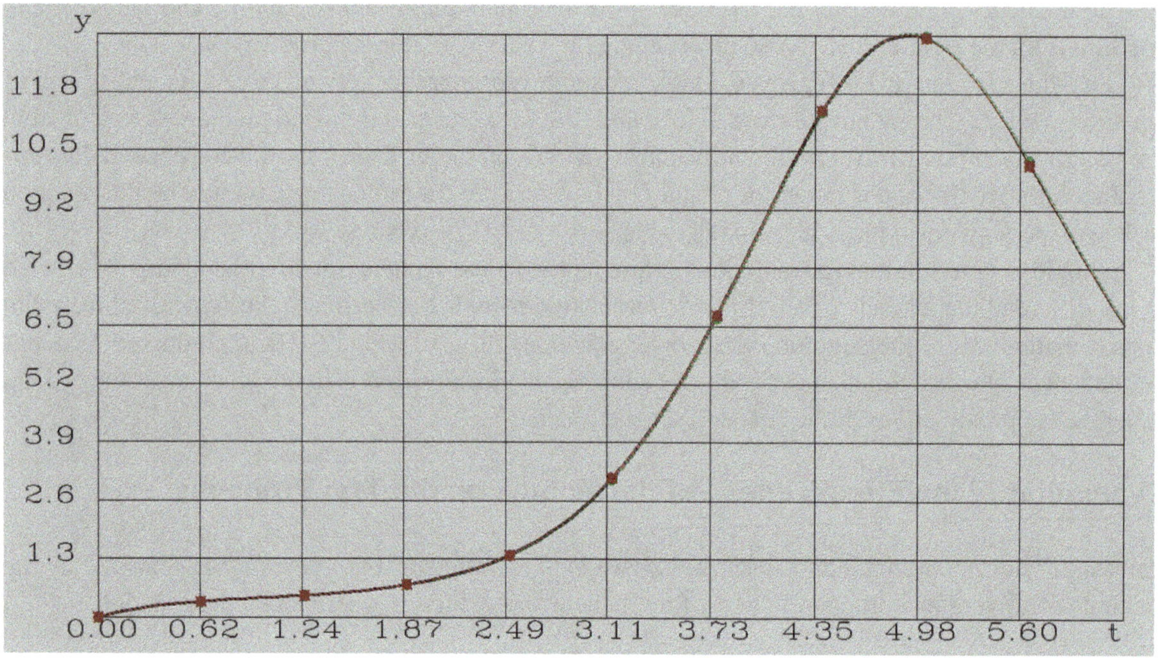

Table 1. Comparing the errors of MS1S and Runge-Kutta methods

Number of Nodes in the Grid *N*	Error of the Method for Given Task by Method	
	MS1S	Runge-Kutta
25	0,03317900	0,04566500
50	0,00656670	0,02208500
100	0,00093927	0,01078000

$$\frac{dv}{dt} = -u - \zeta \cdot v \cdot \left(u^2 + v^2 - 1\right), 0 \leq t \leq 2\pi \tag{8.1.78}$$

with the initial conditions

$$u\left(0\right) = 0, v\left(0\right) = 1.$$

Regardless of the value of the rigidity of ζ this test problem has an exact solution $u = \sin t, v = \cos t$ (see Figure 2).

The problem has been solved by the proposed explicit MS1S on the grid with a constant step $h = 2\pi \cdot 10^{-6}$. It calculates the maximum absolute error over the entire interval of integration. Comput-

Figure 2. The results of the numerical solution for the problem (8.1.78): ■ stands for $v(u,v,t)$; ● stands for (u,v,t); ▬ stands for the exact solution

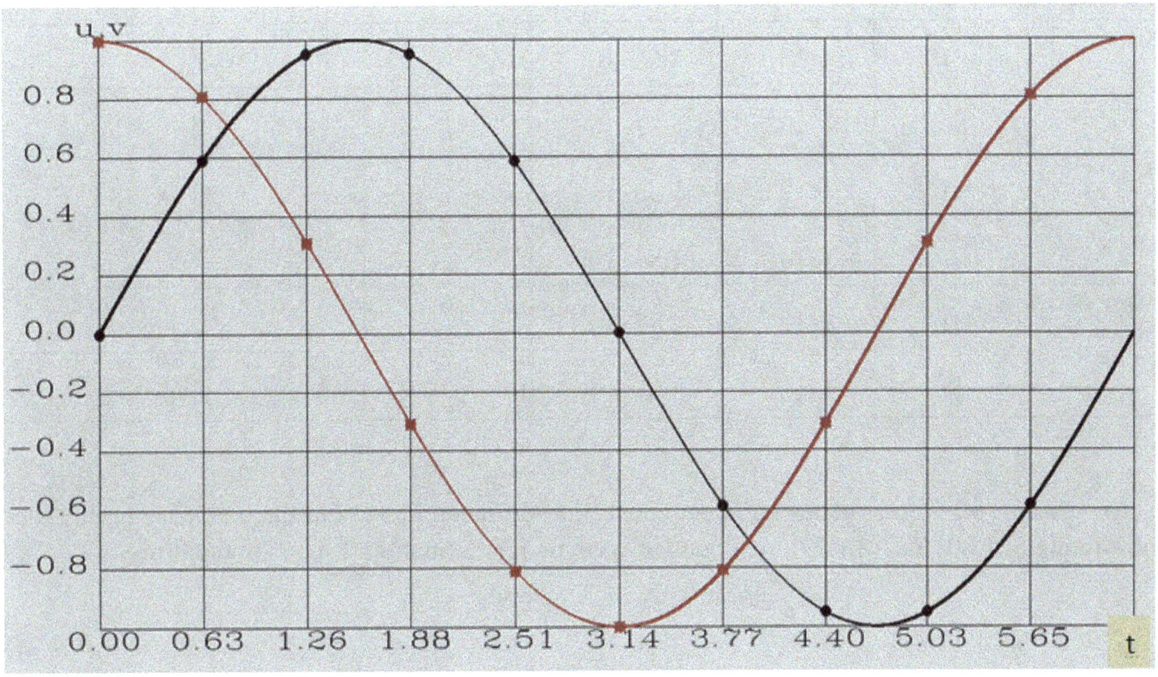

ing costs *Nf* was estimated by amount of computation of the right sides of differential equations (8.1.78). It was compared with realized on the finer grid with explicit adaptive method (Skvortsov, 2007) and the most effective implicit solver for stiffness tasks, MATLAB ode 15s system (Hairer, Vanner, 1999), for which it is also given of the matrix computation of Jacobi NJ and the number of expansions NLU. The results are presented in Table 2. Low efficiency of the implicit method MATLAB ode 15s is due to the rapid change of the Jacobian in the solution process.

To assess the accuracy of the proposed above methods for calculating the derivatives of the first and second order for functions defined on a discrete set of points, we solve the following test problem. Let there be given expression

$$A(y) = -w\frac{\partial u}{\partial y} + \frac{\partial}{\partial y}\left(\nu\frac{\partial u}{\partial y}\right), \in [0,1], \tag{8.1.79}$$

where

$$u = y^{1/n}, w = 0.001u = 0.001y^{1/n}, \nu = 0.16y^2\frac{\partial u}{\partial y}. \tag{8.1.80}$$

Table 2. The accuracy of MS1S, Adaptive and MATLAB ode 15s methods

Method	Grid Step h	Stiffness Parameter ζ	The Absolute Maximum Error	*Nf*	*NJ*	*NLU*
MS1S	$2\pi \cdot 10^{-6}$	$1 \cdot 10^{-2}$	$2.05 \cdot 10^{-3}$	6	0	0
		$1 \cdot 10^{-4}$	$4.41 \cdot 10^{-4}$	6	0	0
Adaptive	$1 \cdot 10^{-6}$	$1 \cdot 10^{-2}$	$1.53 \cdot 10^{-5}$	203	0	0
		$1 \cdot 10^{-4}$	$4.88 \cdot 10^{-5}$	246	0	0
MATLAB ode 15s	$1 \cdot 10^{-6}$	$1 \cdot 10^{-2}$	$2.83 \cdot 10^{-4}$	330	33	94
		$1 \cdot 10^{-4}$	$3.34 \cdot 10^{-4}$	1379	184	443

Expression (8.1.79) simulates turbulent convection-diffusion process in the boundary layer. After substituting (8.1.80) into (8.1.79), we obtain an analytic representation (8.1.79) in the form

$$A\left(y\right) = -\frac{0.001}{n}\left(1 - \frac{320}{n^2}\right)y^{(2/n)-1}, \tag{8.1.81}$$

since

$$\frac{\partial u}{\partial y} = \frac{y^{(2/n)-1}}{n}, \frac{\partial}{\partial y}\left(\nu \frac{\partial u}{\partial y}\right) = 0.32\frac{y^{(2/n)-1}}{n^2}. \tag{8.1.82}$$

It is needed to get discrete values of the derivatives $\partial u / \partial y$, $\partial\left(\nu \partial u / \partial y\right) / \partial y$ and convection-diffusion expression $A\left(y\right)$ by the method of approximation of derivatives of the first (8.1.70) and the second (8.1.71) order on the segment [0,1] at the nodes

$$\left\{y_k; k = 0,1,\ldots M; y_0 = 0, y_M = 1\right\},$$

forming a non-uniform grid $\varpi_{h_k}\left[0,1\right]$ with grid steps $h_k = y_k - y_{k-1}$.

The results are shown in the Figures 3a – 3d when *n*=2, *n*=4, *n*=8 and *n*=12, respectively. In these figures, the following designations are applied:

Figure 3. The results of the implementation of the methods (8.1.70) and (8.1.71) for the problem (1.8.79)

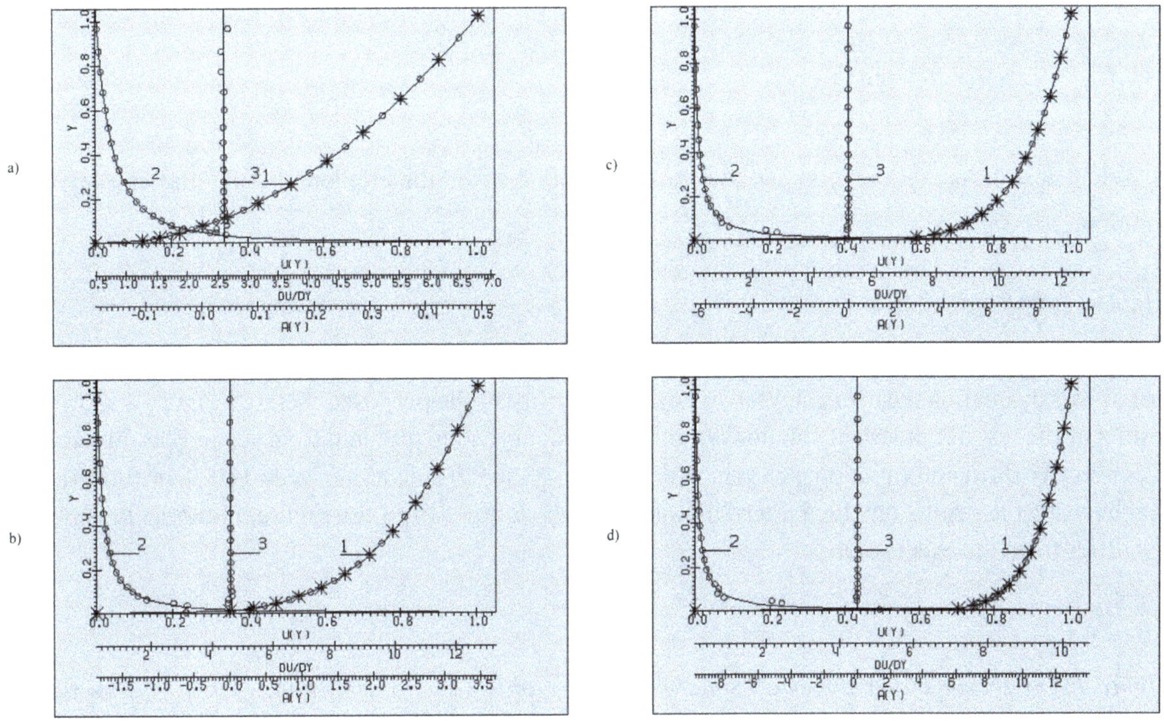

- "1" stands for the function *u*, "2" stands for derivative $\partial u / \partial y$, "3" stands for convection-diffusion term $\partial\left(\nu \partial u / \partial y\right) / \partial y$.
- "■" stands for the exact values obtained from the formulas (8.1.79), (1.8.80).
- "*", "o" stand for the results of the numerical solution for calculation of first and second order and convection-diffusion term.

As follows from the above figures, the accuracy of the calculation of derivatives and convection-diffusion expression even in the high gradient is acceptable. This gives reason to hope for improving the efficiency and accuracy of approximation of derivatives in the differential equations of atmospheric circulation models.

Finally, we turn to the problem

$$\frac{\partial q}{\partial t} + \mathrm{v}\frac{\partial q}{\partial x} = \frac{\partial}{\partial x}\left(\mu \frac{\partial q}{\partial x}\right), \qquad (8.1.83)$$

$$q\left(x,t\right)\Big|_{t=0} = \eta\left(x\right),$$

$$q(x,t)\Big|_{x=a} = \exp(-\mu t)\cdot\sin(a - \mathrm{v}t),$$

$$q(x,t)\Big|_{x=b} = \exp(-\mu t)\cdot\sin(b - \mathrm{v}t).$$

When $\mathrm{v} = const$, $\mu = const$, $\eta(x) = \sin(x)$, a=0, $b = 2\pi$ the problem (8.1.83) has an analytical solution

$$q(x,t) = \exp(-\mu t)\cdot\sin(x - \mathrm{v}t),$$

that allows evaluating the efficiency, accuracy and stability of the proposed method. Figure 4 shows the results of the test of numerical solutions of initial-boundary value problem for a scalar one-dimensional convective diffusion equation on a grid with steps $h = 2\pi / J$ and $\tau = h$ at J=100. The digits on the graphs mark the results obtained after 20 time steps. As follows from the Figure, numerical results do not differ from the exact solution.

Figure 4. The results of the numerical solution for the problem (8.1.83): ● stands for (u,v,t); ■ stands for the exact solution

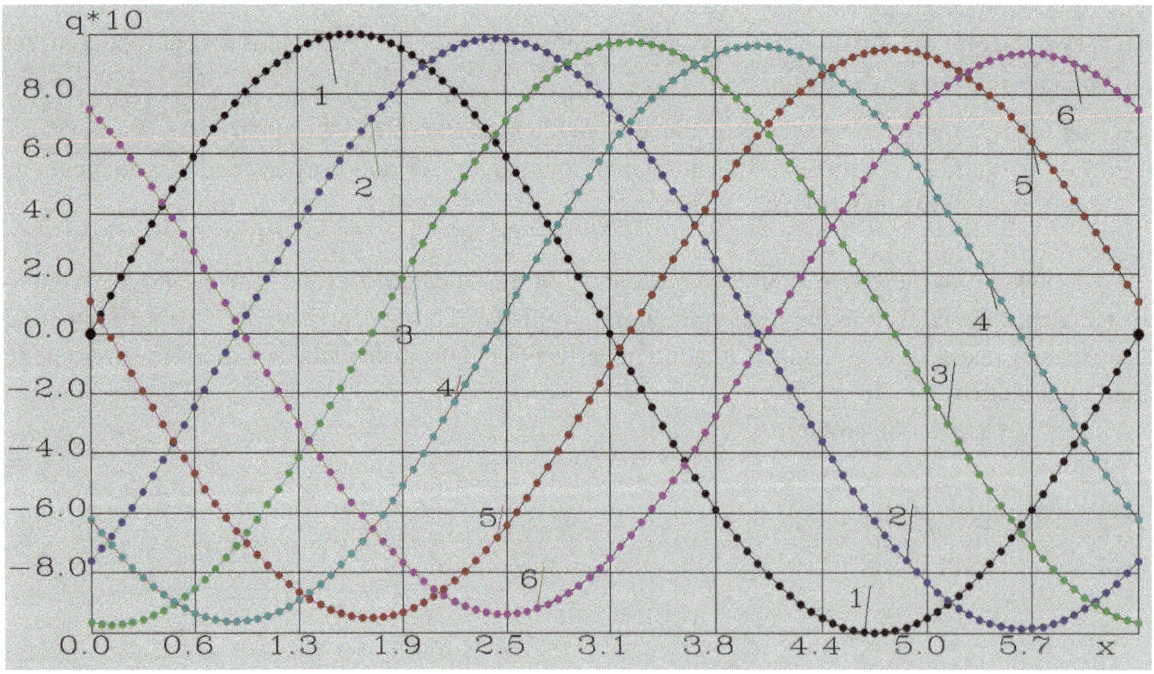

IMPLICIT DIFFERENCE SCHEME WITH AN EXPLICIT SOLUTION OF ONE-DIMENSIONAL CONVECTION-DIFFUSION EQUATION

The Technique Splitting Results in One-Dimensional Advection-Diffusion Equations

The atmospheric flows and heat transfer processes and concentration changes of pollutants are commonly described by a set of partial differential equations, which are mathematical formulations of one or more of the conservation laws of physics. These include the equations of momentum, mass and energy conservation (8.2.2.28) - (8.2.2.34), which involve advection and diffusion terms as the main constituent. Advection-diffusion equations take the form

$$\frac{\partial \Re}{\partial t} + v_1 \frac{\partial \Re}{\partial x_1} + v_2 \frac{\partial \Re}{\partial x_2} + v_3 \frac{\partial \Re}{\partial x_3} = F + \frac{\partial}{\partial x_1}\left(\nu_1 \frac{\partial \Re}{\partial x_1}\right) + \frac{\partial}{\partial x_2}\left(\nu_2 \frac{\partial \Re}{\partial x_2}\right) + \frac{\partial}{\partial x_3}\left(\nu_3 \frac{\partial \Re}{\partial x_3}\right), \tag{8.2.1}$$

where ν_i and v_i are the eddy viscosity and the velocity component of the fluid, respectively, in the direction $x_i\left(i = 1, 2, 3\right)$, and F is the source/sink of the quantity $\Re$. The terms $v_i\left(\partial \Re / \partial x_i\right)$ are usually called advection (or sometimes convection) terms, and describe the transportation of the quantity $\Re$ by the velocity field. The terms $\partial\left(\nu_i \partial \Re / \partial x_i\right) / \partial x_i$ are called diffusion (or sometimes viscous) terms, and express the spreading of quantity $\Re$ by the process of turbulent diffusion. The equation is usually provided with appropriately defined initial and boundary conditions.

Since analytical solutions of advection-diffusion problems cannot usually be found, we have to solve them numerically. However, the numerical treatment of equations of the form (8.2.1) is a highly complicated task. The main reasons for this are the following:

- The equations are nonlinear.
- Due to changes in the input parameters the equations can change the type (hyperbolic, parabolic or elliptic).
- The size of the discretized problem can be very large in a real-life physical model.

Therefore, choosing a sufficiently accurate as well as efficient numerical method for solving advection-diffusion problems is not an easy task. In this case the application of operator splitting seems a good alternative. In (Prusov, Doroshenko, 2008) it is suggested a new finite-difference method for the one-dimensional convection-diffusion equation. Our aim is to extend this method to higher dimensions by using operator splitting based on directional decomposition.

Solving the Three-Dimensional Advection-Diffusion Equation by Operator Splitting

The aim of operator splitting is to replace an initial value problem with a sequence of simpler problems, for which accurate as well as efficient solvers available in standard program packages exist (Faragó, 2005; Lanser, Verwer, 1999). The mathematical background of operator splitting can be sketched as follows. Let S denote a normed space and consider the abstract initial value problem

$$\frac{dw(t)}{dt} = Aw(t) = (A_1 + A_2)w(t), t \in [0,T] \tag{8.2.2}$$

$$w(0) = w_0,$$

where $w(t) \in \boldsymbol{S}$, $t \in [0,T]$ is the unknown function, and A is a given operator $\boldsymbol{S} \rightarrow \boldsymbol{S}$, which can be decomposed into a sum of two "simpler" operators A_1 and A_2. (By "simpler" we mean that the corresponding initial-value problems are easier to treat numerically than the original problem.) The simplest kind of operator splitting is the so-called sequential splitting, where we solve the following sequence of initial value problems:

$$\frac{dw_k^{(1)}(t)}{dt} = A_1 w_k^{(1)}(t), t \in ((k-1)\tau, k\tau] \tag{8.2.3}$$

$$w_k^{(1)}((k-1)\tau) = w_{sp}((k-1)\tau),$$

and

$$\frac{dw_k^{(2)}(t)}{dt} = A_2 w_k^{(2)}(t), t \in ((k-1)\tau, k\tau] \tag{8.2.4}$$

$$w_k^{(2)}((k-1)\tau) = w_k^{(1)}(k\tau),$$

$$w_{sp}(k\tau) = w_k^{(2)}(k\tau)$$

for $k = 1, 2, \ldots, n$, where $\tau = T / n$ is the splitting time step, and $w_{sp}(0) = w_0$. This scheme can be extended to more than two sub-operators in a natural way.

Another possibility is the Marchuk-Strang splitting (Strang, 1968), defined by the following algorithm:

$$\frac{dw_k^{(1)}(t)}{dt} = A_1 w_k^{(1)}(t), t \in ((k-1)\tau, (k-0.5)\tau] \tag{8.2.5}$$

$$w_k^{(1)}((k-1)\tau) = w_{sp}((k-1)\tau),$$

$$\frac{dw_k^{(2)}(t)}{dt} = A_2 w_k^{(2)}(t), t \in ((k-1)\tau, k\tau] \tag{8.2.6}$$

$$w_k^{(2)}\left(\left(k-1\right)\tau\right)=w_k^{(1)}\left(\left(k-0.5\right)\tau\right),$$

$$\frac{dw_k^{(3)}\left(t\right)}{dt}=A_1 w_k^{(3)}\left(t\right), t\in\left(\left(k-0.5\right)\tau, k\tau\right] \tag{8.2.7}$$

$$w_k^{(3)}\left(\left(k-0.5\right)\tau\right)=w_k^{(2)}\left(k\tau\right)$$

$$w_{sp}\left(k\tau\right)=w_k^{(3)}\left(k\tau\right)$$

for $k=1,2,\ldots,n$, where $w_{sp}\left(0\right)=w_0$.

Obviously, there are several ways to define the sub-operators A_i in a splitting procedure. We can choose the sub-operators on a physical base, e.g., we can separate the advection and diffusion terms in Eq. (8.2.2) (physical decomposition) (Dimov et al., 2004). Another possibility is to separate the x_1-, x_2- and x_3- derivatives in the equation (directional decomposition) (Lanser, Blom, Verwer, 2001). For the three-dimensional advection-diffusion problem the physical decomposition would lead to advection problems, which are of hyperbolic type, and diffusion problems, which are of parabolic type. This would cause us difficulties in defining appropriate boundary conditions for the sub-problems. Therefore, we recommend the latter one, which results in three one-dimensional advection-diffusion problems at each time step. Particularly, if we apply the sequential splitting, the detailed algorithm will read as follows:

$$\frac{\partial \Re_k^{(1)}}{\partial t}=-\upsilon_1\frac{\partial \Re_k^{(1)}}{\partial x_1}+\frac{\partial}{\partial x_1}\left(\nu_1\frac{\partial \Re_k^{(1)}}{\partial x_1}\right)+F_{x_1},\ t\in\left(\left(k-1\right)\tau, k\tau\right) \tag{8.2.8}$$

$$\Re_k^{(1)}\left(\left(k-1\right)\tau\right)=\Re_{sp}\left(\left(k-1\right)\tau\right),$$

$$\frac{\partial \Re_k^{(2)}}{\partial t}=-\upsilon_2\frac{\partial \Re_k^{(2)}}{\partial x_2}+\frac{\partial}{\partial x_2}\left(\nu_2\frac{\partial \Re_k^{(2)}}{\partial x_2}\right)+F_{x_2},\ t\in\left(\left(k-1\right)\tau, k\tau\right) \tag{8.2.9}$$

$$\Re_k^{(2)}\left(\left(k-1\right)\tau\right)=\Re_k^{(1)}\left(k\tau\right),$$

$$\frac{\partial \Re_k^{(3)}}{\partial t}=-\upsilon_3\frac{\partial \Re_k^{(3)}}{\partial x_3}+\frac{\partial}{\partial x_3}\left(\nu_3\frac{\partial \Re_k^{(3)}}{\partial x_3}\right)+F_{x_3},\ t\in\left(\left(k-1\right)\tau, k\tau\right) \tag{8.2.10}$$

$$\Re_k^{(3)}\left(\left(k-1\right)\tau\right)=\Re_k^{(2)}\left(k\tau\right),$$

$$\Re_{sp}\left(k\tau\right)=\Re_k^{(3)}\left(k\tau\right)$$

for $k = 1,2,\ldots,n$, where $\Re_{sp}\left(0\right) = \Re_0$. If the original problem is defined over a bounded spatial domain, the equations (8.2.8)-(8.2.10) are also provided with appropriately defined boundary conditions.

Several numerical methods have been constructed for the solution of the resulting one-dimensional advection-diffusion problems (Anderson, Tannehill, Pletcher, 1984; Zlatev, 1995; Datta-Gupta et al., 1991; Orszag, Israeli, 1974; Hirsh, 1975; Pracht, 1971). Also, finite element (Sündermann, 1976) and spectral methods (Markovich, 1986) are very popular. There are also many finite difference schemes that can be considered according to the number of spatial grid points involved, the number of time-levels used, and whether they are explicit or implicit in nature (Verwer, Hundsdorfer, Blom, 2002; Fanchi, 1983; Peaceman, 1977; Celia, Gray, 1992; Sharif, Busnaina, 1988; Leonard, 1995; Liu, Pope, Sepehrnoori, 1995; Radwan, 1996; Harten, 1984; Yee, 1987; Krause, Hirschel, Kordulla, 1973; Weinberg, Leventhal, Ciment, 1977; Ciment, Leventhal, Weinberg, 1978; Berger et al., 1980; Peters, 1975; Rubin & Graves, 1975; Rubin and Graves, 1975; Rubin, Khosla, 1976; Adam, 1975; Adam, 1977; Gupta, Manohar, Stephenson, 1984; Gupta, 1991; White, 1974; Liang, Zhao, 1997; Garbey, Kaper, Romanyukha, 2002; Ropp, Shadid, Ober, 2004; Wang, Shi, Ewing, 2001; Hundsdorfer, Verwer, 2003).

Standard three-point finite difference methods of approximating spatial derivatives can work well for smooth solutions, but they fail when severe gradients or discontinuities are present, which are common in the shock wave problems (Peaceman, 1977; Celia, Gray, 1992; Sharif, Busnaina, 1988; Leonard, 1995; Liu, Pope, Sepehrnoori, 1995). Lower-order accurate finite difference methods, such as upstream-type finite differences, can be a remedy for the numerical oscillations and dispersions. However, they have a large amount of "numerical viscosity" that smooth the solution in much the same way that physical viscosity would, but to an extent that is unrealistic by several orders of magnitude (Leonard, 1995). Standard four-point finite difference methods are good in their higher-order accuracy and in reducing numerical smearing effects (Liu, Pope, Sepehrnoori, 1995). But, they are plagued by their generation of spurious oscillations or overshoots in the neighbourhood discontinuities and lack accuracy (Peaceman, 1977; Celia, Gray, 1992). Total variation stable finite difference schemes (TVD) (Orszag, Israeli, 1974; Hirsh, 1975) guarantee oscillation-free solutions but they are limited to second-order accuracy. Higher-order accurate TVD schemes are attractive for problems with long computational time or with required higher accuracy solutions (Hirsh, 1975). But, the objection to the standard higher-order schemes comes from the additional nodes necessary to achieve the higher-order accuracy. This precludes the use of implicit methods since the obtained matrix is not of three-diagonal form, and it is necessary to use fictitious nodes for the boundary conditions. Also, they do not allow easily for non-uniform grids, unless at the expense of the order of accuracy. On the other hand, the compact schemes that treat functions and their derivatives as unknowns at the grid nodes, like the scheme of (Rubin, Graves, 1975), are fourth-order accurate, and compact in the sense that they reduce to three-diagonal form. The compact schemes generally consist of finite difference schemes which involve two or three grid points. The three-point schemes fall into two classes. The first class consists of methods which are fourth-order accurate for uniform grids, such as schemes (Weinberg, Leventhal, Ciment, 1977; Ciment., Leventhal, Weinberg, 1978; Berger et al., 1980), the operator compact implicit scheme (Weinberg, Leventhal, Ciment, 1977; Ciment, Leventhal, Weinberg, 1978; Berger et al., 1980) and the Hermite finite difference method (Peters, 1975). The second class consists of methods that allow variable grids such as the cubic spline methods (Rubin & Graves, 1975; Rubin and Graves, 1975; Rubin, Khosla, 1976; Adam, 1975; Adam, 1977), and the Hermite finite difference method (Gupta, Manohar, Stephenson, 1984; Gupta, 1991). In (White, 1974; Liang, Zhao, 1997) a compact fourth-order finite difference scheme was introduced with three nodal points for the convection-diffusion equations. This scheme does not seem to suffer excessively from spurious oscillatory behaviour or numerical viscosity.

The disadvantage of the above higher-order compact schemes involving three nodal points is that the boundary conditions are no longer sufficient and they do not allow easily for non-uniform grids, unless at the expense of the order of accuracy. Another disadvantage of some compact schemes is the complexity of the resulting nonlinear finite difference equations and the associated difficulty in solving them efficiently. On the other hand, the compact scheme with two nodal points is fourth-order accurate even for non-uniform spatial grids, and no fictitious points, neither extra formulas are needed for Dirichlet boundary conditions (Liang, Zhao, 1997). The discretization of the convective term might be done in a number of ways (Garbey, Kaper, Romanyukha, 2002; Ropp, Shadid, Ober, 2004). The Ellam scheme is probably one of the best known convective schemes (Wang, Shi, Ewing, 2001).

Coordinate Splitting Three-Dimensional Problem of Convective Diffusion

As it follows from the above references, splitting scheme have been used since the late 70's of the last century. But the computer technology of that time needed low-cost algorithms for the solution. One of the critical resources was main memory of computers. Therefore, it was considered that the use by the algorithm of this resource in the minimum volume is an important positive quality. As a result, almost all splitting algorithms that have been developed at a time looks like conveyor of subtasks, i.e. the results of solving one subtask are the initial data for the other.

The development of parallel computing puts new demands on the numerical algorithms. So, it is natural that some algorithms are revised or will be revised in the future. For example, in (Gnandt, 2005) the additive approach has been refined for its parallel implementation. The modification is to eliminate sub connection through the initial data. Then use the same initial conditions for all subtasks. Such a modification, in the case of using the additive method of splitting with the scheme of the numerical solution of one-dimensional problem, has been proposed in (Prusov et al., 2006; Prusov et al., 2007). Briefly describe the process of this modification additive splitting scheme.

Let us return to the three-dimensional equation (8.2.1) and formulate for it the initial boundary value problem of convective diffusion

$$\frac{\partial \Re}{\partial t}+\sum_{k=1}^{3} v_k \frac{\partial \Re}{\partial x_k}=F+\sum_{k=1}^{3}\frac{\partial}{\partial x_k}\left(\nu_k \frac{\partial \Re}{\partial x_k}\right) \quad (8.2.11)$$

$$\Re\left(0,x_1,x_2,x_3\right)=\Re_0\left(x_1,x_2,x_3\right) \text{ if } \left(x_1,x_2,x_3\right)\in \mathbf{\Omega},$$

$$\Re\left(t,x_1,x_2,x_3\right)=\Re_b \text{ if } \left(x_1,x_2,x_3\right)\in \mathcal{B},$$

where

$\mathcal{B}=\partial\mathbf{\Omega}$ is the boundary of solutions area $\mathbf{\Omega}$,
$\Re=\Re\left(t,x_1,x_2,x_3\right)$ is unknown function,
$F=F\left(t,x_1,x_2,x_3\right)$ is source or sink for $\Re$,

$v_k = v_k(t, x_1, x_2, x_3)$ is the convection coefficient of the k-th coordinate direction,
$\nu_k = \nu_k(t, x_1, x_2, x_3) \geq 0$ is the diffusion coefficient of the k-th coordinate direction.

For certainty, we assume that $v_k > 0$ for k=1,2,3.

Now carry out a space-time discretization of the equation (8.2.11) in accordance with the one-dimensional scheme (Prusov et al., 2006; Prusov et al., 2007). As a result, we obtain the following matrix-vector equation:

$$\frac{\Re^{n+1} - \Re^{n}}{\tau} + \sum_{k=1}^{3}\left(B_k \Re^{n+1} - A_k \Re^{n}\right) = F^{n}, \tag{8.2.12}$$

where

A_k is a matrix operator of the explicit layer of the k-th coordinate direction,
B_k is a matrix operator of the implicit layer of the k-th coordinate direction,
τ is the time step.

The exact solution of equation (8.2.12) has the form

$$\Re^{n+1} = \left(I + \tau\sum_{k=1}^{3} B_k\right)^{-1}\left[\left(I + \tau\sum_{k=1}^{3} A_k\right)\Re^{n} + \tau F^{n}\right]. \tag{8.2.13}$$

For the expression (8.2.13), we use the approximation

$$\left(I + \tau\sum_{k=1}^{3} B_k\right)^{-1} = I - \tau\sum_{k=1}^{3} B_k + O\left[\tau^2\right],$$

$$\left(I + \tau\sum_{k=1}^{3} B_k\right)^{-1}\left(I + \tau\sum_{k=1}^{3} A_k\right) = I + \tau\sum_{k=1}^{3} A_k - \tau\sum_{k=1}^{3} B_k + O\left[\tau^2\right].$$

The result is that up to terms of $O\left[\tau\right]$ the solution of equation (8.2.12) will be

$$\Re^{n+1} = \left(I + \tau\sum_{k=1}^{3} A_k - \tau\sum_{k=1}^{3} B_k\right)\Re^{n} + \tau F^{n}. \tag{8.2.14}$$

The proposed additive splitting of the equation (8.2.12) will be sought in the form

$$\Re_1^{n+1} = \left(I + \tau B_1\right)^{-1}\left[\left(I + \tau A_1\right)\Re^{n} + \tau F_1^{n}\right],$$

$$\Re_2^{n+1} = \left(I + \tau B_2\right)^{-1}\left[\left(I + \tau A_2\right)\Re^n + \tau F_2^n\right], \tag{8.2.15a}$$

$$\Re_3^{n+1} = \left(I + \tau B_3\right)^{-1}\left[\left(I + \tau A_3\right)\Re^n + \tau F_3^n\right],$$

$$\Re^{n+1} = a_1\Re_1^{n+1} + a_2\Re_2^{n+1} + a_3\Re_3^{n+1} + b_1F^n + b_2\Re^n,$$

where a_1, a_2, a_3, b_1, b_2 are unknown coefficients, $F^n = F_1^n + F_2^n + F_3^n$.

We define the coefficients a_1, a_2, a_3, b_1, b_2 so as to ensure the accuracy of the solution of equation (8.2.12) of the order of τ. Substituting approximation

$$\Re_1^{n+1} = \left(I + \tau A_1 - \tau B_1\right)\Re^n + \tau F_1^n + O\left[\tau^2\right],$$

$$\Re_2^{n+1} = \left(I + \tau A_2 - \tau B_2\right)\Re^n + \tau F_2^n + O\left[\tau^2\right],$$

$$\Re_3^{n+1} = \left(I + \tau A_3 - \tau B_3\right)\Re^n + \tau F_3^n + O\left[\tau^2\right]$$

in the expression (8.2.14a) for $\Re^{n+1}$ and comparing with (8.2.14), we find that $a_1 = a_2 = a_3 = 1$ and $b_1 = 0, b_2 = -2$. Consequently, the proposed additive splitting scheme in the directions for (8.2.12) will have the following final form:

$$\Re_1^{n+1} = \left(I + \tau B_1\right)^{-1}\left[\left(I + \tau A_1\right)\Re^n + \tau F_1^n\right],$$

$$\Re_2^{n+1} = \left(I + \tau B_2\right)^{-1}\left[\left(I + \tau A_2\right)\Re^n + \tau F_2^n\right], \tag{8.2.15}$$

$$\Re_3^{n+1} = \left(I + \tau B_3\right)^{-1}\left[\left(I + \tau A_3\right)\Re^n + \tau F_3^n\right],$$

$$\Re^{n+1} = \Re_1^{n+1} + \Re_2^{n+1} + \Re_3^{n+1} - 2\Re^n.$$

The splitting algorithm (8.2.15) allows to make the transition to the next time layer in two stages. The first stage is the decision of three similar problems in each direction. Moreover, these solutions can be carried out in parallel by virtue of their independence from the initial data. In addition, the boundary conditions do not need to be adapted to calculate on any coordinate direction. The second phase is a summation of the results. So, these are only two stages related to the initial data regardless of the dimension of the problem.

We investigate the approximation of the equation (8.2.11) by the scheme (8.2.15). To do this, we will take the immediate approximation error ψ_d that it is the approximation error of the equation (8.2.11) by the finite-difference analogue, which will be discussed below, and the splitting and factorization error ψ_s that is an error introduced in the difference scheme at the transition to the scheme (8.2.15). Then the total equation approximation error will be $\psi = \psi_d + \psi_s$.

To estimate ψ_s use the expression (8.2.14), which is the transformation of the finite-difference equation (8.2.12) up to terms of $O\left[\tau\right]$. Therefore, it does not impair the overall order of the error ψ. We reduce the expression (8.2.14) to the scheme (8.2.15) in order to determine the order of the ψ_s error. We make the transformation to a term *c* by manipulating the terms of higher order of magnitude $O\left[\tau^2\right]$:

$$\begin{aligned}\tau F^n &= \tau\left(I - \tau B_1 - \tau B_2 - \tau B_3\right)\sum_{k=1}^{3} F_k^n + O\left[\tau^2\right] = \tau\sum_{k=1}^{3}\left(I - \tau B_1 - \tau B_2 - \tau B_3\right)F_k^n + O\left[\tau^2\right] \\ &= \tau\sum_{k=1}^{3}\left(I - \tau B_k\right)F_k^n + O\left[\tau^2\right] = \tau\sum_{k=1}^{3}\left(I + \tau B_k\right)^{-1} F_k^n + O\left[\tau^2\right]\end{aligned}.$$

In the same way, we transform matrix operator at $\Re^n$ in (8.2.14)

$$\begin{aligned}\left(I + \tau\sum_{k=1}^{3} A_k - \tau\sum_{k=1}^{3} B_k\right)\Re^n &= \sum_{k=1}^{3}\left(I + \tau A_k - \tau B_k\right)\Re^n - 2\Re^n \\ &= \sum_{k=1}^{3}\left(I - \tau B_k\right)\left(I + \tau A_k\right)\Re^n - 2\Re^n + O\left[\tau^2\right] \\ &= \sum_{k=1}^{3}\left(I + \tau B_k\right)^{-1}\left(I + \tau A_k\right)\Re^n - 2\Re^n + O\left[\tau^2\right]\end{aligned}.$$

As a result of these two changes, we get directly the scheme (8.2.15). Consequently, $\psi_s = O\left[\tau^2\right]$ and the complete approximation error will be $\psi = \psi_d$.

The Method and Algorithm of Explicit Solution a Finite-Difference Scheme for the One-Dimensional Advection-Diffusion Problem

We present here a non-standard finite-difference method for the solution of the sub-problems obtained by splitting (8.2.8) - (8.2.10). Consider the one-dimensional advection-diffusion equation

$$\frac{\partial\Re}{\partial t} + v\frac{\partial\Re}{\partial x} = \frac{\partial}{\partial x}\left(\nu\frac{\partial\Re}{\partial x}\right) + F,\ \nu > 0, 0 \le x \le l, t > 0 \qquad (8.2.16)$$

with initial condition

$$\Re\left(x,0\right) = \eta\left(x\right), 0 \le x \le l \qquad (8.2.17)$$

and Dirichlet boundary conditions

$$\Re\left(0,t\right) = \alpha\left(t\right),\ \Re\left(l,t\right) = \beta\left(t\right), t > 0, \qquad (8.2.18)$$

where $v(x,t)$, $\nu(x,t)$, $\eta(x)$, $\alpha(t)$ and $\beta(t)$ are known functions, while the function $\Re(x,t)$ is unknown.

Assume that the parameters of the problem $v(x,t)$, $\nu(x,t)$, $\eta(x)$, $\alpha(t)$ and $\beta(t)$ are sufficiently smooth, and the initial and boundary conditions are matched so that the problem (8.2.16) - (8.2.18) has a unique solution $\Re(x,t)$.

A parabolic equation (8.2.16) includes terms describing the same physical processes as the members included in the fluid mechanics equations (3.2.29) - (3.2.35), namely:

- Transient member $\partial\Re / \partial t$ describing the rate of change of $\Re$ in time;
- Convective term $v\partial\Re / \partial x$ describing the convection of $\Re$ due to the movement of the surrounding environment with the speed *v*;
- Diffusion (or dissipative) member describing the physical diffusion of in the environment along the -axis.

What prevails, diffusion or convection, depends on the ratio of the quantities and . Obviously, the formation processes and phenomena in the atmosphere in each specific area is influenced by both convection and diffusion. However, the significance of each of them in different circumstances is not the same. Thus, the formation processes at high altitudes (in free atmosphere) is mainly influenced by convection; the turbulent diffusion thus makes a specific, but very little contribution. The presence in the atmosphere of inactive significant horizontal gradients of temperature and humidity contributes to the intensification of diffusion processes. Therefore, the equation (8.2.16) can take type (hyperbolic or parabolic) depending on the ratio and quantities and, i.e. depending on traffic conditions in the atmosphere. Changing the type of equation may lead to incorrectness in the formulation of the problem and computational problems to solve, and the simplification of the original system of equations by neglecting the members of convection or diffusion terms of the equation requires proof of their smallness within the entire region under consideration and implementation time, what is today an insurmountable task.

Let us divide the spatial interval [0,*l*] of the problem into *J* equal parts with division points $x_0 < x_1 < \ldots < x_{J-1} < x_J$, and denote the length of the *j*-th sub-interval by h_j. Besides, divide [0,*T*] into *N* equal parts by points $t^n = nTN^{-1}$, $n = 0,1,\ldots,N$, with time step τ. We define the grid

$$\mathbf{\Omega} = \left\{\left(x_j, t^n\right); j = 0,1,\ldots,J; n = 0,1,\ldots,N\right\},$$

and denote by $\Re_j^n$ the approximation of $\Re\left(x_j, t^n\right)$.

Due to the excessive bulkiness of the mathematical model of atmospheric circulation its numerical implementation is carried out, as a rule, by explicit or semi-implicit difference schemes. Among the existing dual-layer explicit finite difference schemes for solving problems (8.2.16) - (8.2.18), preference is given to schemes in which the convective term is approximated by upwind differences. They have the error of approximation of the equation (8.2.16) of the order $O\left[\tau, h\right]$. The stability condition for them is imposed by time step restrictions $\tau \leq \left[\left(\left|v\right| + 2\nu / h\right) / h\right]^{-1}$. When $\tau\left|v\right| / h < 1$, these schemes make artificial damping. In other words, upwind scheme introduces an artificial viscosity the value of which is equal to $\nu_s = h|v|\left(1 - \tau|v| / h\right) / 2$. This additional dissipation suppresses oscillations characteristic of the scheme with central differences, but at the same time, as a rule, its accuracy is greatly reduced. Explicit schemes (with a finite τ) correctly simulate the physical characteristics of partial differential equations of parabolic type, since the decision at an interior point M $\left(0 < M < l\right)$ located on the characteristic $t = t^{n+1}$, does not depend on the boundary conditions. It seems that for the solution of the parabolic type of equations is better to use implicit methods since they take into account all the information known to the characteristic $t = const$ and under it.

The main shortcomings of the existing two-layer implicit schemes to be mentioned are:

- Implementing them needs to solve a linear system of algebraic equations with three diagonal matrix at each new time step.
- In the case of using the nonlinear equations of motion (8.2.16), i.e. when $\Re = v$ and $\nu = 0$, there may be non-linear instability, and, in order to prevent it for smoothing the solutions, it is necessary to introduce the artificial viscosity.

In addition, if an uncontrolled change of the type of equation with parabolic to the hyperbolic occurs, the use of implicit schemes can cause computational problems with the infinite speed of propagation of disturbances in modeling real atmosphere. Indeed, while for the model equation (8.2.16) at $\nu = 0$, a perturbation of $\Re$ is distributed for a time τ and distance $s = v\tau$, for implicit scheme, since all $\Re^{n+1}$ are involved in the calculation at the same time, the disturbance extends over a distance $s = \infty$ (or to the limits of computational grid). In order to simulate qualitatively the correct behavior of the equation (8.2.16) for the speed of propagation of disturbances, sometimes (e.g., (Anderson, Tannehill, Pletcher, 1990) one used an explicit scheme for the convective term, and implicit scheme for the diffusion term.

Here, we will focus on the construction of unconditionally stable economical two-layer difference schemes for the numerical solution of the problem (8.2.16) - (8.2.18) a generalization of which is to the multidimensional case in the framework of the splitting method can be carried out simply enough.

Integrating Eq. (8.2.16) at x_j from t^n to t^{n+1} yields

$$\Re_j^{n+1} = \Re_j^n - \int\limits_{t^n}^{t^{n+1}} \left[v \frac{\partial \Re}{\partial x} - \frac{\partial}{\partial x}\left(\nu \frac{\partial \Re}{\partial x} \right) - F \right] dt . \tag{8.2.19}$$

Approximating the integral on the right-hand side by the mean-value theorem, we obtain

$$\Re_j^{n+1} = \Re_j^n - \tau \left[v \frac{\partial \Re}{\partial x} - \frac{\partial}{\partial x}\left(\nu \frac{\partial \Re}{\partial x} \right) - F \right]_j^{t=\theta} , \tag{8.2.20}$$

where $t^n < \theta < t^{n+1}$.

For the approximation of the derivatives $\left(\partial \Re / \partial x\right)\Big|_j^{t=\theta}$ and $\partial\left(\nu \partial \Re / \partial x\right) / \partial x\Big|_j^{t=\theta}$ we will use the following difference relations:

$$\left.\left(\frac{\partial \Re}{\partial x}\right)\right|_j^{t=\theta} = \frac{1}{h_{j-1}+h_j}\left[h_{j-1}\frac{\Re_{j+1}-\Re_j}{h_j}+h_j\frac{\Re_j-\Re_{j-1}}{h_{j-1}}\right]^{t=\theta} - \frac{h_{j-1}h_j}{6}\left(\frac{\partial^3\Re}{\partial x^3}\right)^{t=\theta}, \tag{8.2.21a}$$

$$\left.\left[\frac{\partial}{\partial x}\left(\nu\frac{\partial \Re}{\partial x}\right)\right]\right|_j^{t=\theta} = \frac{1}{h_{j-1}+h_j}\left[\begin{array}{l}\left(\nu_{j+1}+\nu_j\right)\dfrac{\Re_{j+1}-\Re_j}{h_j}\\ -\left(\nu_j+\nu_{j-1}\right)\dfrac{\Re_j-\Re_{j-1}}{h_{j-1}}\end{array}\right]^{t=\theta} - \frac{h_j-h_{j-1}}{3}\left(\frac{\partial^3\Re}{\partial x^3}\right)^{t=\theta}, \tag{8.2.22a}$$

The unilateral difference expressions $\left(\Re_{j+1}-\Re_j\right)/h_j$ and $\left(\Re_j-\Re_{j-1}\right)/h_{j-1}$ in (8.2.21a) and (8.2.22a) will be taken at different time levels (n and n+1). For construction of approximations only by two points it is natural for physical reasons to have on the (n+1)-th layer a point x_j as central, and to select the second one from that side from where $\Re$ is transferred by advection to the central point. It is easy to see that the created difference scheme (8.2.21a) and (8.2.22a) has an approximation error of the first order in τ. In this manner we gain the following form:

- For $v > 0$:

$$\left.\left(\frac{\partial \Re}{\partial x}\right)\right|_j^{t=\theta} \approx \frac{1}{h_{j-1}+h_j}\left[h_{j-1}\frac{\Re^n_{j+1}-\Re^n_j}{h_j}+h_j\frac{\Re^{n+1}_j-\Re^{n+1}_{j-1}}{h_{j-1}}\right]+\tau\left.\frac{\partial^2\Re}{\partial t\partial x}\right|_j^{t=\theta}, \tag{8.2.21b}$$

$$\left.\left[\frac{\partial}{\partial x}\left(\nu\frac{\partial \Re}{\partial x}\right)\right]\right|_j^{t=\theta} \approx \frac{1}{h_{j-1}+h_j}\left[\left(\nu_{j+1}+\nu_j\right)\frac{\Re^n_{j+1}-\Re^n_j}{h_j}-\left(\nu_j+\nu_{j-1}\right)\frac{\Re^{n+1}_j-\Re^{n+1}_{j-1}}{h_{j-1}}\right]+\tau\left.\frac{\partial^2\Re}{\partial t\partial x}\right|_j^{t=\theta}. \tag{8.2.22b}$$

- For $v < 0$:

$$\left.\left(\frac{\partial \Re}{\partial x}\right)\right|_j^{t=\theta} \approx \frac{1}{h_{j-1}+h_j}\left[h_{j-1}\frac{\Re^{n+1}_{j+1}-\Re^{n+1}_j}{h_j}+h_j\frac{\Re^n_j-\Re^n_{j-1}}{h_{j-1}}\right]+\tau\left.\frac{\partial^2\Re}{\partial t\partial x}\right|_j^{t=\theta}, \tag{8.2.21c}$$

$$\left.\left[\frac{\partial}{\partial x}\left(\nu\frac{\partial \Re}{\partial x}\right)\right]\right|_j^{t=\theta} \approx \frac{1}{h_{j-1}+h_j}\left[\left(\nu_{j+1}+\nu_j\right)\frac{\Re^{n+1}_{j+1}-\Re^{n+1}_j}{h_j}-\left(\nu_j+\nu_{j-1}\right)\frac{\Re^n_j-\Re^n_{j-1}}{h_{j-1}}\right]+\tau\left.\frac{\partial^2\Re}{\partial t\partial x}\right|_j^{t=\theta}. \tag{8.2.22c}$$

Substituting (8.2.21b), (8.2.22b) or (8.2.21c), (8.2.22c) in (8.2.15) we will receive a difference scheme for the one-dimensional advection-diffusion problem (8.2.16) – (8.2.18) in the following form:

- For $v > 0$:

$$\frac{\Re_j^{n+1}-\Re_j^n}{\tau}+\frac{v_j^n}{h_{j-1}+h_j}\left[h_{j-1}\frac{\Re_{j+1}^n-\Re_j^n}{h_j}+h_j\frac{\Re_j^{n+1}-\Re_{j-1}^{n+1}}{h_{j-1}}\right]-\frac{1}{h_{j-1}+h_j}\left[\left(\nu_{j+1}+\nu_j\right)\frac{\Re_{j+1}^n-\Re_j^n}{h_j}-\left(\nu_j+\nu_{j-1}\right)\frac{\Re_j^{n+1}-\Re_{j-1}^{n+1}}{h_{j-1}}\right]-F_j^n=0; \tag{8.2.23a}$$

$j=0,1,\ldots,J, n=0,1,\ldots,N$

$\Re_j^0=\eta\left(x_j\right), j=0,1,\ldots,J$,

$\Re_0^n=\alpha\left(t^n\right), \Re_J^n=\beta\left(t^n\right), n=0,1,\ldots,N$.

- For $v<0$:

$$\frac{\Re_j^{n+1}-\Re_j^n}{\tau}+\frac{v_j^n}{h_{j-1}+h_j}\left[h_{j-1}\frac{\Re_{j+1}^{n+1}-\Re_j^{n+1}}{h_j}+h_j\frac{\Re_j^n-\Re_{j-1}^n}{h_{j-1}}\right]-\frac{1}{h_{j-1}+h_j}\left[\left(\nu_{j+1}+\nu_j\right)\frac{\Re_{j+1}^{n+1}-\Re_j^{n+1}}{h_j}-\left(\nu_j+\nu_{j-1}\right)\frac{\Re_j^n-\Re_{j-1}^n}{h_{j-1}}\right]-F_j^n=0; \tag{8.2.23b}$$

$j=J-1,J-2,\ldots,2,1, n=0,1,\ldots,N$

$\Re_j^0=\eta\left(x_j\right), j=0,1,\ldots,J$,

$\Re_0^n=\alpha\left(t^n\right), \Re_J^n=\beta\left(t^n\right), n=0,1,\ldots,N$.

In both cases, the differences directed at the (n + 1) th time layer in the direction from where the convection goes to the point x_j. Typically, such differences are called the difference against the flow (upstream differences). Thus, the single difference at the (n+1)-th time layer in the scheme (8.2.23) is always taken against the flow, i.e. in a direction upstream of the point at which $\partial\Re/\partial t$ is calculated. Due to this obstacle, the disturbance cannot propagate in the opposite direction to the physical convection. Therefore, any parasitic wave should not distort the numerical solution.

Patterns corresponding to the scheme (8.2.23) for $v>0$ and $v<0$ are shown in Figure 5a and 5b respectively.

The individual terms of this scheme are implicit, i.e. the values of the grid function $\Re$ occur in it from the top (n+1)-th time layer. However, despite this, the difference scheme (8.2.23) can easily be

Figure 5. Templates for difference schemes: a) the scheme (8.2.23a); b) the scheme (8.2.23b)

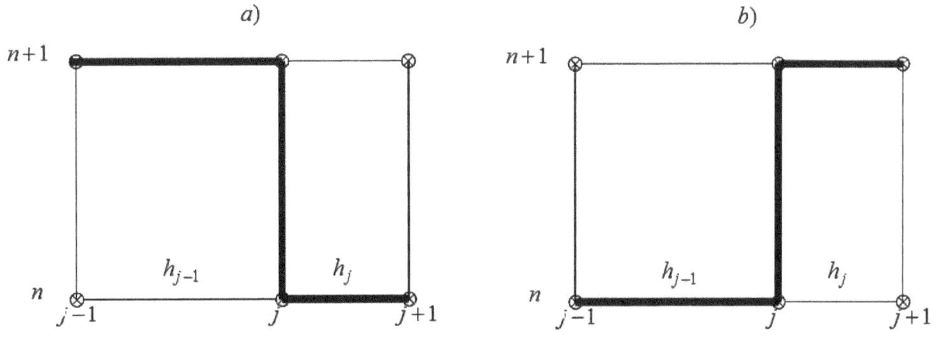

solved explicitly. To demonstrate the difference algorithm for solving differential problem (8.2.16) - (8.2.18), we use the example of the scheme (8.2.23*a*). The set of finite difference equations (8.2.23*a*), recorded for all indices *j* and *n* in the area $\mathbf{\Omega}$, is a system of linear algebraic equations, the number of which is less than the number of unknown grid functions $\Re_j^{n+1}$. The missing equations are derived from the initial and boundary conditions (8.2.13).

Let the values of the grid function $\Re$ of the solution at the *n*-th time layer are known as $\Re_j^n$, $j = 0,1,\ldots,J$. "By applying" a template of Figure 2 for some *j*-th node of the grid (Figure 3), we calculate, by the known values $\Re_j^n$, $\Re_{j+1}^n$ and $\Re_{j-1}^{n+1}$ and using equation (8.2.23a), the value

$$\Re_j^{n+1} = \frac{p_j \Re_{j-1}^{n+1} - d_j \Re_{j+1}^n + \left(1 + d_j\right) \Re_j^n + \tau F_j^n}{1 + p_j}, \tag{8.2.24a}$$

where

$$p_j = \tau b_j = \frac{\tau}{h_{j-1} + h_j} \left(\frac{h_j}{h_{j-1}} v_j^n + \frac{\nu_j + \nu_{j-1}}{h_{j-1}} \right),$$

$$d_j = \tau a_j = \frac{\tau}{h_{j-1} + h_j} \left(\frac{h_{j-1}}{h_j} v_j^n - \frac{\nu_{j+1} + \nu_j}{h_j} \right).$$

"Moving" the template to the right by one grid interval and repeating this procedure, we find the value $\Re_{j+1}^{n+1}$ by the known values of $\Re_{j+1}^n$, $\Re_{j+2}^n$ and the determined value $\Re_j^{n+1}$ (see Figure 6). Consistently "moving" the template, we find the whole sequence of values $\Re_j^{n+1}$ $\left(j = 0,1,\ldots,J-1\right)$ by the formula (8.2.24*a*). Note that to determine solutions using equations (8.2.24*a*) at the time level *n*+1 at the point *j*=1 it is required boundary condition at the point *j*=0. The boundary condition at the point *j*=*J* is not used, as for determining the solution at the point *j*=*J*-1 at the level *n*+1 it is necessary the value of the grid function at the point *j*=*J*-2, and at the boundary point *J*, it is not necessary. This is due to the approximation of the derivative by *x* by a directional difference.

The equation similar to equation (8.2.24*a*) for $v < 0$ is of the form:

Figure 6. A graph of the algorithm for solving the problem (8.2.16) – (8.2.18) by the scheme (8.2.23a)

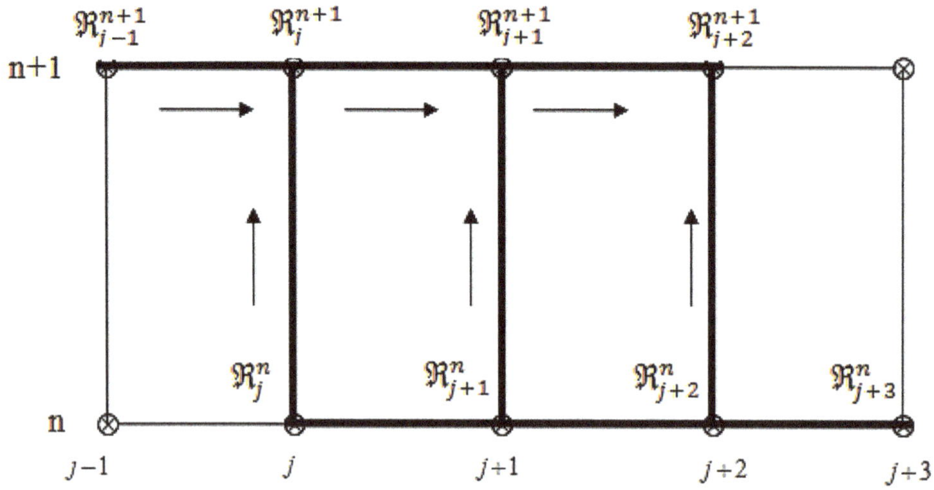

$$\Re_j^{n+1} = \frac{p_j \Re_{j-1}^n - d_j \Re_{j+1}^{n+1} + \left(1 - p_j\right) \Re_j^n + \tau F_j^n}{1 - d_j}. \tag{8.2.24b}$$

Thus, the difference scheme (8.2.23) of the numerical solution of the convective diffusion problem (8.2.16) - (8.2.18) allows consistent building solutions for grid points (in ascending order of the index j at $v > 0$, and in descending order with the index j at $v < 0$) by the algorithm of "running counting". Consequently, the scheme (8.2.23) is an explicit and single-ended.

The resulting finite-difference scheme is similar in appearance to other existing upwind scheme: explicit upwind scheme (Filler, Ludloff, 1961), an explicit scheme of the method of alternating directions (Saul'ev, 1960), explicit patterns with diagonal differences (Roberts, Weiss, 1966), and others. The basic difference between these schemes and proposed one is that they use for approximating the convective term the upstream differences either only at the n-th or only at the $(n+1)$-th time level, and use the classical three-point difference operator at the n-th time level for the approximation of the diffusion term.

Adaptation of a Difference Scheme for the Case of Changes in the Convection Speed Sign

The simplicity of the solution of (8.2.16) - (8.2.18) through the scheme (8.2.23a) or (8.2.23b) is maintained only provided that the is no change of the sign of convection speed in the field, or it occurs only once with convection changing from positive to negative direction of the speed. Otherwise, if the negative convection speed changes to positive one at some point, then there will be problems with implementing the proposed difference scheme, since the problem (8.2.23a), (8.2.23b) are formally "not closed".

Indeed, it follows from the condition $v > 0$ that the difference problem (8.2.23a) should be solved by passing calculation points up the grid, starting from the first point. The limit value at the first point in this case is determined by a given boundary condition $\Re_0^{n+1} = \alpha\left(t^{n+1}\right)$. On the other hand, from the condition $v < 0$, it follows that the problem (8.2.23b) should be solved by searching the calculated grid points down from the last point. In this case, the limit value at the last point is determined by a given boundary condition $\Re_J^{n+1} = \beta\left(t^{n+1}\right)$. Accordingly, if at some internal point the negative convection speed is changed to the positive (see Figure 7), the solution must be carried out by the scheme (8.2.23b); otherwise – by the scheme (8.2.23a). In this case, for the scheme (8.2.23b), an unknown at the (n+1)-th step value $\Re_J^{n+1}$ acts as a "boundary" conditions, and for the scheme (8.2.23a), - an unknown value $\Re_{J-1}^{n+1}$.

This problem is absolutely natural for dynamic meteorology, since it is not difficult to imagine some types of atmospheric circulation (cyclones and anticyclones), in which even at a fairly short distances occur numerous changes in wind direction to the opposite direction. In this regard, the question arises: why the implicit difference schemes "against the flow" in practical use for the problems of gas dynamics of the existence of this problem has not been discussed at all?

For the proposed scheme (8.2.23), we specify the following path, eliminating the problem of changes in the convection speed sign in the calculation of value $\Re_{j-1}^{n+1}$ and $\Re_j^{n+1}$. We will adhere to the requirement that the calculation accuracy of the values $\Re_{j-1}^{n+1}$ and $\Re_j^{n+1}$ consistent with the finite-difference scheme used to solve the problem (8.2.16) - (8.2.18) in general. This requirement corresponds to the solution of differential equations (8.2.23a) and (8.2.23b) with two unknowns $\Re_{j-1}^{n+1}$ and $\Re_j^{n+1}$, when the finite-difference scheme (8.2. 23a) is recorded for the point j, and the scheme (8.2.23b) is recorded for the point j-1

$$\frac{\Re_j^{n+1} - \Re_j^n}{\tau} + \frac{v_j^n}{h_{j-1} + h_j}\left[h_{j-1}\frac{\Re_{j+1}^n - \Re_j^n}{h_j} + h_j\frac{\Re_j^{n+1} - \Re_{j-1}^{n+1}}{h_{j-1}} \right] - \frac{1}{h_{j-1} + h_j}\left[\left(\nu_{j+1} + \nu_j\right)\frac{\Re_{j+1}^n - \Re_j^n}{h_j} - \left(\nu_j + \nu_{j-1}\right)\frac{\Re_j^{n+1} - \Re_{j-1}^{n+1}}{h_{j-1}} \right] - F_j^n = 0, \tag{8.2.25}$$

Figure 7. Geometric illustration of the difference solution when the change of the sign of speed of the convection occurs

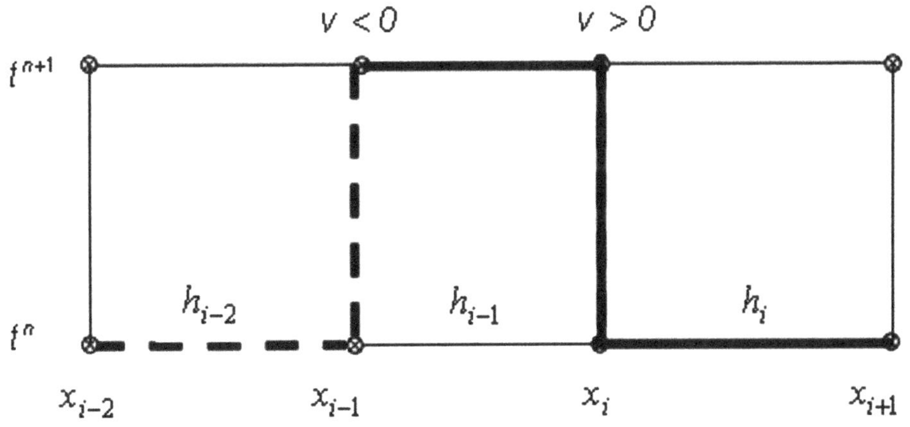

$$\frac{\Re_{j-1}^{n+1}-\Re_{j-1}^{n}}{\tau}+\frac{v_{j-1}^{n}}{h_{j-2}+h_{j-1}}\left[h_{j-2}\frac{\Re_{j}^{n+1}-\Re_{j-1}^{n+1}}{h_{j-1}}+h_{j-1}\frac{\Re_{j-1}^{n}-\Re_{j-2}^{n}}{h_{j-2}}\right]$$
$$-\frac{1}{h_{j-2}+h_{j-1}}\left[\left(\nu_{j}+\nu_{j-1}\right)\frac{\Re_{j}^{n+1}-\Re_{j-1}^{n+1}}{h_{j-1}}-\left(\nu_{j-1}+\nu_{j-2}\right)\frac{\Re_{j-1}^{n}-\Re_{j-2}^{n}}{h_{j-2}}\right]-F_{j-1}^{n}=0 \quad . \tag{8.2.26}$$

In the so obtained system of differential equations the number of unknowns is equal to the number of equations. Solving it in respect to the unknown its $\Re_{j-1}^{n+1}$ and $\Re_{j}^{n+1}$, we obtain

$$\Re_{j-1}^{n+1}=\frac{\left(1+p_{j}\right)\mathcal{B}_{j}-d_{j-1}\mathcal{H}_{j}}{\Im_{j}}, \tag{8.2.27}$$

$$\Re_{j}^{n+1}=\frac{p_{j}\mathcal{B}_{j}+\left(1-d_{j-1}\right)\mathcal{H}_{j}}{\Im_{j}}, \tag{8.2.28}$$

where

$$\mathcal{B}_{j}=\left(1-p_{j-1}\right)\Re_{j-1}^{n}+p_{j-1}\Re_{j-2}^{n}+\tau F_{j-1}^{n},$$

$$\mathcal{H}_{j}=\left(1-d_{j}\right)\Re_{j}^{n}-d_{j}\Re_{j+1}^{n}+\tau F_{j}^{n},$$

$$\Im_{j}=1+p_{j}-d_{j-1},$$

and p_j and d_j have the same form as in (8.2.24).

Condition (8.2.27) and (8.2.28) are quite reliable; the accuracy of their solution meets the basic accuracy of the solution of the difference scheme (8.2.23).

So far, we have demonstrated an algorithm for solving a differential problem (8.2.16) - (8.2.18) by means of a difference scheme (8.2.23) by the example of the first kind boundary conditions (Dirichlet conditions). Also, it will be helpful to demonstrate the behavior of difference scheme for other types of boundary conditions. We choose for this the boundary conditions of the most common type – the conditions of the third kind.

The Case of the Degenerate Equation

In the free atmosphere, i.e. above the boundary layer, there are areas of uniform motion in which the eddy viscosity coefficient is so small that its effect on the change of meteorological variables almost disappears. In fact, the movement of air masses becomes purely convective, although formally described by the equation of convective diffusion. In this case we say that the equation degenerates, or change your style from parabolic to hyperbolic:

$$\frac{\partial\Re}{\partial t}+v\frac{\partial\Re}{\partial x}=F, v>0, 0\leq x\leq l, t>0 \tag{8.2.29}$$

with initial condition

$$\Re(x,0)=\eta(x), 0\le x\le l \tag{8.2.30}$$

and Dirichlet boundary conditions

$$\Re(0,t)=\alpha(t), \Re(l,t)=\beta(t), t>0, \tag{8.2.31}$$

where $v(x,t)$, $\eta(x)$, $\alpha(t)$ and $\beta(t)$ are known functions, while the function $\Re(x,t)$ is unknown.

For the equation (2.8.29), the scheme (8.2.23a), (8.2.23b) takes the following form:

- For $v > 0$:

$$\frac{\Re_j^{n+1}-\Re_j^n}{\tau}+\frac{v_j^n}{h_{j-1}+h_j}\left[h_{j-1}\frac{\Re_{j+1}^n-\Re_j^n}{h_j}+h_j\frac{\Re_j^{n+1}-\Re_{j-1}^{n+1}}{h_{j-1}}\right]-F_j^n=0; \tag{8.2.32a}$$

$$j=0,1,\ldots,J, n=0,1,\ldots,N,$$

$$\Re_j^0=\eta(x_j), j=0,1,\ldots,J,$$

$$\Re_0^n=\alpha(t^n), \Re_J^n=\beta(t^n), n=0,1,\ldots,N.$$

- For $v < 0$:

$$\frac{\Re_j^{n+1}-\Re_j^n}{\tau}+\frac{v_j^n}{h_{j-1}+h_j}\left[h_{j-1}\frac{\Re_{j+1}^{n+1}-\Re_j^{n+1}}{h_j}+h_j\frac{\Re_j^n-\Re_{j-1}^n}{h_{j-1}}\right]-F_j^n=0; \tag{8.2.32b}$$

$$j=J-1, J-2,\ldots,2,1, n=0,1,\ldots,N,$$

$$\Re_j^0=\eta(x_j), j=0,1,\ldots,J,$$

$$\Re_0^n=\alpha(t^n), \Re_J^n=\beta(t^n), n=0,1,\ldots,N.$$

Calculation by scheme (8.2.32a), (8.2.32b) is similar to the scheme (8.2.23a), (8.2.23b) and does not change the general algorithm for solving the problem. Thus, changing the type of equation in the middle of the region does not create computational problems during implementation of the method of explicit calculation and does not require special attention. In fact, the method may work in cases where the original equation repeatedly changes the type several times at a time step.

The Difference Representation of Boundary Conditions of the Third Kind

Let the boundary condition at $x = 0$ for the equation (8.2.15) has the form

$$\varepsilon_0 \frac{\partial \Re(t,0)}{\partial x} + \gamma_0 \Re(t,0) = a(t). \tag{8.2.33}$$

We approximate it as follows:

$$\gamma_0 \Re_0^{n+1} + \phi_0^n \frac{\Re_1^{n+1} - \Re_0^{n+1}}{h_0} - \frac{\varepsilon_0 h_0}{2\nu_0^n}\left(\frac{\Re_0^{n+1} - \Re_0^n}{\tau} - F_0^n\right) = a^{n+1},$$

$$\phi_0^n = \varepsilon_0\left[1 - \frac{h_0 v_0^n - \left(\nu_1^n - \nu_0^n\right)}{2\nu_0^n}\right].$$

and hence

$$\left(h_0\gamma_0 - \phi_0^n - \frac{\varepsilon_0 h_0^2}{2\nu_0^n}\right)\Re_0^{n+1} + \phi_0^n \Re_1^{n+1} = a^{n+1}h_0 - \frac{\varepsilon_0 h_0^2}{2\nu_0^n}\left(\frac{\Re_0^n}{\tau} + F_0^n\right). \tag{8.2.34}$$

Since the left boundary condition is used in the case of $v > 0$, then approximating equation the scheme (8.2.23a) at in the border point we have:

$$\left(1 + p_1\right)\gamma_0 \Re_1^{n+1} = p_1 \Re_0^{n+1} + \left(1 + d_1\right)\Re_1^n - d_1 \Re_2^n + \tau F_1^n, \tag{8.2.35}$$

From the system of equations (8.2.33) and (2.8.34), we find the “initial” value necessary to start the calculation scheme (8.2.23a):

$$\Re_0^{n+1} = \frac{\varkappa_0}{\varkappa_+}, \tag{8.2.36}$$

where

$$\varkappa_0 = \phi_0^n\left[\left(1+d_1\right)\Re_1^n - d_1\Re_2^n + \tau F_1^n\right] - \left(1+p_1\right)\left[a^{n+1}h_0 - \frac{\varepsilon_0 h_0^2}{2\nu_0^n}\left(\frac{\Re_0^n}{\tau} + F_0^n\right)\right],$$

$$\varkappa_+ = \phi_0^n - \left(1+p_1\right)\left(h_0\gamma_0 - \frac{\varepsilon_0 h_0^2}{2\nu_0^n\tau}\right).$$

Similar transformations are carried out in the case where the right boundary condition is of the form:

$$\varepsilon_1\frac{\partial\Re\left(t,l\right)}{\partial x} + \gamma_1\Re\left(t,l\right) = a\left(t\right). \tag{8.2.37}$$

We approximate (8.2.37)

$$\gamma_1\Re_N^{n+1} + \phi_N^n\frac{\Re_N^{n+1} - \Re_{N-1}^{n+1}}{h_{N-1}} - \frac{\varepsilon_1 h_{N-1}}{2\nu_N^n}\left(\frac{\Re_N^{n+1} - \Re_N^n}{\tau} - F_N^n\right) = b^{n+1}, \tag{8.2.38}$$

$$\phi_N^n = \varepsilon_1\left[1 - \frac{h_{N-1}v_N^n - \left(\nu_N^n - \nu_{N-1}^n\right)}{2\nu_N^n}\right],$$

and, at the border point, the equation (8.2.15) can be approximated by the scheme (8.2.23b)

$$\Re_{N-1}^{n+1}\left(1-d_{N-1}\right) = -d_{N-1}\Re_N^{n+1} + \left(1-p_{N-1}\right)\Re_{N-1}^n + p_{N-1}\Re_{N-2}^n + \tau F_{N-1}^n. \tag{8.2.39}$$

From the system (2.8.38), (2.8.39), we obtain the value necessary to start the calculation by the scheme (8.2.23b):

$$\Re_N^{n+1} = \frac{\varkappa_1}{\varkappa_-}, \tag{8.2.40}$$

where

$$\varkappa_1 = \phi_N^n\left[\left(1-p_{N-1}\right)\Re_{N-1}^n + p_{N-1}\Re_{N-2}^n + \tau F_{N-1}^n\right] + \left(1-d_{N-1}\right)\left[b^{n+1}h_{N-1} - \frac{\varepsilon_1 h_{N-1}^2}{2\nu_N^n}\left(\frac{\Re_N^n}{\tau} + F_N^n\right)\right],$$

$$\varkappa_{-} = \phi_N^n + \left(1 - d_{N-1}\right)\left(h_{N-1}\gamma_1 - \frac{\varepsilon_1 h_{N-1}^2}{2\nu_N^n \tau}\right).$$

THEORETICAL STUDY OF APPROXIMATION, ACCURACY, STABILITY AND CONVERGENCE OF THE DIFFERENCE SCHEME

Let us study the properties of the scheme (8.2.23) in order to determine the possibility of its application for solving problems of the forecast of meteorological variables fields. The study will take into account the characteristics associated with the problem to be solved.

First, atmospheric circulation model equations are nonlinear what causes serious difficulties to determine the stability of the numerical scheme, since studying the stability of the scheme for linearized equation or an equation with "frozen" coefficients provide only the necessary conditions for its stability for a nonlinear equation. Second, in order to describe adequately the wave processes in the atmosphere, it should be taken into account two main requirements for the numerical solution: it must correctly reproduce the changes in wave amplitude and correctly describes the change in the oscillation phase, phase and group velocity of the wave perturbations (Belov, Borisenkov, Panin, 1989; Prusov, Doroshenko, 2006; Manabe, 1988). And third, the specifics of the problem, i.e., forecasting meteorological variables, sharply raises the requirement for the efficiency of the numerical algorithm – a reasonable time spent at an appropriate accuracy of the solution.

For concretization of the theoretical studies of the properties of the scheme (8.2.23), restrict ourselves to the case of $v = \text{const} \geq 0$, $\nu = \text{const} \geq 0$ and consider it in the form (8.2.24a).

Approximation

Since the initial condition (8.2.17) and the boundary conditions (8.2.18) in the finite-difference representation (8.2.23) are defined precisely, the study of approximation problem (8.2.16) - (8.2.18) is reduced to the study of approximation of the equation (8.2.16) by the discrete equation (8.2.23).

Given that difference grid on geopotential surfaces of macroscale models are usually uniform, we confine ourselves below to the spatial and temporal grids with uniform steps h and τ respectively, as well as we take $v \geq 0$ and $\nu > 0$, what does not limit the generality. A difference scheme for the one-dimensional problem of convective diffusion (8.2.16) - (8.2.18) at $v \geq 0$, having the form of (8.2.23a), is implemented by the recurrence relation (8.2.24a).

It is known to determine the order of approximation (Samarskii, 1971), the difference equation (8.2.23) approximates equation (8.2.16) with the order (m,n) or has an approximation $O\left[h^m, \tau^n\right]$ on the solution $\Re = \Re\left(x,t\right)$ of equation (8.2.16), if $\psi_d\left(x,t\right) = O\left[h^m + \tau^n\right]$ or $\psi_d \leq M\left(h^m + \tau^n\right)$ for all $t \in \varpi_\tau$ (M is a positive constant not depending on h and τ, $\left\|\cdot\right\|$ is a norm on the grid ϖ_τ), where ψ_d is a local error of the approximation scheme (8.2.23) on the solution of equation (8.2.16), equal to

$$\psi_d = F_j^n - \frac{1}{\tau}\left[\left(1 + p_j\right)\Re_j^{n+1} - p_j \Re_{j-1}^{n+1} + d_j \Re_{j+1}^n - \left(1 + d_j\right)\Re_j^n\right]. \tag{8.2.41}$$

Estimate the order of local approximation error ψ_d. For this purpose, expand the function $\Re(x,t)$ by Taylor's formula in a neighborhood of the point (x_j, t^n):

$$\Re_j^{n+1} = \Re_j^n + \frac{\tau}{1!}\left(\frac{\partial \Re}{\partial t}\right)_j^n + \frac{\tau^2}{2!}\left(\frac{\partial^2 \Re}{\partial t^2}\right)_j^n + \frac{\tau^3}{3!}\left(\frac{\partial^3 \Re}{\partial t^3}\right)_j^n + O[\tau^4],$$

$$\Re_{j+1}^n = \Re_j^n + \frac{h}{1!}\left(\frac{\partial \Re}{\partial x}\right)_j^n + \frac{h^2}{2!}\left(\frac{\partial^2 \Re}{\partial x^2}\right)_j^n + \frac{h^3}{3!}\left(\frac{\partial^3 \Re}{\partial x^3}\right)_j^n + O[h^4], \tag{8.2.42}$$

$$\Re_{j-1}^{n+1} = \Re_j^n + \frac{\tau}{1!}\left(\frac{\partial \Re}{\partial t}\right)_j^n - \frac{h}{1!}\left(\frac{\partial \Re}{\partial x}\right)_j^n + \frac{\tau^2}{2!}\left(\frac{\partial^2 \Re}{\partial t^2}\right)_j^n - \frac{\tau h}{1!}\left(\frac{\partial^2 \Re}{\partial t \partial x}\right)_j^n + \frac{h^2}{2!}\left(\frac{\partial^2 \Re}{\partial t^2}\right)_j^n + \frac{\tau^3}{3!}\left(\frac{\partial^3 \Re}{\partial t^3}\right)_j^n.$$
$$-\frac{\tau^2 h}{2!}\left(\frac{\partial^3 \Re}{\partial t^2 \partial x}\right)_j^n + \frac{\tau h^2}{2!}\left(\frac{\partial^3 \Re}{\partial t \partial x^2}\right)_j^n - \frac{h^3}{3!}\left(\frac{\partial^3 \Re}{\partial x^3}\right)_j^n + O[\tau^4 + \tau^3 h + \tau^2 h^2 + \tau h^3 + h^4]$$

Substituting the values p_j and d_j in the form of (8.2.24) into the expansion (8.2.42) to (2.8.41), we obtain

$$\psi_d = F_j^n - \left(\frac{\partial \Re}{\partial t}\right)_j^n - \left(\upsilon \frac{\partial \Re}{\partial x}\right)_j^n + \frac{\nu_{j+1}^n - \nu_{j-1}^n}{2h}\left(\frac{\partial \Re}{\partial x}\right)_j^n + \frac{\nu_{j+1}^n + 2\nu_j^n + \nu_{j-1}^n}{4}\left(\frac{\partial^2 \Re}{\partial x^2}\right)_j^n$$
$$+\frac{2\upsilon_j^n h^2 + \left(\nu_{j+1}^n - \nu_{j-1}^n\right)}{12}\left(\frac{\partial^3 \Re}{\partial x^3}\right)_j^n + \left[\upsilon_j^n h + \left(\nu_j^n + \nu_{j-1}^n\right)\right]\left(\frac{\tau}{2h}\left(\frac{\partial^2 \Re}{\partial t \partial x}\right)_j^n + \frac{\tau^2}{4h}\left(\frac{\partial^3 \Re}{\partial t^2 \partial x}\right)_j^n - \frac{\tau}{4}\left(\frac{\partial^3 \Re}{\partial t \partial x^2}\right)_j^n\right).$$
$$+O\left[\frac{\tau^3}{h} + \tau + \tau h + h^2\right]$$

Given that

$$F_j^n = \left(\frac{\partial \Re}{\partial t} + \upsilon\frac{\partial \Re}{\partial x} - \nu\frac{\partial^2 \Re}{\partial x^2}\right)_j^n,$$

$$\nu_{j+1}^n + \nu_j^n = 2\nu_j^n + h\left(\frac{\partial \nu}{\partial x}\right)_j^n + O[h^2],$$

$$\nu_j^n + \nu_{j-1}^n = 2\nu_j^n - h\left(\frac{\partial \nu}{\partial x}\right)_j^n + O[h^2], \tag{8.2.43}$$

$$\frac{\nu_{j+1}^{n} - \nu_{j-1}^{n}}{2h} = \left(\frac{\partial \nu}{\partial x}\right)_j^n + O\left[h^2\right],$$

$$\frac{\nu_{j+1}^{n} + 2\nu_j^n + \nu_{j-1}^{n}}{4} = \nu_j^n + O\left[h^2\right],$$

$$h\frac{\nu_{j+1}^{n} - \nu_{j-1}^{n}}{12} = \frac{h^2}{6}\left(\frac{\partial \nu}{\partial x}\right)_j^n + O\left[\mathrm{h}^2\right],$$

local problem error of approximation (8.2.16) - (8.2.18) by the scheme (8.2.23) is estimated as follows:

- On an even spatial grid:

$$\psi_d = O\left[\frac{\tau}{h} + \tau + h^2\right].$$

- On an uneven spatial grid:

$$\psi_d = O\left[\tau\right] + O\left[\sum_{j=1}^{3}\left(\max_j\left\{h_j^{(k)} - h_{j-1}^{(k)}\right\} + \frac{\tau}{\max_j\left\{h_j^{(k)}\right\}}\right)\right].$$

Since the error of factorization and splitting are related as $\psi_s = \psi_d + O\left[\tau^2\right]$, the total approximation error is $\psi = \psi_d$.

Turning to the grid norm *C*, get that scheme (8.2.23) is of the order of approximation

- On an even spatial grid:

$$\left\|\psi_d\left(x,t\right)\right\|_C = \max_{i=1,N-1}\left|\psi_d\right| = O\left[\frac{\tau}{h} + \tau + h^2\right]. \tag{8.2.44a}$$

- On an uneven spatial grid:

$$\left\|\psi_d\left(x,t\right)\right\|_C = \max_{i=1,N-1}\left|\psi_d\right| = O\left[\tau\right] + O\left[\sum_{j=1}^{3}\left(\max_j\left\{h_j^{(k)} - h_{j-1}^{(k)}\right\} + \frac{\tau}{\max_j\left\{h_j^{(k)}\right\}}\right)\right]. \tag{8.2.44b}$$

Thus, the scheme (8.2.23) there is a first-order approximation in space and time. In addition, the approximation of equation (8.2.43) is conditional

- On an even spatial grid:

$$\frac{\tau}{h} \to 0 \text{ when } \tau \to 0, h \to 0 .$$

- On an uneven spatial grid:

$$\frac{\tau}{\max_j \left\{h_j^{(k)}\right\}} \to 0 \text{, when } \tau \to 0, \max_j \left\{h_j^{(k)}\right\} \to 0, k = 1,2,3 .$$

Consequently, the problem (8.2.16) - (8.2.18) and its discrete analogue (8.2.23) are conditionally agreed.

Effect of Changes in the Convection Speed Sign Inside of the Computational Domain on the Order of Approximation Schemes

Show that in the case of the speed sign change of convection inside the computational domain from minus to plus, while solving the of system (8.2.25), (2.8.26), the approximation order of the scheme does not change.

Let ψ_1 is a local error of the condition (8.2.27), and ψ_2 is a local error of the condition (8.2.28)

$$\psi_1 = \left(1 + p_j\right)\mathcal{B}_j - d_{j-1}\mathcal{H}_j - \Im_j \Re_{j-1}^{n+1} ,$$

$$\psi_2 = \Im_j \Re_j^{n+1} - p_j \mathcal{B}_j - \left(1 - d_{j-1}\right)\mathcal{H}_j .$$

The total local error is

$$\psi_1 + \psi_2 = \Im_j \left(\Re_j^{n+1} - \Re_{j-1}^{n+1}\right) + \mathcal{B}_j - \mathcal{H}_j . \qquad (8.2.45)$$

Considering in (2.8.45) expressions $\mathcal{B}_j, \mathcal{H}_j$ and $\Im_j$, we get

$$\begin{aligned}\psi_1 + \psi_2 = &\left[\left(\Re_j^{n+1} - \Re_j^n\right) + p_j\left(\Re_j^{n+1} - \Re_{j-1}^{n+1}\right) + d_j\left(\Re_{j+1}^n - \Re_j^n\right) - \tau F_j^n\right] \\ &+ \left[\tau F_j^n - \left(\Re_{j-1}^{n+1} - \Re_{j-1}^n\right) - d_{j-1}\left(\Re_j^{n+1} - \Re_{j-1}^{n+1}\right) - p_{j-1}\left(\Re_{j-1}^n - \Re_{j-2}^n\right)\right]\end{aligned} .$$

Substituting in the last equality the Taylor series expansion (8.2.42) and the relation (8.2.43), we obtain the order of local approximation scheme (8.2.23) at the points j-1 and j when $v_{j-1} < 0$ and $v_j > 0$.

- On an even spatial grid:

$$\psi_1 + \psi_2 = O\left[\frac{\tau}{h} + \tau + h^2\right].$$

- On an uneven spatial grid:

$$\psi_1 + \psi_2 = O\left[\tau\right] + O\left[\sum_{j=1}^{3}\left(\max_j\left\{h_j^{(k)} - h_{j-1}^{(k)}\right\} + \frac{\tau}{\max_j\left\{h_j^{(k)}\right\}}\right)\right].$$

Therefore, the order of approximation of the problem (8.2.16) - (8.2.18) by the difference scheme (8.2.23) for the case when in the middle of the field there is a change of speed convection sign from positive to negative, does not change.

Stability and Convergence of the Difference Scheme

The solution of the difference scheme (8.2.23) must be closer to the exact solution of the differential problem (8.2.16) - (8.2.18); and the difference between them should decrease with decreasing a grid size. This property of the difference scheme is called the convergence of the scheme.

Definition 8.2.1: The solution of the difference problem (8.2.23) converges to the exact solution of the problem (8.2.16) - (8.2.18) in the grid norm if for an arbitrary time t_n, n=1,2,…, the following holds

$$\left\|\Re\left(x_j, t^n\right) - \Re_j^n\right\| \to 0 \text{ when } h_j \to 0, \tau \to 0, j = 1, 2, \ldots, N-1.$$

If the estimate holds:

$$\left\|\Re\left(x_j, t^n\right) - \Re_j^n\right\| = O\left[h_j^m + \tau^p\right],$$

then the solution has the m-th order of convergence in space and p-th order in time.

Typically, the difference scheme convergence analysis is based on the theoretical study of the properties of its approximation and stability, as the necessary and sufficient conditions (Bahvalov, 1975; Godunov, 1959; Lyashko, Makarov, Skorobagatko, 1977). Stability of a difference scheme refers to its property when small perturbations in the input data are not amplified in the process of computing and do not bend the numerical solution.

Definition 8.2.2: If the inequality holds for the difference scheme (8.2.23)

$$\left\|\hat{\Re}(t)-\Re(t)\right\|_{(1)} \le M_1\left\|\hat{\eta}(t)-\eta(t)\right\|_{(1)} + M_2 \max_{0\le t'\le t}\left\|\hat{F}(t')-F(t')\right\|_{(2)},$$

where $\hat{\Re}$, $\Re$ are solutions of the difference problem (8.2.23) with input data $\hat{\eta}$, η and $\hat{F}$, F, respectively; $\|\cdot\|_{(1)}$, $\|\cdot\|_{(2)}$ are some grid norms; $M_1 > 0$, $M_2 > 0$ are the constants, independent of space and time steps and the choice of input data, then the scheme is called stable.

Inequality

$$\left\|\hat{\Re}(t)-\Re(t)\right\|_{(1)} \le \left\|M_1\hat{\eta}(t)-\eta(t)\right\|_{(1)}$$

expresses the stability of the scheme with respect to the source data $(F \equiv 0)$, and the inequality

$$\left\|\hat{\Re}(t)-\Re(t)\right\|_{(1)} \le M_2 \max_{0\le t'\le t}\left\|\hat{F}(t')-F(t')\right\|_{(2)}$$

means stability with respect to the right side $(\eta \equiv 0)$.

The issue of stability of finite difference schemes is crucial for problems of modeling the general circulation of the atmosphere, as well as for problems of numerical weather prediction (especially, long-term). However, the scheme stability investigation for the numerical solution of the equations included in the circulation model of the atmosphere, are facing considerable difficulties due to their non-linearity. So, often the stability analysis is carried out for a system of linear equations obtained by linearization of nonlinear equations (Samarskii, Popov, 1994).

Here, in the study of stability, convergence, monotony and the dispersion properties of the scheme (8.2.23), the coefficients of the equation (8.2.16) are assumed to be constant.

Consider the space H_h of grid functions $\Re = \{\Re_j\}$, defined on a grid $\bar{\omega}_h$ and equal to zero at j=0 $j = 0$ and j=N. The operation of scalar product and the norm in H_h we define by formulas

$$(\Re, v) = \sum_{j=1}^{N-1} \Re_j v_j h, \tag{8.2.46}$$

$$\|\Re\| = \sqrt{(\Re, \Re)}. \tag{8.2.47}$$

Define the following operators acting in the space H_h:

$$(\Lambda\Re)_j = \Re_{\bar{x},j} = \frac{\Re_j - \Re_{j-1}}{h},$$

$$\left(\Lambda^*\Re\right)_j = -\Re_{x,j} = -\frac{\Re_{j+1} - \Re_j}{h},$$

$$\left(\Lambda_0\Re\right)_j = -0.5h\Re_{\bar{x}x,j} = -0.5h\frac{\Re_{j+1} - 2\Re_j + \Re_{j-1}}{h},$$

$$\left(\Lambda_1\Re\right)_j = \Re_{0x,j} = \frac{\Re_{j+1} - \Re_{j-1}}{2h},$$

$j = 1,2,\ldots,N-1$.

The difference problem (2.8.46) can be written in the canonical form:

$$B\Re_t + A\Re = \varphi, \tag{8.2.48}$$

$$\Re_j^0 = \eta_j, j = 0,1,\ldots,N,$$

and the vector $\varphi = \varphi\left(t^n\right) \in H_h$ is defined as follows

$$\varphi_0 = 0, \varphi_1 = F_1 + \left(\frac{v}{2} + \frac{\nu}{h}\right)\frac{\alpha^{n+1}}{h}, \varphi_j = F_j, j = 2,\ldots,N-2,$$

$$\varphi_{N-1} = F_{N-1} - \left(\frac{v}{2} - \frac{\nu}{h}\right)\frac{\beta^n}{h}, \varphi_N = 0.$$

Now prove

Statement 8.2.1: The operator defined by (8.2.50) is positively defined.

We show that $\left(\Re,\Re\right) > 0$ for $\forall\Re \in H_h$ except $\Re \equiv 0$. The operator *A* can be represented as follows:

$$A = \left(\frac{v}{2} + \frac{\nu}{h}\right)\Lambda - \left(\frac{v}{2} - \frac{\nu}{h}\right)\Lambda^* = \frac{v}{2}\left(\Lambda - \Lambda^*\right) + \frac{\nu}{h}\left(\Lambda + \Lambda^*\right) = v\Lambda_1 + 2\frac{\nu}{h}\Lambda_0.$$

It is shown in (Samarskii, Vabishchevich, 1999) that Λ_0 is positively defined operator and Λ_1 is skew-symmetric, i.e. $\left(\Lambda_1\Re,\Re\right) = 0$. Having this in mind, we have

$$\left(A\Re,\Re\right)=\left(2\frac{\nu}{h}\Lambda_0\Re,\Re\right)>0\,,$$

what proves the statement 8.2.1.

To prove the stability with respect to the initial data, set the right side of (8.2.48) to zero. Then we can write the scheme (8.2.46) in the form

$$\Re^{n+1}=S\Re^{n}\,,$$

where $S=E-\tau B^{-1}A$ is the operator of transition from a time layer to the next one; operators *B* and *A* are defined by (8.2.49) and (8.2.50).

Let us give a definition of the stability with respect to the initial data (Samarskii, Popov, 1994). Difference scheme (2.8.46) will be stable with respect to the initial data, if the operator *S* is bounded by the norm:

$$\left\|S\right\|\le 1\,. \tag{8.2.51}$$

Obviously, if the condition (8.2.51) holds then

$$\left\|\Re^{n+1}\right\|=\left\|S\Re^{n}\right\|\le\left\|S\right\|\left\|\Re^{n}\right\|\le\left\|\Re^{n}\right\|\le\ldots\le\left\|\Re^{0}\right\|,$$

the uniform stability with respect to the initial data is provided Samara, Popov (1994).

Theorem 8.2.1: For the scheme (8.2.23) of the approximation problem (8.2.16) - (8.2.18), the estimation holds

$$\left\|\Re^{n+1}\right\|\le\left\|\Re^{0}\right\|,$$

That means that the scheme is evenly stable with respect to the initial data in the grid norm (8.2.47)

Proof: It follows formthe definition of the norm that

$$\left\|S\right\|=\left\|E-\tau B^{-1}A\right\|=\sup_{\left\|x\right\|=1}\left\|\left(E-\tau B^{-1}A\right)x\right\|,$$

from this, we have:

$$\left\|S\right\|^2=\left\|E-\tau B^{-1}A\right\|^2=\sup_{\left\|x\right\|=1}\left\|\left(E-\tau B^{-1}A\right)x,\left(E-\tau B^{-1}A\right)x\right\|,$$

$$\left\|S\right\|^2 \le 1 + \sup_{\|x\|=1}\left[\left(\tau B^{-1}Ax, \tau B^{-1}Ax\right) - 2\left(\tau B^{-1}A\right)x, x\right]. \tag{8.2.52}$$

Statement 8.2.2. Operators A and B^{-1} are commutative.

We first show that the commutativity of A and B implies commutativity A and B^{-1}:

$$AB^{-1} = B^{-1}BAB^{-1} = B^{-1}ABB^{-1} = B^{-1}A.$$

Then, on one hand,

$$\begin{aligned} BA &= \left[E + \tau\left(\frac{v}{2} + \frac{\nu}{h}\right)\Lambda\right]\left[\left(\frac{v}{2} + \frac{\nu}{h}\right)\Lambda - \left(\frac{v}{2} - \frac{\nu}{h}\right)\Lambda^*\right] \\ &= \left(\frac{v}{2} + \frac{\nu}{h}\right)\Lambda + \tau\left(\frac{v}{2} + \frac{\nu}{h}\right)^2\Lambda^2 - \left(\frac{v}{2} - \frac{\nu}{h}\right)\Lambda^* - \tau\left(\frac{v}{2} + \frac{\nu}{h}\right)\left(\frac{v}{2} - \frac{\nu}{h}\right)\Lambda\Lambda^* \end{aligned}, \tag{8.2.53}$$

on another

$$\begin{aligned} AB &= \left[\left(\frac{v}{2} + \frac{\nu}{h}\right)\Lambda - \left(\frac{v}{2} - \frac{\nu}{h}\right)\Lambda^*\right]\left[E + \tau\left(\frac{v}{2} + \frac{\nu}{h}\right)\Lambda\right] \\ &= \left(\frac{v}{2} + \frac{\nu}{h}\right)\Lambda + \tau\left(\frac{v}{2} + \frac{\nu}{h}\right)^2\Lambda^2 - \left(\frac{v}{2} - \frac{\nu}{h}\right)\Lambda^* - \tau\left(\frac{v}{2} + \frac{\nu}{h}\right)\left(\frac{v}{2} - \frac{\nu}{h}\right)\Lambda^*\Lambda \end{aligned}, \tag{8.2.54}$$

Since the operators Λ and Λ^* commute in the space of functions with zero boundary conditions (Samarskii, Gulin, 1973), then it follows from (8.2.53), (8.2.54) that the operators A and B also commute in the space H_h. This implies the commutativity of A and B^{-1}, and thus, the assertion is proved.

Using the commutativity of A and B^{-1}, we can write the equation (8.2.52) as follows:

$$\left\|S\right\|^2 \le 1 + \sup_{\|x\|=1}\left[\left(\tau AB^{-1}x, \tau B^{-1}Ax\right) - 2\left(\tau AB^{-1}\right)x, x\right]. \tag{8.2.55}$$

Let $B^{-1}x = \Re$, then $x = B\Re$. Taking into account the input replacement, rewrite equation (8.2.55)

$$\left\|S\right\|^2 \le 1 + \sup_{\|x\|=1}\left[\left(\tau A\Re, \tau A\Re\right) - 2\left(\tau A\Re, B\Re\right)\right]$$

From the last inequality, one can be see that for the condition (8.2.51) is necessary and sufficient for

$$\left(\tau A\Re,\tau A\Re\right)-2\left(\tau A\Re,B\Re\right)\le 0 \text{ for } \left\|B\Re\right\|=1\,. \tag{8.2.56}$$

We write the operators *A* and *B* as follows

$$B\Re=\Re+\tau\left(\beta+\alpha\right)\Lambda\Re\,, \tag{8.2.57}$$

$$A\Re=\left(\beta+\alpha\right)\Lambda\Re-\left(\beta-\alpha\right)\Lambda^{*}\Re\,, \tag{8.2.58}$$

$$\beta=\frac{v}{2}\ge 0,\alpha=\frac{\nu}{h}>0\,.$$

Substituting (8.2.57) and (8.2.58) in (8.2.56), we obtain

$$\left(\tau A\Re,\tau A\Re\right)-2\left(\tau A\Re,B\Re\right)=\tau^{2}\begin{bmatrix}\left(\beta+\alpha\right)\Lambda\Re-\left(\beta-\alpha\right)\Lambda^{*}\Re,\\ \left(\beta+\alpha\right)\Lambda\Re-\left(\beta-\alpha\right)\Lambda^{*}\Re\end{bmatrix}-2\tau\begin{bmatrix}\left(\beta+\alpha\right)\Lambda\Re-\left(\beta-\alpha\right)\Lambda^{*}\Re,\\ \Re+\tau\left(\beta+\alpha\right)\Lambda\Re\end{bmatrix}.$$

$$=\tau^{2}\left(\beta+\alpha\right)^{2}\left\|\Lambda\Re\right\|^{2}+\tau^{2}\left(\beta-\alpha\right)^{2}\left\|\Lambda^{*}\Re\right\|^{2}-2\tau\left(A\Re,\Re\right)$$

Given that $\left\|\Lambda\Re\right\|^{2}=\left\|\Lambda^{*}\Re\right\|^{2}$ and $\left(A\Re,\Re\right)>0$, we obtain

$$\left\|S\right\|^{2}\le 1+\sup_{\left\|x\right\|=1}\left[\left(\tau A\Re,\tau A\Re\right)-2\left(\tau A\Re,B\Re\right)\right]=1-\inf_{\left\|B\Re\right\|=1}\left[4\tau^{2}\beta\alpha\left\|\Lambda\Re\right\|^{2}+2\tau\left(A\Re,\Re\right)\right]<1,$$

What proves the Theorem 8.2.1.

The scheme (8.2.23) stability with respect to the right side is established by the following

Theorem 8.2.2: For the scheme (8.2.23) of the approximation problem (8.2.16) - (8.2.18), the estimation holds

$$\left\|\Re^{n+1}\right\|_{A_0}\le\exp\left[\frac{v^{2}}{4\nu}\left(t^{n+1}+1\right)\right]\times\left[\left\|\Re^{0}\right\|_{A_0}+\sqrt{\frac{t^{n}+1}{2}}\left(\sum_{k=0}^{n}\tau\left\|\varphi\left(t^{k}\right)\right\|^{2}\right)^{1/2}\right], \tag{8.2.59}$$

$$A_0\Re=-\Re_{\bar{x}x}$$

that guarantees the absolute stability of the difference scheme (8.2.23) with respect to the initial data and the right side in the norm of the operator A_0.

Proof: The operator A can be represented as the sum of the self-adjoint and skew-symmetric operators (Samarskii, Gulin, 1973):

$$A = \nu A_0 + \upsilon A_1,$$

where the symmetric part of the operator A has the form

$$A_0 \Re = \frac{2}{h} \Lambda_0 \Re = -\Re_{\bar{x}x}, \tag{8.2.60}$$

and skew-symmetric part has the form

$$A_1 \Re = \Lambda_1 \Re = \Re_{0x}. \tag{8.2.61}$$

Using the theorem on the stability of schemes with a positively defines nonself-adjoint operator (Samarskii, Gulin, 1973) according to which the solution of the problem (8.2.48) satisfies

$$\left\|\Re^{n+1}\right\|_{A_0} \le \exp\left[\frac{c_0}{4\varepsilon}\left(t^{n-1}+1\right)\right] \times \left[\left\|\Re^0\right\|_{A_0} + \sqrt{\frac{t^n+1}{2\varepsilon}}\left(\sum_{k=0}^{n} \tau \left\|\varphi\left(t^k\right)\right\|^2\right)^{1/2}\right], \tag{8.2.62}$$

if the following conditions are met:

$$A_0 > 0, \tag{8.2.63}$$

$$\left\|A_1 \Re\right\|^2 \le c_0 \left(A_0 \Re, \Re\right) \text{for all } \Re \in H_h, \tag{8.2.64}$$

$$B \ge \varepsilon E + \frac{\tau}{2} \nu A_0, \tag{8.2.65}$$

where $c_0 > 0$, $\varepsilon > 0$ are constant independent of τ and h.

We will show the conditions (8.2.63) - (2.8.65) of this theorem for the operators (8.2.60), (8.2.61), (8.2.49).

In Samarskii, Gulin (1973), it is shown that

$$\left(\Lambda_0\mathfrak{R},\mathfrak{R}\right)\geq\frac{h}{l^2}\left\|\mathfrak{R}\right\|^2.$$

Hence, in view of the operator A_0 (8.2.60) we obtain the estimate

$$\left(A_0\mathfrak{R},\mathfrak{R}\right)\geq\frac{h}{l^2}\left\|\mathfrak{R}\right\|^2,$$

i.e. $A_0>0$ and inequality (8.2.63) holds.

To prove (8.2.64), we use the equality

$$\left\|A_1\mathfrak{R}\right\|^2=\frac{1}{4}\left\|\left(\Lambda-\Lambda^*\right)\mathfrak{R}\right\|^2. \tag{8.2.66}$$

Using the parallelogram equality for norms (8.2.47), we get

$$\left\|\left(\Lambda-\Lambda^*\right)\mathfrak{R}\right\|^2\leq 2\left(\left\|\Lambda\mathfrak{R}\right\|^2+\left\|\Lambda^*\mathfrak{R}\right\|^2\right).$$

Since Samarskii, Gulin (1973)

$$\left\|\Lambda\mathfrak{R}\right\|^2=\left\|\Lambda^*\mathfrak{R}\right\|^2\leq\frac{2}{h}\left(\Lambda_0\mathfrak{R},\mathfrak{R}\right),$$

then

$$\left\|\left(\Lambda-\Lambda^*\right)\mathfrak{R}\right\|^2\leq\frac{8}{h}\left(\Lambda_0\mathfrak{R},\mathfrak{R}\right). \tag{8.2.67}$$

Using (8.2.56), (8.2.57), as well as the definition of the operator (8.2.61), we find

$$\left\|A_1\mathfrak{R}\right\|^2=\frac{1}{4}\left\|\left(\Lambda-\Lambda^*\right)\mathfrak{R}\right\|^2\leq\frac{2}{h}\left(\Lambda_0\mathfrak{R},\mathfrak{R}\right)=\left(A_0\mathfrak{R},\mathfrak{R}\right).$$

That is, the inequality (8.2.64) holds with a constant c_0=1.

Let us now consider the inequality (8.2.65). According to the definition of the operator B (8.2.49), we have

$$\left(B\Re,\Re\right)=\left(\Re,\Re\right)+\tau\left(\frac{v}{2}+\frac{\nu}{h}\right)\left(\Lambda\Re,\Re\right).$$

But, as noted above, $\Lambda=\Lambda_0+\Lambda_1$ and $\left(\Lambda_1\Re,\Re\right)=0$, so

$$\left(B\Re,\Re\right)=\left(\Re,\Re\right)+\tau\left(\frac{v}{2}+\frac{\nu}{h}\right)\left(\Lambda_0\Re,\Re\right).$$

This implies inequality

$$\begin{aligned}\left(B\Re,\Re\right)&=\left(E\Re,\Re\right)+\tau\frac{\nu}{h}\left(\frac{v}{2}+\frac{\nu}{h}\right)\nu\left(A_0\Re,\Re\right)\\&=\left(E\Re,\Re\right)+\tau\left(1+\frac{\nu h}{2\nu}\right)\nu\left(A_0\Re,\Re\right)\geq\left(E\Re,\Re\right)+\frac{\tau}{2}\nu\left(A_0\Re,\Re\right)\end{aligned},$$

which proves (8.2.65) for $\varepsilon=1$.

This proves the theorem.

Note 8.2.1: In (Samaraskii, Gulin, 1973), there is provided even stability with respect to the initial data of the degenerate scheme (8.2.32) $\nu=0$

$$\left\|\Re^{n+1}\right\|_D\leq\left\|\Re^n\right\|_D,$$

where

$$D=\Re-\frac{v\tau h}{4}\Re_{\bar{x}x}.$$

By the theorem on the connection of stability with respect to the initial data and to the right part of the scheme (8.2.32) the a priori estimation holds

$$\left\|\Re^{n+1}\right\|_D\leq\left\|\Re^n\right\|_D+\sum_{k=0}^{n}\tau\left\|B^{-1}\varphi_k\right\|_D.$$

Convergence

From conditional consistency (approximation) and the stability of the scheme (8.2.23) by the theorem on the equivalence of Lax (Hoffman, 2001), it follows the conditional convergence solutions $\Re^*$ of the difference scheme (8.2.23) to the solution $\Re$ of the problem (8.2.16) - (8.2.18) with the first order in τ and h

$$\left\|\Re - \Re^{*}\right\|_{h} = O\left[\tau + h^{2} + \tau / h\right]$$

in the norm $\left\|\cdot\right\|_{C}$ at conditions $\tau / h \to 0$, $\tau \to 0$, $h \to 0$.

Monotone

We put in (8.2.24) $F\left(x,t\right) \equiv 0$ and write the resulting expression

$$\Re_{j}^{n+1} = \frac{\tau b_{j}}{1+\tau b_{j}} \Re_{j-1}^{n+1} + \frac{1+\tau a_{j}}{1+\tau b_{j}} \Re_{j}^{n} - \frac{\tau a_{j}}{1+\tau b_{j}} \Re_{j+1}^{n}$$

explicitly

$$\Re_{j}^{n+1} = \sum_{k=0}^{+\infty} \frac{\tau^{2}\left(a_{k} b_{k-1} - a_{k-1} b_{k}\right) + \tau\left(a_{k} + b_{k-1}\right) + 1}{1+\tau b_{k-1}} \psi\left(j-k,j\right) \Re_{j-k}^{n} - \tau a_{j} \psi\left(j,j\right) \Re_{j+1}^{n}, \tag{8.2.68}$$

where

$$\psi\left(k,m\right) = \tau^{m-k} \left(\sum_{n=k+1}^{m} b_{n}\right) / \left[\sum_{n=k+1}^{m} \left(1+\tau b_{n}\right)\right] \geq 0 .$$

According to a feature of monotony, find conditions (Kalitkin, 1978) in which all the coefficients in the right-hand side of (8.2.68) will be non-negative

$$\begin{cases} \tau^{2}\left(a_{k} b_{k-1} - a_{k-1} b_{k}\right) + \tau\left(a_{k} + b_{k-1}\right) + 1 \geq 0 \\ -a_{k} \geq 0 \end{cases} \tag{8.2.69}$$

Suppose for certainty that spatial grid is defined by the relation

$$h_{j} = \gamma h_{j-1} = \gamma^{j} h_{0}, \gamma \geq 1 . \tag{8.2.70}$$

Consider the first inequality in (8.2.69). We write the formula for computing the roots of a quadratic equation

$$\tau_{1,2} = \frac{-\left(a_{k} + b_{k-1}\right) \pm \sqrt{\left(a_{k} + b_{k-1}\right)^{2} - 4 a_{k-1} b_{k}}}{2\left(a_{k} b_{k-1} - a_{k-1} b_{k}\right)} \tag{8.2.71}$$

and evaluate these roots. It's obvious that

$$a_k = \frac{h_{k-1}v - 2\nu}{\gamma h_{k-1}\left(h_{k-1} + h_k\right)} = \frac{\gamma h_{k-2}v - 2\nu}{\gamma^3 h_{k-2}\left(h_{k-2} + h_{k-1}\right)} \le b_k = \frac{\gamma h_{k-1}v + 2\nu}{h_{k-1}\left(h_{k-1} + h_k\right)} = \frac{\gamma^2 h_{k-2}v + 2\nu}{\gamma^2 h_{k-2}\left(h_{k-2} + h_{k-1}\right)},$$

$$a_{k-1} = \frac{h_{k-2}v - 2\nu}{\gamma h_{k-2}\left(h_{k-2} + h_{k-1}\right)} \le b_{k-1} = \frac{\gamma h_{k-2}v + 2\nu}{h_{k-2}\left(h_{k-2} + h_{k-1}\right)}.$$

From this, we get

$$a_k + b_{k-1} = \frac{\gamma h_{k-2}v\left(1 + \gamma^3\right) + 2\nu + \left(\gamma^3 - 1\right)}{\gamma^3 h_{k-2}\left(h_{k-2} + h_{k-1}\right)} \ge 0,$$

$$a_k b_{k-1} = \frac{\left(\gamma h_{k-2}v\right)^2 - \left(2\nu\right)^2}{\gamma^3 h_{k-2}\left(h_{k-2} + h_{k-1}\right)},$$

$$a_{k-1} b_k = \frac{\left(\gamma h_{k-2}v\right)^2 - \left(2\nu\right)^2 + 2\nu v h_{k-2}\left(1 - \gamma^2\right)}{\gamma^3 h_{k-2}\left(h_{k-2} + h_{k-1}\right)^2}.$$

From the last two equations, it follows that $a_k b_{k-1} \ge a_{k-1} b_k$. For radicand of (8.2.71) we have the inequality:

$$\left(a_k + b_{k-1}\right)^2 - 4a_{k-1}b_k = \left(a_k - b_{k-1}\right)^2 + 4\left(a_k b_{k-1} - a_{k-1}b_k\right) \ge 0.$$

Consequently, $\tau_{1,2} < 0$ and quadratic inequality in (8.2.69) holds for $\forall \tau \ge 0$.

Note 8.2.2: For any grid of the type $h_0 < h_1 < \ldots < h_{J-1}$, this result is not the case.

Consider the second inequality in (8.2.69) and write it in the form, using the expression (8.2.24) for a_k, allowing to assess the impact of the difference grid settings

$$2\nu - h_{j-1}v = 2\nu - h_0 v\gamma^j \ge 0, or\ \gamma \le \left(\frac{2\nu}{h_0 v}\right)^{1/j}, \forall j.$$

From this, when *j*=*J*-1 and the inequality $\gamma \ge 1$, obtain the condition of monotonicity scheme associated with the grid parameters (8.2.70)

$$h_0 \le \frac{2\nu}{v} \, and \; \gamma \le \left(\frac{2\nu}{h_0 v}\right)^{\frac{1}{J-1}}.$$

Note 8.2.3: For a even grid, the scheme monotony condition has the form

$$h \le \frac{2\nu}{v}.$$

Numerical Viscosity

We find the first differential approximation (Samarskii, Popov, 1992) by substituting the series (8.2.24a) in the scheme (8.2.23a) and making the necessary transformations. But, as the result is rather cumbersome, we restrict ourselves to only the part that contains all the members from $\partial^2\Re / \partial x^2$ and $\partial^3\Re / \partial x^3$:

$$\frac{\partial \Re}{\partial t} + v\frac{\partial \Re}{\partial x} - \nu\frac{\partial^2 \Re}{\partial x^2} \approx \frac{\tau v}{2}\frac{v\left(h_j - h_{j-1}\right) + 4\nu}{h_j + h_{j-1}}\frac{\partial^2 \Re}{\partial x^2}$$
$$+2\nu\left[\frac{h_j - h_{j-1}}{6} + \tau\frac{\tau v\left(vh_{j-1} - 2\nu\right) - \nu\left(h_j + h_{j-1}\right)}{\left(h_j + h_{j-1}\right)^2}\right]\frac{\partial^3 \Re}{\partial x^3}$$

From this, we find the numerical evaluation of viscosity

$$\nu_c = \frac{\tau v}{2}\frac{v\left(h_j - h_{j-1}\right) + 4\nu}{h_j + h_{j-1}}. \tag{8.2.72}$$

Consequently, the value of the numerical viscosity ν_c directly proportional to the time step τ, convection velocity v and physical viscosity ν. In particular, it is absent at all in the diffusion equation $\left(v = 0\right)$.

Let us find the condition of correctness of the scheme (8.2.23a), i.e. when $\left(\nu + \nu_c\right) \ge 0$. After simple transformations, we obtain the condition on the steps of the spatial grid

$$h_j \ge \frac{\tau v^2 - 2\nu}{\tau v^2 + 2\nu}h_{j-1} - \frac{4\tau v\nu}{\tau v^2 + 2\nu}.$$

However, since

$$\left|\frac{\tau v^2 - 2\nu}{\tau v^2 + 2\nu}\right| \le 1 \text{ and } \frac{4\tau v\nu}{\tau v^2 + 2\nu} \ge 0,$$

it is a sufficient condition for the correctness is $h_j \geq h_{j-1}$.

Note that for even spatial grid $\left(h_j = h_{j-1} = h\right)$, as follows from (8.2.71), the scheme is absolutely correct.

Phase and Module Change in a Point

Consider the initial-boundary value problem (8.2.16) - (8.2.18) for the homogeneous equation of convective diffusion (8.2.16). If $F\left(x\right) \equiv 0$ it has an analytical solution (Prusov, Doroshenko, 2006) of the form

$$\Re\left(x,t\right) = \exp\left(-\nu k^2 t\right) \cdot \exp\left[ik\left(x - vt\right)\right]. \tag{8.2.73}$$

Obviously, at some fixed point in x_j at time instances t^n and $t^{n+1} = t^n + \tau$, the value of the exact solution will differ (see Figure 8) by module (the first factor) and by phase (the second factor).

This change is given by the relation

$$\Re^{n+1} = \rho \Re^n,$$

where

$$\rho = \exp\left(-\nu k^2 t\right) \cdot \exp\left(ikvt\right).$$

Therefore, the exact solution module (8.2.73) at a fixed point in space during the time inteval τ will change in $\left|\rho\right| = \exp\left(-\nu k^2 t\right) < 1$ times and the phase – by the angle

Figure 8. The change of phase and module in one time step

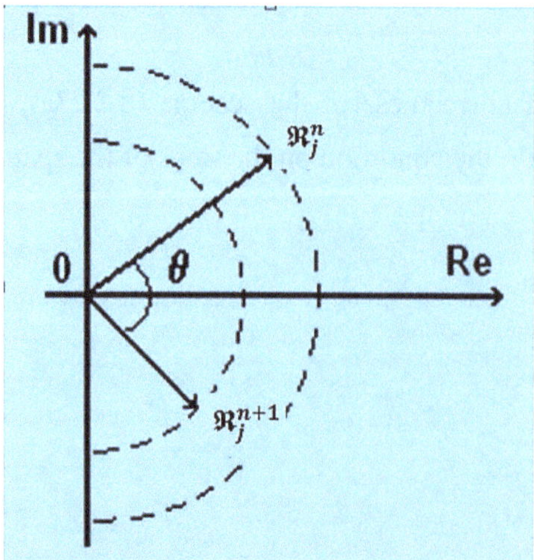

$$\theta = \operatorname{arctg}\frac{\operatorname{Im}\rho}{\operatorname{Re}\rho} = -kvt\,.$$

Find the same changes for the numerical solution

$$\Re_j^{n+1} = \lambda \Re_j^n\,.$$

The solution will be sought in the form of function

$$\Re_j^n = \Re_j^0 \lambda^n \exp\left(ikx_j\right) = \Re_j^0 \lambda^n \exp\left[ik\left(x_0 + \sum_{s=0}^{j-1} h_s\right)\right], \qquad (8.2.74)$$

where k is a wave number of the harmonics; $i = \sqrt{-1}$ is the imaginary unit; j and n are numbers of spatial and temporal grid point respectively (Prusov et al., 2007).

Substituting this solution into (8.2.24) and conducting the necessary simple algebraic transformations, we obtain

$$\begin{aligned}\lambda\left(k\right) = &\frac{\left[1+2q_j \sin^2\left(\frac{k}{2}h_j\right)\right]\left[1+2p_j \sin^2\left(\frac{k}{2}h_{j-1}\right)\right]}{1+4p_j\left(1+p_j\right)\sin^2\left(\frac{k}{2}h_{j-1}\right)} - \frac{p_j q_j \sin\left(kh_j\right)\sin\left(kh_{j-1}\right)}{1+4p_j\left(1+p_j\right)\sin^2\left(\frac{k}{2}h_{j-1}\right)} \\ &-i\frac{p_j \sin\left(kh_{j-1}\right)\left[1+2q_j \sin^2\left(\frac{k}{2}h_j\right)\right]}{1+4p_j\left(1+p_j\right)\sin^2\left(\frac{k}{2}h_{j-1}\right)} + \frac{q_j \sin\left(kh_j\right)\left[1+2p_j \sin^2\left(\frac{k}{2}h_{j-1}\right)\right]}{1+4p_j\left(1+p_j\right)\sin^2\left(\frac{k}{2}h_{j-1}\right)}\,.\end{aligned} \qquad (8.2.75)$$

Select the real and imaginary parts from the resulting expression

$$\operatorname{Re}\lambda = \frac{\left[1+2q_j \sin^2\left(\frac{k}{2}h_j\right)\right]\left[1+2p_j \sin^2\left(\frac{k}{2}h_{j-1}\right)\right]}{1+4p_j\left(1+p_j\right)\sin^2\left(\frac{k}{2}h_{j-1}\right)} - \frac{p_j q_j \sin\left(kh_j\right)\sin\left(kh_{j-1}\right)}{1+4p_j\left(1+p_j\right)\sin^2\left(\frac{k}{2}h_{j-1}\right)},$$

$$\operatorname{Im}\lambda = -\frac{p_j \sin\left(kh_{j-1}\right)\left[1+2q_j \sin^2\left(\frac{k}{2}h_j\right)\right]}{1+4p_j\left(1+p_j\right)\sin^2\left(\frac{k}{2}h_{j-1}\right)} - \frac{q_j \sin\left(kh_j\right)\left[1+2p_j \sin^2\left(\frac{k}{2}h_{j-1}\right)\right]}{1+4p_j\left(1+p_j\right)\sin^2\left(\frac{k}{2}h_{j-1}\right)}\,.$$

Then the change the module λ of the numerical solution is equal to

$$|\lambda| = \sqrt{\frac{1+4q_j\left(1+q_j\right)\sin^2\left(\frac{k}{2}h_j\right)}{1+4p_j\left(1+p_j\right)\sin^2\left(\frac{k}{2}h_{j-1}\right)}}, \tag{8.2.76}$$

and the change in the numerical solution phase, because of the cumbersome, yet leave in the formal form

$$\varphi = \mathrm{arctg}\frac{\mathrm{Im}\,\lambda}{\mathrm{Re}\,\lambda}. \tag{8.2.77}$$

In order to facilitate further discussion, we make two assumptions. First, we assume that $h_j = h_{j-1} = \text{const} = h$. Second, as short waves $\left(k > \pi / 4h\right)$ may not be correctly displayed on the grid, it will be assumed that $kh < 1$. Then the formula (8.2.76) takes the form

$$|\lambda| = \sqrt{\frac{1+4q\left(1+q\right)\sin^2\left(\frac{k}{2}h\right)}{1+4p\left(1+p\right)\sin^2\left(\frac{k}{2}h\right)}} < 1,$$

where

$$p = \frac{\tau}{2h^2}\left(hv+2\nu\right) \geq 0, q = \frac{\tau}{2h^2}\left(hv-2\nu\right), |q| < p.$$

Using assessment for $|\lambda|$ by the expansion in Taylor's formula, we obtain

$$|\lambda| \approx 1 - \nu k^2\tau - \frac{\left(k\tau\right)^2 v\nu}{2h}.$$

Thus, the circuit (8.2.23a) is dissipative $|\lambda| \leq 1$. And when one consider that

$$|\rho| \approx 1 - \nu k^2\tau + \frac{\left(k^2\tau\nu\right)^2}{2} - \ldots,$$

we get the accuracy of the order of τ^2 for the transmission of module change in the fixed point in space. The difference is other terms of the expansion points to the need for a separate study of the dissipative properties of the scheme. One such study was conducted above for the numerical viscosity.

Now estimate the change in the solution phase at x_j. According to the accepted assumptions, the simplified formula (2.8.77) takes the form

$$\varphi = -\mathrm{arctg}\left\{\frac{p\sin(kh)\left[1+2q\sin^2\left(\frac{k}{2}h\right)\right]}{\left[1+2q\sin^2\left(\frac{k}{2}h\right)\right]\left[1+2p\sin^2\left(\frac{k}{2}h\right)\right]-pq\sin^2(kh)} + \frac{q\sin(kh)\left[1+2p\sin^2\left(\frac{k}{2}h\right)\right]}{\left[1+2q\sin^2\left(\frac{k}{2}h\right)\right]\left[1+2p\sin^2\left(\frac{k}{2}h\right)\right]-pq\sin^2(kh)}\right\}.$$

Using the approach of Taylor's formula, we can estimate the change in phase of the numerical solution

$$\varphi \approx -k\tau v + \frac{k\tau v}{6}(kh)^2 = \theta\left[1-\frac{(kh)^2}{6}\right].$$

Consequently, the difference scheme (8.2.23a) is decelerating $(\varphi/\theta \le 1)$. Slowing down is the larger, the larger is the spatial step or wave number.

The Phase Velocity

The phase velocity characterizes the rate of the phase change in time at the fixed point in space. For the exact solution (8.2.73), it is $v_p^A = -v$ for all wavelengths and points in space (Belov, Borisenkov, Panin, 1989). For the numerical solution, the evaluation of phase velocity is defined as

$$v_p = \frac{\varphi}{k\tau} \approx -v\left[1-\frac{(kh)^2}{6}\right] = v_p^A\left[1-\frac{(kh)^2}{6}\right].$$

Thus, the numerical solution of the phase velocity is too low for all waves. The reason for this is in the finite-difference representation of the spatial derivatives. In addition, the phase velocity is different for different kinds of waves: long wavelengths have a greater speed. Therefore, the difference scheme (8.2.23a) numerical dispersion is present.

The Group Velocity

The group velocity is the speed of propagation of the wave with the maximum amplitude, which is generated by the superposition of a wave group with close wave numbers. This velocity characterizes the rate of energy transfer. We define the group velocity for the exact solution (8.2.73)

$$v_g^A = \frac{1}{\tau}\frac{d\theta}{dk} = -v.$$

It is constant for all wavelengths, and is equal to a true phase velocity (Belov, Borisenkov, Panin (1989)). We find numerical group velocity as

$$v_g = \frac{1}{\tau}\frac{d\varphi}{dk} \approx -v + \frac{v\left(kh\right)^2}{2} = v_g^A\left[1 - \frac{\left(kh\right)^2}{2}\right].$$

As expected, it was understated for all wavelengths due to the finite-difference representation of the spatial derivatives. For different kinds of waves, the group velocity is different: with increasing wavelength (decreasing wave number *k*), it is close to the true value. So, energy transmission by long waves is better expressed by the scheme (8.2.23a) than the shortwave energy spreading.

Computational Complexity

Consider the diagram (8.2.24). For simplicity, assume that $h_j = h_{j-1} = \text{const} = h$, $j = 0,1,\ldots,J-1$. Then the number of operations required to move to the next time layer is equal $M_s = 8\left(J-1\right)+11$. To attach a specific concept to the value of this number, we compare the number of operations for problem solving algorithm (8.2.16) - (8.2.18) by means of explicit and implicit schemes (all other conditions are equal). Simple calculations give the following number of transactions:

- The explicit scheme $M_s = 7\left(J-1\right)+8$.
- The implicit scheme $M_s = 10\left(J-1\right)$.

Consequently, the difference scheme (8.2.24), being unconditionally stable as implicit one, requires almost the same operations during the transition to the next time layer as explicit scheme. But when you consider that the problem of atmospheric circulation modeling is solved numerically always for some long period of time, the efficiency of the scheme (8.2.24) over existing both explicit and implicit scheme becomes apparent. It is expected that the numerical solution of nonlinear multidimensional equations of hydrodynamics using the finite difference scheme (8.2.23) will require less computer time in comparison with known methods.

AN ANALYSIS OF THE EFFECTIVENESS OF THE METHOD BY COMPARING KNOWN DIFFERENCE SCHEMES BASED ON MODEL PROBLEMS

To confirm theoretical results obtained in the previous section and verify the effectiveness of the difference scheme (8.2.23), we will test this scheme on model problems with analytical solutions.

Define a difference between the numerical and exact solution by the maximum deviation from the exact numerical solution

$$\delta_M = \max_{0\le j\le N}\left|\Re\left(x_j,T\right)-\Re_j^K\right|,$$

where $\Re\left(x_j,T\right)$ is the exact solution at x_j at a time instance T, $\Re_j^K$ is the numerical solution of a test problem at x_j at a time instance T, $T = K\tau$ is the time interval of calculation, M is the grid node number, in which the maximum deviation of the numerical solution of the exact solution is obtained.

As a first example, take the test task of spreading a physical quantity $\Re$ in a viscous homogeneous medium, moving at a constant speed v. The function

$$\Re\left(x,t\right)=e^{-k^2\mu t}\sin\left[k\left(x-vt\right)\right] \tag{8.2.78}$$

is the exact solution of the problem

$$\frac{\partial\Re}{\partial t}+v\frac{\partial\Re}{\partial x}=\mu\frac{\partial^2\Re}{\partial x^2},0\le x\le l,0<t\le T, \tag{8.2.79}$$

with the initial condition

$$\Re\left(x,0\right)=\sin\left(kx\right),0\le x\le l,$$

and the boundary conditions

$$\Re\left(0,t\right)=e^{-k^2\mu t}\sin\left(kvt\right),t>0,$$

$$\Re\left(l,t\right)=e^{-k^2\mu t}\sin\left[k\left(l-vt\right)\right],t>0.$$

Numerical calculations of the scheme (8.2.23) were carried out on the even grid $x_j = jlN^{-1}, j=0,1,\dots,N$, with an increment of h; τ is a time step. The results of these calculations are shown in Figure 9 and Table 2 for the following parameters: k=1, v=20 m/s, $\mu = 0.01$ m²/s, $l=\pi\approx 3.14$ m, $T=\pi\approx 3.14s$. The results in the Table for δ_M are shown at different ratios in space and time steps.

Table 2. The maximum deviation δ_M of the solution (8.2.79) of the scheme (8.2.23) of the exact solution

τ / h	$\pi \cdot 10^{-2}$	$2\pi \cdot 10^{-2}$	$5\pi \cdot 10^{-2}$	$\pi \cdot 10^{-3}$	$\pi \cdot 10^{-4}$	$\pi \cdot 10^{-5}$	$\pi \cdot 10^{-6}$
$\pi \cdot 10^{-2}$	0.2106	0.0571	0.0136	0.0056	0.0013	0.0010	0.0010
$2\pi \cdot 10^{-2}$	0.1810	0.0483	0.0101	0.0042	0.0004	0.0003	0.0002
$5\pi \cdot 10^{-2}$	0.1497	0.0526	0.0189	0.0092	0.0010	0.0001	0.0000
$\pi \cdot 10^{-3}$	0.1877	0.0901	0.0355	0.0177	0.0018	0.0002	0.0000
$\pi \cdot 10^{-4}$	0.6648	0.4845	0.2681	0.1528	0.0174	0.0018	0.0002

One can see from the Table 2 that there is a convergence of the numerical solution to the exact solution ($\delta_M \to 0$) for some fixed h and $\tau \to 0$. Being reduced to a certain value of magnitude $O\left[h^2\right]$, the deviation δ_M then stabilizes. For a fixed τ and $h \to 0$, while $\tau < h$, there is a slight approximation to the exact numerical solutions; when $\tau \geq h$ the error δ_M increases. This behavior of the numerical solution is explained by the conditional convergence of a difference scheme (8.2.23).

In the case when $h \to 0$, $\tau \to 0$ (changing error across the diagonal of the Table) and $\tau / h \leq 1$, the error is reduced to a certain critical value δ_{cr}, which tends to zero as $\tau / h \to 0$. When $\tau / h > 1$, $h \to 0$ and $\tau \to 0$, the error increases, but the decision remains stable.

Visualization of the results for J=500, N=2000, T=10hN/v s. is shown in Figure 9 for six consecutive temporary layers. It is seen that the solution moves from left to right with a speed v, and, at the same time, the front edge spreads at a rate determined by the value of μ. Numerical solution coincides with the exact solution (8.2.78), which is applied here (they are indistinguishable in the Figure).

The compliance with the numerical solution to the exact solution is more clearly seen in Figure 10 which shows the time behavior of the values of $\Re\left(x,t\right)$ at the space point x=l/2. There are also indistinguishable.

As a further test problem that has an exact solution, we choose a one-dimensional problem for nonlinear equation Burgers

$$\frac{\partial u}{\partial t} + u\frac{\partial u}{\partial x} = \nu\frac{\partial^2 u}{\partial x^2}, \tag{8.2.80}$$

which simulates the simplest case of shock waves formation in gas dynamics ($\nu \to 0$) or a one-dimensional motion of a viscous fluid ($\nu > 0$). The transformation (Cole, 1951; Hopf, 1950):

$$u\left(x,t\right) = -2\nu\frac{\partial}{\partial x}\left(\ln \Re\right) \tag{8.2.81}$$

Figure 9. Calculation results for the problems (8.2.79) at successive instants: t=0.2; 2) t=0.4; 3) t=0.6; 4) t=0.8; 5) t=1.0; 6) t=1.2

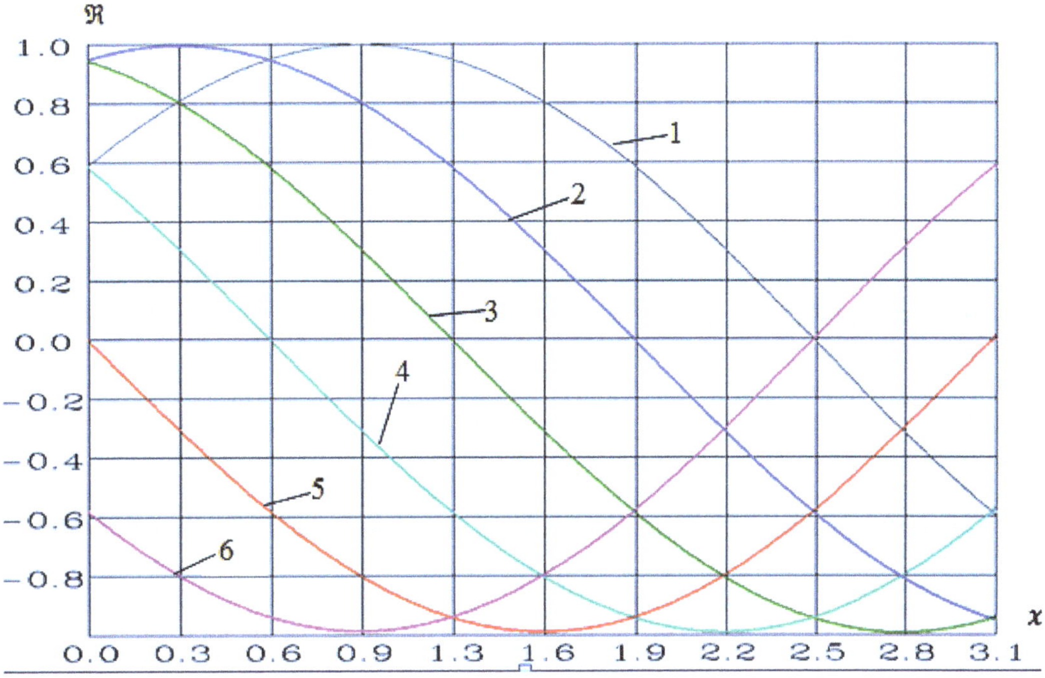

Figure 10. The results of numerical solution of the problem (8.2.79) at the point $x = l / 2$

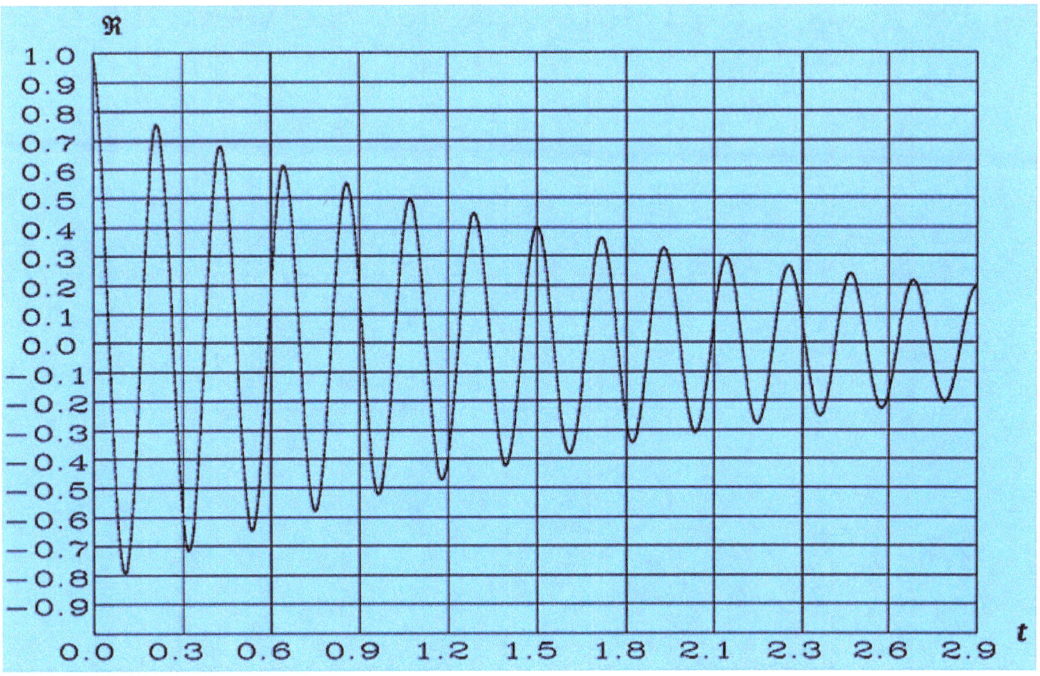

for the Burgers equation (8.2.80) simplifies significantly the task of finding analytical solutions, because with the help of (8.2.81) it is reduced to the linear heat equation,

$$\frac{\partial \Re}{\partial t} = \nu \frac{\partial^2 \Re}{\partial x^2}, \tag{8.2.82}$$

for which a solution can be found by the classical methods (Arsenin, 1984). Thus, for the equation (8.2.80) we can get the exact analytical solutions for some initial and boundary conditions. These solutions are useful for assessing the quality of the difference scheme (8.2.23) for solving nonlinear equations.

In (Anderson, Tannehill, Pletcher, 1990), there is proposed to combine linear convective diffusion equation with nonlinear Burgers equation in the generalized equation

$$\frac{\partial u}{\partial t} + \left(c + bu\right)\frac{\partial u}{\partial x} = \nu \frac{\partial^2 u}{\partial x^2}, \tag{8.2.83}$$

where c and b are free parameters. When c=1/2 and b=-1, generalized Burgers equation has an exact stationary solution

Figure 11. Current (8/2/84) and numerical (8.2.23) of the solution of equation (8.2.80)

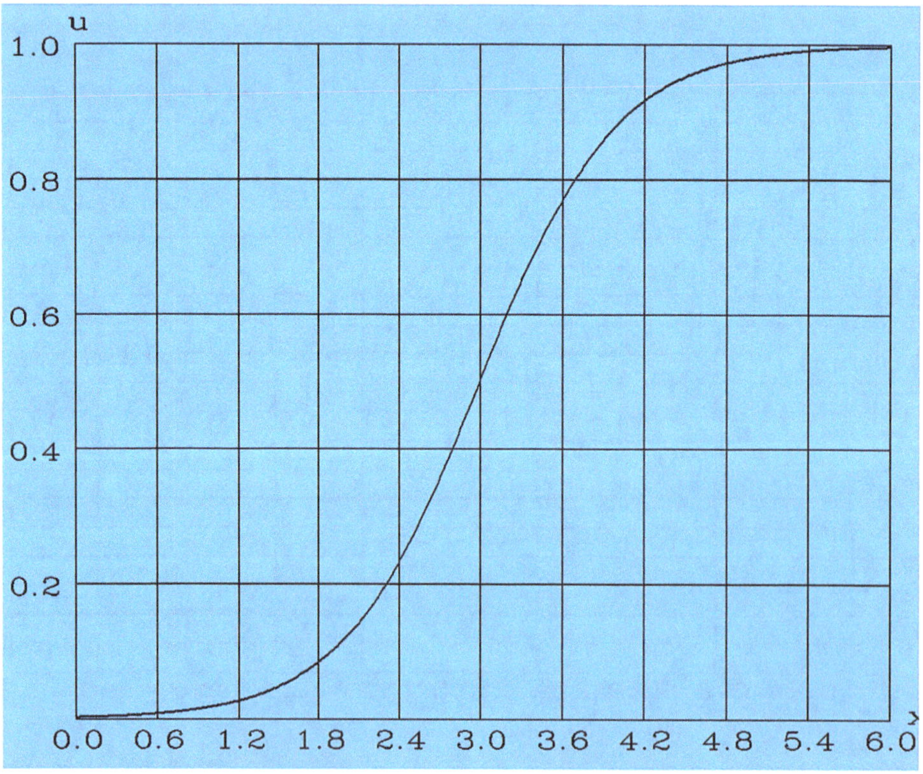

$$u = -\frac{c}{b}\left[1 + \text{th}\frac{c\left(x - x_0\right)}{2\nu}\right]. \tag{8.2.84}$$

Therefore, if the initial distribution of the function u is given by (8.2.84), the exact solution does not change in time, and remains the same at a given initial distribution. Such behavior of the solution of equation (8.2.83) at x_0=3, $\nu = 1/4$ is shown in Figure 11. It applied in 2000 successive numerical results obtained with the scheme (8.2.23) at $\tau = h$ and the exact solution of (8.2.84). As seen in the Figure, all of these are placed on one line.

To practically assess the accuracy of the difference scheme (8.2.23), we compare the obtained results with the help of the solutions obtained by other numerical methods, which are widely used in solving various problems of applied fluid dynamics. A comparative analysis on the next test problem

$$\frac{\partial \Re}{\partial t} + v\frac{\partial \Re}{\partial x} = \mu\frac{\partial^2 \Re}{\partial x^2} + F\left(x,t\right), 0 \le x \le 1, t > 0, \tag{8.2.85}$$

where

$$v \ge 0, \mu > 0 - \text{const},$$

$$F\left(x,t\right) = \left(1 + Av\right)\cos\left(Ax + t\right) + \mu A^2 \sin\left(Ax + t\right).$$

exact solution has the form:

$$\Re\left(x,t\right) = \sin\left(Ax + t\right), \tag{8.2.86}$$

and the boundary and initial conditions are formulated in accordance tothis decision.

Consider the following finite difference schemes for the approximation of the equation (8.2.85):

1. The proposed difference scheme (8.2.23):

$$\frac{\Re_j^{n+1} - \Re_j^n}{\tau} + \frac{v}{2}\left(\frac{\Re_{j+1}^n - \Re_j^n}{h} + \frac{\Re_j^{n+1} - \Re_{j-1}^{n+1}}{h}\right) = \mu\left(\frac{\Re_{j+1}^n - \Re_j^n}{h^2} - \frac{\Re_j^{n+1} - \Re_{j-1}^{n+1}}{h^2}\right) + F_j^n.$$

2. Monotone scheme of (Samarskii, Vabishchevich, 1999):

$$\begin{gathered}\frac{\Re_j^{n+1} - \Re_j^n}{\tau} + \sigma v\frac{\Re_j^{n+1} - \Re_{j-1}^{n+1}}{h} + \left(1 - \sigma\right)v\frac{\Re_j^n - \Re_{j-1}^n}{h} \\ = \sigma\mu\frac{\Re_{j+1}^{n+1} - 2\Re_j^{n+1} + \Re_{j-1}^{n+1}}{h^2} + \left(1 - \sigma\right)\mu\frac{\Re_{j+1}^n - 2\Re_j^n + \Re_{j-1}^n}{h^2} + F_j^n\end{gathered}.$$

3. Crank-Nicolson scheme (Paskonov, Polezhaev, Chudov, 1984)

$$\frac{\Re_j^{n+1} - \Re_j^n}{\tau} + \frac{v}{2}\left(\frac{\Re_{j+1}^{n+1} - \Re_{j-1}^{n+1}}{2h} + \frac{\Re_{j+1}^n - \Re_{j-1}^n}{2h}\right) =$$

$$= \frac{\mu}{2}\left(\frac{\Re_{j+1}^{n+1} - 2\Re_j^{n+1} + \Re_{j-1}^{n+1}}{h^2} + \frac{\Re_{j+1}^n - 2\Re_j^n + \Re_{j-1}^n}{h^2}\right) + \frac{F_j^{n+1} + F_j^n}{2}.$$

4. A scheme of linear modified method of characteristics

$$\frac{\Re_j^{n+1} - \Re^n(P)}{\tau} = \mu\frac{\Re_{j+1}^{n+1} - 2\Re_j^{n+1} + \Re_{j-1}^{n+1}}{h^2} + F_j^{n+1},$$

$$\Re^n(P) = \alpha\Re_{j-m-1}^n + (1-\alpha)\Re_{j-m}^n, C\Re = \tau v / h,$$

$$m = |C\Re|, \alpha = C\Re - m, 0 \le \alpha \le 1.$$

The results of calculations using all four schemes for the following values of the parameters *A*=8, *v*=10 m/s, *h*=0.005, *T*=30 s. are shown in Table 3 This example shows the behavior of the δ_M in all schemes for large Peclet numbers $Pe = hv / \mu$ and Courant $Cu = \tau v / h$.

It is seen from the Table that when $\tau / h < 1$ the difference scheme (8.2.23) is not inferior in accuracy to other methods considered. When $\tau / h = 1$, in contrast to other methods, the accuracy of the scheme (8.2.23) somewhat worse, due to the conditional convergence of the method. Increased number *Pe* does not affect largely on the accuracy δ_M of these methods.

The results of numerical experiment showed that the solving the problem by the scheme (8.2.23) is not inferior to the Crank-Nicholson scheme and is the most accurate among the considered schemes.

As noted above, the complexity of models is combined with restrictions on the time to implement them in dynamic meteorology problems. That is why when choosing a method for solving such problems, an important role play not only accuracy, but also the efficiency of the implementation of the algorithm. Since the scheme (8.2.23) and the Crank-Nicholson have the same accuracy on the test task, preference should be given to the scheme (8.2.23), as the decision procedure is faster. Crank-Nicholson method requires solving a system of equations with three diagonal matrices, and the implementation of the scheme (8.2.23), as noted in previous sections, is reduced to an explicit account and requires less computer time. In support of this conclusion, we present Table 4, which shows the time of the decision the test problem with each of these methods in the performance of 720,000 time steps under the same experimental conditions.

Table 3. Comparing calculation results for the problem (8.2.85) with schemes 1-4 of the exact solution of problem (8.2.86)

		$Cu = 1$	$Cu = 2$	$Cu = 5$	$Cu = 10$
$\mu = 10^{-2}$ $Pe = 5 \cdot 10^{0}$	1	0.0014	0.0017	0.0027	0.0044
	2	0.0283	0.0286	0.0294	0.0309
	3	0.0009	0.0010	0.0016	0.0028
	4	0.0286	0.0549	0.1336	0.2646
$\mu = 10^{-3}$ $Pe = 5 \cdot 10^{1}$	1	0.0017	0.0021	0.0031	0.0047
	2	0.0281	0.0284	0.0292	0.0307
	3	0.0011	0.0011	0.0016	0.0028
	4	0.0284	0.0538	0.1308	0.2705
$\mu = 10^{-4}$ $Pe = 5 \cdot 10^{2}$	1	0.0015	0.0018	0.0030	0.0056
	2	0.0281	0.0284	0.0292	0.0306
	3	0.0011	0.0012	0.0017	0.0035
	4	0.0284	0.0543	0.1356	0.2844
$\mu = 10^{-6}$ $Pe = 5 \cdot 10^{4}$	1	0.0009	0.0016	0.0033	0.0075
	2	0.0281	0.0284	0.0292	0.0306
	3	0.0009	0.0010	0.0018	0.0035
	4	0.0284	0.0550	0.1368	0.2874

Table 4. Solving test tasks with methods 1-4

Time Solution			
The Serial Number of the Difference Scheme in Experiment			
1	**2**	**3**	**4**
44 sec	1 min 21 sec	1 min 30 sec	1 min 35 sec

An Example of the Practical Use of the Method for Solving the Problem of Planetary Weather Forecast

In order to challenge the practical implementation of the comprehensive method (splitting method and the difference scheme) in the numerical solution of the equations of hydrodynamic model of the atmosphere does not obscure by very complex physical and mathematical problems inherent in the process of atmospheric circulation modeling, we consider the simplified hydrodynamic model of the atmosphere.

We confine ourselves to the non-stationary air movement in the horizontal plane at a height (in the numerical experiment adopted geopotential height of the 500 hPa), where the influence of the thermal and orographic heterogeneity of the underlying surface on atmospheric motion can be neglected. We introduce the assumption that air is vertically homogeneous medium in which the particle trajectories lie in horizontal planes. We neglect the influence of external and internal sources/sinks of heat and moisture. Then the equation of the system (3.2.29) - (3.2.35) will be simplified and take the following form:

$$\frac{\partial \mathcal{R}}{\partial t}+\frac{u}{r\cos\varphi}\frac{\partial \mathcal{R}}{\partial \lambda}+\frac{v}{r}\frac{\partial \mathcal{R}}{\partial \varphi}=-\left[\frac{1}{r\cos\varphi}\frac{\partial u}{\partial \lambda}+\frac{1}{r\cos\varphi}\frac{\partial}{\partial \varphi}\left(v\cos\varphi\right)\right], \tag{8.2.87}$$

$$\begin{aligned}\frac{\partial u}{\partial t}&+\frac{u}{r\cos\varphi}\frac{\partial u}{\partial \lambda}+\frac{v}{r}\frac{\partial u}{\partial \varphi}+\frac{uv}{r}\mathrm{tg}\varphi=-e^{-\mathcal{R}}\frac{1}{r\cos\varphi}\frac{\partial p}{\partial \lambda}\\&+\frac{2}{r\cos\varphi}\frac{\partial}{\partial \lambda}\left[\nu\left(\frac{1}{r\cos\varphi}\frac{\partial u}{\partial \lambda}+\frac{w}{r}+\frac{v}{r}\mathrm{tg}\varphi\right)\right]\\&+\frac{1}{r}\frac{\partial}{\partial \varphi}\left[\nu\left(\frac{1}{r\cos\varphi}\frac{\partial v}{\partial \lambda}+\frac{1}{r}\frac{\partial u}{\partial \varphi}-\frac{u}{\mathrm{r}}\mathrm{tg}\varphi\right)\right]+2\Omega v\sin\varphi\end{aligned}, \tag{8.2.88}$$

$$\begin{aligned}\frac{\partial v}{\partial t}&+\frac{u}{r\cos\varphi}\frac{\partial v}{\partial \lambda}+\frac{v}{r}\frac{\partial v}{\partial \varphi}+\frac{u^2}{r}\mathrm{tg}\varphi=-e^{-\mathcal{R}}\frac{1}{r}\frac{\partial p}{\partial \varphi}\\&+\frac{1}{r\cos\varphi}\frac{\partial}{\partial \lambda}\left[\nu\left(\frac{1}{r\cos\varphi}\frac{\partial v}{\partial \lambda}+\frac{1}{r}\frac{\partial u}{\partial \varphi}-\frac{u}{r}\mathrm{tg}\varphi\right)\right]\\&+\frac{2}{r}\frac{\partial}{\partial \varphi}\left[\nu\left(\frac{1}{r}\frac{\partial v}{\partial \varphi}\right)\right]-2\Omega u\sin\varphi\end{aligned}, \tag{8.2.89}$$

$$\frac{\partial \theta}{\partial t}+\frac{u}{r\cos\varphi}\frac{\partial \theta}{\partial \lambda}+\frac{v}{r}\frac{\partial \theta}{\partial \varphi}=\frac{1}{r^2\cos^2\varphi}\frac{\partial}{\partial \lambda}\left(k_T\frac{\partial \theta}{\partial \lambda}\right)+\frac{1}{r^2}\frac{\partial}{\partial \varphi}\left(k_T\frac{\partial \theta}{\partial \varphi}\right), \tag{8.2.90}$$

$$p=e^{-\mathcal{R}}RT. \tag{8.2.91}$$

Recall the notation adopted earlier that are repeated in the system of equations (8.2.87) - (8.2.91): u, v are horizontal components of the wind velocity vector $\mathbf{V}$; $\mathcal{R}=\ln\left(\rho\right)$, ρ is air density; $T=t+273.15$ K° is the absolute temperature; $\theta=T\left(p_{00}/p\right)^{R/c_p}$ is the potential temperature; p is air pressure in millibars, p_{00} is the pressure at sea level; R=287.04 is the universal gas constant for dry air; c_p=3.5R is the specific heat at constant pressure; Ω is the angular speed of rotation of the Earth; ν is a coefficient of horizontal turbulent kinematic viscosity with dimension $m^2 \cdot s^{-1}$; $k_T = \nu / \Pr = \nu$ is a coefficient of turbulent thermal diffusivity.

Take the coefficient of horizontal turbulent exchange ν by Smagorinskiy formula (Smagorinskiy, 1963)

$$\nu = \sigma\left[k_0 + 0.08r\left(\Delta\lambda^2 \cos^2\varphi + \Delta\varphi^2\right)\sqrt{D_c^2 + D_d^2}\right],$$

where $k_0 = 5 \cdot 10^4$ m²/s is a constant, $\Delta\lambda$, $\Delta\varphi$ are parameters of the finite-difference grid, σ is a parameter varying levels of dissipation in the model Marchuk et al. (1984) D_c and D_d are strain rate of averaged flux having the following form:

$$D_c = \frac{\partial u}{\partial\left(\lambda\cos\varphi\right)} - \frac{\partial v}{\partial\varphi} - v\mathrm{tg}\varphi,$$

$$D_d = \frac{\partial \mathrm{v}}{\partial\left(\lambda\cos\varphi\right)} + \frac{\partial \mathrm{u}}{\partial\varphi} + u\mathrm{tg}\varphi. \tag{8.2.92}$$

Numerical experiments on forecasting the state of the atmosphere at 500 hPa of the geopotential surface were carried out for the territory of Europe. As a source of information for the initial conditions and the non-stationary boundary conditions, data of objective analysis at 00 GMT and forecast at 12 GMT from German national service Deutscher Wetterdienst, Offenbach, (DWD) in binary GRIB format for the territory ($0^o - 90^0 N, 0^o - 90^0 E$) in the nodes of a regular latitude-longitude grid with steps $\Delta\lambda = \Delta\varphi = 1.5^o$.

Implementation of the constructed model was carried out using the additive-average and splitting the difference scheme (8.2.23).

In the space-time domain

$$\mathfrak{H} = \left\{\left(\lambda,\varphi\right): 0 \le \lambda \le \pi/2, 0 \le \varphi \le \pi/2\right\}, \Im_\lambda = \left\{\left(\lambda,\varphi\right) \in \mathfrak{H} : \lambda \in \left[0,\pi/2\right]\right\},$$

$$\Im_\varphi = \left\{\left(\lambda,\varphi\right) \in \mathfrak{H} : \varphi \in \left[0,\pi/2\right]\right\}, \Im = \Im_\lambda \bigcup \Im_\varphi, t \in \mathfrak{H}_t = \left[0;T\right],$$

$$T = 12 \cdot 3600\,s,$$

we introduce the grid sets:

$$\omega_\tau = \left\{t^n : t^n = n\tau, n = \overline{0,N}, \tau = T/N\right\},$$

$$\omega = \left\{ \left(\lambda_i, \varphi_j \right) : \lambda_i = i\Delta\lambda, \varphi_j = j\Delta\varphi, i = 0,1,\ldots,J_\lambda, j = 0,1,\ldots,J_\varphi \right\},$$

$$\gamma_\lambda = \left\{ \left(\lambda_i, \varphi_j \right) \in \omega : i \in \left[0, J_\lambda \right] \right\}, \gamma_\varphi = \left\{ \left(\lambda_i, \varphi_j \right) \in \omega : j \in \left[0, J_\varphi \right] \right\}, \gamma = \gamma_\lambda \bigcup \gamma_\varphi .$$

Let

$$\Re = \begin{Bmatrix} \mathcal{R} \\ u \\ v \\ \theta \end{Bmatrix}, \; F = \begin{Bmatrix} -\left(r\cos\varphi \right)^{-1} \left[\partial u / \partial\lambda + \partial\left(v\cos\varphi \right) / \partial\varphi \right] \\ -e^{-\mathcal{R}} \left(r\cos\varphi \right)^{-1} \partial p / \partial\lambda + 2\mathbf{\Omega} v \sin\varphi \\ -e^{-\mathcal{R}} r^{-1} \partial p / \partial\varphi - 2\mathbf{\Omega} u \sin\varphi \\ 0 \end{Bmatrix}.$$

Present the system of equations (8.2.87) - (8.2.91) in vector form:

$$\frac{\partial\Re}{\partial t} + \frac{u}{r\,\cos\varphi}\frac{\partial\Re}{\partial\lambda} + \frac{v}{r}\frac{\partial\Re}{\partial\varphi} = \frac{1}{r^2\,\cos^2\varphi}\frac{\partial}{\partial\lambda}\left(\nu\frac{\partial\Re}{\partial\lambda} \right) + \frac{1}{r^2}\frac{\partial}{\partial\varphi}\left(\nu\frac{\partial\Re}{\partial\varphi} \right) + F .$$

Rewrite it in operator form:

$$\frac{\partial\Re}{\partial t} + A_\lambda\Re + A_\varphi\Re = F , \tag{8.2.93}$$

where

$$A_\lambda\Re = \frac{1}{r\cos\varphi}\left[u\frac{\partial\Re}{\partial\lambda} - \frac{1}{r\cos\varphi}\frac{\partial}{\partial\lambda}\left(\nu\frac{\partial\Re}{\partial\lambda} \right) \right],$$

$$A_\varphi\Re = \frac{1}{r}\left[v\frac{\partial\Re}{\partial\varphi} - \frac{1}{r}\frac{\partial}{\partial\varphi}\left(\nu\frac{\partial\Re}{\partial\varphi} \right) \right].$$

Splitting of the equation (8.2.93) into subtasks on the interval $t \in \left[t^n, t^{n+1} \right]$. is as follows:

$$\frac{1}{2}\frac{\Re_{\lambda,i} - \Re_i}{\tau} + \frac{1}{2r\cos\varphi}\left\{ u_i \begin{pmatrix} \dfrac{\Re_{i+1} - \Re_i}{h_\lambda} \\ + \dfrac{\Re_{\lambda,i} - \Re_{\lambda,i-1}}{h_\lambda} \end{pmatrix} - \frac{1}{r\cos\varphi}\frac{1}{h_\lambda}\begin{bmatrix} \left(\nu_{i+1} + \nu_i \right)\dfrac{\Re_{i+1} - \Re_i}{h_\lambda} \\ -\left(\nu_i + \nu_{i-1} \right)\dfrac{\Re_{\lambda,i} - \Re_{\lambda,i-1}}{h_\lambda} \end{bmatrix} \right\} + \frac{1}{2}F_i ,$$

$$\frac{1}{2}\frac{\Re_{\varphi,i}-\Re_{i}}{\tau}+\frac{1}{2r}\left\{v_{i}\left(\frac{\Re_{i+1}-\Re_{i}}{h_{\varphi}}+\frac{\Re_{\varphi,i}-\Re_{\varphi,i-1}}{h_{\varphi}}\right)-\frac{1}{r}\frac{1}{h_{\varphi}}\left[\left(\nu_{i+1}+\nu_{i}\right)\frac{\Re_{i+1}-\Re_{i}}{h_{\varphi}}-\left(\nu_{i}+\nu_{i-1}\right)\frac{\Re_{\varphi,i}-\Re_{\varphi,i-1}}{h_{\varphi}}\right]\right\}+\frac{1}{2}F_{i},$$

$$\Re^{n+1}=\frac{\Re_{\lambda}+\Re_{\varphi}}{2},i=1,2,\ldots,N,N=J_{\lambda}J_{\varphi}-1.$$

Calculation of the coefficients of horizontal turbulent exchange by the formula (8.2.92) for the points of the grid ω as follows:

$$D_{c,ij}=\frac{u_{i+1j}-u_{i-1jk}}{2\Delta\lambda\cos\varphi_{j}}-\frac{v_{ij+1}-v_{ij-1}}{2\Delta\varphi}-v_{ij}\mathrm{tg}\varphi_{j},$$

$$D_{d,ij}=\frac{v_{i+1j}-v_{i-1jk}}{2\Delta\lambda\cos\varphi_{j}}+\frac{u_{ij+1}-u_{ij-1}}{2\Delta\varphi}-u_{ij}\mathrm{tg}\varphi_{j}.$$

Numerical results of the experiment for the forecast the state of the atmosphere at 500 hPa geopotential surface at 12 GMT are shown in Figure 11 and Figure 13. Figure 12 shows the results obtained using the proposed method with the step $\tau=3600s$. Figure 13 for comparison shows the field of meteorological variables derived from prognostic data of the German national service Deutscher Wetterdienst (DWD) Offenbach.

Isolines of temperature field are drawn through 1°C. Field of wind are shown by vectors; the direction indicates where the wind blows at a given point from, and their unit value of a cross-dashes. One such dash corresponds the speed of 5 m/s.

As one can see from the figures, the solution obtained describes qualitatively enough the picture of the distribution of fields of meteorological variables, despite the very serious adopted simplifications in the formulation of the problem.

SUMMARY

This chapter is devoted to the analysis of numerical methods for reproducing the specific physical processes. The problems of numerical solution of equations for predictive models of regional atmospheric circulation are considered.

The weather forecasting models are mainly classified according to the type of meteorological processes, specifics of their mathematical realization of the forecast period, as well as the size of the territory for which they are formulated. To implement a global model with a local high resolution with limited computing resources requires the use of efficient numerical methods. Choosing the best solution method is a difficult task as the purpose (accuracy) and means (cost) are contradictive. Currently, various

Figure 12. Weather forecast (12 GMT, 31.07.09) for the isobaric surface of 500 hPa, calculated on the basis of a simplified atmospheric circulation model using additively-averaged and splitting the difference scheme (8.2.23) at increments of $\tau = 3600\ s$

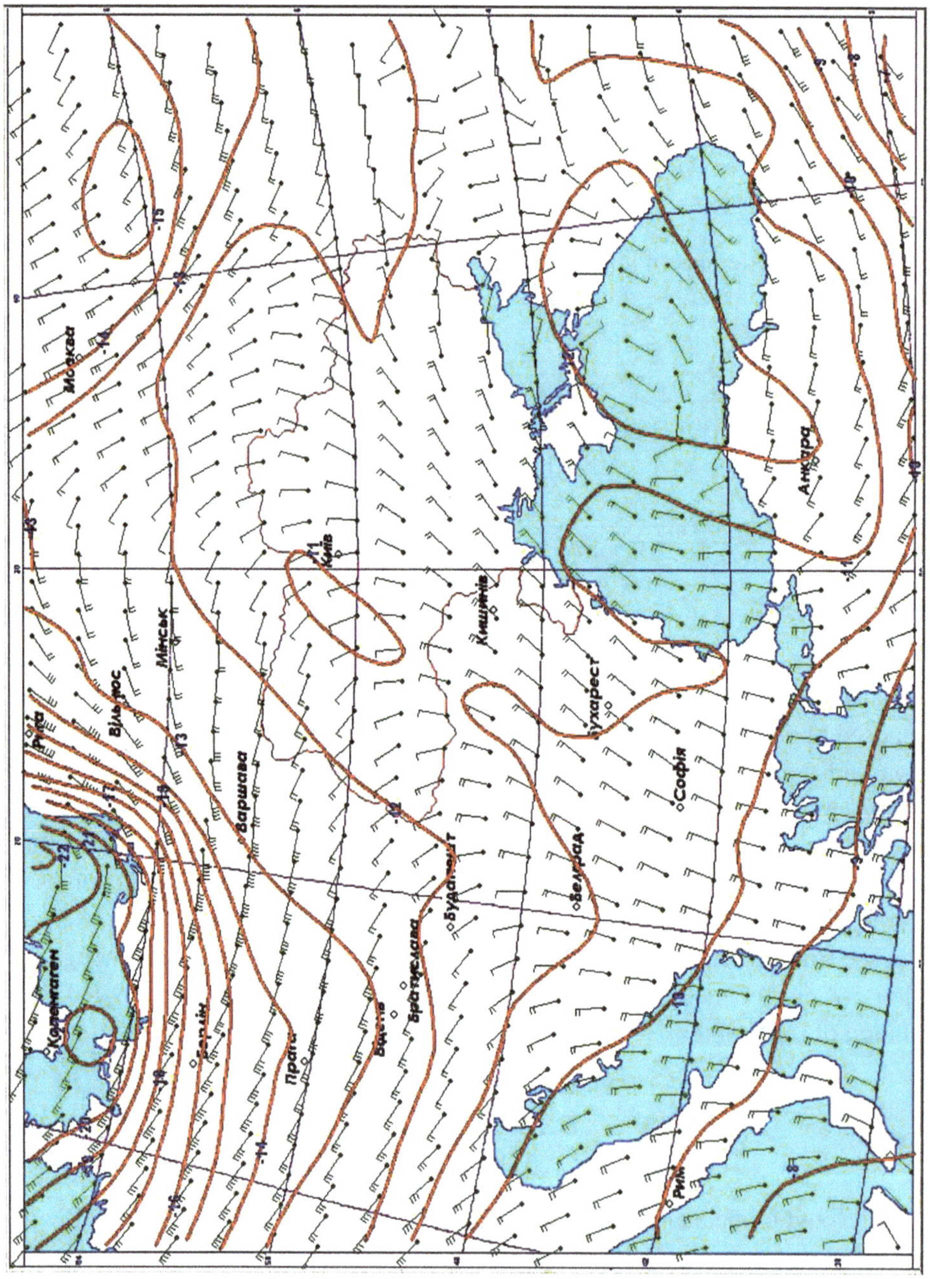

modifications of approaches to the numerical solution of the Navier-Stokes equations are widely known that lay the basis of hydrodynamic models of atmospheric circulation. Among them are: direct methods (mainly Galerkin methods and their modifications), finite element and finite difference methods, and semi-Lagrangian methods.

Figure 13. Weather forecast (12 GMT 31.07.09) for the isobaric surface of 500 hPa got from the service of Deutscher Wetterdienst (DWD), Offenbach (Germany)

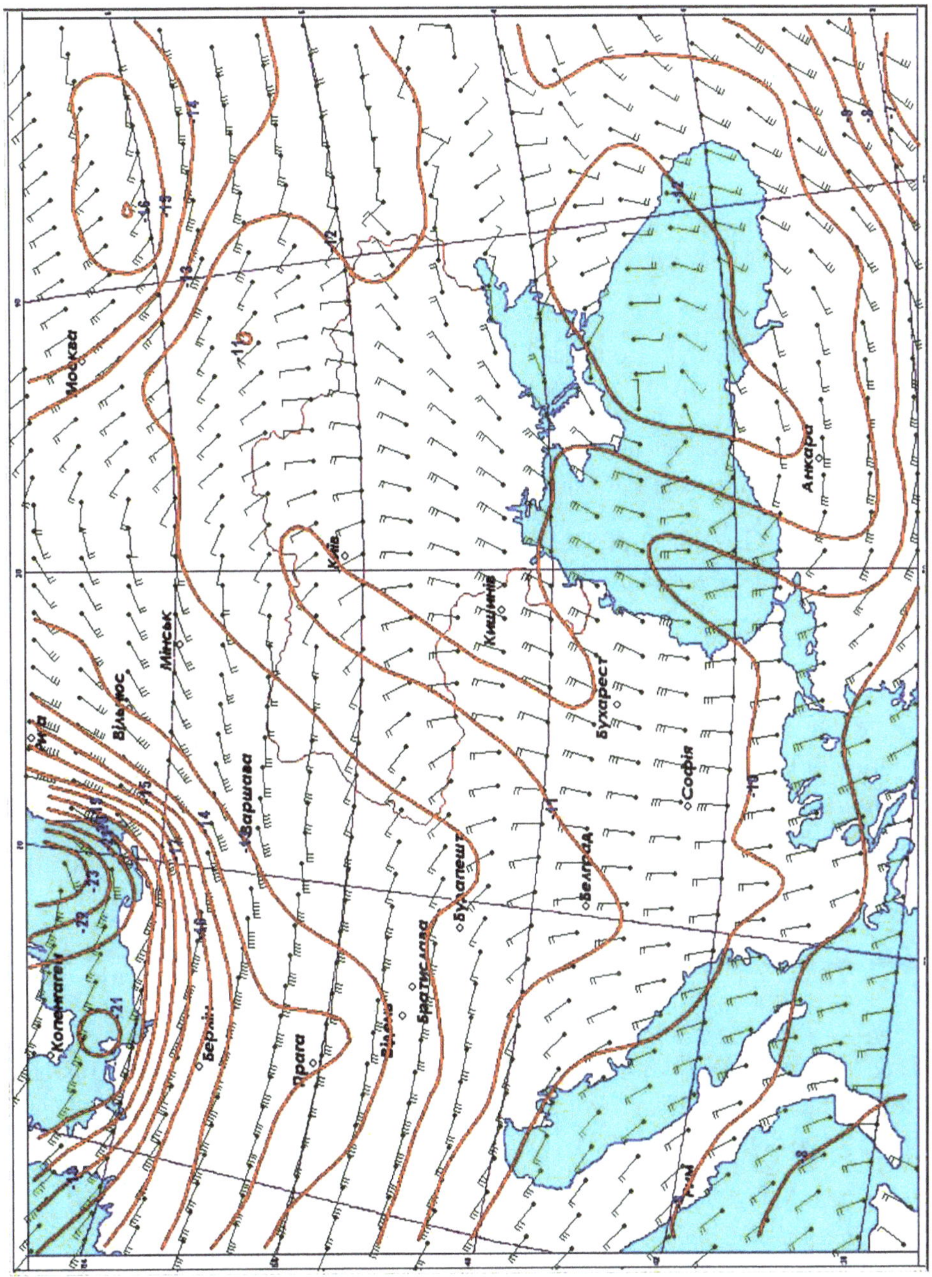

After brief overview of the existing methods, we develop new numerical techniques used to solve the non-stationary problem of general circulation of the atmosphere with a prehistory. We propose a family of strongly resistant linear *l*-step finite difference schemes on a uniform and irregular computational grid and obtain fourth-order approximation of partial derivatives in the right side of the atmosphere model equations. Then we construct an implicit difference scheme with an explicit solution of one-dimensional convection-diffusion equation and extend the general case with the technique of splitting. We give an analysis of the effectiveness of the method by comparing with known difference schemes based on model problems and apply the method with an example of the practical use of the method for solving the problem of planetary weather forecast.

REFERENCES

Adam, Y. (1975). A Hermitian Finite Difference Method for the Solution of Parabolic Equations. *Computers & Mathematics with Applications (Oxford, England)*, *1*(3-4), 393–406. doi:10.1016/0898-1221(75)90041-3

Adam, Y. (1977). Highly Accurate Compact Implicit Methods and Boundary Conditions. *Journal of Computational Physics*, *24*(1), 10–22. doi:10.1016/0021-9991(77)90106-1

Anderson, D., Tannehill, J., & Pletcher, R. (1984). *Computational fluid mechanics and heat transfer* (Vols. 1-2). New York: Hemisphere Publishing Corporation.

Argyris, J. H., Mareczek, G., & Scharpf, D. V. (1970). Two- and three-dimensional flow analysis using finite elements. *Nuclear Engineering and Design*, *11*(2), 230–236. doi:10.1016/0029-5493(70)90147-0

Arsenin, V. Y. (1984). *Methods of mathematical physics and special functions*. Moscow: Nauka. (in Russian)

Artemiev, S. S., & Demidov, D. S. (1975). A-stable method of Rosenbrock-type and fourth order accuracy of the solution of the Cauchy problem for stiff systems of ordinary differential equations. In *Some problems of computational and applied mathematics* (pp. 214–220). Novosibirsk: Nauka. (in Russian)

Babenko, K. I. (1986). *Fundamentals of numerical analysis*. Moscow: Nauka. (in Russian)

Bakhvalov, N. S. (1975). *Numerical methods*. Moscow: Nauka. (in Russian)

Bakhvalov, N. S., Zhidkov, N. P., & Kobelkov, G. M. (1987). *Numerical methods*. Moscow: Nauka. (in Russian)

Barrett, J. W., & Morton, K. W. (1980). Optimal finite element approximation for diffusion-convection problems in one dimension. *International Journal for Numerical Methods in Engineering*, *15*(10), 1457–1474. doi:10.1002/nme.1620151004

Bates, J. R., & Chen, M. (1996). A comparison of climate simulations from a semi-Lagrangian and Eulerian GCM. *Journal of Climate*, *9*(5), 1126–1149. doi:10.1175/1520-0442(1996)009<1126:ACOCSF>2.0.CO;2

Bates, J. R., Li, Y., Brandt, A., McCormick, S. F., & Ruge, J. (1995). A global shallow-water numerical model based on the semiLagrangian advection of potential vorticity. *Quarterly Journal of the Royal Meteorological Society*, *121*(528), 1981–2005. doi:10.1002/qj.49712152810

Belov, P. N., Borisenkov, E. P., & Panin, B. D. (1989). *Numerical methods of weather forecasting*. Leningrad: Gidrometeoizdat Publisher. (in Russian)

Berezin, I. S., & Zhidkov, N. P. (1962). *Calculation methods*. Moscow: Fizmatgiz. (in Russian)

Berger, A. E., Solomom, J. M., Ciment, M., Leventhal, S. H., & Wineberg, B. C. (1980). Generalized OCI Schemes for Boundary Layer Problems. *Mathematics of Computation*, *35*(151), 695–731. doi:10.1090/S0025-5718-1980-0572850-8

Bermejo, R., & Staniforth, A. (2002). The conversion of semi-Lagrangian advection schemes to quasi-monotone schemes. *Monthly Weather Review*, *130*, 423–430. doi:10.1175/1520-0493(2002)130<0423:ACQMSL>2.0.CO;2

Boer, G. J., McFarlane, N. A., Laprise, R., Henderson, J. D., & Blanchet, J.-P. (1984). The Canadian Climate Centre spectral atmospheric general circulation model. *Atmosphere-ocean*, *22*(4), 397–429. doi:10.1080/07055900.1984.9649208

Carpenter, R. L. Jr, Droegemeier, K. K., Woodward, P. R., & Hane, C. E. (1990). Application of the Piecewise Parabolic Method (PPM) to meteorological modelling. *Monthly Weather Review*, *118*(3), 586–612. doi:10.1175/1520-0493(1990)118<0586:AOTPPM>2.0.CO;2

Celia, M. A., & Gray, W. G. (1992). *Numerical Methods for Differential Equations*. Englewood Cliffs, NJ: Prentice Hall.

Ciment, M., Leventhal, S. H., & Weinberg, B. C. (1978). The Operator Compact Implicit Method for Parabolic Equations. *Journal of Computational Physics*, *28*(2), 135–166. doi:10.1016/0021-9991(78)90031-1

Cole, J. D. (1951). On a quasilinear parabolic equation occurring in aerodynamics. *Quarterly of Applied Mathematics*, *9*(3), 225–236. doi:10.1090/qam/42889

Courtier, P., & Geleyn, J. F. (1988). A global numerical weather prediction model with variable resolution: Application to the shallow water equations. *Quarterly Journal of the Royal Meteorological Society*, *114*(483), 1321–1346. doi:10.1002/qj.49711448309

Cullen, M. J. P. (1973). A simple finite-element method for meteorological problems. *Journal of the Institute of Mathematics and Its Applications*, *11*(1), 15–31. doi:10.1093/imamat/11.1.15

Datta-Gupta, A., Lake, L. W., Pope, G. A., Sepehrnoori, K., & King, M. J. (1991). High Resolution Monotonic Schemes for Reservoir Fluid Flow Simulation. *In Situ*, *15*(3), 289–317.

Dimov, I., Faragó, I., Havasi, A., & Zlatev, Z. (2004). Operator splitting and commutativity analysis in the Danish Eulerian Model. *Mathematics and Computers in Simulation*, *67*(3), 217–233. doi:10.1016/j.matcom.2004.06.017

Ducrocq, V., & Bougeault, P. (1995). Simulations of an observed squall line with a meso-beta scale hydrostatic model. *Weather and Forecasting*, *10*(2), 380–399. doi:10.1175/1520-0434(1995)010<0380:SOAOSL>2.0.CO;2

Fanchi, J. R. (1983). Multidimensional Numerical Dispersion. *SPE Journal*, *23*(01), 143–151. doi:10.2118/9018-PA

Faragó, I. (2005). Splitting methods for abstract Cauchy problems. In Z. Li, L. Vulkov, & J. Was'niewski (Eds.), *Numerical Analysis and Its Application* (pp. 35–45). Berlin: Springer Verlag. doi:10.1007/978-3-540-31852-1_4

Filler, L., & Ludloff, H. F. (1961). Stability analysis and integration of the viscous equations o motion. *Mathematics of Computation*, *15*, 261–274.

Fletcher, C. A. J. (1984). *Computational Galerkin methods*. New York: Spinger-Verlag. doi:10.1007/978-3-642-85949-6

Garbey, M., Kaper, H. G., & Romanyukha, N. (2002) A Some Fast Solver for System of Reaction-Diffusion Equations. In *Proc. 13*[th] *Int. Conf. on Domain Decomposition DD13, Domain Decomposition Methods in Science and Engineering* (pp. 397-394). Barcelona: CIMNE.

Gear, S. W. (1971). *Numerical initial value problems in ordinary differential equations*. Englewood Cliffs, NJ: Prentice-Hall.

Geleyn, J.-F., & Hollingsworth, A. (1979). An economical analytical method for the computation of the interaction between scattering and line absorption of radiation. *Beitrage zur Physik der Atmosphare*, *52*, 1–16.

Giraldo, F. X. (2000). Lagrange-Galerkin methods on spherical geodesic grids: The shallow water equations. *Journal of Computational Physics*, *160*(1), 336–368. doi:10.1006/jcph.2000.6469

Girard, C., & Delage, Y. (1990). Stable schemes for nonlinear vertical diffusion in atmospheric circulation models. *Monthly Weather Review*, *118*(3), 737–745. doi:10.1175/1520-0493(1990)118<0737:SSFNVD>2.0.CO;2

Gnandt, B. (2005). A new operator splitting method and its numerical investigation. In I. Faragó, K. Georgiev, & Á. Havasi (Eds.), *Advances in Air Pollution Modeling for Environmental Security, 54* (pp. 229–241). Groningen: Springer Nethelands. doi:10.1007/1-4020-3351-6_21

Godunov, S.K. (1959) Difference method for numerical calculation of discontinuous solutions of the equations of hydrodynamics. *Mathematical Digest, 47*(3), 271-306.

Godunov, S. K., & Ryaben'kii, V. S. (1973). *Difference schemes (introduction to the theory)*. Moscow: Science. (in Russian)

Gorbunov, A. D. (1973). *Difference methods for solving the Cauchy problem for systems of ordinary differential equations*. Moscow: MGU. (in Russian)

Gospodinov, I. G., Spiridonov, V. G., Bénard, P., & Geleyn, J.-F. (2002). A refined semi-Lagrangian vertical trajectory scheme applied to a hydrostatic atmospheric model. *Quarterly Journal of the Royal Meteorological Society*, *128*(579), 323–336. doi:10.1256/00359000260498905

Gravel, S., Staniforth, A., & Côté, J. (1993). A stability analysis of a family of baroclinic semi-Lagrangian forecast models. *Monthly Weather Review*, *121*(3), 815–824. doi:10.1175/1520-0493(1993)121<0815:ASAOAF>2.0.CO;2

Griffiths, D. F., & Mitchel, A. R. (1979). *Finite elements for convection dominated flows*. New York: ASME.

Gupta, M. M. (1991). High Accuracy Solutions of Incompressible Navier-Stokes Equations. *Journal of Computational Physics*, *93*(2), 342–359. doi:10.1016/0021-9991(91)90188-Q

Gupta, M. M., Manohar, R. P., & Stephenson, J. W. (1984). A Single Cell High Order Scheme for the Convection-Diffusion Equation with Variable Coefficients. *International Journal for Numerical Methods in Fluids*, *4*(7), 641–651. doi:10.1002/fld.1650040704

Hairer, E., & Wanner, G. (1996). *Solving Ordinary Differential Equations II. Stiff and Differential-Agebraic Problems*. Berlin: Springer. doi:10.1007/978-3-642-05221-7

Hall, J., & Watt, J. (1976). *Modern numerical methods for solving ordinary differential equations*. Oxford, UK: Oxford University Press.

Hamming, R. W. (1973). *Numerical Methods for Scientists and Engineers* (2nd ed.). New York: McGraw-Hill.

Harten, A. (1984). On a Class of High Resolution Total Variation Stable Finite Difference Schemes. *SIAM Journal on Numerical Analysis*, *21*(1), 1–23. doi:10.1137/0721001

Henrici, P. (1962). *Discrete variable methods in ordinary differential equations*. New York: J. Wiley & Sons.

Hirsh, R. S. (1975). Higher-Order Accurate Difference Solutions of Fluid Mechanics Problems by a Compact Differencing Technique. *Journal of Computational Physics*, *19*(1), 90–109. doi:10.1016/0021-9991(75)90118-7

Hoffman, J. D. (2001). *Numerical methods for engineers and scientists*. New York: CRC Press.

Hopf, E. (1950). The partial differential equation $u_t + uu_x = \mu u_{xx}$. *Communications on Pure and Applied Mathematics*, *3*(3), 201–230. doi:10.1002/cpa.3160030302

Hortal, M. (2002). The development and testing of a new two-time-level semi-Lagrangian scheme (SETTLS) in the ECMWF forecast model. *Quarterly Journal of the Royal Meteorological Society*, *128*(583), 1671–1687. doi:10.1002/qj.200212858314

Hortal, M., & Simmons, A. (1991). Use of reduced Gaussian grids in spectral models. *Monthly Weather Review*, *119*(4), 1057–1074. doi:10.1175/1520-0493(1991)119<1057:UORGGI>2.0.CO;2

Hundsdorfer, W., & Verwer, J. (2003). *Numerical solution of time-dependent advection-diffusion-reaction equations*. Berlin: Springer. doi:10.1007/978-3-662-09017-6

Jakob-Chien, N. R., Hack, J. J., & Williamson, D. L. (1995). Spectral transform solutions to the shallow water test set. *Journal of Computational Physics*, *119*(1), 164–187. doi:10.1006/jcph.1995.1125

Kalitkin, N. N. (1978). *Numerical methods*. Moscow: Nauka. (in Russian)

Krause, E., Hirschel, E. H., & Kordulla, W. (1976). Fourth-Order Mehrstellen Integration for Three-Dimensional Turbulent Boundary Layers. *Computers & Fluids*, *4*(2), 77–92. doi:10.1016/0045-7930(76)90014-1

Krylov, V. I., Bobkov, V. V., & Monastyrnyi, P. I. (1977). *Computational methods*. Moscow: Nauka. (in Russian)

Lanser, D., Blom, J. G., & Verwer, J. G. (2001). Time integration of the shallow water equations in spherical geometry. *Journal of Computational Physics*, *171*(1), 373–393. doi:10.1006/jcph.2001.6802

Lanser, D., & Verwer, J. G. (1999). Analysis of operator splitting for advection-diffusion-reaction problems in air pollution modelling. *Journal of Computational and Applied Mathematics*, *111*(1-2), 201–216. doi:10.1016/S0377-0427(99)00143-0

Lele, S. K. (1992). Compact finite difference schemes with spectral-like resolution. *Journal of Computational Physics*, *103*(1), 16–42. doi:10.1016/0021-9991(92)90324-R

Leonard, B. P. (1995). Order of Accuracy of Quick and Related Convection-Diffusion Schemes. *Applied Mathematical Modelling*, 19.

Li, Y., & Bates, J. R. (1996). A study of the behavior of semi-Lagrangian models in the presence of orography. *Quarterly Journal of the Royal Meteorological Society*, *122*(535), 1675–1700. doi:10.1002/qj.49712253510

Liang, D., & Zhao, W. (1997). A high-order upwind method for the convection-diffusion problem. *Computer Methods in Applied Mechanics and Engineering*, *147*(1-2), 105–115. doi:10.1016/S0045-7825(97)00004-2

Liu, J., Pope, G. A., & Sephernoori, K. A. (1995). High-Resolution Finite-Difference Scheme for Non-uniform Grids. *Applied Mathematical Modelling*, *19*(3), 162–172. doi:10.1016/0307-904X(94)00020-7

Lyashko, I. I., Makarov, V. L., & Skorobagatko, A. A. (1977). *Calculation methods (Numerical analysis. Methods for solving problems of mathematical physics)*. Kyiv: Vyshcha Shkola. (in Ukrainian)

Manabe, S., Smagorinsky, J., & Strickler, R. F. (1965). Simulated climatology of a general circulation model with a hydrologic cycle. *Monthly Weather Review*, *93*(12), 769–798. doi:10.1175/1520-0493(1965)093<0769:SCOAGC>2.3.CO;2

Markovich, S. A. (1986). *Spectral Models for General Circulation of Atmosphere and Numerical Weather Forecasting*. Leningrad: Hydrometeoizdat. (in Russian)

Mesinger, F., & Arakava, A. (1976). Numerical models used in atmospheric models. *GARP WMO-ICSU Publications Series*, *17*, 1–64.

Milne, V. E. (1955). *The numerical solution of differential equations*. Moscow: IL.

Mitchel, A. R., & Griffiths, D. F. (1980). *The finite difference method in partial differential equations*. Chichester, UK: Wiley.

Moorthi, S., & Higgins, R. W. (1993). Application of fast Fourier transforms to the direct solution of a class of two-dimensional separable elliptic equations on the sphere. *Monthly Weather Review*, *121*(1), 290–296. doi:10.1175/1520-0493(1993)121<0290:AOFFTT>2.0.CO;2

Morton, K. W., & Parrott, A. K. (1980). Generalised Galerkin methods for first-order hyperbolic equations. *Journal of Computational Physics*, *36*(2), 249–270. doi:10.1016/0021-9991(80)90187-4

Naughton, M., & Bourke, W. (1996). Experiments with the semi-Lagrangian version of the BMRC global spectral atmospheric model. In *Procs. of the ECMWF Seminar on Semi-Lagrangian methods 1995*. Reading, UK: ECMWF.

Orszag, S. A., & Patterson, G. S. (1972). Numerical simulation of three-dimension homogeneous isotropic turbulence. *Physical Review Letters*, *28*(2), 76–79. doi:10.1103/PhysRevLett.28.76

Ortega, J., & Poole, W. (1986). *Introduction to numerical methods for solving differential equations.* Moscow: Nauka.

Paskonov, V. M., Polezhaev, V. I., & Chudov, L. A. (1984). *Numerical modeling of heat and mass transfer*. Moscow: Science.

Peaceman, D. W. (1977). *Fundamentals of Numerical Reservoir Simulations.* Amsterdam: Elsevier Science Publishers.

Peters, N. (1975) Boundary Layer Calculation by a Hermitian Finite Difference Method. In R.D. Richtmyer (Ed.), *Proc. of the Fourth International Conference on Numerical Methods in Fluid Mechanics, Lecture Notes in Physics*. New York: Springer-Verlag. doi:10.1007/BFb0019767

Pracht, W. E. (1971). A numerical method for calculating transient creep flows. *Journal of Computational Physics*, *7*(1), 46–60. doi:10.1016/0021-9991(71)90048-9

Prusov, V., & Doroshenko, A. (2003). Modeling and Forecasting Atmospheric Pollution over Region. *Annales Univ. Sci.*, *46*, 55–73.

Prusov, V., Doroshenko A., Faragó I., Havasi Á. (2006). On the numerical solution of the three-dimensional advection-diffusion equation. *Problems of Programming, 2-3*, 641-647.

Prusov, V. A., & Doroshenko, A. E. (2015). Multistep method for the numerical solution of the problem of modeling the circulation of the atmosphere in the formulation of the Cauchy problem. *Cybernetics and Systems Analysis*, *51*(4), 547–555. doi:10.1007/s10559-015-9745-6

Prusov, V. A., Doroshenko, A. E., Chernysh, R. I., & Guk, L. N. (2007). Effective difference schemes for the numerical solution of the convective diffusion problem. *Cybernetics and Systems Analysis*, *43*(3), 368–376. doi:10.1007/s10559-007-0058-2

Prusov, V. A., & Doroshenko, A. Yu. (2005). Methods of efficient modeling and forecasting regional atmospheric processes. In *Advances in Air Pollution Modeling for Environmental Security* (pp. 143–152). Leiden: Springer-Netherlands.

Prusov, V. A., & Doroshenko, A. Yu. (2006). *Modeling Natural and Technogenic Atmospheric Processes.* Kyiv: Naukova Dumka. (in Ukrainian)

Prusov, V. A., & Doroshenko, A. Yu. (2008). On efficient numerical solution of one-dimensional convection–diffusion equations in modelling atmospheric processes. *International Journal of Environment and Pollution*, *32*(2), 231–249. doi:10.1504/IJEP.2008.017104

Prusov, V.A., & Doroshenko, Y.U., & Tyrchak. (2005). Highly efficient methods for regional weather forecasting. *System Analysis and Information Technologies*, *4*, 312–319.

Radwan, S. F. (1996). On the Higher-Order Accurate Scheme for Solving Two-Dimensional Unsteady Burgers' Equation. In *Proc. Sixth International Congress on Fluid Dynamics and Propulsion*. Cairo, Egypt: Cairo Univ.

Ritchie, H., & Tanguay, M. (1996). A comparison of spatially averaged Eulerian and semi-Lagrangian treatments of mountains. *Monthly Weather Review*, *124*(1), 167–181. doi:10.1175/1520-0493(1996)124<0167:ACOSAE>2.0.CO;2

Roberts, K. V., & Weiss, W. O. (1966). Convective difference schemes. *Mathematics of Computation*, *20*(94), 272–299. doi:10.1090/S0025-5718-1966-0198702-6

Ropp, D. L., Shadid, J. N., & Ober, C. C. (2004). Studies of the Accuracy of Time Integration Methods for Reaction-Diffusion Equations. *Journal of Computational Physics*, *194*(2), 544–574. doi:10.1016/j.jcp.2003.08.033

Rubin, G., & Graves, R. A. Jr. (1975). Viscous Flow Solutions with a Cubic Spline Approximation. *Computers & Fluids*, *3*(1), 1–36. doi:10.1016/0045-7930(75)90006-7

Rubin, S. G., & Graves, R. A. (1975). *A Cubic Spline Approximation for Problems in Fluid Mechanics*. NASA Tech. Rep. R-436. Washington, DC: NASA.

Rubin, S. G., & Khosla, P. K. (1976). Higher-Order Numerical Methods Derived from Three-Point Polynomial Interpolation. *NASA CR*, *2735*, TR-228.

Samarskii, A. A. (1971). *Introduction to the theory of difference schemes*. Moscow: Nauka. (in Russian)

Samarskii, A. A. (1982). *Introduction to Numerical Methods*. Moscow: Nauka. (in Russian)

Samarskii, A. A., & Gulin, A. V. (1973). *Stability of difference scheme*. Moscow: Nauka. (in Russian)

Samarskii, A. A., & Popov, Y. P. (1994). *Difference methods for solving problems of gas dynamics* [in Russian]. Moscow: Nauka.

Samarskii, A. A., & Vabishchevich, P. N. (1999). *Numerical methods for solving convection-diffusion problems* [in Russian]. Moscow: Editorial URSS.

Saul'ev, V. K. (1960). *The integration of parabolic equations by the method of nets* [in Russian]. Moscow: Fizmatgiz.

Shampine, L. F., & Gordon, M. K. (1975). *Computer solution of ordinary differential equations*. W. H. Freeman.

Sharif, M. A. R., & Busnaina, A. A. (1988). Assessment of Finite Difference Approximations for the Advection Terms in the Simulation of Practical Flow Problems. *Journal of Computational Physics*, *74*(1), 143–176. doi:10.1016/0021-9991(88)90074-5

Skvortsov, L.M. (2007). Explicit multi-step method for the numerical solution of stiff differential equations [in Russian]. *J. of Numerical Methods and Mathematical Physics, 47*(6), 959-967.

Smagorinsky, J. (1963). General circulation experiments with the primitive equations.1: The basic experiment. *Monthly Weather Review*, *91*(3), 99–164. doi:10.1175/1520-0493(1963)091<0099:GCEWTP>2.3.CO;2

Steger, J. L., & Warming, R. F. (1981). Flux-vector splitting of the inviscid gasdynamics equations with application to finite difference methods. *Journal of Computational Physics*, *40*(2), 263–293. doi:10.1016/0021-9991(81)90210-2

Stetter, H. J. (1973). *Analysis of discretization methods for ordinary differential equations*. Berlin: Springer. doi:10.1007/978-3-642-65471-8

Strang, G. (1968). On the construction and comparison of difference schemes. *SIAM Journal on Numerical Analysis*, *5*(3), 507–517. doi:10.1137/0705041

Sündermann, J. (1976). The application of finite element and finite difference techniques in hydrodynamical-numerical models. In Formualtions and computational algorithms in finite element analysis. Cambridge, MA: MIT Press.

Taylor, M., Tribbia, J., & Iskandarani, M. (1997). The spectral element method for the shallow water equations on the sphere. *Journal of Computational Physics*, *130*(1), 92–108. doi:10.1006/jcph.1996.5554

Thuburn, J., & Yong, L. (2000). Numerical simulations of Rossby-Haurwitz waves. *Tellus*, *52A*(2), 181–189. doi:10.1034/j.1600-0870.2000.00107.x

Van Leer, B. (1974). Towards the ultimate conservative difference scheme. Pt. 2. Monotonicity and conservative in a second-order scheme. *Journal of Computational Physics*, *14*(4), 361–370. doi:10.1016/0021-9991(74)90019-9

Verwer, J. G., Hundsdorfer, W., & Blom, J. G. (2002). Numerical time integration for air pollution models. *Surveys Math. Indust.*, *10*, 107–174.

Wang, H., Shi, X., & Ewing, R. E. (2001). An Ellam Scheme for multidimensional advection-reaction equations and its optimal error estimate. *SIAM Journal on Numerical Analysis*, *38*(6), 1846–1885. doi:10.1137/S0036142999362389

Weinberg, B.C., Leventhal, S.H. & Ciment, M. (1977). *The Operator Compact Implicit Scheme for Viscous Flow Problems.* AIAA paper N 77.

White, A. B. (1974). *Numerical Solution of Two-Point Boundary-Value Problems* (Ph. D. Thesis). California Inst. of Technology.

Williamson, D. L., & Olson, J. G. (1998). A comparison of semi-Lagrangian and Eulerian polar climate simulation. *Monthly Weather Review*, *126*(4), 991–1000. doi:10.1175/1520-0493(1998)126<0991:ACOSLA>2.0.CO;2

Yee, H. C. (1987). Construction of Explicit and Implicit Symmetric TVD Schemes and Their Applications. *Journal of Computational Physics*, *68*(1), 151–179. doi:10.1016/0021-9991(87)90049-0

Young, J. A. (1968). Comparative properties of some time differencing schemes for linear and nonlinear oscillations. *Monthly Weather Review*, *96*(6), 357–364. doi:10.1175/1520-0493(1968)096<0357:CPOSTD>2.0.CO;2

Zalëtkin, S. F. (1986). *The numerical solution of ordinary differential equations, multi-step method.* Moscow: Lomonosov Moscow State University. (in Russian)

Zlatev, Z. (1995). *Computer treatment of large air pollution models*. Dordrecht: Kluwer Academic Publishers. doi:10.1007/978-94-011-0311-4

Chapter 9
Numerical Methods for Solving Regional and Mesoscale Problems

ABSTRACT

Regular grids with even steps of the spatial coordinates in the whole computational domain are the most convenient for implementing numerical methods for integration of equations of weather forecasts. However, computing a local numerical weather forecast based on the global general circulation models of the atmosphere will need enormous increase in computation time exceeding reasonable limits. Moreover, as some regional weather details are well localized it is reasonable to apply high-resolution grids locally. In this chapter, we study how to use the high-resolution grids in the numerical methods for solving regional and mesoscale weather forecast problems.

BRIEF INTRODUCTION

When using numerical methods for integration of equations for fluid flow, heat and the mass transfer, the regular grid with even steps of the spatial coordinates in the whole computational domain are most convenient conditions. However, computing associated with the numerical weather forecast for the global general circulation models of the atmosphere, even for the most powerful modern computers, limits the size of the bottom of the grid used. Breaking grids to a size suitable for describing the regional and small-scale processes and phenomena in the atmosphere causes an increase in computation time exceeding reasonable limits. As some of the regional weather details are well localized (atmospheric disturbances associated with orographic and thermal heterogeneity of the underlying surface), of course, there appears the idea of a local field of the high-resolution grid.

Separate modeling of regional atmospheric processes, in which the boundary conditions are obtained from the initial to the forecast point of time according to synoptic observations are considered unchanged for the time interval calculation, or extrapolated at each time step to the side edges of the inner part of the solution region, now is only of scientific interest. Stated thus lateral boundary conditions lead to

DOI: 10.4018/978-1-5225-2636-0.ch009

large errors, and sometimes to a false numerical solution because the errors generating fictitious wave with large amplitude at the borders and in the vicinity of them, extends deep into the area of solutions (Asselin, 1972; Davies, 1976; Gustafsson, 1975; Hill, 1968; Kreiss, 1970; Kreiss, 1980; Miyakoda, Rosati, 1977; Oliger, Sundstrom, 1978; Perkey, Kreitzberg, 1976; Shapiro, O'Brien, 1970; Sundstrom, 1986; Williamson, Browning, 1974). It is obvious that to reduce the forecast error in the restricted area, one must add a buffer zone to expand the boundaries of the area to such an extent that generated the disturbance at the boundaries do not reach the computational domain. Increasing the area of solutions entails increasing computer power requirements and, thus, leads to difficulties in the practical realization of such models.

In recent years, regional modeling of atmospheric processes is implemented in view of the fact that the field of meteorological variables in a bounded domain is formed under the influence of macro-scale atmospheric circulation. Therefore, a bounded domain of solutions is seen as part of a whole and non-stationary boundary conditions at its lateral boundaries are formulated on the basis of data obtained from a bordering region. Thus, in practice, when solving numerical weather prediction problem in a limited area, they make the grid more condensed to achieve the desired accuracy for solving the problem in the field of large gradients of dependent functions.

As the use of variable resolution grids in the spectral methods causes considerable inconvenience, we will analyze the problems arising from the application of non-uniform grids to difference schemes of numerical integration of the equations of hydrodynamics and heat and mass transfer, members of regional weather prediction model.

The model with variable resolution can be realized in two ways (Prusov, Doroshenko, 2006). In the first one, you can use a common mathematical model with continuously or stepwise changing density grid (the algorithm of "bilateral interaction"). In this case, there are a number of problematic issues related to the change of the grid step. They belong to the local and global effects of reflection, transmission and conversion of different waves at the boundaries of the variable resolution grid. In addition, most of the difference schemes are the most accurate in the constant grid steps. Even with the gradually changes of grid step in the transition zone, there are errors exceeding the error in the rest of the grid. If this transition occurs abruptly, the error value can be increased by an order of magnitude.

All of the explicit difference schemes for the numerical solution of equations of fluid flow, heat and mass transfer are conditionally stable, i.e., there is an upper limit for the magnitude of the integration step by time. If the change of the spatial grid steps is smooth, the maximum time step is proportional to the ratio of the maximum grid step to the local maximum of the characteristic velocity. However, as the experience of work with nested grids shows, there is still not enough studied the possibility of converting a stable scheme in an unstable scheme at a local approximation.

In the second method, the problem of weather forecasting for the restricted area is formulated by combining two models of different completeness, assuming that the field of meteorological variables of a limited area is formed on the background of the general circulation of the atmosphere. Combined model consists of:

- The global atmospheric general circulation model, including simplified equations numerically implemented on the coarse grid.
- The regional model, which includes the complete equations of fluid flow, heat, and mass transfer, numerically implemented on a fine grid.

Necessary for the regional model boundary conditions are identified with the solution to the global model that can be integrated together with the regional model, or in advance. This method of forecasting for nested grids method was called the method of " one-way interaction" because the numerical results of the internal model do not affect the integration of the equations of the external model.

THE METHOD OF "ONE-WAY INTERACTION"

In the analysis of the essence of the method of "one-way interaction", it may seem that from a computational point of view, there is no upper limit on the term of the regional forecast on embedded grids. Therefore, it was considered that the regional model, containing the complete equations of fluid flow, heat, and mass transfer solved numerically on a fine grid, will be the basis of operational weather prediction schemes in the near future. But in practice, non-stationary boundary conditions are determined by extrapolation in the models of "one-way interaction". This leads to the emergence of errors spread into the area of solutions with the speed of external gravitational waves that reduces significantly possible terms of a weather forecast for a limited area. Consequently, in order to make the regional weather forecasts real, it is necessary to continue the study of the interaction of grids with different resolutions.

Theoretical foundations of the method of "one-way interaction" are not yet well understood. For models in which the non-stationary boundary conditions are taken from the forecast model with a coarse resolution, the question of the formulation of the boundary conditions has been considered first as a minor problem. However, it was soon discovered that the redefined boundary conditions give rise to unacceptably large errors, especially in the case of models with complete hydrodynamic equations in hydrostatic approximation. These errors are caused by the generation of computing modes, gravitational waves and other disturbances at the boundaries of the solution area. The strongest disturbance of the motion picture is observed from that part of the waves that propagate at a high speed in the form of external gravitational waves.

The issue of correct formulation of boundary conditions in the case of the method of "one-way interaction" is difficult enough. In the books (Issues, 1985) devoted to the study of the effect of different types of boundary conditions on numerical simulation results for non-uniform grids, it is shown that the optimal mathematical and computational boundary conditions for the scheme with the "one-way interaction" should have the following properties:

- A boundary conditions must specify the main quasi-geostrophic part of the solution when the movement is directed into the region; the outlet of the solutions region should not contain reflection in the form of computing mode or gravitational waves.
- Fast gravitational waves generated inside the area should, if possible, go across the border or should not be reflected and reinforced by the boundary conditions.

Most of the boundary conditions applied in practice, do not meet not only both requirements, but each of them separately.

Here, we present only some of the problems arising in the problem of the numerical forecast of the atmospheric state for a limited area, but they are enough to see what multilateral is this challenge and how much of difficulties arise in its numerical solution.

Since the traditional numerical solution of the problem of regional weather forecasting in the formulation of classical initial-boundary value problem is associated with a lot of difficulties discussed above, we replace it with a similar problem satisfying the special requirements arising from the concept of "one-way interaction" (Prusov, Doroshenko, 2006). Let's explain this with a simple example.

Let it is needed to find a function $\Re(t)$ at the interval $t_{j-1} \le t \le t_j$ satisfying the differential equations solving

$$\frac{d\Re}{dt} = F(t) \tag{9.1.1}$$

and additional conditions from prehistory

$$\Re(t_j) = \Re_j, j = 1, 2, \ldots, N, \tag{9.1.2}$$

where $\Re_j$ are the discrete value of the function $\Re(t)$, given in the nodes $t=t_j$ (j=1,2,…,N) of large-scale grid ϖ_h with the grid step pitch $h_j=t_j-t_{j-1}$. In general, F is a nonlinear operator. It is algebraic for ordinary differential equations, and is a spatial differential operator for partial differential equations.

At first, we restrict ourselves in the problem (9.1.1) – (9.1.2) to the case of an ordinary differential equation. The natural way to solve this problem is as follows. On the segment $[t_{j-1},t_j]$, we introduce a small-scale grid ϖ_h of points ζ_k(k=01,..,M) with the increment of $\tau_{k-1}=\zeta_k-\zeta_{k-1}$ provided that $\zeta_0=t_{j-1}$, $\zeta_M=t_j$, and consider the differential equation not everywhere on t, but only on the interval $[t_{j-1},t_j]$. Replacing the derivative $d\Re(\zeta_k)/dt$ its approximate value, for example,

$$d\Re(\zeta_k)/dt = \left[\Re(\zeta_{k+1}) - \Re(\zeta_k)\right]/\tau_k \equiv (\Re_{k+1} - \Re_k)/\tau_k,$$

we move to the difference scheme

$$\Re_{k+1} = \Re_k + \tau_k F(\zeta_k), k = 0, 1, \ldots, M-1. \tag{9.1.3}$$

The initial value of $\Re(\zeta_0) = \Re(t_{j-1}) = \Re_{j-1}$ is defined. Other successive values $\Re(\zeta_0) = \Re(t_{j-1}) = \Re_{j-1}$ are found by solving the differential equation (9.1.3). That is, to solve the problem (9.1.1) - (9.1.2) one can apply a difference scheme for the Cauchy problem.

It can be done also in different way (Prusov, Doroshenko, 2006). Taking a closer look at the statement of the problem (9.1.1) - (9.1.2), one can see that finding the value of $\Re(t)$ in the interval $[t_{j-1},t_j]$ is a task of interpolation. Indeed, at the points $t=t_j$ (j=1,2,…,N), there are known the values of the function $\Re_j = \Re(t_j)$ and values of the right side of the equation (9.1.1) $F(t_j)$, i.e., Its first derivative $d\Re/dt|_{t=t_j} = F(t_j)$. Under the terms of the problem (9.1.1) - (9.1.2), one must find $\Re(t)$ for the value of t, different from the values of the nodes $t=t_j$ ($j=1,2,\ldots,N$). Hence, we come to the method known as the interpolation "with multiple nodes" or Hermit polynomial interpolation.

A Numerical Method Based on a Prehistory and Interpolation with Multiple Nodes for Solving the Non-Stationary Problem

The following will consider the general problem of interpolation functions by a set of its several values (pre-history) and a derivative of the first order with a help of an algebraic polynomial. An appropriate representation of the polynomial interpolation will be obtained, its uniqueness proved, and interpolation error estimated.

Suppose that in addition to the values of the function $\Re_j$, there are given at the nodes of the grid under consideration its first order derivatives $\Re'_j$ so that the total of known data is $n+1=2N$. Suppose, furthermore, the values of t does not coincide with the grid points t_j ($j=1,2,\ldots,N$), where the values $\Re_j^{(\alpha)}$, $\alpha=0,1$ are given. It is required, using the source data, to find a value $\Re(t)$, i.e. solve the problem of interpolation. To do this, you must specify a rule by which, based on given values t_j and $\Re^{(\alpha)}(t_j)$, $\alpha=0,1$, $j=1,2,\ldots,N$, we can calculate the approximate value of $\Re(t)$. It is known that if the function $\Re$ on the interval $a\leq t\leq b$ of small length has derivatives of sufficiently high order, then their behavior on a small portion of it is not very different from the geometric representation of an algebraic polynomial. This follows from the fact that in the considered interval function $\Re$ can be represented by the formula of Taylor with a small error term. Therefore, interpolation of the function by an algebraic polynomial should give good accuracy by taking a sufficient number of sites close to the point of interpolation t.

Construct an algebraic polynomial $P_n(t)$ of degree n, satisfying the conditions

$$P_n^{(\alpha)}(t_j) = \Re^{(\alpha)}(t_j), \alpha = 0,1, j = 1,2,\ldots,N . \qquad (9.1.4)$$

We prove that there is only one interpolation polynomial $P_n(t)$ satisfying the conditions (9.1.4). The interpolation conditions (9.1.4) represent a system of linear algebraic equations for the coefficients $a_0,a_1,\ldots,a_n$ of the polynomial

$$P_n(t) = a_0 t^n + a_1 t^{n-1} + \ldots + a_n . \qquad (9.1.5)$$

The number of equations of this system is equal to the number of unknowns and is equal to $2N$. Therefore, it suffices to show that the homogeneous system

$$P_n^{(\alpha)}(t_j) = 0, \alpha = 0,1, j = 1,2,\ldots,N \qquad (9.1.6)$$

has only the trivial solution $a_0=a_1=\ldots=a_n=0$. The condition (9.1.6) for a fixed j and $\alpha=0,1$ means that the number t_j is a root of multiplicity n of the polynomial $P_n(t)$. Thus, the polynomial $P_n(t)$ given the multiplicity has just no less than $n+1$ roots at the [a, b]. Since the degree of $P_n(t)$ is equal to n, this polynomial is identically zero, all its coefficients are equal to zero, the homogeneous system of equations (9.1.6) has a unique solution $a_0=a_1=\ldots=a_n=0$, and there is only one polynomial (9.1.5) satisfying the conditions (9.1.4).

Heterogeneous system (9.1.4) is uniquely solvable for any right-hand side. Since the values of $\Re^{(\alpha)}\left(t_j\right)$, $\alpha=0,1$, $j=1,2,\ldots,N$ includes only the right side of the system (9.1.4), then the coefficients a_j of the polynomial $P_n(t)$ are expressed linearly through the value of $\Re^{(\alpha)}\left(t_j\right)$, i.e.

$$P_n\left(t\right)=\sum_{j=1}^{N}L_j\left(t\right)\Re'\left(t_j\right). \tag{9.1.7}$$

Find explicit representation of the formula (9.1.7) with the help of the residual

$$r_n\left(\Re,t\right)=\Re\left(t\right)-P_n\left(t\right). \tag{9.1.8}$$

It occurs when replacing the function $\Re\left(t\right)$ interpolation polynomial $P_n(t)$ and is called the error of interpolation or residual member of interpolation formula. Obviously, this error is zero at the nodes of the interpolation. Estimate the error at an arbitrary point $t\in[a,b]$.

Theorem 9.1.1: Let nodes t_j ($j=1,2,\ldots,N$) and a point t belong to the interval $[a,b]$ and function $\Re\left(t\right)$ has a continuous derivative of order $n+1$ on $[a,b]$. Then there exists a point ξ on the interval $[a,b]$, which for the error $r_n\left(\Re,t\right)$ of interpolation $r_n\left(\Re,t\right)$ the following equality holds

$$r_n\left(\Re,t\right)=\frac{\Psi_n\left(t\right)}{\left(n+1\right)!}\Re^{\left(n+1\right)}\left(\xi\right),_n\left(t\right)=\prod_{j=1}^{N}\left(t-t_j\right)^2. \tag{9.1.9}$$

Proof: Consider the auxiliary function

$$\varrho\left(s\right)=\Re\left(s\right)-P_n\left(s\right)-K\Psi_n\left(s\right), \tag{9.1.10}$$

where $s\in[a,b]$, K is a constant.

Suppose you want to evaluate $r_n\left(\Re,t\right)$ at a given point $t\in[a,b]$, which is not a node of interpolation. We choose the constant K from the condition $\varrho(s)=0$. It's enough to put

$$K=\frac{\Re\left(t\right)-P_n\left(t\right)}{\Psi_n\left(t\right)}.$$

The function $\varrho(s)$ has at least $n+2$ zeros on $[a,b]$, namely, at the points t and t_j $(j=1,2,\ldots,N)$. The derivative of the first order from $\varrho'(s)$ by the theorem of Rolle has at least one zero within each interval between adjacent points $t_1,t_2,\ldots,t_N$, t. The number of zeros will be N. Moreover, $\varrho'(s)$ has at each point t_j $(j=1,2,\ldots,N)$ n roots of the multiplicity one. Thus, $\varrho'(s)$ has, given the multiplicity, of at least $n+1$ zeros on $[a,b]$. Similarly, the second order derivative of $\varrho''(s)$ is at least n roots, etc. Derivative $\varrho^{(n+1)}(s)$ at least once vanishes on $[a,b]$. Thus, there exists a point $\xi\in[a,b]$, where $\varrho^{(n+1)}(s)=0$.

From (9.1.10) we have

$$\varrho^{(n+1)}(s) = \Re^{(n+1)}(s) - K\Psi_n(s).$$

Since $\Psi_n(s)$ is the polynomial of degree $n+1$ with leading coefficient 1, we have $\Psi_n(s)=(n+1)!$. Therefore, from the condition $\varrho^{(n+1)}(s)=0$ and taking into account the expression for the K, we obtain the following representation for the error of interpolation

$$\Re(t) - P_n(t) = \frac{\Psi_n(s)}{(n+1)!}\Re^{(n+1)}(\xi).$$

From this, it follows that (9.1.9) holds.

Without loss of generality (this will become clear below) assume that the function $\Re(t)$ is analytic in the closed end area of $\mathfrak{H}$, bounded by $\wp$ and containing $[a,b]$ inside. Then the formula of Cauchy

$$\Re(t) = \frac{1}{2\pi i}\int_{\wp}\frac{\Re(t)}{\xi - t}d\xi \tag{9.1.11}$$

allows to reduce the construction of interpolation polynomial $P_n(t)$ to the problem of finding a polynomial $P_n(1/(\xi-t),t)$ for a fairly simple linear fractional functions $1/(\xi-t)$, where ξ is taken as a parameter by which the integration is carried out in (9.1.11).

We investigate the dependence of the error of interpolation

$$r_n\left(\frac{1}{\xi - t},t\right) = \frac{1}{\xi - t} - P_n\left(\frac{1}{\xi - t},t\right) = \frac{1}{\xi - t} - \sum_{j=1}^{N} L_j(t)\frac{1}{(\xi - t)^2} \tag{9.1.12}$$

from ξ. Equation (9.1.12) is an expansion of a rational function into simple fractions. It is obvious that the point $\xi=t$ for the error is a first-order pole with residue equal to one. The common denominator of all the terms in the expression for the error (9.1.12) is equal to $(\xi-t)\Psi_n(\xi)$. Thus, the error $r_n(1/(\xi-t),t)$ may be represented as

$$r_n\left(\frac{1}{\xi - t}, t\right) = \frac{\Omega(\xi, t)}{(\xi - t)\Psi_n(\xi)}, \tag{9.1.13}$$

where $\mathbf{\Omega}(\xi,t)$ is a polynomial on ξ of degree not higher than n–1.

Statement 9.1.1: The polynomial $\mathbf{\Omega}(\xi,t)$ on the right side of (9.1.13) does not depend on ξ and is equal to $\Psi_n(\xi)$.

Proof: When $|\xi| \gg |t|$, the equality holds

$$\frac{1}{\xi - t} = \sum_{I=1}^{\infty} \frac{t^I}{\xi^{I+1}},$$

and it follows from (9.1.13) that the error is linearly dependent on the interpolated function. Consequently, we have

$$r_n\left(\frac{1}{\xi - t}, t\right) = \sum_{I=1}^{\infty} \frac{r_n\left(t^I, t\right)}{\xi^{I+1}}. \tag{9.1.14}$$

Since the function $\varphi = t^k (k=0,1,\ldots,n)$ are interpolated exactly, then

$$r_n\left(t^I, t\right) = 0, I = 0, 1, \ldots, n.$$

Thus, the first n+1 terms in the sum (9.1.14) disappear, and the expansion (9.1.14) begins with a member $r_n\left(s^{I+1}, s\right) / \xi^{I+2}$. It follows that the degree of the numerator $\mathbf{\Omega}(\xi,s)$ in (9.1.13) with respect to ξ is by n+2 units less than the denominator of. But the degree of the denominator is equal to n+2. Thus, $\mathbf{\Omega}(\xi,s)$ is of zero degree and does not depend on ξ, i.e., $\mathbf{\Omega}(\xi,t) = \mathbf{\Omega}(t)$.

Finally, since the value $\xi = t$ for $r_n(1/(\xi - t), t)$ is a simple pole, minus 1, then $\mathbf{\Omega}(t) = \Psi n_(t)$ and

$$r_n\left(\frac{1}{\xi - t}, t\right) = \frac{\Psi_n(t)}{(\xi - t)\Psi_n(\xi)}. \tag{9.1.15}$$

Taking into account (9.1.15) to (9.1.11), we obtain a representation of error $r_n(\Re, t)$ in the form of a contour integral

$$r_n(\Re, t) = \frac{1}{2\pi i}\int_{\wp} \Re(\xi) r_n\left(\frac{1}{\xi - t}, t\right) d\xi = \frac{\Psi_n(t)}{2\pi i}\int_{\wp} \frac{\Re(\xi)}{(\xi - t)\Psi_n(\xi)} d\xi. \tag{9.1.16}$$

The resulting integral expression is the basis for finding the required interpolation polynomial

$$P_n(t) = \Re(t) - r_n(\Re, t). \tag{9.1.17}$$

The residue function $\Psi_n(t)\Re(\xi)/\left[(\xi - t)\Psi_n(\xi)\right]$ at the point $\xi=t$ is obviously, $\Re(t)$. Let us calculate the residue at the pole $\xi=t_j$, $j=1,2,\ldots,N$. For values of ξ closed to the values of t_j, the following expansions in power series hold:

$$\Re(\xi) = \sum_{I=0}^{\infty} \frac{1}{I!} \Re^{(I)}(t_j)(\xi - t_j)^I,$$

$$\frac{1}{\xi - t} = -\frac{1}{(t - t_j) - (\xi - t_j)} = \sum_{I=0}^{\infty} \frac{(\xi - t)^I}{(t - t_j)^{I+1}}, \tag{9.1.18}$$

$$\frac{\xi - t_j}{\Psi_n(\xi)} = \sum_{I=0}^{\infty} C_I^{(j)} (\xi - t_j)^I.$$

Representing the function $\Re(\xi)/\left[(\xi - t)\Psi_n(\xi)\right]$ in the form of:

$$\frac{\Re(\xi)}{(\xi - t)\Psi_n(\xi)} = \frac{1}{\xi - t_j} \cdot \frac{\xi - t_j}{\Psi_n(\xi)} \cdot \frac{\Re(\xi)}{\xi - t}, \tag{9.1.19}$$

find the rest of it by multiplying the power series (9.1.18).

The expression (9.1.17) considered with (9.1.18), (9.1.19) and the equality $\Re'(t_j) = F(t_j)$ according to (9.1.1.) Leads to the desired interpolation polynomial

$$P_n(t) = \sum_{j=1}^{N} \frac{\aleph^2(t)}{(t - t_j)^2 \left[\aleph'(t_j)\right]^2} \times \left\{ \left[1 - \frac{\aleph''(t_j)}{\aleph'(t_j)}(t - t_j) \right] \Re(t_j) + (t - t_j) F(t_j) \right\}, \tag{9.1.20}$$

where

$$\aleph(t) = \sqrt{\Psi_n(t)} = \prod_{j=1}^{N} (t - t_j).$$

Above, we assumed that the function $\Re(t)$ is analytic in the closed end area of $\mathfrak{H}$, bounded by $\wp$ and containing $[a,b]$ inside. But (9.1.20) contains only the values $\Re^{(\alpha)}(t_j)$, j=1,2,…,N, α=0,1. Thus, the formula (9.1.20) remains true for any function $\Re(t)$ with finite values of $\Re^{(\alpha)}(t_j)$, j=1,2,…,N, α=0,1.

The interpolation polynomial (9.1.20) for N=3 and $\forall t\in[t_{j-1},t_j]$ is simplified to a polynomial of the fifth degree

$$\begin{aligned}P_3(t)=&\left(\frac{t_j-t}{t_j-t_{j-1}}\right)^2\left(\frac{t-t_{j-1}}{t_j-t_{j-1}}\right)^2\left[\left(1+\frac{3}{4}\frac{t-t_{j-1}}{t_j-t_{j-1}}\right)\Re_{j-2}+\frac{1}{4}\left(t-t_{j-2}\right)F_{j-2}\right]\\&+\left(\frac{t_j-t}{t_j-t_{j-1}}\right)^2\left(\frac{t-t_{j-2}}{t_{j-1}-t_{j-2}}\right)^2\left[\Re_{j-1}+\left(t-t_{j-1}\right)F_{j-1}\right]\\&+\left(\frac{t-t_{j-2}}{t_{j-1}-t_{j-2}}\right)^2\left(\frac{t-t_{j-1}}{t_j-t_{j-1}}\right)^2\left[\left(1-\frac{3}{4}\frac{t-t_{j-1}}{t_j-t_{j-1}}\right)\Re_j-\frac{1}{4}\left(t_j-t\right)F_j\right]\end{aligned}\qquad(9.1.20a)$$

and finally, when N=2 we will have $\forall t\in[t_{j-1},t_j]$ the third degree polynomial

$$\begin{aligned}P_2(t)=&\left(\frac{t_j-t}{t_j-t_{j-1}}\right)^2\left[\left(1+2\frac{t-t_{j-1}}{t_j-t_{j-1}}\right)\Re_{j-1}+\left(t-t_{j-1}\right)F_{j-1}\right]\\&+\left(\frac{t-t_{j-1}}{t_j-t_{j-1}}\right)^2\left[\left(1+2\frac{t_j-t}{t_j-t_{j-1}}\right)\Re_j+\left(t_j-t\right)F_j\right].\end{aligned}\qquad(9.1.20b)$$

One can make sure that the last formula gives correct results at $\Re=1,t,t^2,t^3$. At $\Re=t^4$ the maximal error $r_{\max}$ in the range (0,1) is $r_{\max}$=-0.0625. At the same time, by the formula of Bessel, $r_{\max}$=-0.5625 with the same number of given $\Re(t_j)$ function values.

Note the main advantages of the interpolation formulas (9.1.20), built on the basis of predetermined values of the function and its derivative $\Re^{(\alpha)}(t_j)$, j=1,2,…,N, α=0,1:

- They are more accurate than any of the formulas that use only the value of $\Re(t_j)$.
- When interpolating inside the interval (t_{N-1},t_N), they do not need the data outside of the right border of the segment of interpolation, so they can be used for the rightmost interval.
- Values of the function $\Re(t_j)$ and its derivative $\Re(t_j)$ can be set on an irregular grid t_j.

The results obtained in this paragraph indicate that solving problems of the type (9.1.1) - (9.1.2) for ordinary differential equations can be carried out effectively by interpolation (Prusov, Doroshenko, 2006).

The Test Problem 9.1.1: The algorithm "one-way interaction" suggests the joint implementation of macro scale modeling of atmospheric processes and regional solution to these problems, and as a rule needs different computers. Transferring detailed tables of macro scale meteorological fields with small time steps used as unsteady boundary conditions in the regional model is associated with the large amounts of memory and is time-consuming in the computer networks. Consequently, the efficiency of the numerical method for solving the regional problem is determined, above all, the ability to use the most acceptable time discreteness.

The purpose of the test problem is to show that numerical simulation of meteorological processes that are usually of diurnal variation, is possible by setting the macro scale meteorological data as unsteady boundary conditions in the regional model with a time step $\tau \leq 12$ hours.

The numerical study of wave propagation is modeled by the differential equation

$$\frac{du}{dt} = \beta\pi \cos\left(\beta\pi t\right) - \left[u - \sin\left(\beta\pi t\right)\right] \times \left[\frac{m}{\alpha}\left(1 - \frac{t}{\alpha}\right)^{m-1} - \frac{c\pi}{\mathrm{tg}\left(c\pi t + \varphi\right)}\right] \tag{9.1.21}$$

with varying the coefficients α,β and c therein.

The results of its numerical solutions are compared with the exact solution

$$u = \sin\left(\beta\pi t\right) + \sin\left(c\pi t + \varphi\right)\exp\left(1 - t / \alpha\right)^{m}. \tag{9.1.22}$$

An algorithm for solving the test problem is as follows. First, according to (9.1.22), the initial values u_n are calculated for $n=0,1,\ldots,19$ on a grid with a large step $\tau=1$ (macro scale data). The general behavior of the function u_n (solid line) and its discrete values with a step of τ (points line) in the solution is shown in Figure 1a and Figure 2a. The values obtained for the coarse grid are used to fill the values of the function u_n by interpolation formula (9.1.20b) inside of the sequence of segments $[t_{n-5},t_{n-4}]$, $[t_{n-4},t_{n-3}]$,…, $[t_{n-1},t_n]$. To demonstrate the behavior of the solution within each segment, the problem is repeatedly solved by using the equation (9.1.20b) for the solution on a fine grid in increments 0.1τ. The results of these tasks are shown in Figure 1b - 1e and Figure 2b - 2e.

For comparison, dashed lines of Figure 1a and 2a indicate the results of interpolation for the same time intervals using cubic splines. Analysis of the test results shown in Figure 1b - 1e and Figure 2b - Figure 1a shows that the numerical modeling of wave processes using interpolation method "with multiple nodes" is possible even in cases where the macro scale data are set on a grid, a step of which is commensurate with the half-life changes in these processes.

Application of Interpolation with Multiple Nodes for Modeling Regional Atmospheric Processes in a "One-Way Interaction"

Let the general circulation of the atmosphere in the macro-scale area $\wp\left(r\right)$ is determined by the vector $\Re\left(r,t\right) = \left(\mathcal{R}, u, v, w, \mathfrak{H}\right)$ of discrete values obtained from the analysis and forecast on the basis of the macro-scale model. Recall that function $\mathfrak{H}$ refers to scalar quantities θ,q,δ,s.

*Figure 1. The results of solving the problem for α=7, β=1, c=1, m=2, φ=0.5; 1 – stands for the exact solution; o – stands for the initial values; * – stands for the calculated values*

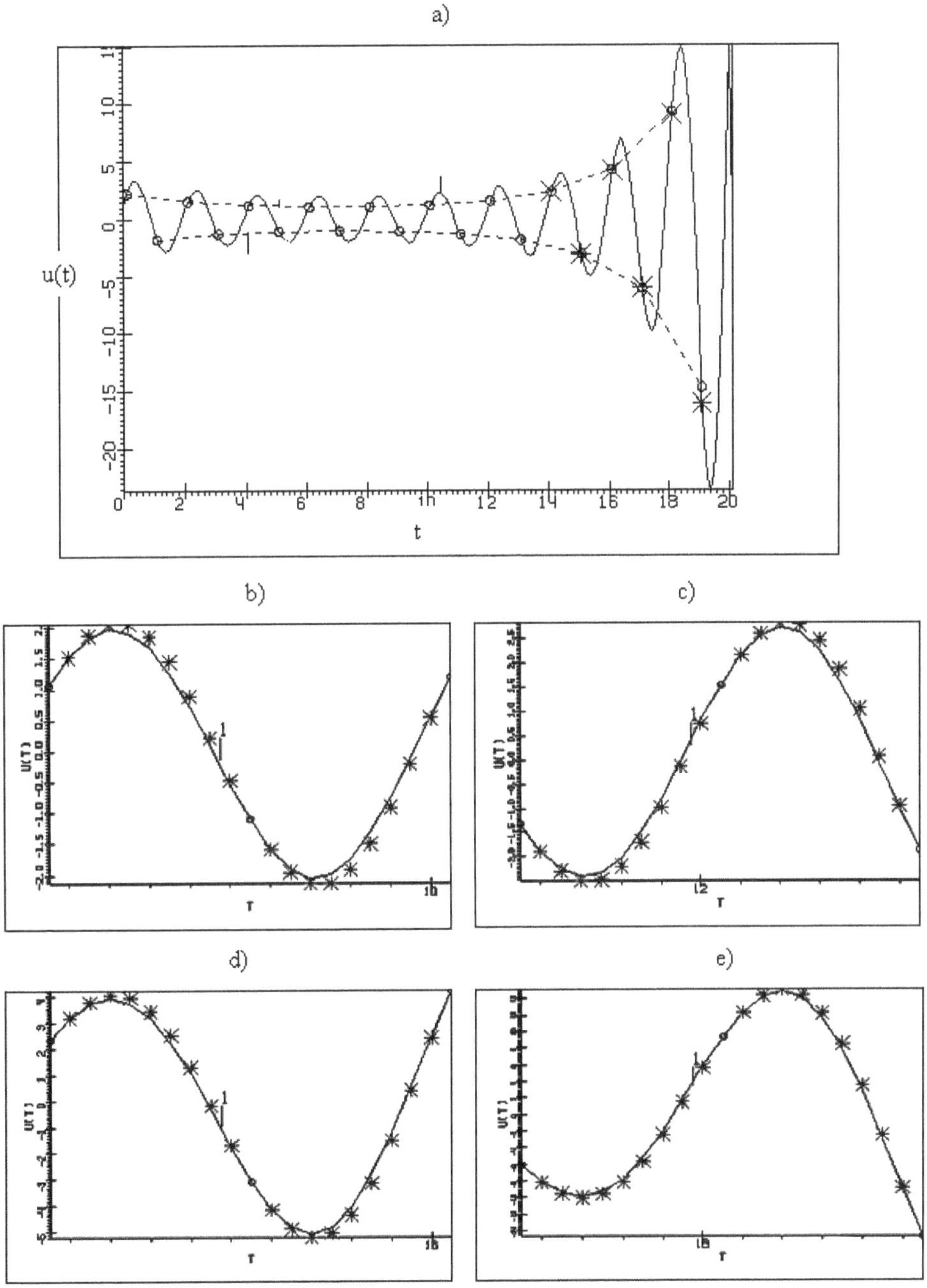

To distinguish the time grid and the spatial grid, quantities relating to specific points of time, as in Chapter 8, we will mark by a superscript, and the values specified in the nodes of the spatial grid – by a subscript. The state vector $\Re\left(r,t\right)$ in the area $\wp\left(r\right)$ is determined by a vector of discrete values $\Re\left(r,t^n\right)=\Re^n\left(r\right)$, in predetermined time instants $t=t^n$ (n=0,1,2,…) with the step $\tau = t^n - t^{n-1}$.

*Figure 2. The results of solving the problem for α=10, β=0.5, c=1, m=2, φ=0.5; 1 – stands for the exact solution; o – stands for the initial values; * – stands for the calculated values*

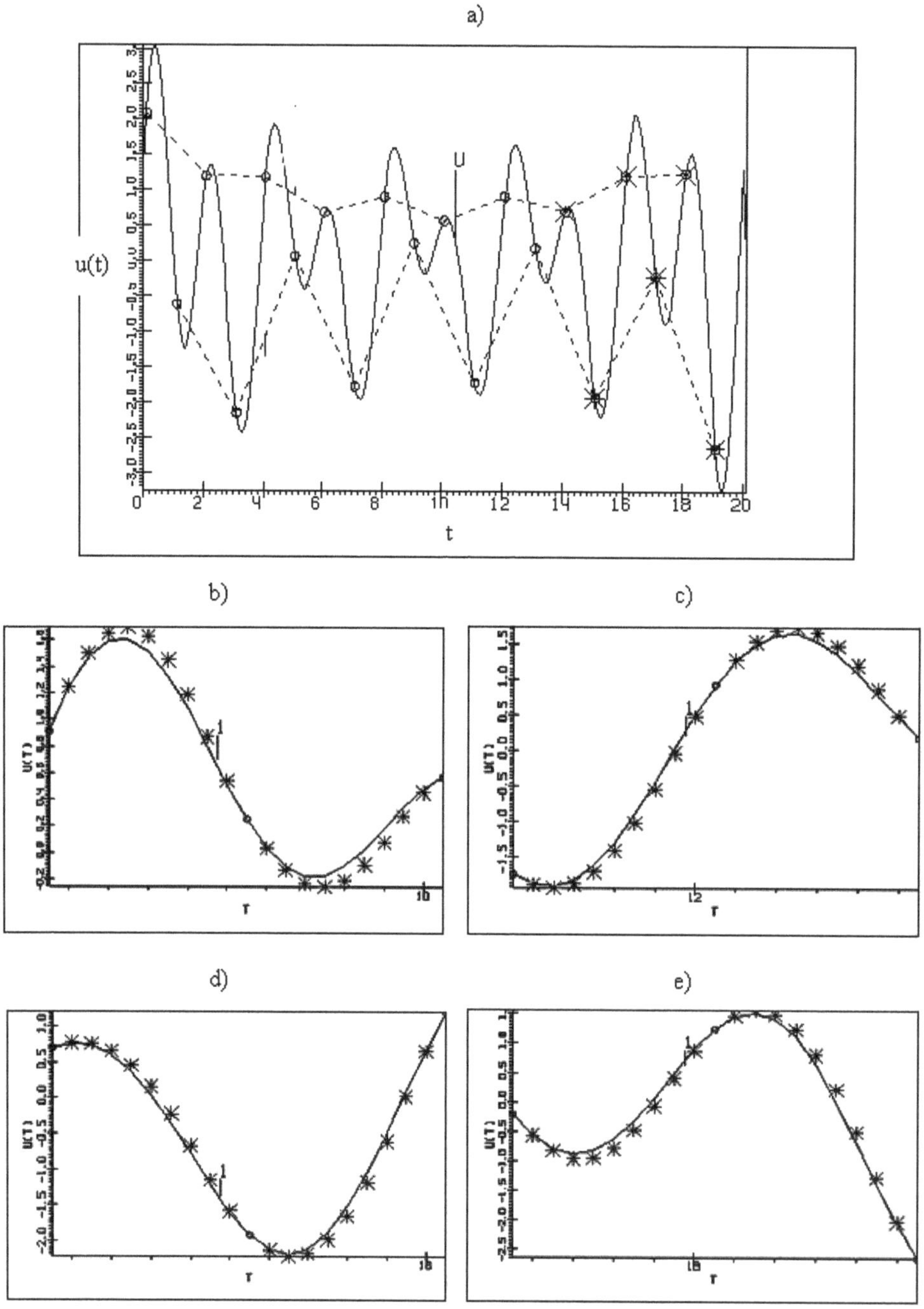

Let in $\wp(r)$ independent continuous variables λ, φ, σ of the coordinate system run through the values: $\alpha_1 \leq \lambda \leq \alpha_2$, $\beta_1 \leq \varphi \leq \beta_2$ и $\varsigma_1 \leq \sigma \leq \varsigma_2$. The spatial grid ϖ_h in this area is formed by dividing the segments $[\alpha_1, \alpha_2]$, $[\beta_1, \beta_2]$ and $[\varsigma_1, \varsigma_2]$ into the set of *I*–1 elements $\Delta\lambda_i$, *J*-1 elements $\Delta\varphi_j$ and *K*-1 elements $\Delta\sigma_k$ respectively. Designate the value of the vector $\Re\left(\lambda_i,\varphi_j,\sigma_k,t\right)$ in the nodes of the grid ϖ_h

$$\lambda_i = \alpha_1 + \sum_{\zeta=1}^{i-1}\Delta\lambda_\zeta, 1 \le i \le I,$$

$$\varphi_i = \beta_1 + \sum_{\zeta=1}^{i-1}\Delta\varphi_\zeta, 1 \le j \le J,$$

$$\sigma_i = \varsigma_1 + \sum_{\zeta=1}^{i-1}\Delta\sigma_\zeta, 1 \le k \le K$$

at time $t = t^n$ $\left(n = 0,1,2,\ldots\right)$ as $\Re^n_{ijk}$.

Define a mesoscale region of solutions $\bar{\wp}\left(r\right)$ in the macro scale area $\wp\left(r\right)$. In the region $\bar{\wp}\left(r\right)$, if it uses α or β mesoscale (defined in the Chapter 5), under x_1,x_2,x_3 we mean the rectangular Cartesian coordinate system *x,y,z*; if it considers γ mesoscale, we mean under x_1,x_2,x_3 the spherical coordinate system λ,φ,σ.

Leaving the vertical size of the mesoscale region $\bar{\wp}\left(r\right)$ equal to the vertical size of the macro scale area $\wp\left(r\right)$, we define the horizontal dimensions of $\bar{\alpha}_1 \le x_1 \le \bar{\alpha}_2$ and $\bar{\beta}_1 \le x_2 \le \bar{\beta}_2$ in such a way that $\bar{\alpha}_1 \ge \alpha_1$, $\bar{\alpha}_2 \le \alpha_2$, $\bar{\beta}_1 \ge \beta_1$, $\bar{\beta}_2 \le \beta_2$. Hence, $\bar{\wp}\left(r\right) \subset \wp\left(r\right)$. Construct in the region $\bar{\wp}\left(r\right)$ a small-scale grid $\bar{\varpi}_h$ of points by breaking the area $\bar{\wp}$ into a set of I_1-1 elements Δx_{1i}, J_1-1 elements Δx_{2j} and K_1-1 elements Δx_{3k} respectively. Construct a vector $\left\{r_{ijk}\right\}$, defining a continuous variable *r* only at the points $i\left(1 \le i \le I_1\right)$, $j\left(1 \le j \le J_1\right)$, $k\left(1 \le k \le K_1\right)$. As the result, we have:

$$x_i = \bar{\alpha}_1 + \sum_{\eta=1}^{I_1-1}\Delta x_{1\eta}, 1 \le i \le I_1;$$

$$y_j = \bar{\beta}_1 + \sum_{\eta=1}^{J_1-1}\Delta x_{2\eta}, 1 \le j \le J_1; \tag{9.1.23}$$

$$z_k = \bar{\varsigma}_1 + \sum_{\eta=1}^{K_1-1}\Delta x_{3\eta}, 1 \le k \le K_1;$$

Suppose that one need to solve the problem of determining the state of the atmosphere in a limited region $\bar{\wp}(r)$ at $\forall t \in \left[t^{n-1}, t^{n}\right]$ that in the vector representation is of the form:

$$\frac{\partial \Re(r,t)}{\partial t} = D\Re(r,t), \forall t \in \left[t^{n-1}, t^{n}\right], n = 1,2,\ldots, \forall r \in \bar{\wp},$$

$$\Re(r,t^{n}) = \Re^{n}(r), n = 1,2,\ldots, \tag{9.1.24}$$

where, according to (5.1.7) – (5.1.9), $D\Re(r,t) =$

- For the equation of continuity:

$$= -u\frac{\partial \mathcal{R}}{\partial x_1} - v\frac{\partial \mathcal{R}}{\partial x_2} - \tilde{w}\frac{\partial \mathcal{R}}{\partial x_3} - \left(\frac{\partial u}{\partial x_1} + \frac{\partial v}{\partial x_2} + \frac{\partial \tilde{w}}{\partial x_3}\right).$$

- For the equations of motion for the atmosphere:

$$= -u\frac{\partial u}{\partial x_1} - v\frac{\partial u}{\partial x_2} - \tilde{w}\frac{\partial u}{\partial x_3} - e^{-\mathcal{R}}\frac{\partial p}{\partial x_1} - \frac{Hg}{H - F(x_1,x_2)}\left(1 - \frac{x_3}{H}\right)\frac{\partial F(x_1,x_2)}{\partial x_1}$$
$$+2\frac{\partial}{\partial x_1}\left(\nu\frac{\partial u}{\partial x_1}\right) + \frac{\partial}{\partial x_2}\left[\nu\left(\frac{\partial v}{\partial x_1} + \frac{\partial u}{\partial x_2}\right)\right] + \frac{H^2}{\left[H - F(x_1,x_2)\right]^2}\frac{\partial}{\partial x_3}\left\{\nu\left[\frac{\partial u}{\partial x_3} + \left(1 - \frac{F(x_1,x_2)}{H}\right)\frac{\partial w}{\partial x_1}\right]\right\},$$
$$-2\Omega\left(\mathrm{w}\cos\varphi - v\sin\varphi\right)$$

$$= -u\frac{\partial v}{\partial x_1} - v\frac{\partial v}{\partial x_2} - \tilde{w}\frac{\partial v}{\partial x_3} - e^{-\mathcal{R}}\frac{\partial p}{\partial x_2} - \frac{H g}{H - F(x_1,x_2)}\left(1 - \frac{x_3}{H}\right)\frac{\partial F(x_1,x_2)}{\partial x_2}$$
$$+\frac{\partial}{\partial x_1}\left[\nu\left(\frac{\partial v}{\partial x_1} + \frac{\partial u}{\partial x_2}\right)\right] + 2\frac{\partial}{\partial x_2}\left(\nu\frac{\partial v}{\partial x_2}\right) + \frac{\partial}{\partial x_3}\left\{\nu\left[\left(1 - \frac{F(x_1,x_2)}{H}\right)\frac{\partial w}{\partial x_2} + \frac{\partial v}{\partial x_3}\right]\right\} - 2\Omega u\sin\varphi,$$

$$= -u\frac{\partial w}{\partial x} - v\frac{\partial w}{\partial y} - w\frac{\partial w}{\partial z} - g\left[1 - \frac{H}{H - F(x_1,x_2)}\right] + \frac{\partial}{\partial x_1}\left[\nu\left(\frac{H}{H - F(x_1,x_2)}\frac{\partial u}{\partial x_3} + \frac{\partial w}{\partial x_1}\right)\right]$$
$$+\frac{\partial}{\partial x_2}\left[\nu\left(\frac{H}{H - F(x_1,x_2)}\frac{\partial v}{\partial x_3} + \frac{\partial w}{\partial x_2}\right)\right] + \frac{2H^2}{\left[H - F(x_1,x_2)\right]^2}\frac{\partial}{\partial x_3}\left(\nu\frac{\partial w}{\partial x_3}\right) + 2\Omega u\cos\varphi.$$

- For the equations of distribution in the atmosphere of scalar values $\mathfrak{H} = (\theta, q, \delta, s)$

$$= -u\frac{\partial \mathfrak{H}}{\partial x_1} - v\frac{\partial \mathfrak{H}}{\partial x_2} - \left(\tilde{w} - w_g\right)\frac{\partial \mathfrak{H}}{\partial x_3} - \sigma\mathfrak{H} + f + \frac{\partial}{\partial x_1}\left(d_T\frac{\partial \mathfrak{H}}{\partial x_1}\right)$$
$$+\frac{\partial}{\partial x_2}\left(d_T\frac{\partial \mathfrak{H}}{\partial x_2}\right) + \frac{H^2}{\left[H - F\left(x_1, x_2\right)\right]^2}\frac{\partial}{\partial x_3}\left(d_T\frac{\partial \mathfrak{H}}{\partial x_3}\right).$$

By approximating the operations of differentiation of the first and second orders, members of the $D\Re\left(r,t\right)$, with respect to the spatial variables x_1, x_2, x_3 with formulas (1.8.70) and (8.1.71), we thus approximate the partial differential equation by the differential difference equations, which represent a system of large number of ordinary differential equations

$$\frac{\partial \Re_{ijk}}{\partial t} = \Lambda\Re_{ijk}, 1 \le i \le I_1, 1 \le j \le J_1, 1 \le k \le K_1, \tag{9.1.25}$$

where Λ is the difference representation of the differential operator *D*.

To determine the state of the atmosphere at the node r_{ijk} of the grid $\bar{\varpi}_h$ in a bounded domain $\bar{\wp}\left(r\right)$ at $\forall t \in \left[t^{n-1}, t^n\right]$ we will solve the problem

$$\frac{\partial \Re_{ijk}}{\partial t} = \Lambda\Re_{ijk}, \forall t \in \left[t^{n-1}, t^n\right], n = 1, 2, \ldots, \tag{9.1.26}$$

$$\Re_{ijk}\left(t^n\right) = \Re_{ijk}^n, \forall t \in \left[t^{n-1}, t^n\right], n = 1, 2, \ldots, \tag{9.1.27}$$

where $\Lambda\Re_{ijk}$ is the numerical value of the right-hand side of equation (9.1.25) at r_{ijk} of the grid $\bar{\varpi}_h$ in $\bar{\wp}\left(r\right)$.

We take without a proof the following obvious statement.

Statement 9.1.1: The initial problem (9.1.1) - (9.1.2) for an ordinary differential equation and the initial problem (1.9.26) - (1.9.27) for the differential-difference equation for specific values of *i,j,k* are equivalent.

Now, if you take the designation $\Re_n = \Re_{ijk}^n$ and $F\left(t\right) = \Lambda\Re_{ijk}$, the problem (1.9.26) – (9.1.27), can be solved with the help of interpolation formulas with multiple nodes (01.09.20).

Recall the basic requirements of the approximation: given an order of accuracy, convergence $\Lambda\Re \to D\Re$ as $h \to 0$, where $h = \left(\Delta x_{1i}, \Delta x_{2j}, \Delta x_{3k}\right)$ is the step of the spatial grid; transportability (disturbance imposed on a function $\Re$ is transferred by convection only in the direction of the velocity). If we put in line the differential operator *D* in (1.9.24) to a grid operator Λ, satisfying these requirements and solve the problem (1.9.26) - (1.9.27) for all specific values of indices $i \in \left(\overline{1, I_1}\right)$, $j \in \left(\overline{1, J_1}\right)$ and $k \in \left(\overline{1, K_1}\right)$, we obtain the solution of the problem (9.1.24).

APPROXIMATION OF FUNCTIONS IN THE MACRO SCALE GRID NODES INTO A MESOSCALE GRID NODES

To use the results of the forecast obtained in the macro-scale model of atmospheric circulation as binding conditions in the regional model, it is necessary to move from discrete values specified in the nodes of large-scale spatial grid ϖ_h of the area $\wp\left(r\right)$, to the discrete values on a fine grid $\bar{\varpi}_h$ of the region $\bar{\wp}\left(r\right)$. That is, we need using the original data in the nodes $\lambda_i, \varphi_j, \sigma_k$ ($1 \le i \le I$, $1 \le j \le J$, $1 \le k \le K$) of the large-scale spatial grid ϖ_h at time points $=t^n$ (n=0,1,2…), to find the value of the function $\Re\left(r,t\right) = \left(\mathcal{R}, \mathrm{u}, \mathrm{v}, \mathrm{w}, \mathfrak{H}\right)$ at the same time points $t=t^n$ at the nodal points r_{ijk}

$$\left(1 \le i \le I_1, 1 \le j \le J_1, 1 \le k \le K_1\right)$$

of the grid $\bar{\varpi}_h$ in a bounded domain $\bar{\wp}\left(r\right)$. The rule, that allows calculating an approximate value of $\Re^n\left(r\right)$ for given discrete values of the function $\Re^n\left(\lambda_i, \varphi_j, \sigma_k\right)$, we call the completion or interpolation (Prusov, Doroshenko, 2006).

The issues of completion or interpolation of functions of one variable have been developed in detail. Currently, based on the results of theoretical studies, there are created interpolation methods characterized by high precision and efficiency (Bahvalov, Zhidkov, Kobel'kov, 1987; Bahvalov, 1975). Therefore, a significant part of the results of theoretical research for the one-dimensional case is automatically transferred to the case of functions of two or more variables, though faced with the fact that there appear not enough effective methods. Besides, complexity problems of solutions dramatically increase with its dimension, so it is generally not possible to develop standard methods for solving a wide range of multidimensional problems with accuracy as high as for one-dimensional case.

In our multi-dimensional problem of the interpolation, the location of grid nodes ϖ_h and $\bar{\varpi}_h$ is regular, so the interpolation functions of several variables can be reduced to the consistent implementation of the operations functions of one variable. Let have η to be one of the axes of the coordinate system $r = \left(\lambda, \varphi, \sigma\right)$ and the segment $a \le \eta \le b$ of the linear dimension of the macro scale area $\wp\left(r\right)$ for this coordinate. Let be given discrete values $f_j = f\left(\eta_j\right)$ of a function $f\left(\eta\right) = \Re^n\left(\lambda, \varphi, \sigma\right)$ in the points $a = \eta_0 < \eta_1 < \ldots < \eta_N = b$, forming a regular macro scale grid $\varpi_h\left[a, b\right]$ with the grid step $h = \left(b - a\right) / \left(N - 1\right)$.

We confine ourselves to a way of constructing the function $f\left(\eta\right)$ by interpolation $\bar{f} = f_j$, $j = \left(\widehat{1, N}\right)$, in the form of a polynomial of degree N

$$f(\eta) = S\overline{f} = \sum_{j=0}^{N} f_j q_j(\eta), \tag{9.2.1}$$

where S is the operator of completion, $q_j(\eta)$ is a polynomial of degree N; particularly, for the Lagrange interpolation formula, the polynomial has the form

$$q_j(\eta) = \frac{(\eta - \eta_0)\ldots(\eta - \eta_{j-1})(\eta - \eta_{j+1})\ldots(\eta - \eta_N)}{(\eta_j - \eta_0)\ldots(\eta_j - \eta_{j-})(\eta_j - \eta_{j+1})\ldots(\eta_j - \eta_N)}.$$

Accuracy of interpolation using an algebraic polynomial (9.2.1) obviously depends on the properties of the function $f(\eta)$, the table of nodes (grid) and finally, on the number of nodes N. The main problem of convergence is the question of how to be interconnected the properties of the function $f(\eta)$ and a table of nodes to provide in the adopted approach a convergence of polynomial (9.2.1) to $f(\eta)$. To answer this question, it is necessary a thorough study of the tendency to zero of the error measure $\varepsilon_n(f,\eta)$. To date, in the published works devoted to the issue of convergence of interpolation processes, the necessary and sufficient conditions are found to be satisfied by a table of nodes to provide the convergence process on the set of functions of any fixed order of smoothness of (Nicholas, 1977; Reinsch, 1967). Likewise, for certain types of grids, classes of functions are found in which these grids lead to convergent interpolation process. We don't give even a basic system analysis of published results, not so much because of their large numbers, but because of the impossibility of their application. This is due to the fact that in finding unknown numerical values of any magnitude in the mesoscale grid nodes from the known values of the nodes of its macro-scale grid, interpolation parameters remain uncontrolled. Indeed, the macro scale grid (size of elements and the number of nodes), as well as the location of mesoscale region in the macro scale area defined by a general problem and cannot be changed by the interpolation task. In addition, the properties of the interpolated function are unknown a priori.

For surely convergent interpolation process, we construct the piecewise polynomial interpolation operator S^*, putting in line the vector $\overline{f} = (f_j)_{j=0}^{N}$ on $[\eta_j, \eta_{j+1}]$, $j = \left(\overline{1, N-1}\right)$ and the function $f(\eta) = S^*\overline{f}$, equal to

$$f(\eta) = S^*\overline{f} = \sum_{j=0}^{n} f_j q_j^*(\eta), \tag{9.2.2}$$

where $q_j^*(\eta)$ is a polynomial of degree $n \ll N$.

The splines are the simplest example of the piecewise polynomial function (9.2.2) (Bahvalov, Zhidkov, Kobel'kov, 1987).

Approximation of Functions of the Macro Scale Grid Nodes into the Horizontal Mesoscale Grid Nodes

Let *F* is an initial complete normed (Banach) space (Richtmyer, Morton, 1967; Trenogin, 1980), and $\bar{F}$ is another Banach space, approximating *F*. Let relationship between *F* and $\bar{F}$ is performed by a linear operator $T \in \lambda\left(F, \bar{F}\right)$ with the range of values $R\left(T\right) = \bar{F}$. The operator *T* will be called the narrowing operator, as in this situation, the typical case is $H\left(T\right) \neq \left\{0\right\}$ (zeros subspace of *T* is not trivial). Otherwise, *T* would have both a right inverse $S = T_r^{-1}$ (what follows from $TS = I_{\bar{F}}$), and left inverse $S = T_L^{-1}$ (what follows from $TS = I_F$) and would continuously invertible by the Banach Theorem (Banach, 1932).

The operator $S \in \lambda\left(\bar{F}, F\right)$ will be called the interpolation operator (completion operator) if $H\left(S\right) = \left\{0\right\}$. In the general case, this terminology is justified by the fact that the operator *S* acts of a "narrow" space $\bar{F}$ to "broad" space *F* (Prusov, Doroshenko, 2006).

The condition $R\left(T\right) = \bar{F}$ means that the operator *T* has a right inverse $S = T_r^{-1}$ that is not, however, unique. It is, therefore, natural to raise the question of the optimal, in one sense or another, choice of the operator *S*.

Let us consider further the Hilbert space *Y* and the operator ($A \in \lambda\left(F, Y\right)$. We call $s \in F$ the interpolation spline ((*T*,*A*)-spline) corresponding to the vector input $\bar{f} \in \bar{F}$, if

$$As_Y^2 = \inf_{f \in T^{-1}\left(\bar{f}\right)} Af^2. \tag{9.2.3}$$

Here, $T^{-1}\left(\bar{f}\right)$ denotes the set of preimages of the element $\bar{f}$ when mapping *T*, i.e., a set of solutions of the equation

$$Tf = \bar{f}. \tag{9.2.4}$$

Denote as usual, by *H*(*T*) and *H*(*A*) null spaces of the operators *T* and *A*. Note that *H*(*T*) and *H*(*A*) lie in *F*.

Theorem 9.2.1: If the linear manifold *AH*(*T*) is closed, and $H\left(A\right) \cap H\left(T\right) = \left\{0\right\}$, then for each $\bar{f} \in F$ there exists a unique interpolation (*T*,*A*)-spline *s*.

Proof: Since $R\left(T\right) = \bar{F}$ by the theorem's hypothesis, then for every element $\bar{f} \in \bar{F}$ there is an element f_0 such that $Tf_0 = \bar{f}$. But then the general solution of equation (9.2.4) has the form $f = f_0 + \phi$, where ϕ is an arbitrary element of *H*(*T*). Thus, the problem consists in determining the distance in the Hilbert space *Y* from zero to affine manifold

$$M = \left\{Af_0 + A\phi : \phi \in H\left(T\right)\right\}.$$

This diversity is closed by the hypothesis:

$$M = AT^{-1}(\bar{f}) = Af_0 + AH(T).$$

Then, as it is easy to show (Riesz's theorem) there exists a unique vector $y \in Y$ equal to the distance from the point 0 to the manifold M, i.e. that $y = \rho(0, M)$.

Since $y \in AT^{-1}(\bar{f})$ is a closed set, then according to (9.2.3) and (9.2.4), the spline s, if it exists, is the general solution of two equations:

$$As = y, Ts = \bar{f}. \tag{9.2.5}$$

Let us prove the uniqueness of the spline s. Let some $\bar{f}$ matches the two splines s_1 and s_2. According to (9.2.5), for the difference s_1-s_2 there is a system of equations

$$A(s_1 - s_2) = 0, T(s_1 - s_2) = 0.$$

Consequently, $s_1 - s_2 \in H(A)$ and $s_1 - s_2 \in H(T)$. But by the hypothesis, $H(A) \cap H(T) = \{0\}$. It follows that s_1=s_2 what was necessary to prove.

For particular applications of this theorem, the following statements will be useful.

Lemma 9.2.1: Let F is a unit ball that is weakly compact (e.g., F is reflexive or is Hilbert space); let also A is normally solvable, its zero subspace $H(A)$ is finite, and $H(A) \cap H(T) = \{0\}$. Then $AH(T)$ is closed.

Proof: Let $\{y_j\} \subset AH(T)$, and $y_j \to y_0$ when $j \to \infty$. This means that there are elements $f_j \in H(T)$, $Ax_j = y_j$, $j = 1,2,\ldots$. We expand the space F into a direct sum $F = H(A) + \tilde{F}$. Then the restriction $\tilde{A}$ of the operator A to $\tilde{F}$ is a continuously invertible operator. Assuming $f_j = \phi_j + \bar{f}_j$, where $\phi_j \in H(A)$, $f_j \in \tilde{F}$, and noting that $\tilde{A}f_j = y_j$, $j = 1,2,\ldots$, we obtain $\bar{f}_j = \tilde{A}^{-1}y_j$, i.e., $\{\tilde{f}_j\}$ is limited.

Next, $T\phi_j = -T\bar{f}_j$, since Tf_j=0. Let $\tilde{T}$ be the restriction of T to $H(A)$, find $\phi_j = -\tilde{T}^{-1}T\bar{f}_j$, from where $\{\phi_j\}$ is also limited. Consequently, $\{f_j\}$ is also limited. Let $\{f_n\}$ is its weakly convergent subsequence and f_0 is its limit. Then at $n \to \infty$ we have

$$Af_n = y_n \to Af_0 = y_0 \text{ weakly}$$

$$0 = Tf_n \to Tf_0 \text{ weakly.}$$

Consequently, $y_0 \in AH(T)$ and Lemma 9.2.1 is proved.

Lemma 9.2.2: The rank of the matrix $(\omega_i \psi_j)$, $i = 1,2,\ldots,m, j = 1,2,\ldots,n$ equals *m*.
Proof: Consider a homogeneous system of *n* linear equations with *m* unknowns

$$\sum_{i=1}^{m} \alpha_i \omega_i \psi_j = 0, j = 1,2,\ldots,n. \tag{9.2.6}$$

The rank of its matrix is *r*≤*m*. Suppose that *r*<*m*. Then the system (9.2.6) has a nontrivial solution $(\alpha_1^0, \alpha_2^0, \ldots, \alpha_m^0)$. Consider the vector

$$\phi = \sum_{i=1}^{m} \alpha_i^0 \omega_i \in H(A).$$

Obviously, $\phi \neq 0$ and $\phi \in H(T)$, for $\phi, \psi_j = 0$, $j = 1,2,\ldots,n$, and $(\alpha_1^0, \alpha_2^0, \ldots, \alpha_m^0)$ is a solution of the system of linear equations (9.2.6). If the condition of Theorem $H(A) \cap H(T) = \{0\}$, we arrive at a contradiction. Therefore, *r*=*m*, and Lemma 9.2.2 is proved.

Lemma 9.2.3: Let $\Im$, a linear span of the elements of the system $\{\psi_j\}_1^n$, and $H(A)^\perp$, an orthogonal complement of zero subspace *H*(*A*) are subspaces of F^*. Then $\dim\left\{\Im \cap H(A)^\perp\right\} = n - m$.
Proof: Let

$$\psi = \sum_{j=1}^{n} \lambda_j \psi_j \in H(A)^\perp,$$

where λ_j is scalar. This means that $\omega_i \psi_j = 0$, $j = 1,2,\ldots,n$. To determine $\lambda_1, \lambda_2, \ldots, \lambda_n$, obtain the following homogeneous system of *m* equations in *n* unknowns:

$$\sum_{j=1}^{n} \omega_i \psi_j \lambda_j = 0, \quad i = 1,2,\ldots,m.. \tag{9.2.7}$$

The matrix of the system (9.2.7) is the transpose of the matrix system (9.2.6), and therefore, its rank is equal to *m*. But then the system (9.2.7) has *n-m* linearly independent solutions. Lemma 9.2.3 is proved.

We fix in following that $\{\zeta_k\}_1^{n-m}$ is a basis in $\Im \cap H(A)^\perp$. Notice, that

$$\zeta_k = \sum_{j=1}^{n} \lambda_{kj}\psi_j, k = 1,2,\ldots,n-m, \tag{9.2.8}$$

where $\left(\lambda_{k1},\lambda_{k2},\ldots,\lambda_{kn}\right)$, $k = 1,2,\ldots,n-m$ is a fixed set of linearly independent solutions of the system (9.2.7). We have seen above that if *s*-spline, then $As \in \left\{AH\left(A\right)\right\}^{\perp}$.

Lemma 9.2.4: Between the manifolds $\left\{AH\left(A\right)\right\}^{\perp}$ and $\Im \cap H\left(A\right)^{\perp}$, there is a link $\left\{AH\left(A\right)\right\}^{\perp} = \left(A^{*}\right)^{-1}\left\{\Im \cap H\left(A\right)^{\perp}\right\}$.

Proof: The lemma is equivalent to the following equation $A^{*}\left\{AH\left(A\right)\right\}^{\perp} = \Im \cap H\left(A\right)^{\perp}$. It is easy to show that $H\left(A\right)^{\perp} = R\left(A^{*}\right)$. If $\omega \in \left(A^{*}\right)\left\{AH\left(A\right)\right\}^{\perp}$, then there exists $y \in \left\{AH\left(T\right)\right\}^{\perp}$ such that $\omega = A^{*}y$. Consequently, $y \in H\left(A\right)^{\perp} = R\left(A^{*}\right)$. We show that $\psi \in \Im$. Let $\phi \in H\left(T\right)$, then $\phi,\psi = \phi, A^{*}y = A\phi, y = 0$ since $y \in \left\{AH\left(T\right)\right\}^{\perp}$, and the inclusion $A^{*}\left\{AH\left(T\right)\right\}^{\perp} \subset \Im \cap H\left(A\right)^{\perp}$ holds.

Conversely, if $\psi \in \Im$ and $\psi \in H\left(T\right)^{\perp}$, then $\psi \in R\left(A^{*}\right)$, i.e. $\psi = A^{*}y$. Herewith, $\phi,\psi = 0$ for $\forall \phi \in H\left(T\right)$, i.e. $0 = \phi, A^{*}y = A\phi, y$, $y \in \left\{AH\left(T\right)\right\}^{\perp}$, and the reverse inclusion is proved, and along with Lemma 9.2.4.

This theorem and Lemmas allow to specify a method for constructing interpolation splines. According to Theorem 9.2.1 and Lemma 9.2.1, if the following conditions are implemented:

1. $R\left(A\right) = Y$ (if *A* is normally solvable, it is sufficient to replace *Y* by *R*(*A*));
2. *H*(*A*) is finite;
3. $\bar{F}$ is finite;
4. $\dim H\left(A\right) = m < n = \dim \bar{F}$;
5. $H\left(A\right) \cap H\left(T\right) = \left\{0\right\}$;
6. The unit ball *F* is weakly compact, the problem of spline interpolation has a unique solution.

Lemmas 9.2.2-9.2.4 allow to formulate the following procedure of calculation of the spline *s*. Let $\left\{\theta_j\right\}_1^m$ is the basis in *H*(*A*) and $\left\{\psi_j\right\}_1^n \subset F^{*}$ is linearly independent system of functional, such that

$$H\left(T\right) = \left\{f : f, \psi_j = 0, j = 1,2,\ldots,n\right\}. \tag{9.2.9}$$

If $\left\{\bar{e}_j\right\}_1^n$ denotes a basis of *F*, then we can write

$$Tf = \sum_{j=1}^{n} f, \psi_j \bar{e}_j . \tag{9.2.10}$$

Let $\{\zeta_k\}_1^{n-m}$ is a basis in $\Im \cap H(A)^{\perp}$. Then, according to Lemma 9.2.4, $\{\vartheta_k\}_1^{n-m} = \left\{(A^*)^{-1} \zeta_k\right\}_1^{n-m}$ is a basis in $\{AH(T)\}^{\perp}$, i.e. in the subspace where the vector *As* lies. Then the element *As*=*y* can be found in the form of an expansion in the basis $\{\vartheta_k\}_{k=1}^{n-m}$:

$$As = \sum_{k=1}^{n-m} \beta_k \vartheta_k . \tag{9.2.11}$$

Multiplying this equation by the vector ϑ_L, we obtain, using (9.2.8)

$$\sum_{k=1}^{n-m} \beta_k (\vartheta_k, \vartheta_L) = (As, \vartheta_L) = s, A^* \vartheta_L = s, \zeta_L = s, \sum_{j=1}^{n} \lambda_{Lj} \psi_j = \sum_{j=1}^{n} \lambda_{Lj} s, \psi_j .$$

Since, according to the formula (9.2.10)

$$Ts = \sum_{j=1}^{n} s, \psi_j \bar{e}_j = \bar{f} , \tag{9.2.12}$$

then s, ψ_j, $j = 1, 2, \dots, n$ are the known coordinates η_j of the vector of input data

$$\bar{f} = \sum_{j=1}^{n} \eta_j \bar{e}_j .$$

Thus, to determine the number β_k, we obtain the system of equations

$$\sum_{k=1}^{n-m} \beta_k (\vartheta_k, \vartheta_L) = \sum_{j=1}^{n} \lambda_{Lj} \eta_j, L = 1, 2, \dots, n-m \tag{9.2.13}$$

with Gram matrix, which has a unique solution, and then find a vector

$$As = C\bar{f} , \tag{9.2.14}$$

where C is a linear operator acting from F to $\left\{AH\left(T\right)\right\}^{\perp}$. For the definition of the spline s it is necessary to solve simultaneously the equations (9.2.12) and (9.2.14). This problem has a unique solution (see Eq. (9.2.6)). Thus, each element $\bar{f} \in F$ is associated with a specific element $s \in F$, and, hence, it is defined a bounded linear operator $s = S^{*}\bar{f}$, i.e. the construction of the extension optimal operator is completed.

The specific polynomial splines are treated similarly. Below, we will need the representation of private polynomial splines to fill functions having on the segment of interpolation generalized derivatives of first and second order squared integral (Prusov, Doroshenko, 2006).

Example 9.2.1: Let $F = H^{1}\left(0, L\right)$, $\bar{F} = c^{n+1}$, $Tf\left(\eta\right) = \left\{f\left(\eta_{j}\right)\right\}_{j=0}^{n}$, where $\left\{\eta_{j}\right\}_{0}^{n}$ is a grid on $\left[0, L\right]$.

As the extension operator S^{*}, take the piecewise linear interpolation operator, assigning to a vector $\bar{f} = \left\{f_{j}\right\}_{j=0}^{n} \in c^{n+1}$ the function $s_{1}\left(\eta\right) = S^{*}\bar{f}$ equal at $\left[\eta_{j}, \eta_{j+1}\right]$, $j = 1, 2, \ldots, n-1$, to

$$s_{1}\left(\eta\right) = \frac{\eta - \eta_{j+1}}{\eta_{j} - \eta_{j+1}} f_{j} + \frac{\eta - \eta_{j}}{\eta_{j+1} - \eta_{j}} f_{j+1} \equiv f_{j} + \frac{\eta - \eta_{j}}{\eta_{j+1} - \eta_{j}}\left(f_{j+1} - f_{j}\right). \tag{9.2.15}$$

We show that piecewise linear interpolation in S^{*} (9.2.15) is optimal in the following sense of the minimum of a quadratic functional

$$J = \int_{0}^{L} f'^{2}\left(\eta\right) d\eta \tag{9.2.16}$$

with the proviso that $f\left(\eta_{j}\right) = f_{j}$, $j = 1, 2, \ldots, n$, is attained at the function $s_{1}\left(\eta\right) = S^{*}\bar{f}$ and only at it (see. (9.2.15)). Since

$$s_{1}'\left(\eta\right) = \left(f_{j+1} - f_{j}\right) / \left(\eta_{j+1} - \eta_{j}\right) \text{ на } \left[\eta_{j}, \eta_{j+1}\right],$$

then for any function $\phi \in H^{1}\left(0, L\right)$ such that $T\phi = 0$, i.e. $\phi\left(\eta_{j}\right) = 0$, $j = 1, 2, \ldots, n$, we have

$$\int_{0}^{L} s_{1}'\left(\eta\right)\phi\left(\eta\right) d\eta = \sum_{j=0}^{n-1} \frac{f_{j+1} - f_{j}}{\eta_{j+1} - \eta_{j}} \int_{0}^{L} \phi'\left(\eta\right) d\eta = 0. \tag{9.2.17}$$

However, for any function $f\left(\eta\right) = s_{1}\left(\eta\right) + \phi\left(\eta\right)$ inequality holds

$$J\left(f\right) = \int_{0}^{L} f'^{2}\left(\eta\right) d\eta = \int_{0}^{L} \left(s_{1}' + \phi'\right)^{2}\left(\eta\right) d\eta = \int_{0}^{L} s_{1}'^{2}\left(\eta\right) d\eta + \int_{0}^{L} \phi'^{2}\left(\eta\right) d\eta \geq \int_{0}^{L} s_{1}'^{2}\left(\eta\right) d\eta = J\left(s\right).$$

This means that the infimum d^2 of the functional $J(f)$ is achieved at function $s_1(\eta)$. Now assume that this lower bound is also achieved on the function $\hat{s}(\eta)$, where $T\hat{s}=0$. Take $\phi(\eta)=\hat{s}_1(\eta)+s_1(\eta)$. Then, according to (9.2.17), we obtain

$$\left(s_1',\hat{s}_1'\right)=\int_0^L s_1'(\eta)\hat{s}'(\eta)\,d\eta=\int_0^L s_1'^2(\eta)\,d\eta=d^2=\int_0^L \hat{\phi}'^2(\eta)\,d\eta,$$

or shorter $\left(s_1',\hat{s}_1'\right)=s_1'^2=\hat{s}_1'^2$. It follows $\hat{s}_1(\eta)=s_1(\eta)$, and since, for example, $\hat{s}_1(\eta)=s_1(\eta)$, then $\hat{s}_1(\eta)\equiv s_1(\eta)$ on [0,*L*]. Thus, the piece-wise linear function $s_1(\eta)$ is the only function in $H^1(0,L)$ interpolating $\bar{f}$ and minimizing $J(f)$.

We introduce the characteristic function $\sigma_j(\eta)$ such that on the interval $\left[\eta_j,\eta_{j+1}\right]$ and outside $\sigma_j=0$. Then the optimal extension operator S^* can be written as follows:

$$s_1(\eta)=S^*\bar{f}=\sum_{j=0}^{n-1}\left[f_j+\frac{\eta-\eta_j}{\eta_{j+1}-\eta_j}\left(f_{j+1}-f_j\right)\right]\sigma_j(\eta). \tag{9.2.18}$$

Example 9.2.2: Let $F=H^2(0,L)$, $Y=\lambda_2(0,L)$ $\bar{F}=I_2^{n+1}$, $Tf(\eta)=\left\{f(\eta_j)\right\}_{j=0}^n$ $Af(\eta)=f''(\eta)$, where $\left\{\eta_j\right\}_0^n$ is a grid on [0,*L*]. We state the problem of finding the function $s(\eta)\in H^2(0,L)$, implementing the minimum of the functional

$$J_2(f)=\int_0^L\left[f''(\eta)\right]^2 d\eta. \tag{9.2.19}$$

Verify the implementation of the set of conditions 1) -5) formulated above. The range of values of the operator $d^2/d\eta^2$ is $\lambda_2(0,L)$; *H*(*A*) is two-dimensional subspace of polynomials $c_0+c_1\eta$, i.e., of the dimension 2, and $\bar{F}$ is $(n+1)$-dimensional. For *n*>1 the conditions 4) and 5), as *H*(*T*) consists of functions that vanish at *n* points, what is impossible for a non-zero polynomial of the first degree. Finally, condition 6) holds, since $H^2(0,L)$ is a Hilbert space. Therefore, the variational problem (9.2.19) has a unique solution $s_2(\eta)$, which is called interpolation cubic spline corresponding to the input vector $\bar{f}$.

Also suitable is the procedure for constructing spline function described above. Functional $\left\{\psi_j\right\}_{j=0}^n$ on the elements $f(\eta)\in H^2(0,L)$ are set in this case by the formulas

$$f,\psi_j=f(\eta_j), j=1,2,\ldots,n.$$

Further, it is easy to check that the basis $\left\{\zeta_k\right\}_{k=0}^{n-2}$ in $\Im \cap H\left(A\right)^{\perp}$ can be selected as follows:

$$\zeta_s = \frac{2}{\eta_{s+1}-\eta_{s-1}}\left(\frac{\psi_{s+1}-\psi_s}{\eta_{s+1}-\eta_s}+\frac{\psi_s-\psi_{s-1}}{\eta_s-\eta_{s-1}}\right),$$

$$s = 1,2,\ldots,n-1.. \tag{9.2.20}$$

The operator $A^* = \left(d^2/d\eta^2\right)^*$ is a bounded linear operator from $\lambda_2\left(0,L\right)$ in $H^2\left(0,L\right)$; $H\left(A^*\right)=\left\{0\right\}$ and $R\left(A^*\right)$ consists of functions that are orthogonal to 1 and η. Therefore, $\zeta_s \in R\left(A^*\right)$ and the functions

$$\vartheta_s\left(\eta\right)=\left(A^*\right)^{-1}\zeta_s, s=1,2,\ldots,n-1,$$

forming the basis of $\left\{AH\left(T\right)\right\}^{\perp}$. Since for any $f\left(\eta\right)\in H^2\left(0,L\right)$ we have the equality

$$\left(\vartheta_s, Af\right)_{\lambda_2(0,L)} = \left(A^*\vartheta_s, f\right)_{\lambda_2(0,L)} = \left(\zeta_s, f\right),$$

the function $\vartheta_s\left(\eta\right)$ is given by the following explicit formulas

$$\vartheta_s = \frac{2}{\eta_{s+1}-\eta_{s-1}}\left[\frac{\left(\zeta_{s+1}-\zeta\right)_+-\left(\zeta_s-\zeta\right)_+}{\eta_{s+1}-\eta_s}+\frac{\left(\zeta_s-\zeta\right)_+-\left(\zeta_{s-1}-\zeta\right)_+}{\eta_s-\eta_{s-1}}\right], \tag{9.2.21}$$

where $\left(*\right)_+=\left(*\right)$ with $\left(*\right)>0$, and $\left(*\right)_+=0$ for $\left(*\right)\leq 0$.

It is easy to show that for any function $f\left(\eta\right)\in H^2\left(0,L\right)$ and functions given by formulas (9.2.21) the following equality holds

$$\int\limits_0 \vartheta_s\left(\eta\right)f''\left(\eta\right)d\eta = \frac{2}{\eta_{s+1}-\eta_{s-1}}\left[\frac{f\left(\eta_{s+1}\right)-f\left(\eta_s\right)}{\eta_{s+1}-\eta_s}+\frac{f\left(\eta_s\right)-f\left(\eta_{s-1}\right)}{\eta_s-\eta_{s-1}}\right].$$

Finally, one can calculate scalar products $\left(\vartheta_k,\vartheta_L\right)_{\lambda_2(0,L)}$, and hence the system matrix (9.2.13). If we denote $h_j=\eta_{j+1}-\eta_j$, $j=1,2,\ldots,n-1$ the step grid, it is easy to show that $\left(\vartheta_k,\vartheta_k\right)=\left(h_{k+1}+h_k\right)/3$, $\left(\vartheta_k,\vartheta_{k-1}\right)=h_k/6$, $\left(\vartheta_k,\vartheta_{k+1}\right)=h_{k+1}/6$, and $\left(\vartheta_k,\vartheta_l\right)=0$, for $l\neq k-1,k,k+1$. Thus, the system (9.2.13) has the form

$$\begin{bmatrix} (h_0+h_1)/3 & h_1/6 & 0 \dots & & \dots 0 \\ h_1/6 & (h_0+h_1)/3 & h_1/6 \dots & & \dots 0 \\ & \dots & \dots & & \\ 0 \dots & & \dots (h_{n-2}+h_{n-1})/3 & h_{n-1}/6 & \end{bmatrix} \begin{bmatrix} \beta_1 \\ \beta_2 \\ \dots \\ \beta_{n-1} \end{bmatrix} = \begin{bmatrix} (f_2-f_1)/h_1-(f_1-f_0)/h_0 \\ (f_3-f_2)/h_2-(f_2-f_1)/h_1 \\ \dots\dots\dots\dots\dots\dots \\ (f_n-f_{n-1})/h_{n-1}-(f_{n-1}-f_{n-2})/h_{n-2} \end{bmatrix}. \quad (9.2.22)$$

Since the coefficients of the system (9.2.22) form the tridiagonal matrix, the system (9.2.13) can be easily solved by the sweep method.

In accordance with (9.2.11), we have

$$s''(\eta)=\sum_{k=1}^{n-1}\beta_k\vartheta_k(\eta), \quad (9.2.23)$$

and since $\vartheta_k(\eta_l)=\delta_{kl}$, $k,l=1,2,\dots,n-1$, then

$$s''(\eta_j)=\beta_j;\ j=1,2,\dots,n-1;\ s''(0)=s''(L)=0. \quad (9.2.24)$$

Formulas (9.2.21) and (2.9.23) allow to find for the segment $[\eta_j,\eta_{j+1}]$ the dependence

$$s''(\eta)=\frac{1}{h_j}\left[(\eta_{j+1}-\eta)\beta_j+(\eta-\eta_j)\beta_{j+1}\right],$$

which, considering the boundary conditions (2.9.24), leads to the formula that defines a spline $s(\eta)$ on the interval $[\eta_j,\eta_{j+1}]$

$$s(\eta)=\frac{1}{6h_j}\left[(\eta_{j+1}-\eta)^3\beta_j+(\eta-\eta_j)^3\beta_{j+1}\right]+\frac{1}{h_j}\left[(\eta_{j+1}-\eta)\left(f_j-\frac{\beta_jh_j^2}{6}\right)+(\eta-\eta_j)\left(f_{j+1}-\frac{\beta_{j+1}h_j^2}{6}\right)\right]. \quad (9.2.25)$$

Similarly, the splines can be considered that are generated by a variety of differential operators.

RESTORING FUNCTIONS, DEFINED IN THE MACRO SCALE GRID NODES, INTO THE VERTICAL MESOSCALE GRID NODES

While solving problems of the weather forecast in a limited area, to ensure the desired accuracy near the underlying surface, where the vertical change meteorological quantities have the greatest gradients, it is desirable to use a grid with fine vertical steps. On the other hand, in order to save calculation time and computer memory, it is advisable to use a coarser grid away from the solid surface. Therefore, solving these problems should apply the mesoscale irregular (variable steps) vertical grid. However, the vertical grid nodes of the macro scale predictive models are usually found in standard geopotential surfaces: z_0, 850, 700, 500, … hPa. Obviously, this grid is too coarse for mesoscale applications. So, we meet again with the need to fill the data given on the macro-scale grid nodes into the irregular mesoscale grid. We have already noted that there is no interpolation formula that would ensure both the uniqueness and the required accuracy of the solution interpolation problem especially on the segment [z_0, 850], i.e. in the boundary layer of the atmosphere, where all the processes of the turbulent exchange occur. Therefore, to compensate for the data given on the macro-scale vertical grid nodes in the irregular vertical mesoscale grid, it is necessary to use a special procedure for computing interpolation of the macro scale forecast data in the computational grid nodes.

We take in the field $\bar{\wp}\left(r\right)$ of the regional task the condition of hydrostatic stability and horizontal homogeneity of meteorological fields in the free atmosphere, i.e. above the boundary layer. The main factors that affect the interaction of the atmosphere and the underlying surface are:

- The turbulent exchange.
- Forced convective motions generated by the horizontal movement of air over the uneven topography of the underlying surface.
- Free convective motions caused by the lifting force of Archimedes in the thermal heterogeneity of the underlying surface.

Let the components u,v,w of the velocity $\mathbf{V}$ are associated with the coordinates

$$x = x_1, y = x_2, \sigma = H\left[x_3 - F\left(x,y\right)\right] / \left[H - F\left(x,y\right)\right]$$

as follows: u is in the direction of growth of x (East direction); v is in the direction of increasing of y (North direction); w is in the direction of growth of σ (the direction opposite to the gravitational force, that is, upward). In view of the simplifying assumptions made above, the system of equations (5.1.7) – (5.1.9) as a model of vertical atmospheric column takes the form (Prusov, Doroshenko, 2006):

$$\frac{\partial u}{\partial t} + \bar{w}\frac{\partial u}{\partial \sigma} = \frac{H^2}{\left(H-F\right)^2}\frac{\partial}{\partial \sigma}\left(\nu\frac{\partial u}{\partial \sigma}\right) + f\left(v - v_g\right) - g\frac{H-\sigma}{H}\frac{\partial F}{\partial x},$$

$$\frac{\partial v}{\partial t}+\bar{w}\frac{\partial v}{\partial \sigma}=\frac{H^2}{\left(H-F\right)^2}\frac{\partial}{\partial \sigma}\left(\nu\frac{\partial v}{\partial \sigma}\right)-f\left(u-u_g\right)-g\frac{H-\sigma}{H}\frac{\partial F}{\partial y},$$

$$\frac{\partial w}{\partial t}+\bar{w}\frac{\partial w}{\partial \sigma}=\frac{H^2}{\left(H-F\right)^2}\frac{\partial}{\partial \sigma}\left(\nu\frac{\partial w}{\partial \sigma}\right)-\frac{g}{\theta}\frac{H}{H-F}\frac{\partial \theta}{\partial \sigma}\cdot\sigma, \qquad (9.2.26)$$

$$\frac{\partial \theta}{\partial t}+\bar{w}\frac{\partial \theta}{\partial \sigma}=\frac{H^2}{\left(H-F\right)^2}\frac{\partial}{\partial \sigma}\left(\nu\frac{\partial \theta}{\partial \sigma}\right)+L\left(\alpha\frac{\partial q}{\partial t}\right)+Q_{rad},$$

$$\frac{\partial q}{\partial t}+\bar{w}\frac{\partial q}{\partial \sigma}=\frac{H^2}{\left(H-F\right)^2}\frac{\partial}{\partial \sigma}\left(\nu\frac{\partial q}{\partial \sigma}\right)-\alpha\frac{\partial q}{\partial t},$$

$$\frac{\partial \delta}{\partial t}+\bar{w}\frac{\partial \delta}{\partial \sigma}=\frac{H^2}{\left(H-F\right)^2}\frac{\partial}{\partial \sigma}\left(\nu\frac{\partial \delta}{\partial \sigma}\right)+\alpha\frac{\partial q}{\partial t},$$

$$\frac{\partial s}{\partial t}+\bar{w}\frac{\partial s}{\partial \sigma}=\frac{H^2}{\left(H-F\right)^2}\frac{\partial}{\partial \sigma}\left(\nu\frac{\partial s}{\partial \sigma}\right)+Q_s.$$

where t is the time, performing the role of the iteration parameter; u_g and v_g are values of the horizontal components of the wind speed at the height $\sigma = H$ (geostrophic wind components), $f = 2\Omega\sin\varphi$ is the Coriolis parameter; φ is the latitude; L is the latent heat of condensation; $\alpha = \left(0,1\right)$ is a pointer to condensation in an atmosphere of sub-grid scale; Q_{rad} is the rate of radiative cooling or heating; Q_s is the source/sink impurity concentration in the atmosphere; ν is a coefficient of the turbulent viscosity and diffusion at the Prandtl and Schmidt numbers $Pr = Sc = 1$.

To close the system of equations of the averaged motion (9.2.26) containing terms with the turbulent viscosity coefficient ν, use additional equation (6.5.13) that models the turbulence. Model (6.5.13) is applicable in this case, since, first, the background of the process is not important, and, second, it is conceptually very simple, very convenient to use, and, as shown in Section 8.1.2, does not cause computational difficulties.

In view of the above simplifications made in the task of the mesoscale model of the vertical atmospheric column, the equation (6.5.13)

$$\nu=\chi\frac{\left|\mathrm{rot}\mathbf{V}\right|^2}{\left[\nabla\left(\nabla\cdot\mathbf{V}\right)-\Delta\mathbf{V}\right]^2}\sqrt{\left(\mathrm{rot}\mathbf{V}\right)^2},$$

takes the form

$$\nu = \chi\left(1-\frac{F}{H}\frac{\left(\frac{\partial u}{\partial \sigma}\right)^2+\left(\frac{\partial v}{\partial \sigma}\right)^2}{\left(\frac{\partial^2 u}{\partial \sigma^2}\right)^2+\left(\frac{\partial^2 v}{\partial \sigma^2}\right)^2}\times\left[\left(\frac{\partial u}{\partial \sigma}\right)^2+\left(\frac{\partial v}{\partial \sigma}\right)^2\right]^{\frac{1}{2}}\right). \tag{9.2.27}$$

In the stratified atmosphere, the coefficient of turbulent viscosity ν, which depends on the vertical gradient of the horizontal component of the wind velocity vector, is also subject to buoyancy forces. We reflect the influence of buoyancy forces on turbulence by inclusion to the equation (9.2.27) additional term f_b describing the buoyancy force (Prusov, Doroshenko, 2006)

$$f_b = \begin{cases}\left(1-\varsigma_1 \mathrm{Ri}\right)^2 & on\ condition\ Ri > 0,\\ \left(1-\varsigma_2 \mathrm{Ri}\right)^{-1/2} & on\ condition\ Ri < 0,\end{cases} \tag{9.2.28}$$

where $\varsigma_1 \approx 7$, $\varsigma_2 \approx 14$ are empirical constants; Ri is the Richardson number, defined by the expression

$$\mathrm{Ri} = -\frac{g}{\rho}\frac{\partial \rho / \partial \sigma}{\left(\frac{\partial^2 u}{\partial \sigma^2}\right)^2+\left(\frac{\partial^2 v}{\partial \sigma^2}\right)^2}.$$

As a result, the equation for the turbulent viscosity coefficient ν will take the final shape:

$$\nu = \chi f_b\left(1-\frac{F}{H}\frac{\left(\frac{\partial u}{\partial \sigma}\right)^2+\left(\frac{\partial v}{\partial \sigma}\right)^2}{\left(\frac{\partial^2 u}{\partial \sigma^2}\right)^2+\left(\frac{\partial^2 v}{\partial \sigma^2}\right)^2}\times\left[\left(\frac{\partial u}{\partial \sigma}\right)^2+\left(\frac{\partial v}{\partial \sigma}\right)^2\right]^{\frac{1}{2}}\right). \tag{9.2.29}$$

An effective method of high order of accuracy for the approximation of partial derivatives of the first (8.1.70) and the second (8.1.71) orders, which are an integral part of the equation (9.2.29), using the boundary conditions (8.1.76a) and (8.1.76b), gives the ability to determine the value of $\nu(\sigma)$ with a third order of accuracy on an irregular grid.

The above system of equations (9.2.26) – (9.2.29) can be integrated numerically for the atmospheric layer from the Earth's surface $\sigma = z_0$ to the upper boundary of the solution area $\sigma = H$.

The boundary conditions are taken as follows:

- With $\sigma = H, u = u_g, v = v_g, w = 0, \mathfrak{H} = \mathfrak{H}_H$.
- When $\sigma = z_0, u = v = w = 0, \mathfrak{H} = \mathfrak{H}_0$.

To increase the accuracy and effectiveness of the solutions of equations (9.2.26) – (2.9.29), we use a vertical grid (Prusov, Doroshenko, 2006)

$$\frac{\sigma}{H} = 1 - \frac{\ln\left\{\left[\beta + 1 - \left(x_3 / H\right)\right] / \left[\beta - 1 + \left(x_3 / H\right)\right]\right\}}{\ln\left[\left(\beta + 1\right) / \left(\beta - 1\right)\right]}, \tag{9.2.30}$$

that includes K counting levels, with the uneven distribution of the value of the grid steps ς_k $\left(k = 1, 2, \ldots, K\right)$.

The values of β are set in the range of $1 < \beta < \infty$. The proposed formula allows to place the greater number of nodes near $\sigma = 0$, the closer the value of the parameter β to one.

The values of the meteorological variables in the nodes of an irregular vertical computing grid (9.2.30), located at an altitude of at least 100 m above the terrain, i.e. for $\forall \sigma_k > 100$ m, are taken on the basis of the data located at the nodes of the regional horizontal grid. They are obtained by filling the macro scale data in the horizontal components of the regional grid, that are given on geopotential surfaces with interpolating by polynomial splines (9.2.25). In the nodes $\sigma_k \leq 100$ m (k=1,2,…), the meteorological variables are calculated based on the numerical solution of hydrodynamic equations of simplified vertical atmospheric column (9.2.26) - (9.2.29). Solving the first task of interpolation by splines is described above. Let us consider the solution of the second problem.

The equations (9.2.26) – (2.9.29) apply to all interior points of the vertical layer $0 < \sigma < H$, and the boundary conditions are imposed at the boundary points $\sigma = 0$ and $\sigma = H$.

At the level of $\sigma = 0$, the boundary conditions for the velocity are formulated on the basis of the properties of impermeable surface $u=u_b$, $v=v_b$, w=0, where u_b, v_b are horizontal components of the velocity vector of the surface movement. When air flows along the stationary underlying surface, $u_b = v_b = 0$. For temperature, humidity and conductivity, the boundary conditions are formulated on the basis of balance equations defining the different exchange regimes.

At the level of $\sigma = H$, the boundary conditions of the third kind are set to ensure fulfillment of the conditions matching the calculated data from the macro scale data.

Equations (9.2.26) – (2.9.29) are solved numerically on a grid (9.2.30) using the known difference unconditionally stable scheme (Crank, Nicolson, 1947)

$$\frac{\Re_k^{n+1} - \Re_k^n}{\tau} + \frac{1}{2}\left[\begin{array}{l} \overline{w}_k^{n+1} \dfrac{\varsigma_{k-1}^2 \left(\Re_{k+1}^{n+1} - \Re_k^{n+1}\right) + \varsigma_k^2 \left(\Re_k^{n+1} - \Re_{k-1}^{n+1}\right)}{\varsigma_{k-1}\varsigma_k\left(\varsigma_{k-1} + \varsigma_k\right)} \\ + \overline{w}_k^n \dfrac{\varsigma_{k-1}^2 \left(\Re_{k+1}^n - \Re_k^n\right) + \varsigma_k^2 \left(\Re_k^n - \Re_{k-1}^n\right)}{\varsigma_{k-1}\varsigma_k\left(\varsigma_{k-1} + \varsigma_k\right)} \end{array}\right]$$
$$= \frac{1}{2}\frac{H^2}{\left(H - F\right)^2}\left[\begin{array}{l} \dfrac{\left(\nu_{k+1}^{n+1} + \nu_k^{n+1}\right)\left(\Re_{k+1}^{n+1} - \Re_k^{n+1}\right)}{\varsigma_k\left(\varsigma_{k-1} + \varsigma_k\right)} - \dfrac{\left(\nu_k^{n+1} + \nu_{k-1}^{n+1}\right)\left(\Re_k^{n+1} - \Re_{k-1}^{n+1}\right)}{\varsigma_{k-1}\left(\varsigma_{k-1} + \varsigma_k\right)} \\ + \dfrac{\left(\nu_{k+1}^n + \nu_k^n\right)\left(\Re_{k+1}^n - \Re_k^n\right)}{\varsigma_k\left(\varsigma_{k-1} + \varsigma_k\right)} - \dfrac{\left(\nu_k^n + \nu_{k-1}^n\right)\left(\Re_k^n - \Re_{k-1}^n\right)}{\varsigma_{k-1}\left(\varsigma_{k-1} + \varsigma_k\right)} \end{array}\right] + \frac{1}{2}\left(Q_k^{n+1} + Q_k^n\right)$$

(9.2.31)

using the method of successive approximations, in which the role of the iteration parameter is performed by t. As an initial vertical profiles of meteorological variables u, v, θ, q, δ and s in the nodes of the vertical grid $\sigma_k > 100$ m, the values of meteorological variables are taken on geopotential surfaces of macro scale models obtained with the completion in the nodes of shallow horizontal grid by interpolating with polynomial splines (9.2.25). Initial vertical profiles of vertical velocity w are calculated based on the incompressibility equation integration.

In the nodes of $\sigma_k \leq 100$ m ($k = 1,2,\ldots$), initial vertical profiles of meteorological variables $\Re = \left(u, v, \theta, q, \delta, s\right)$ are calculated based on the formula

$$\Re_k = \Re_{k-1} + \left(\Re_{k+1} - \Re_{k-1}\right)\frac{\ln\left(\sigma_k / \sigma_{k-1}\right)}{\ln\left(\sigma_{k+1} / \sigma_{k-1}\right)},$$

derived from the well-known logarithmic distribution of meteorological variables with height in the boundary layer of the atmosphere (Oke, 1982).

Here, it should be particularly noted that all the calculated values of u, v, θ, q, δ and s in the grid points of $\sigma_k > 100$ m, after each iteration, are returned to its original value obtained in macro scale models. Such an algorithm, on the one hand, establishes a connection between the unperturbed regional characteristics of the underlying surface, the macro scale and calculated data, and, on the other hand, not only delivers, but also accelerates the convergence of the iterative process.

Thus, the combination of interpolation method of the completion (9.2.14) for $\sigma_k > 100$ m and the numerical solution of equations (9.2.26) – (2.9.29) for $\sigma_k \leq 100$ m (k=1,2,...) with the above-stated boundary conditions led to a vicious iterative process. It allows to determine the vertical profiles of meteorological variables on the computational grid according to their values known at macro scale grid. As shown by conducted numerical experiments, to achieve convergence, four iterations are usually sufficient.

Completion quality for the vertical profiles of meteorological variables on the limited input data is shown in Figure 3. The points indicate data of radiosonde meteorological quantities at standard levels (at ground level, 850, 700, 500 hPa); the solid lines show the results of the completion in the mesoscale grid nodes. The data radiosonde meteorological quantities, which are measured at intermediate levels, are compared with their calculated data.

Thus, to use the results of the forecast from the macro scale model of atmospheric circulation as binding conditions for the regional model, the transition from discrete values specified in the nodes of large-scale spatial grid ϖ_h of the area $\wp\left(r\right)$ to the discrete values on a fine grid $\bar{\varpi}_h$ of the region $\bar{\wp}\left(r\right)$ should be made. To do this, using the baseline data in the nodes $\lambda_i, \varphi_j, \sigma_k$ ($1 \leq i \leq I$, $1 \leq j \leq J$, $1 \leq k \leq K$) of the large-scale spatial grid ϖ_h at time instants t=t^n (n=0,1,2,...) and combining the interpolation method of completion (9.2.14), and a numerical method for solving the system of equations (9.2.26) – (2.9.29), the values of $\Re = \left(\mathbf{V}, p, \theta, \rho, q, \delta, s\right)$ are determined at the same time points $t = t^n$ at the nodal points

$$r_{ijk}\left(1 \leq i \leq I_1, 1 \leq j \leq J_1, 1 \leq k \leq K_1\right)$$

Figure 3. Results fill in the vertical column of atmospheric values: a) horizontal projection of wind speed vector u; b) - horizontal projection of wind speed vector v; c) vertical projection of wind speed vector w; d) temperature T; e) specific humidity q; f) specific water content δ ; g) turbulent viscosity coefficient ν in the computational grid nodes on the radiosonde data in Kyiv for (18.00) 03.01.89 year.

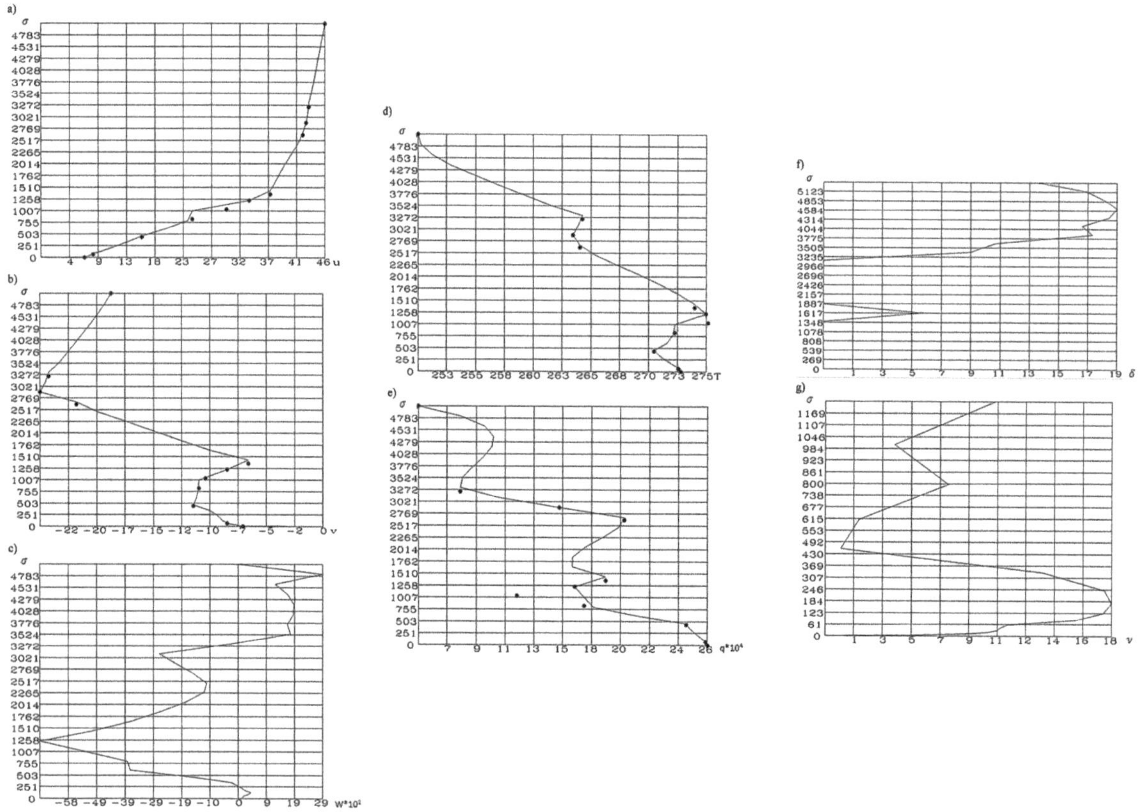

of the grid $\bar{\varpi}_h$ in a bounded domain $\bar{\wp}(r)$.

Next, determine the state of the atmosphere at the nodal points r_{ijk}

$$\left(1 \le i \le I_1, 1 \le j \le J_1, 1 \le k \le K_1\right)$$

of the grid $\bar{\varpi}_h$ in a bounded domain $\bar{\wp}(r)$ for $\forall t \in \left[t^{n-1}, t^n\right]$, n=1,2,…,N by solving $L = I_1 \cdot J_1 \cdot K_1$ problems (9.1.26) – (01.09.27) for ordinary differential equations (9.1.25) by means of interpolation formulas with multiple nodes (9.1.20).

Testing of Sharing Methods

For the numerical solution of the problem

$$\frac{\partial \Re(r,t)}{\partial t} = D\Re(r,t), \forall t \in \left[t^{n-1}, t^{n}\right], n = 1,2,\ldots, \forall r \in \bar{\wp},$$

$$\Re(r,t^{n}) = \Re^{n}(r), n = 1,2,\ldots, \tag{9.3.1}$$

of predicting the atmospheric state $\Re(r,t) = (\mathcal{R}, \mathrm{u}, \mathrm{v}, \mathrm{w}, \mathfrak{H})$ in a limited area $\bar{\wp}(r)$ at time instants $t \in \left[t^{n-1}, t^{n}\right]$, $n = 1,2,\ldots$, where $D\Re(r,t)$ are convection-diffusion terms based on the algorithm of "one-way interaction", some new, more accurate and effective methods were offered in the previous chapters:

- Smooth completion of discrete values of functions, defined on a large-scale grid ϖ_h, in the nodes of a finer grid $\bar{\varpi}_h$ with the spline interpolation (9.2.25) having a good approximation and algorithmic properties that ensure the uniqueness of the interpolation process for a wide range of functions $\Re(\eta)$ at all interval $[a,b]$ with a third order of smoothness.
- Approximation of high (the fourth) order of accuracy of the first (8.1.70) and the second (8.1.71) order included in the convection-diffusion terms of differential equations with partial derivatives of the hydrodynamic model of the atmosphere (5.1.7) – (5.1.9); since the solution obtained in this system of algebraic equations at each node of the grid depends on the values of a function in the rest of the nodes, it is due to the global dependence, rather than local, as in the standard three-point difference approximation.
- Efficient numerical solution of hydrodynamic equations of atmospheric model based on the method of interpolation with multiple nodes (9.1.20) that has the properties of a smooth completion for $\forall t \in \left[t^{n-1}, t^{n}\right]$.

To evaluate the results of applying these methods, numerical weather prediction experiments have been carried out for the territory of Ukraine and its administrative regions. Ukraine is located in the south-west of the East European Plain. In the west of the territory of the Ukrainian Carpathian Mountains rise, in the south along the Southern coast of the Crimean Peninsula, the Crimean Mountains are located. From the West to the East (from 22° to 40° of the East longitude) of its territory stretches for 1300 km from, from the North to the South (from 52° to 45° of the North latitude) it stretches almost for 900 km. On the territory of Ukraine, large and small rivers, high mountains, hills, lowland plains, artificial reservoirs placed, it is surrounded by warm seas. Complex physical and geographical conditions cause a variety of weather conditions, what was the reason for choosing this area as a testing ground for experiments. The main objectives of the experiments were:

- Study of the properties of numerical methods used in the interpolation, approximation, and integration.
- Recreation of general laws of interaction of global and regional atmospheric processes.
- Analysis of the influence of the vertical grid structure on the dynamics and heat transfer in the boundary layer of the atmosphere.

- Estimation of values of parameters in the model of turbulence heat exchange process, moisture and momentum between the surface and free atmosphere.

Regardless of the area of solving the problem, it was filled with the spatial difference grid, consisting of homogeneous horizontal cells formed by dividing the horizontal dimensions of the field of solutions into 100 pieces. The height of the top boundary of the solution area was assumed constant at H=15000 m. 41 levels has been chosen to build the vertical grid. These levels are located at heights, the value of which is calculated as the dimensionless quantities σ / H by formula (9.2.30). Their values are shown in Table 1.

The data of objective analysis and forecast obtained from the German national service Deutscher Wetterdienst (DWD Offenbach) in the binary GRIB format for the territory (0°-90°N, 0°-90°E) are used as background information for the task of the regional forecast with the step 1.5° in the nodes of the regular latitude-longitude grid. One complete set of data consisting of 255 files containing data of objective analysis for 00 or 12 hours GMT of the current day and forecast data for the time periods of 12, 24, 36, 48 hours GMT. The composition and structure of each data set are shown in Table 2 (the code "98" denotes the level of the sea, and the code „89" denotes the ground level). In the table, the symbol "+" marks meteorological values present in the data set at an appropriate level, and the symbol "-" denotes missing ones.

The obtained weather data (pressure, temperature, humidity, and horizontal components of wind) based on the method of completion with the spline interpolation (9.2.25) are translated into the nodes of a mesoscale horizontal grid spaced in both latitude and longitude for significantly smaller distances than in the original grid. Then, the data of mesoscale horizontal grid, by solving numerically the system of equations (9.2.26) - (2.9.29), taking into account the topography of the underlying surface on the

Table 1. The structure of the vertical grid

N°	Height σ / H	N°	Height σ / H	N°	Height σ / H	N°	Height σ / H
1	0.0000000	11	$1.48381 \cdot 10^{-2}$	21	0.2366667	31	0.5750000
2	$6.86644 \cdot 10^{-5}$	12	$2.34671 \cdot 10^{-2}$	22	0.2666000	32	0.6136000
3	$1.82167 \cdot 10^{-4}$	13	$3.61051 \cdot 10^{-2}$	23	0.2974000	33	0.6530667
4	$3.69678 \cdot 10^{-4}$	14	$5.34639 \cdot 10^{-2}$	24	0.3290667	34	0.6934000
5	$6.79145 \cdot 10^{-4}$	15	$7.53174 \cdot 10^{-2}$	25	0.3616000	35	0.7346000
6	$1.18906 \cdot 10^{-3}$	16	0.1000000	26	0.3950000	36	0.7766666
7	$2.02703 \cdot 10^{-3}$	17	0.1256000	27	0.4292667	37	0.8196000
8	$3.39807 \cdot 10^{-3}$	18	0.1520667	28	0.4644000	38	0.8634000
9	$5.62527 \cdot 10^{-3}$	19	0.1794000	29	0.5004000	39	0.9080667
10	$9.20149 \cdot 10^{-3}$	20	0.2076000	30	0.5372667	40	0.9536000
						41	1.0000000

Table 2. The contents of the data set DWD Offenbach

The Pressure at the Standard Level (hp)	The Height Level	The Pressure at Sea Level	Relative Humidity of Air	Air Temperature	The Components of the Wind Velocity u,v
98	-	+		-	-
89	-	-		+	+
850	+	-	+	+	+
700	+	-	+	+	+
500	+	-	+	+	+
300	+	-	-	+	+
250	+	-	-	+	+
200	+	-	-	+	+
100	+	-	-	+	+
50	+	-	-	+	+

territory of the forecast, are translated into the nodes of mesoscale vertical grid. And, finally, the right parts of the equations of the circulation model of the atmosphere (9.1.24) containing the partial derivatives of the first and second-order were approximated with compact difference schemes (8.1.70) and (8.1.71). As a result of these calculations, the values of difference analogues of the right parts $\Lambda\Re_{ijk}$ ($1 \le i \le I_1, 1 \le j \le J_1, 1 \le k \le K_1$) of equations for $\Re(r,t) = (\mathcal{R}, \mathrm{u}, \mathrm{v}, \mathrm{w}, \mathfrak{H})$ were obtained.

To determine the state of the atmosphere at a node r_{ijk} of the grid $\bar{\varpi}_h$ in a limited region $\bar{\wp}(r)$ at $\forall t \in [t^n, t^{n+1}]$, the problem was solved on the mesoscale grid for each term of the objective analysis and forecast

$$\frac{\partial \Re_{ijk}}{\partial t} = \Lambda\Re_{ijk}, \forall t \in [t^n, t^{n+1}], n = 1, 2, \ldots,$$

$$\Re_{ijk}(t^n) = \Re_{ijk}^n, n = 0, 1, \ldots,$$

$$1 \le i \le I_1, 1 \le j \le J_1, 1 \le k \le K_1.$$

The interpolation method with multiple nodes (9.1.20) was used sequentially $\forall t \in [t^n, t^{n+1}]$, n=1,2,.... Herewith, the source data were taken from the previous time t^{n-1}, $t^n = t^{n-1} + 12$ and $t^{n+1} = t^{n-1} + 24$ hours.

The algorithm for solving the problem of numerical weather prediction for the limited area on the basis of the method of one-way interaction is shown in Figure 4.

Examples of the results of weather problems solutions obtained by the method (9.1.20) with the step of 4 hours are given in Figure 5 in the form of surface maps.

Figure 4. Results of an experimental solution of problems of weather forecast

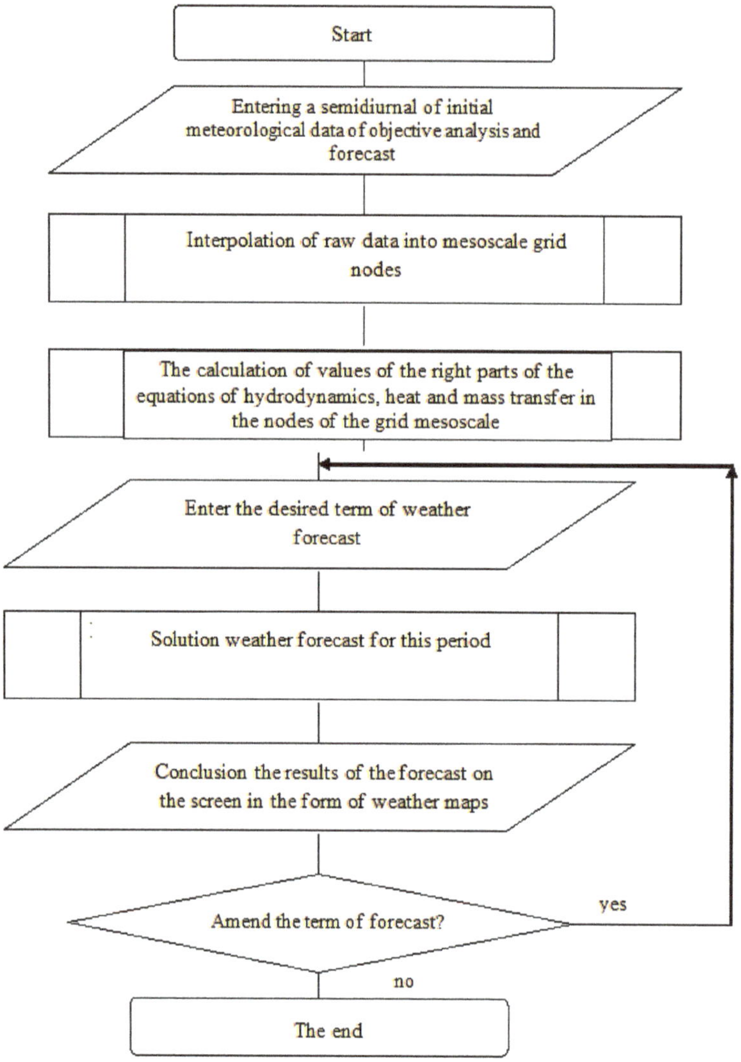

Comparative analysis of the numerical results and actual weather maps shows that the results are consistent qualitatively and quantitatively with the actual data.

Research Problems for the Dissemination of Impurities in the Atmosphere and Methods of Their Resolution

Environmental and meteorological model of the physical processes in the atmosphere, formulated on the basis of the full Navier-Stokes (Reynolds) equations, heat, and mass transfer, describes a wide range of movements: mesoscale acoustic and gravitational waves, barotropic and baroclinic Rossby waves, quasi-stationary jet streams and others. Resources of modern computers still do not allow for the spatial discretization of ecological and meteorological model with high resolution. And spatial discretization with coarse resolution leads to the fact that a lot of mesoscale meteorological processes (including im-

Figure 5. A cloud model in the atmosphere from the salvo out emission put: ☼ is the place salvo emission; a), b), ..., n) are forecasts on 1,2, ..., 14 hours after the salvo emission

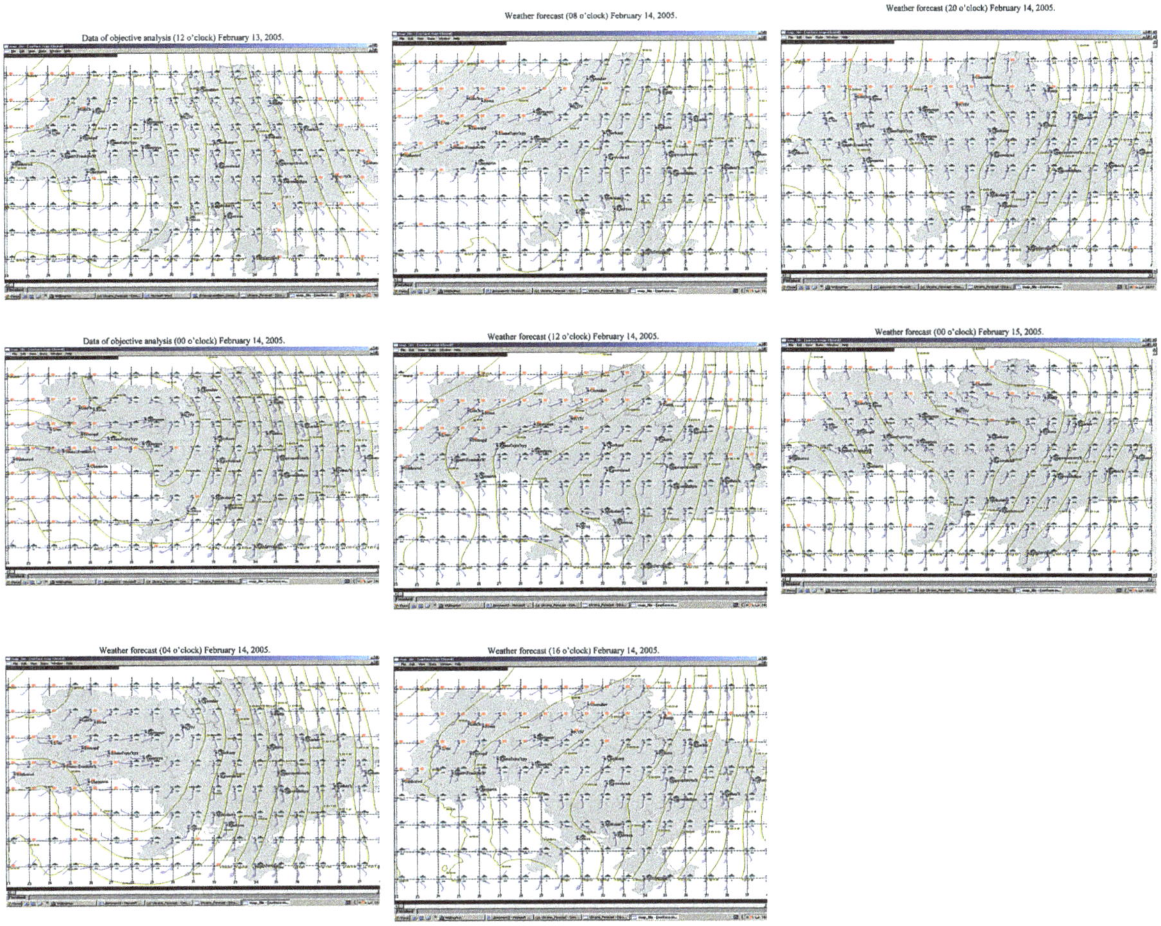

purity scattering from a single source) will not be described in this model, as they will be in the class of sub-grid scale processes in this case.

The problem of numerical implementation of a comprehensive ecological and meteorological models can be viewed from two perspectives. On the one hand, it is a study of atmospheric motions based on the hydrodynamic model with the required level of detail. In our case, we will examine directly the dynamics of the atmosphere and the processes within it. On the other hand, it is a solution of the specific problem of physical transfer and diffusion of impurities in the atmosphere. It is necessary to have a clear idea of the thermodynamic mechanisms of the process under investigation, as well as of the adequacy of the atmosphere model to real mesoscale meteorological conditions and their impact on the quality of the diffusion model. Obviously, the selection of the characteristic spatial scale is dominated in this problem.

Solving the problem is directly related to the construction of a computational algorithm, which allows to solve the problem of modeling the transfer and diffusion of impurities in the atmosphere from a point source at almost arbitrary computational grids, which numerically implements the hydrodynamic model of the atmosphere. There are two ways to describe mechanisms of interaction between processes of different scales. The first is based on discretization of the diffusion problem on an irregular grid with the necessary interpolation of meteorological parameters, defined on the coarse grid, into the fine grid nodes. The complexity of this algorithm is obvious. The second method is to divide the solution area of the diffusion problem to a number of sub-areas, to transform the original formulation of the problem in each sub-region taking into account only the description of the basic processes and the subsequent crosslinking of solutions at the borders. This path seems to be more economical, as the main construction of the following algorithms for solving the diffusion problem allows using both analytical and numerical methods.

In solving evolutionary problems of the transfer and diffusion of atmospheric dust-gas mixture emitted by industrial enterprises during their project operation and with salvo emissions, the pollution zone will conditionally be divided into "near" and "distant" areas. Such a hypothetical division allows the use of both numerical and analytical methods to solve this problem (Prusov, Kopeika, Grymov, 1998; Prusov, 2001). Splitting method is accepted as a basic algorithm for solving the considered problems.

The model of the spreading aerosols in the atmosphere that allows to calculate the concentration of suspended and surface impurities has been formulated in Section 5.3 for the equation (5.3.7) with coefficients that represent the actual value of the meteorological wind speed $\boldsymbol{V} = \left(v_1, v_2, v_3\right)$ and the coefficient turbulent diffusion $\nu = \left(\nu_1, \nu_2, \nu_3\right)$. Obviously, these factors are complex non-linear functions of the coordinates $\boldsymbol{X} = \left(x_1, x_2, x_3\right)$ and the time t, which determines the practical uselessness of analytic methods for the solution of the problem as they require significant styling and simplifications in the formulation of the problem. Therefore, for real weather conditions, only numerical solution is possible together with the implementation of a hydrodynamic model of the atmosphere discussed above. In other words, at first, there to be received the fields $\boldsymbol{V} = \left(v_1, v_2, v_3\right)$, $\nu = \left(\nu_1, \nu_2, \nu_3\right)$ based on the numerical integration of the equations of hydrodynamic model of the atmosphere (5.1.7) – (5.1.9) at the moment of forecast $t_{n+1} = t_n + \Delta t$. Only after substituting these values in the coefficients of the equation (5.3.7) and specifying the initial and boundary conditions, one can get the real numerical values of the impurity concentration at the same time $t_{n+1} = t_n + \Delta t$.

The approximation of equation (5.3.7) with the difference schemes leads, in general, to a system of algebraic equations. Their efficient solution is a difficult problem. Great difficulties arise when approximating the original equation with implicit difference schemes for multidimensional problems. However, it is implicit difference schemes that are cost-effective for the numerical solution of the equation (5.3.7). They either do not dictate restrictions on the ratio of the temporal and spatial grid spacing, or significantly weaken these constraints.

In the construction of implicit difference schemes for the numerical solution of multidimensional equation (5.3.7), a variety of approaches are used based on the notion of a difference scheme in the form of a sequence of simpler difference analogues. One such approach is the method of fractional steps that allows to reduce the solution of a difficult problem to a successive solution of simpler problems. We distinguish the following types of method of fractional steps:

- Analytical splitting that allows to solve various problems analytically at separate fractional steps.
- Geometric splitting that transforms multidimensional problem to a sequence of successive problems of lower dimension, in particular, to the space one-dimensional problems; in this case, we refer to the splitting by directions.

The said types of splitting suggest a different realization of the fractional step method – the method of approximate factorization of the operator. Both methods are equivalent in the case of commutative operators. The splitting and approximate factorization of the operator can be represented in the differential, differential-discrete and discrete forms. We limit the application of the above types of splitting method and the corresponding approximate factorization, at first, to analytical form, in order to move afterwards to the differential and discrete implementation. The differential form is useful for pre-transformation, especially in the case of irregular grids and the curvilinear coordinate system.

Simulation of Impurity Scattering in the "Near Zone" of an Instantaneous Point Source

Let it occurs in the atmosphere an aerosol of an intensity *M* at an altitude $x_3 = x_3^{eff}$ as the result of salvo emission at time $t=t_0$, which is then transported by air masses and diffuses under the influence of atmospheric turbulence. If not to take into account the initial spread of the aerosol contaminants and almost completely insignificant influence of the underlying surface, the propagation of the aerosol cloud can be interpreted as the convection-diffusion process of impurity migration from an instantaneous point source of the power *M* in unlimited space. Assume that the instantaneous point source of an aerosol is in the neighborhood of the point $\boldsymbol{X}_0 = \left(x_1^{eff}, x_2^{eff}, x_3^{eff}\right)$. Then it can be described by the δ-function as $s = \delta\left(\boldsymbol{X} - \boldsymbol{X}_0\right)\delta\left(t - t_0\right)$.

There is a very significant problem in the finite-difference approximation of the initial conditions under which the point source $\boldsymbol{X}_0$ is described by the δ-function. It is related to the fact that the impurity concentration in this description infinitely grows as approaching the source and becomes infinitely large at the source. So, for the corresponding approximation in the vicinity of the source, it should be required unlimited grinding the computational grid. This leads to a major problem of the sharing the numerical solution of the meteorological and environmental problems – their scales differ by several orders of magnitude. Adjustment (in fact, means grinding) of the difference grid for numerical solution of the meteorological problem with cross-geometric dimensions of the impurity clouds spread, not exceeding a few meters close to the emission source, significantly affects the cost of computer time, even taking into account the capabilities of modern computers. Doubling of the number of nodes on the three-dimensional computational grid of meteorological tasks leads to 16-fold increase of computational complexity and the required amount of memory. Consequently, despite the fact that the numerical simulation of the spread of impurities in the atmosphere is completely natural, however, it leaves urgent the search for intelligent ways to simplify the task.

The coefficients of the equation (5.3.7) in the real world are not permanent either in time or in space. Considerable significance is the question of the influence of the length of a time interval during which the meteorological parameters $\boldsymbol{V} = \left(v_1, v_2, v_3\right)$, $\nu = \left(\nu_1, \nu_2, \nu_3\right)$ can be considered constant, as it is important to study the relationship between the calculated and actual concentrations obtained by observation. The estimates show (Dovgiy, Prusov, Kopeika, 2000; Prusov, Grymov, Kopiiko, 1996; Prusov, Kopeika, Grymov, 1998; Prusov, 2001; Prusov, 2002; Prusov, Doroshenko, 2002; Prusov, Doroshenko, 2003; Doroshenko Prusov, 2005; Prusov, Doroshenko, 2006), that the field concentration s, described by the original equation of turbulent diffusion (5.3.7) with constant values of ν, is influenced by a relatively small-scale eddies of $t \approx 2$-3 minutes. The estimates allow to assume that calculation methods based on analytical solutions will be in accordance with a particular experiment for the relatively short distances from the source, or in other words, at low impurity scattering time. Therefore, if we restrict ourselves by the local property impurity movement in the cloud, located on the outer edge of the boundary layer of the atmosphere, one can use the simplest and most affordable for theoretical research approach of the homogeneous and isotropic atmospheric turbulence.

We assume that the boundary of the solution area is at relatively short distance from the source ("near zone") and the dependence of meteorological parameters $\boldsymbol{V} = \left(v_1, v_2, v_3\right)$, $v = \left(v_1, v_2, v_3\right)$ on the spatial coordinates in it can be neglected. Then we arrive at the following model:

$$\frac{\partial s}{\partial t} + v_j \frac{\partial s}{\partial x_j} + \varrho s = \frac{\partial}{\partial x_j}\left(\frac{\nu_j}{\xi}\frac{\partial s}{\partial x_j}\right), j = 1,2,3, t_0 \le t \le \tau, \tag{9.4.1}$$

$$s = s_0\left(\mathbf{X}\right), \text{when } t = t_0, \tag{9.4.2}$$

$$s \to s_v\left(x_1, x_2, x_3\right) \text{when } x_1 \to \pm\infty, x_2 \to \pm\infty, x_3 \to \pm\infty, \tag{9.4.3}$$

$$s = 0, \text{when } x_3 = 0, \tag{9.4.4a}$$

or

$$\frac{\partial s}{\partial x_3} = 0 \text{ when } x_3 = 0, \tag{9.4.4b}$$

The homogeneous boundary condition of the first kind (9.4.4a) corresponds to the propagation of impurities over the land, and the homogeneous boundary condition of the second kind (9.4.4b) – over the water surface.

In solving the problem (9.4.1) – (9.4.4) of the transfer and diffusion of polluting substances concentration s, the components of velocity and turbulent exchange in the atmosphere at a given point are calculated by the methods of mezo-meteorology. After receiving the necessary information about the fields of meteorological parameters, which are assumed to be constant in $\wp$ in the time interval $t_0 \le t \le \tau$, to solve the problem of propagation of the explosion products discarded at a given point $\boldsymbol{X}_0 \in \wp$, one can apply the Fourier transformation method.

The equation (7.7.1) with constant coefficients is reduced by the substitution

$$a = \frac{\nu_j}{\xi}, \beta_j = \frac{v_j}{2a}, \lambda_j = -\left(\frac{v_j^2}{4a} + \varrho\delta_j^3\right), j = 1,2,3,$$

$$s = \varphi\prod_{j=1}^{3} e^{\lambda_j t + \beta_j x_j}$$

to the form

$$\frac{\partial \varphi}{\partial t} = \sum_{j=1}^{3} \frac{\nu_j}{\xi} \frac{\partial^2 \varphi}{\partial x_j^2} = a\Delta\varphi, \tag{9.4.5}$$

where φ is a concentration of the diffusing substance, a is a coefficient of diffusion in the homogeneous isotropic turbulent environment, Δ is the Laplace operator, δ_j^3 is the Kronecker symbol.

At first, we prove the following lemma.

Lemma 9.4.1: If the solution of the equation

$$\Delta\varphi - \frac{1}{a}\frac{\partial \varphi}{\partial t} = 0$$

depends only on r and t, then the function $\psi = r\varphi$ satisfies the equation

$$\frac{\partial^2 \psi}{\partial r^2} = \frac{1}{a}\frac{\partial \psi}{\partial t}. \tag{9.4.6}$$

Proof: Indeed, writing the Laplace operator in spherical coordinate system, we see that the function $\varphi = \varphi(r,t)$ satisfies the equation

$$\frac{\partial^2 \varphi}{\partial r^2} + \frac{2}{r}\frac{\partial \varphi}{\partial r} - \frac{1}{a}\frac{\partial \varphi}{\partial t} = 0$$

or

$$\frac{1}{r}\frac{\partial^2 (r\varphi)}{\partial r^2} - \frac{1}{a}\frac{\partial \varphi}{\partial t} = 0.$$

Assuming then that $\psi = r\varphi$, we obtain for ψ the equation (9.4.6), what completes the proof of the Lemma.

Let a continuously operating constant power source of impurity Q is placed in the origin of coordinates, and in the rest of the space, the background concentration is zero $\varphi(r,0) = 0$ for $r \neq 0$. Obviously, in this case the concentration φ is a function of r and t.

The presence the source of impurities in point r=0 means that the flow of concentration per time unit through the sphere S_ε (centered at r=0 and the radius ε) with $\varepsilon \to 0$ is equal to Q, i.e.

$$\lim_{\varepsilon\to 0}\left(-\iint_{S_\varepsilon} a\frac{\partial\varphi}{\partial n}\,d\varsigma\right) = Q.$$

Since the normal derivative $\partial\varphi / \partial n = \partial\varphi / \partial r$ is constant on the surface S_ε by symmetry, then

$$-a\frac{\partial\varphi}{\partial r}\cdot 4\pi r^2\Big|_{r=\varepsilon} \to Q \text{ when } \varepsilon \to 0,$$

what means that the derivative $\partial\varphi / \partial r$ a singularity of the form $-Q / \left(4\pi a r^2\right)$ at r=0. Therefore, the function itself should have at r=0 a singularity of the form

$$\varphi \sim \frac{Q}{4\pi a r},$$

so the product $r\varphi = \psi$ remains limited at r=0.

The function ψ defined by the conditions

$$\left.\begin{array}{l} \partial^2\psi / \partial r^2 = a^{-1}\partial\psi / \partial t, \\ \psi(0,t) = Q / \left(4\pi a r^2\right) = \psi_0, \\ \psi(r,0) = 0, \end{array}\right\}$$

is expressed by the formula

$$\psi(r,t) = \frac{2}{\sqrt{\pi}}\frac{Q}{4\pi a}\int_{\frac{r}{2\sqrt{at}}}^{\infty} e^{-\alpha^2}\,d\alpha,$$

where

$$\alpha = \frac{\varsigma - r}{2\sqrt{a\left(t - t_0\right)}}.$$

Thus, the solution of the problem of spreading impurities from a permanent source of the power Q, placed at the origin point r=0 has the form

$$\varphi\left(r,t\right) = Q\Phi\left(r,t\right) = \frac{2}{\sqrt{\pi}}\frac{1}{r}\frac{Q}{4\pi a}\int\limits_{\frac{r}{2\sqrt{at}}}^{\infty} e^{-\alpha^2} d\alpha,$$

where $\Phi\left(r,t\right)$ is a concentration corresponding to a single source (Q=1).

To go to the instantaneous source, consider the source Q, placed in point $\boldsymbol{X}_0 = \left(x_1^{eff}, x_2^{eff}, x_3^{eff}\right)$, that is acting for a period of time τ. Such a source can be considered as equivalent to two sources of the power +Q and -Q. The first of them is activated when t=0, the second – at $t = \tau$. The concentration distribution in this case is expressed by the formula

$$\varphi_\tau\left(r,t\right) = Q\left[\Phi\left(r,t\right) - \Phi\left(r,t-\tau\right)\right].$$

The ejected amount of impurity $M = Q\tau$ over the period of time τ, therefore, is

$$\varphi_\tau\left(r,t\right) = \frac{M}{\tau}\left[\Phi\left(r,t\right) - \Phi\left(r,t-\tau\right)\right].$$

Passing to the limit as T → 0 and assuming M to be a constant, we find

$$\varphi_0\left(r,t\right)\lim_{\tau\to 0}\varphi_\tau\left(r,t\right) = M\frac{\partial |}{\partial t} = \frac{2}{\sqrt{\pi}}\frac{M}{4\pi ar}\frac{ra}{4a\sqrt{at^3}}e^{-\frac{r^2}{4at}},$$

or

$$\varphi_0\left(r,t\right) = MG\left(x_1, x_2, x_3, t, x_1^{eff}, x_2^{eff}, x_3^{eff}\right),$$

where

$$G\left(x_1, x_2, x_3, t, x_1^{eff}, x_2^{eff}, x_3^{eff}\right) = \left(\frac{1}{2\sqrt{\pi at}}\right)^3 e^{-\frac{\left(x_1 - x_1^{eff}\right)^2 + \left(x_2 - x_2^{eff}\right)^2 + \left(x_3 - x_3^{eff}\right)^2}{4at}}. \quad (9.4.7)$$

The function $G\left(x_1, x_2, x_3, t, x_1^{eff}, x_2^{eff}, x_3^{eff}\right)$ is a function of the impact of an instant source of impurity. It is the concentration at the point $\left(x_1, x_2, x_3\right)$ at the time instant *t*, caused by a point source of power *M*, placed at the time *t*=0 in the point $\left(x_1^{eff}, x_2^{eff}, x_3^{eff}\right)$.

It is easily seen that

$$\iiint\limits_{-\infty}^{+\infty} G\left(x_1, x_2, x_3, t, x_1^{eff}, x_2^{eff}, x_3^{eff}\right) dx_1^{eff} dx_2^{eff} dx_3^{eff} = 1 .$$

It follows from (9.4.7) that the influence function *G* has the symmetry property

$$G\left(x_1, x_2, x_3, t, x_1^{eff}, x_2^{eff}, x_3^{eff}\right) = G\left(x_1^{eff}, x_2^{eff}, x_3^{eff}, t, x_1, x_2, x_3\right),$$

that is an expression of the principle of reciprocity: the action at the point $\left(x_1, x_2, x_3\right)$ from the source located at the point $\left(x_1^{eff}, x_2^{eff}, x_3^{eff}\right)$, is equal to the action at the point $\left(x_1^{eff}, x_2^{eff}, x_3^{eff}\right)$ from the same source placed in the point $\left(x_1, x_2, x_3\right)$. However, with respect to the variable *t* such the symmetry is absent what is an expression of the irreversibility of diffusion processes in time. If the instantaneous point source is placed at the point $\left(x_1^{eff}, x_2^{eff}, x_3^{eff}\right)$ at the time $t=t_0$, then (9.4.7), respectively, is transformed into

$$G\left(x_1, x_2, x_3, t, x_1^{eff}, x_2^{eff}, x_3^{eff}\right) = \left(\frac{1}{2\sqrt{\pi a\left(t - t_0\right)}}\right)^3 e^{-\frac{\left(x_1 - x_1^{eff}\right)^2 + \left(x_2 - x_2^{eff}\right)^2 + \left(x_3 - x_3^{eff}\right)^2}{4a\left(t - t_0\right)}} . \tag{9.4.8}$$

The function $G\left(x_1, x_2, x_3, t, x_1^{eff}, x_2^{eff}, x_3^{eff}\right)$ satisfies the equation of diffusion

$$\frac{\partial G}{\partial t} = a\sum_{j=1}^{3} \frac{\partial^2 G}{\partial x_j^2}$$

with variable $\left(x_1, x_2, x_3, t\right)$, as it is easily seen by direct differentiation.

Determine the form of the influence function G in the case of two dimensions. Let a straight line parallel to the axis x_3 and passing through the point $\left(x_1^{eff}, x_2^{eff}\right)$, contains an infinite linear source of the power $\bar{M} = const$ per length unit. The influence of the function $G_{1,2}$ of such a source will not depend on x_3 and fully characterized by its values in the plane $\left(x_1, x_2\right)$. We calculate the function $G_{1,2}$. Let the amount of impurity $dM = \bar{M} dx_3$ is formed on the element dx_3; then the concentration distribution in space is given by the integral

$$\bar{\varphi} = \int_{-\infty}^{+\infty} \bar{M} dx_3^{eff} G\left(x_1, x_2, x_3, t, x_1^{eff}, x_2^{eff}, x_3^{eff}, t_0\right).$$

Calculating the integral

$$\int_{-\infty}^{+\infty} e^{-\frac{\left(x_3 - x_3^{eff}\right)^2}{4a\left(t-t_0\right)}} dx_3 = 2\sqrt{a\left(t - t_0\right)} \int_{-\infty}^{+\infty} e^{-\alpha^2} d\alpha = 2\sqrt{\pi a\left(t - t_0\right)},$$

where $\alpha = \dfrac{x_3^{eff} - x_3}{2\sqrt{a\left(t - t_0\right)}}$, we obtain $\bar{\varphi} = \bar{M} G_{1,2}$. Comparing function

$$G_{1,2}\left(x_1, x_2, t, x_1^{eff}, x_2^{eff}, t_0\right) = \left(\frac{1}{2\sqrt{\pi a\left(t - t_0\right)}}\right)^2 e^{-\frac{\left(x_1 - x_1^{eff}\right)^2 + \left(x_2 - x_2^{eff}\right)^2}{4a\left(t-t_0\right)}} \tag{9.4.9}$$

with (9.4.8), we see the similarity in structure.

Similarly, we can derive an expression for the source function in the one-dimensional case. Considering the flat endless source of constant power of $\tilde{M}$ per area unit, we get that

$$\tilde{\varphi} = \int_{-\infty}^{+\infty}\int_{-\infty}^{+\infty} \tilde{M} dx_1^{eff} dx_2^{eff} G\left(x_1, x_2, x_3, t, x_1^{eff}, x_2^{eff}, x_3^{eff}, t_0\right) = \tilde{M} G_3\left(x_3, t, x_3^{eff}, t_0\right),$$

where

$$G_3\left(x_3, t, x_3^{eff}, t_0\right) = \frac{1}{2\sqrt{\pi a\left(t - t_0\right)}} e^{-\frac{\left(x_3 - x_3^{eff}\right)^2}{4a\left(t-t_0\right)}} \tag{9.4.10}$$

is a source function for one measurement.

From (9.4.8), (9.4.9) and (9.4.10), it is seen that

$$G\left(x_1, x_2, x_3, t, x_1^{eff}, x_2^{eff}, x_3^{eff}, t_0\right) = G_{1,2}\left(x_1, x_2, t, x_1^{eff}, x_2^{eff}, t_0\right) \cdot G_3\left(x_3, t, x_3^{eff}, t_0\right). \tag{9.4.11}$$

This equation allows us to apply the method of splitting by directions and reduce a multidimensional problem to a sequence of alternating tasks of smaller dimension.

We state the following problem: to find a bounded function $\varphi\left(x_1, x_2, x_3, t\right)$, defined in the region $-\infty < x_1 < +\infty$, $-\infty < x_2 < +\infty$, $-\infty < x_3 < +\infty$, $t - t_0 \geq 0$ which satisfies the diffusion equation

$$\frac{\partial \varphi}{\partial t} = a\sum_{j=1}^{3}\frac{\partial^2 \varphi}{\partial x_j^2}, \tag{9.4.12}$$

with the initial condition

$$\varphi = \varphi_0\left(x_1, x_2, x_3\right), when\ t = t_0, \tag{9.4.13}$$

and can be represented as the product

$$\varphi\left(x_1, x_2, x_3, t\right) = \varphi_{1,2}\left(x_1, x_2, t\right)\cdot\varphi_3\left(x_3, t\right), \tag{9.4.14}$$

where $\varphi_{1,2}\left(x_1, x_2, t\right)$ is a function only of variables x_1, x_2, t; $\varphi_3\left(x_3, t\right)$ is a function only of variables x_3, t.

If $\varphi_0\left(x_1, x_2, x_3\right)$ is a continuous function, the fulfillment of the initial condition (9.4.13) will be understood in the sense that $\varphi\left(x_1, x_2, x_3, t\right)$ is continuous for $t=t_0$, i.e.

$$\lim_{t\to t_0, X\to X_0}\varphi\left(x_1, x_2, x_3, t\right) = \varphi_0\left(x_1, x_2, x_3\right)$$

As it is known, the required solution of the diffusion equation (9.4.12) is uniquely determined by its initial conditions, if it is limited. Taking into account the influence function (9.4.8), this solution has the form

$$\varphi\left(x_1, x_2, x_3, t\right) = \iiint_{-\infty}^{+\infty} G\left(x_1, x_2, x_3, t, x_1^{eff}, x_2^{eff}, x_3^{eff}\right)\varphi_0\left(x_1^{eff}, x_2^{eff}, x_3^{eff}\right)dx_1^{eff}dx_2^{eff}dx_3^{eff}. \tag{9.4.15}$$

If at the initial time $t=t_0$ at the point $\left(x_1^{eff}, x_2^{eff}, x_3^{eff}\right)$, there is ejected an amount of impurities $\varphi = \varphi_0\left(x_1, x_2, x_3\right) = \text{const} = M$, one can ensure by the direct verification that that the function

$$\varphi\left(x_1, x_2, x_3, t\right) = MG\left(x_1, x_2, x_3, t, x_1^{eff}, x_2^{eff}, x_3^{eff}, t_0\right) \tag{9.4.16}$$

Is the concentration of impurities at the point $\left(x_1, x_2, x_3\right)$ at the time *t* from the instantaneous power of a point source of *M*. From the definition of $\bar{M}$ and $\tilde{M}$ it follows that $M = \bar{M}\cdot\tilde{M}$, then one can write the following chain of expression based on (9.4.11) and (9.4.16):

$$\bar{M}G_{1,2}\left(x_1, x_2, t, x_1^{eff}, x_2^{eff}, t_0\right)\cdot\tilde{M}G_3\left(x_3, t, x_3^{eff}, t_0\right) = \varphi_{1,2}\left(x_1, x_2, t\right)\cdot\varphi_3\left(x_3, t\right).$$

So, the problem (9.4.12), (9.4.13) can be split into two:

1. Find a limited function $\varphi_{1,2}\left(x_1, x_2, t\right)$ defined in the region $-\infty < x_3 < +\infty$, $t - t_0 \geq 0$, that satisfies the diffusion equation

$$\frac{\partial \varphi_{1,2}}{\partial t} = a\sum_{j=1}^{2}\frac{\partial^2 \varphi_{1,2}}{\partial x_j^2}, \tag{9.4.17}$$

and a condition

$$\varphi_3 = \bar{M}\delta\left(x_3 - x_3^{eff}\right)\delta\left(t - t_0\right); \tag{9.4.18}$$

2. Find a limited function $\varphi_3\left(x_3, t\right)$ defined in the region $-\infty < x_3 < +\infty$, $t - t_0 \geq 0$ that satisfies the diffusion equation

$$\frac{\partial \varphi_3}{\partial t} = a\frac{\partial^2 \varphi_3}{\partial x_3^2}, \tag{9.4.19}$$

and a condition

$$\varphi_3 = \bar{M}\delta\left(x_3 - x_3^{eff}\right)\delta\left(t - t_0\right); \tag{9.4.20}$$

The solution of (4.9.17), (9.4.18) with (9.4.9) has the form

$$\varphi_{1,2}\left(x_1, x_2, t\right) = \frac{\bar{M}}{4\pi a\left(t - t_0\right)} e^{-\frac{\left(x_1 - x_1^{eff}\right)^2 + \left(x_2 - x_2^{eff}\right)^2}{4a\left(t - t_0\right)}}.$$

We carry out in the resulting expression a transition back to the original variable *s* according to the rule (9.4.4), taking into account that

$$s\left(x_1, x_2, x_3, t\right) = s_{1,2}\left(x_1, x_2, t\right)\cdot s_3\left(x_3, t\right).$$

The result is:

$$s_{1,2}\left(x_1, x_2, t\right) = \frac{\bar{M}}{4\pi\mu_{1,2}\left(t - t_0\right)} e^{-\frac{\left[x_1 - x_1^{eff} - v_1\left(t - t_0\right)\right]^2}{4\mu_{1,2}\left(t - t_0\right)}} \times e^{-\frac{\left[x_2 - x_2^{eff} - v_2\left(t - t_0\right)\right]^2}{4\mu_{1,2}\left(t - t_0\right)}}, \tag{9.4.21}$$

where it is assumed that

$$\mu_{1,2} = \mu_1 \equiv \nu_1 / \xi = \mu_2 \equiv \nu_2 / \xi = const ,$$

i.e., the atmosphere is a homogeneous isotropic medium in the horizontal plane.

In general, the problem (9.4.17), (4.9.18) needs to be clarified with the additional boundary condition, taking into account the effect of the underlying surface. This leads to the problem of finding a solution of the diffusion equation

$$\frac{\partial \varphi_3}{\partial t} = a \frac{\partial^2 \varphi_3}{\partial x_3^2}, x_3 > 0, t > t_0 \tag{9.4.22}$$

on the semi-infinite line $x_3 > 0$ for $t > t_0$ values, satisfying the initial condition

$$\varphi_3\left(x_3, t\right) = \chi\left(x_3\right), x_3 > 0 \tag{9.4.23}$$

and boundary condition, which, depending on the given boundary regime, is taken in one of the following forms:

$$\varphi_3\left(0, t\right) = 0 \text{ (the first boundary problem)} \tag{9.4.24a}$$

$$\frac{\partial \varphi_3}{\partial x_3}\left(0, t\right) = 0 \text{ (the second boundary problem)} \tag{9.4.24b}$$

Let's start with detailed study of the first boundary problem (9.4.22) – (9.4.24a). To ensure the uniqueness of the solution of the problem, it is necessary to impose certain conditions on the behavior in the infinity. We require as an additional condition to the function $\varphi_3\left(x_3, t\right)$ to be limited everywhere:

$$\left|\varphi_3\left(x_3, t\right)\right| < A \text{ when } 0 < x_3 < \infty \text{ and } t \geq t_0,$$

where A is a constant. It follows that the initial function $\chi\left(x_3\right)$ must also satisfy the condition of limitation $\left|\chi\left(x_3\right)\right| < A$.

We first prove two lemmas regarding the function $\varphi_3\left(x_3, t\right)$, defined by the Poisson integral, that can be easily obtained at $\varsigma = x_3^{eff}$ from (9.4.10):

$$\varphi_3\left(x_3,t\right)=\frac{1}{2\sqrt{\pi}}\int_{-\infty}^{+\infty}\frac{1}{\sqrt{a\left(t-t_0\right)}}e^{-\frac{\left(x_3-\varsigma\right)^2}{4a\left(t-t_0\right)}}\psi\left(\varsigma\right)d\varsigma\,. \tag{9.4.25}$$

Lemma 9.4.2: If the function $\psi\left(x_3\right)$ is an odd function, i.e., $\psi\left(x_3\right)=-\psi\left(-x_3\right)$, then $\varphi_3\left(x_3,t\right)$ from (9.4.25) vanishes at x_3=0.

Proof: Let the integral (9.4.25), which determines the function $\varphi_3\left(x_3,t\right)$, converges. This occurs because $\psi\left(x_3\right)$ is limited. The integrand in the integral

$$\varphi_3\left(0,t\right)=\frac{1}{2\sqrt{\pi}}\int_{-\infty}^{+\infty}\frac{1}{\sqrt{a\left(t-t_0\right)}}e^{-\frac{\varsigma^2}{4a\left(t-t_0\right)}}\psi\left(\varsigma\right)d\varsigma$$

is odd with respect to ς, as it is the product of two odd functions. And the integral of an odd function within a symmetric interval with respect to the origin is zero. Consequently, $\varphi_3\left(0,t\right)=0$ what proves the Lemma 9.4.2.

Lemma 9.4.3: If the function $\psi\left(x_3\right)$ is an even function, i.e., $\psi\left(x_3\right)=\psi\left(-x_3\right)$, then the derivative function of $\varphi_3\left(x_3,t\right)$ from formula (9.4.25) is zero when x_3=0 for all $t>t_0$.

Proof: Differentiating the formula (9.4.25) by x_3 and substituting x_3=0, we get

$$\left.\frac{\partial\varphi_3}{\partial x_3}\right|_{x_3=0}=-\frac{1}{2\sqrt{\pi}}\int_{-\infty}^{+\infty}\frac{\left(x_3-\varsigma\right)}{2\left[a\left(t-t_0\right)\right]^{3/2}}e^{-\frac{\left(x_3-\varsigma\right)^2}{4a\left(t-t_0\right)}}\psi\left(\varsigma\right)d\varsigma\Bigg|_{x_3=0}=0\,,$$

since the integrand is odd at x_3=0, if the function $\psi\left(x_3\right)$ is even. Lemma 9.4.3 is proved. ■

We turn now to the construction of the function $\varphi_3\left(x_3,t\right)$, which satisfies the diffusion equation, homogeneous boundary condition of the first kind and the initial condition

$$\varphi_3\left(x_3,0\right)=\psi\left(x_3\right)=\begin{cases}\chi\left(x_3\right) & if\ x_3>0,\\ -\chi\left(-x_3\right) & if\ x_3<0.\end{cases} \tag{9.4.26}$$

Considering the values of the function $\varphi_3\left(x_3,t\right)$ only in the region of interest x_3≥0 and using the definition (9.4.26) of the function $\psi\left(x_3\right)$ from (9.4.25), we have

$$\varphi_3\left(x_3,t\right)=\frac{1}{2\sqrt{\pi}}\int_{-\infty}^{0}\frac{1}{\sqrt{a\left(t-t_0\right)}}e^{-\frac{\left(x_3-\varsigma\right)^2}{4a\left(t-t_0\right)}}\psi\left(\varsigma\right)d\varsigma+\frac{1}{2\sqrt{\pi}}\int_{0}^{+\infty}\frac{1}{\sqrt{a\left(t-t_0\right)}}e^{-\frac{\left(x_3-\varsigma\right)^2}{4a\left(t-t_0\right)}}\psi\left(\varsigma\right)d\varsigma$$
$$=-\frac{1}{2\sqrt{\pi}}\int_{0}^{+\infty}\frac{1}{\sqrt{a\left(t-t_0\right)}}e^{-\frac{\left(x_3+\varsigma\right)^2}{4a\left(t-t_0\right)}}\psi\left(\varsigma\right)d\varsigma+\frac{1}{2\sqrt{\pi}}\int_{0}^{+\infty}\frac{1}{\sqrt{a\left(t-t_0\right)}}e^{-\frac{\left(x_3-\varsigma\right)^2}{4a\left(t-t_0\right)}}\psi\left(\varsigma\right)d\varsigma.$$

Combining both integral together, we get the desired function

$$\varphi_3\left(x_3,t\right)=\frac{1}{2\sqrt{\pi}}\int_{0}^{+\infty}\frac{1}{\sqrt{a\left(t-t_0\right)}}\left[e^{-\frac{\left(x_3-\varsigma\right)^2}{4a\left(t-t_0\right)}}-e^{-\frac{\left(x_3+\varsigma\right)^2}{4a\left(t-t_0\right)}}\right]\psi\left(\varsigma\right)d\varsigma. \quad (9.4.27a)$$

Note that when x_3=0, the expression in square brackets is converted to zero and $\varphi_3\left(x_3,t\right)=0$; (9.4.27a) is the solution of the diffusion equation with homogeneous boundary conditions of the first kind and the initial condition $\varphi_3\left(x_3,t\right)=\chi\left(x_3\right)$.

Applying Lemma 2, we can easily see that the solution of the diffusion equation with homogeneous boundary condition of the second kind, and the initial condition $\varphi_3\left(x_3,t\right)=\chi\left(x_3\right)$ is represented as

$$\varphi_3\left(x_3,t\right)=\frac{1}{2\sqrt{\pi}}\int_{0}^{+\infty}\frac{1}{\sqrt{a\left(t-t_0\right)}}\left[e^{-\frac{\left(x_3-\varsigma\right)^2}{4a\left(t-t_0\right)}}+e^{-\frac{\left(x_3+\varsigma\right)^2}{4a\left(t-t_0\right)}}\right]\psi\left(\varsigma\right)d\varsigma. \quad (9.4.27b)$$

We use these formulas to solve the problem (7.7.19), (7.7.20) of the impurity diffusion in the direction x_3 from an instantaneous point source of power $\tilde{M}$. The result is:

- For the homogeneous boundary condition of the first kind:

$$\varphi_3\left(x_3,t\right)=\frac{\tilde{M}}{2\sqrt{\pi a\left(t-t_0\right)}}\left[e^{-\frac{\left(x_3-x_3^{eff}\right)^2}{4a\left(t-t_0\right)}}-e^{-\frac{\left(x_3+x_3^{eff}\right)^2}{4a\left(t-t_0\right)}}\right].$$

- For the homogeneous boundary condition of the second kind:

$$\varphi_3\left(x_3,t\right)=\frac{\tilde{M}}{2\sqrt{\pi a\left(t-t_0\right)}}\left[e^{-\frac{\left(x_3-x_3^{eff}\right)^2}{4a\left(t-t_0\right)}}+e^{-\frac{\left(x_3+x_3^{eff}\right)^2}{4a\left(t-t_0\right)}}\right].$$

We carry out the reverse transition to the original variable s in these expressions according to the rule (7.7.4), taking into account that

$$s\left(x_1, x_2, x_3, t\right) = s_{1,2}\left(x_1, x_2, t\right) \cdot s_3\left(x_3, t\right).$$

The result is following:

$$s_3\left(x_3, t\right) = \frac{\tilde{M}}{4\pi\mu_3\left(t - t_0\right)}\left[1 \pm e^{-\frac{x_3 x_3^{eff}}{\mu_3\left(t-t_0\right)}}\right] \times e^{-\left\{\varrho\left(t-t_0\right)+\frac{\left[x_3 - x_3^{eff} - v_3\left(t-t_0\right)\right]^2}{4\mu_3\left(t-t_0\right)}\right\}}. \quad (9.4.28)$$

The square brackets, the "plus" sign corresponds to the spread of impurities over the land, when homogeneous boundary conditions of the first kind hold, and the sign "minus" corresponds to the spread of impurities above the water surface when the homogeneous boundary conditions are of the second kind.

The product of $s_{1,2}\left(x_1, x_2, t\right)$ and $s_3\left(x_3, t\right)$ defined by the formulas (9.4.21) and (9.4.28), respectively, gives a solution of problem (9.4.1) – (9.4.3) of impurities distribution at a relatively short distance from the instantaneous point source ("near zone") where the dependence of meteorological parameters v_j (j=1,2,3) on the spatial coordinates can be neglected.

With the help of analytical solutions of atmospheric diffusion problems, one can get a number of independent interesting and important theoretical results. This applies, above all, to establishing the concentration of functional dependencies on key operating factors. At the same time, the analytical methods may be useful in calculating the concentration value in the cloud in close proximity to the site of the accident (explosion). Cross cloud geometrical dimensions, in which the turbulent scattering of impurities within the time interval τ may increase to such values that the problem of matching difference grids will disappear that occurs when the general numerical solution of the meteorological and environmental problems for $t > \tau$. Therefore, the solution of (9.4.1) – (9.4.3) may be based on analytic formulas (9.4.21) and (9.4.28) until the horizontal geometric dimensions of the cloud will not be comparable, as a minimum, to a five-step spatial grid, on which the numerical solution of the weather forecast problem is realized.

Roughness and thermal heterogeneity of the underlying surface can affect significantly the vertical distribution of wind speed and the eddy diffusion coefficient, and, consequently, the convection and diffusion of the impurities in the atmospheric boundary layer. Therefore, in the simulation of accidental releases at small altitudes, or releases of "mushroom" shaped, the impurity concentration cannot be expressed by the formula (9.4.28), even in a rough approximation. The concentration $s_3\left(x_3, t\right)$ in the boundary layer of the atmosphere can be determined only based on numerical methods for solving the split problem (9.4.1) – (9.4.3), as the vertical component of the wind speed v_3 and turbulent diffusion coefficient ν_3 are here vertical coordinate functions.

For cases where the homogeneous and isotropic approximation of the atmosphere are unacceptable in a vertical plane, we change slightly the problem of splitting scheme (9.4.1) - (9.4.3) discussed above. Let be, as well as above, $s\left(x_1, x_2, x_3, t\right) = \bar{s}\left(x_1, x_2, t\right) \cdot s_3\left(x_3, t\right)$. In accordance with this view, we split the general problem of diffusion and convection from the instantaneous point source emissions (9.4.1) – (9.4.3) into following two:

1. Given the values of $\overline{\mu}(H) = \nu_1(x_1, x_2, t) / \xi = \nu_2(x_1, x_2, t) / \xi = const$, $v_1(H) = v_1(x_1, x_2, H) = const$, $v_2(H) = v_2(x_1, x_2, H) = const$, find a bounded function $\overline{s}(x_1, x_2, t)$ defined in the region $-\infty < x_1 < +\infty$, $-\infty < x_2 < +\infty$, $x_3 = H$, $t - t_0 \geq \tau$ that satisfies the equation of convective diffusion

$$\frac{\partial \overline{s}}{\partial t} + v_j \frac{\partial \overline{s}}{\partial x_j} = \frac{\partial}{\partial x_j}\left(\overline{\mu}\frac{\partial \overline{s}}{\partial x_j}\right), j = 1,2, \tag{9.4.29}$$

the initial condition

$$\overline{s} = M\delta\left(x_1 - x_1^{eff}\right)\delta\left(x_2 - x_2^{eff}\right) \text{when } t = t_0, \tag{9.4.30}$$

and the boundary conditions

$$\overline{s} \to 0 \text{ when } x_1 \to \pm\infty, x_2 \to \pm\infty, t - t_0 \geq 0; \tag{9.4.31}$$

2. Given the values $\mu_3(x_3, t) = \nu_3(a, b, x_3, t) / \xi$, $w(x_3, t) = v_3(a, b, x_3, t) + w_g(a, b)$, find a limited function $s(x_3, t)$ defined in $x_1 = a$, $x_2 = b$, $0 \leq x_3 \leq H$, $t - t_0 \leq \tau$ that satisfies the equation of convective diffusion

$$\frac{\partial s_3}{\partial t} + \overline{w}\frac{\partial s_3}{\partial \sigma} + \varrho s_3 = \left(\frac{H}{H - F}\right)^2 \frac{\partial}{\partial \sigma}\left(\mu_3 \frac{\partial s_3}{\partial \sigma}\right), \tag{9.4.32}$$

the initial condition

$$s_3(\sigma, t) = 1 \; when \; \sigma = x_3^{eff}, t = t_0, \tag{9.4.33}$$

and the boundary conditions

$$\alpha_1 \frac{H}{H - F}\frac{\partial s_3}{\partial \sigma} + \beta_1 s_3 = \chi_1 \; when \; \sigma = 0,$$

$$\alpha_2 \frac{H}{H - F}\frac{\partial s_3}{\partial \sigma} + \beta_2 s_3 = \chi_2 \; when \; \sigma = H, \tag{9.4.34}$$

Here, a and b are given values of the coordinates x_1 and x_2 respectively; H is the known value of the upper boundary of the solution area; $\sigma = H\left[x_3 - F\left(x,y\right)\right] / \left[H - F\left(x,y\right)\right]$ is the vertical coordinate associated with terrain $F\left(x,y\right)$; $\tilde{w}$ is the reduced vertical velocity related to the terrain (see Section 5.1.) and having the form:

$$\tilde{w} = \frac{H}{H - F\left(x_1, x_2\right)} \left\{ w - \left(1 - \frac{x_3}{H}\right) \left[u \frac{\partial F\left(x_1, x_2\right)}{\partial x_1} + v \frac{\partial F\left(x_1, x_2\right)}{\partial x_2} \right] \right\}.$$

The boundary conditions (9.4.34) recorded as total differential form, depends on the given numbers $\alpha_1, \beta_1, \chi_1$ and $\alpha_2, \beta_2, \chi_2$. If $\alpha_1 = 0$, then the value of unknown function s_3 is set on the underlying surface, and if $\beta_1 = 0$, then the value of its derivative is set at the same point. Similarly, it is for the upper boundary of the solution.

The solution (9.4.29) – (9.4.31) for a given horizontal plane $-\infty < x_1 < +\infty$, $-\infty < x_2 < +\infty$ and $\sigma = x_3 = H$ with $t - t_0 \le \tau$ is provided by (9.4.21), i.e.

$$\overline{s}\left(x_1, x_2, t\right) = \frac{M}{4\pi\overline{\mu}\left(t - t_0\right)} e^{-\frac{\left[x_1 - x_1^{eff} - v_1\left(t - t_0\right)\right]^2}{4\overline{\mu}\left(t - t_0\right)}} \times e^{-\frac{\left[x_2 - x_2^{eff} - v_2\left(t - t_0\right)\right]^2}{4\overline{\mu}\left(t - t_0\right)}}, \tag{9.4.35}$$

The problem (9.4.32) - (9.4.34) must be solved only by numerical methods. In the construction of difference schemes for the equations (4.9.33), it is possible to use different approximation (explicit, implicit, symmetrical, single-sided) for the individual terms of the equation. Consider some of the existing difference schemes approximating the problem (9.4.32) - (9.4.34). Build the uneven grid (9.2.30) along $0 \le \sigma \le H$ with steps $\zeta_k = \sigma_k - \sigma_{k-1}$ $\left(k = 1, 2, \ldots, K\right)$. Approximating spatial derivatives by three-point difference relations

$$\overline{w} \frac{\partial s_3}{\partial \sigma} \approx \overline{w}_k \frac{\zeta_k^2 \left[\left(s_3\right)_{k+1} - \left(s_3\right)_k\right] + \zeta_{k+1}^2 \left[\left(s_3\right)_k - \left(s_3\right)_{k-1}\right]}{\zeta_k \zeta_{k+1} \left(\zeta_k + \zeta_{k+1}\right)},$$

$$\frac{\partial}{\partial \sigma}\left(\mu_3 \frac{\partial s_3}{\partial \sigma}\right) \approx \frac{\zeta_k \left[\left(\mu_3\right)_{k+1} + \left(\mu_3\right)_k\right]\left[\left(s_3\right)_{k+1} - \left(s_3\right)_k\right]}{\zeta_k \zeta_{k+1} \left(\zeta_k + \zeta_{k+1}\right)} - \frac{\zeta_{k+1} \left[\left(\mu_3\right)_{k+1} + \left(\mu_3\right)_k\right]\left[\left(s_3\right)_k - \left(s_3\right)_{k-1}\right]}{\zeta_k \zeta_{k+1} \left(\zeta_k + \zeta_{k+1}\right)}, \tag{9.4.36}$$

and the derivative with respect to t – by one-sided difference quotient "forward" with a constant step $\ell = t^{n+1} - t^n$ $\left(n = 0, 1, \ldots\right)$, we obtain for the equation (9.4.32) the following schemes.

Explicit Four-Point Scheme

Substituting the difference expression of first and second order (9.4.36), relating them to the n-th temporal layer, into the equation (9.4.32), will have the resulting scheme:

$$\frac{\left(s_3\right)_k^{n+1}-\left(s_3\right)_k^n}{\ell}=-\overline{w}_k^n\,\frac{\zeta_k^2\left[\left(s_3\right)_{k+1}^n-\left(s_3\right)_k^n\right]+\zeta_{k+1}^2\left[\left(s_3\right)_k^n-\left(s_3\right)_{k-1}^n\right]}{\zeta_k\zeta_{k+1}\left(\zeta_k+\zeta_{k+1}\right)}-\varrho\left(s_3\right)_k^n$$
$$+\left(\frac{H}{H-F}\right)^2\left\{\frac{\zeta_k\left[\left(\mu_3\right)_{k+1}^n+\left(\mu_3\right)_k^n\right]\left[\left(s_3\right)_{k+1}^n-\left(s_3\right)_k^n\right]}{\zeta_k\zeta_{k+1}\left(\zeta_k+\zeta_{k+1}\right)}-\frac{\zeta_{k+1}\left[\left(\mu_3\right)_k^n+\left(\mu_3\right)_{k-1}^n\right]\left[\left(s_3\right)_k^n-\left(s_3\right)_{k-1}^n\right]}{\zeta_k\zeta_{k+1}\left(\zeta_k+\zeta_{k+1}\right)}\right\},\tag{9.4.37}$$

$2\le k\le K-1$.

The number of equations in (9.4.37) is less than the number of unknowns $\left(s_3\right)_k^n$, $1\le k\le K$; missing equations are derived from the boundary conditions (9.4.34):

$$\left(s_3\right)_1^n=\frac{\left(s_3\right)_2^n-\zeta_1\chi_1/\alpha_1}{1-\zeta_1\beta_1/\alpha_1},\left(s_3\right)_K^n=\frac{\left(s_3\right)_{K-1}^n-\zeta_{K-1}\chi_2/\alpha_2}{1-\zeta_{K-1}\beta_2/\alpha_2}.$$

In the internal nodes the scheme (9.4.37) is implemented very simply. Since in the difference equation, there is only one unknown $\left(s_3\right)_k^{n+1}$, solving it at each time step is carried out by the recurrence formula $\left(s_3\right)_k^{n+1}=\left(1+\Lambda\right)\left(s_3\right)_k^n$ $\left(2\le k\le K-1\right)$, where Λ is the difference operator of the right side of (9.4.37). The error of approximation of the scheme (9.4.37) at $\zeta=\zeta_k=\zeta_{k+1}$ is equals to $O\left(\ell,\zeta^2\right)$. In order to study its resistance to the initial data we use the method of separation of variables (Fourier method), using the approach of "freezing" the coefficients of the variables $\overline{w}$ and μ_3.

In accordance with the method of Fourier, a solution of homogeneous linear difference equations with constant coefficients on the grid $t^n=t^0+n\ell$ when $\zeta=\zeta_k=\zeta_{k+1}$ is written as:

$$\left(s_3\right)_k^n=A\exp\left(i\eta k\zeta\right),\left(s_3\right)_k^{n+1}=\lambda\left(s_3\right)_k^n,\tag{9.4.38}$$

where η is an arbitrary real number; A is a constant amplitude; λ is the growth factor; $i=\sqrt{-1}$.

Substituting (9.4.38) in (9.4.37) and dividing by $\lambda\exp\left(i\eta k\zeta\right)$, we get

$$\lambda = 1 - 4\frac{\ell\mu_3}{\zeta^2}\sin^2\left(\frac{\eta\zeta}{2}\right) - i\frac{\ell w}{\zeta}\sin\eta\zeta + \varrho\ell\,. \tag{9.4.39}$$

A necessary and sufficient condition for the limitations of harmonic disturbances (9.4.38) by Neumann is as follows:

$$\left|\lambda\right| \le 1 + \upsilon\ell, \upsilon = \text{const}\,. \tag{9.4.40}$$

Indeed, suppose that (9.4.40). When $t^n = t^0 + n\ell \le \tau$, we have:

$$\left|\lambda^n\right| \le \left(1 + \upsilon\ell\right)^{\tau/\ell} \le \exp\left(\upsilon\ell\right).$$

The sufficient condition (4.9.41) is proved. Now prove necessary. Suppose that for $t^n = t^0 + n\ell \le \tau$ and for some $\mathcal{L} > 1$ we have $\left|\lambda^n\right| \le \mathcal{L}$. Putting $t^0 + n\ell = \tau$, we find

$$\left|\lambda\right| \le \mathcal{L}^{1/n} = \exp\left[\ell\ln\left(\frac{\mathcal{L}}{\tau}\right)\right] \le 1 + \upsilon\ell\,,$$

i.e., the condition (9.4.40) is satisfied.

The Neumann condition (9.4.40) with the conventional (formal) study of the stability, the members of the order of ℓ are insignificant, therefore (9.4.40) can be replaced by:

$$\left|\lambda\right| \le 1, \tag{9.4.41}$$

and the term $\varrho\ell$ in (9.4.39) can be omitted, what we will do.

For further consideration, it is convenient to introduce the following notation:

$$\vartheta = \sin^2\left(\frac{\eta\zeta}{2}\right), k_p = 2\frac{\ell\mu_3}{\zeta^2}, k_g = \frac{\ell\left|w\right|}{\zeta}, \tag{9.4.42}$$

where k_p and k_g are parabolic and hyperbolic Courant numbers, respectively. A sufficient condition for stability of (9.4.41) can be represented as $1 - \left|\lambda\right|^2 \ge 0$ and substitute in (9.4.39). After simple transformations, we obtain

$$4k_p\vartheta - 4k_p^2\vartheta^2 - 4k_g^2\vartheta\left(1 - \vartheta\right) \ge 0\,,$$

from where

$$k_p - k_p^2\vartheta - k_g^2\left(1-\vartheta\right) \geq 0 .$$

The left side of this inequality is a linear function of ϑ, therefore, it suffices to verify the inequality in the extreme points of the interval of ϑ changes, i.e. when $\vartheta = 0$ and $\vartheta = 1$. Taking $\vartheta = 1$, we get $k_p \geq k_p^2$, from where

$$k_p \leq 1, \ell \leq \frac{\zeta^2}{2\mu_3}, \tag{9.4.43a}$$

When $\vartheta = 0$, we have $k_p \geq k_g^2$ or $k_g \geq k_g^2$, from where we have

$$\ell \leq 2\frac{\mu_3}{w^2}. \tag{9.4.43b}$$

Thus, the explicit scheme (9.4.37) is conditionally stable under very onerous conditions (04.09.44). In practical calculations performed at finite values of the time step ℓ, violation of conditions (9.4.43) can lead to unrealistic results.

The Implicit Four-Point Scheme

Transfer the grid analogs of derivatives by σ and member ϱs on the n+1 time layer. The result is

$$\frac{\left(s_3\right)_k^{n+1} - \left(s_3\right)_k^n}{\ell} = -\overline{w}_k^{n+1}\left\{\frac{\zeta_k^2\left[\left(s_3\right)_{k+1}^{n+1} - \left(s_3\right)_k^{n+1}\right]}{\zeta_k\zeta_{k+1}\left(\zeta_k + \zeta_{k+1}\right)} + \frac{\zeta_{k+1}^2\left[\left(s_3\right)_k^{n+1} - \left(s_3\right)_{k-1}^{n+1}\right]}{\zeta_k\zeta_{k+1}\left(\zeta_k + \zeta_{k+1}\right)}\right\} - \varrho\left(s_3\right)_k^{n+1}$$
$$+\left(\frac{H}{H-F}\right)^2\left\{\frac{\zeta_k\left[\left(\mu_3\right)_{k+1}^{n+1} + \left(\mu_3\right)_k^{n+1}\right]\left[\left(s_3\right)_{k+1}^{n+1} - \left(s_3\right)_k^{n+1}\right]}{\zeta_k\zeta_{k+1}\left(\zeta_k + \zeta_{k+1}\right)} - \frac{\zeta_{k+1}\left[\left(\mu_3\right)_k^{n+1} + \left(\mu_3\right)_{k-1}^{n+1}\right]\left[\left(s_3\right)_k^{n+1} - \left(s_3\right)_{k-1}^{n+1}\right]}{\zeta_k\zeta_{k+1}\left(\zeta_k + \zeta_{k+1}\right)}\right\}, \tag{9.4.44}$$

$$2 \leq k \leq K - 1 .$$

The scheme (9.4.44) is a system of algebraic equations with a tridiagonal matrix that at each time step is realized by sweep as described in Sec. 8.1.3. The approximation error at $\zeta = \zeta_k = \zeta_{k+1}$ is equal to $O\left(\ell, \zeta^2\right)$.

Investigating stability based on the Fourier method of harmonic disturbances by Neumann, we find

$$\lambda = \frac{1}{1+4\dfrac{\ell\mu_3}{\zeta^2}\sin^2\left(\dfrac{\eta\zeta}{2}\right)+i\dfrac{\ell w}{\zeta}\sin\eta\zeta-\varrho\ell}. \tag{9.4.45}$$

If $\varrho = 0$, then (9.4.45) implies $\left|\lambda\right| \le 1$, and, consequently, the stability of $\ell \to 0, \zeta \to 0$ is valid for arbitrary values k_p and k_g.

Implicit Six-Point Balanced Scheme

We integrate numerically the equation (9.4.32) in the interval $t = \left[t^n, t^{n+1}\right]$ using the trapezium formula. The result is

$$\begin{aligned}
\frac{\left(s_3\right)_k^{n+1}-\left(s_3\right)_k^{n}}{\ell} = &-\frac{1}{2}\overline{w}_k^{n+1}\left\{\frac{\zeta_k^2\left[\left(s_3\right)_{k+1}^{n+1}-\left(s_3\right)_k^{n+1}\right]}{\zeta_k\zeta_{k+1}\left(\zeta_k+\zeta_{k+1}\right)}+\frac{\zeta_{k+1}^2\left[\left(s_3\right)_k^{n+1}-\left(s_3\right)_{k-1}^{n+1}\right]}{\zeta_k\zeta_{k+1}\left(\zeta_k+\zeta_{k+1}\right)}\right\} \\
&+\overline{w}_k^{n}\left\{\frac{\zeta_k^2\left[\left(s_3\right)_{k+1}^{n}-\left(s_3\right)_k^{n}\right]}{\zeta_k\zeta_{k+1}\left(\zeta_k+\zeta_{k+1}\right)}+\frac{\zeta_{k+1}^2\left[\left(s_3\right)_k^{n}-\left(s_3\right)_{k-1}^{n}\right]}{\zeta_k\zeta_{k+1}\left(\zeta_k+\zeta_{k+1}\right)}\right\}-\frac{1}{2}\left[\varrho\left(s_3\right)_k^{n+1}+\varrho\left(s_3\right)_k^{n}\right] \\
&+\frac{1}{2}\left(\frac{H}{H-F}\right)^2\left\{\begin{aligned}&\frac{\zeta_k\left[\left(\mu_3\right)_{k+1}^{n+1}+\left(\mu_3\right)_k^{n+1}\right]\left[\left(s_3\right)_{k+1}^{n+1}-\left(s_3\right)_k^{n+1}\right]}{\zeta_k\zeta_{k+1}\left(\zeta_k+\zeta_{k+1}\right)}+\frac{\zeta_k\left[\left(\mu_3\right)_{k+1}^{n}+\left(\mu_3\right)_k^{n}\right]\left[\left(s_3\right)_{k+1}^{n}-\left(s_3\right)_k^{n}\right]}{\zeta_k\zeta_{k+1}\left(\zeta_k+\zeta_{k+1}\right)}\\
&-\frac{\zeta_{k+1}\left[\left(\mu_3\right)_k^{n+1}+\left(\mu_3\right)_{k-1}^{n+1}\right]\left[\left(s_3\right)_k^{n+1}-\left(s_3\right)_{k-1}^{n+1}\right]}{\zeta_k\zeta_{k+1}\left(\zeta_k+\zeta_{k+1}\right)}-\frac{\zeta_{k+1}\left[\left(\mu_3\right)_k^{n}+\left(\mu_3\right)_{k-1}^{n}\right]\left[\left(s_3\right)_k^{n}-\left(s_3\right)_{k-1}^{n}\right]}{\zeta_k\zeta_{k+1}\left(\zeta_k+\zeta_{k+1}\right)}\end{aligned}\right\}
\end{aligned} \tag{9.4.46}$$

$2 \le k \le K-1$.

Like (9.4.44), the scheme (9.4.46) is a system of algebraic equations with a tridiagonal matrix and at each time step is realized by the sweep. Its approximation error at $\zeta = \zeta_k = \zeta_{k+1}$, in contrast to (9.4.44), is $O\left(\ell^2, \zeta^2\right)$.

Assuming, as before, $\varrho = 0$, we find

$$\lambda = \frac{1-2\dfrac{\ell\mu_3}{\zeta^2}\sin^2\left(\dfrac{\eta\zeta}{2}\right)-i\dfrac{1}{2}\dfrac{\ell w}{\zeta}\sin\eta\zeta}{1+2\dfrac{\ell\mu_3}{\zeta^2}\sin^2\left(\dfrac{\eta\zeta}{2}\right)+i\dfrac{1}{2}\dfrac{\ell w}{\zeta}\sin\eta\zeta}. \tag{9.4.47}$$

It is obvious that $\left|\lambda\right| \leq 1$ for arbitrary ℓ and ζ, hence, the scheme (9.4.46) is unconditionally stable. Since the error of approximation scheme is of second order, it allows evaluation on smooth solutions with larger time steps than the first-order scheme (9.4.36), (9.4.44). Therefore, the additional cost of machine time for the implementation of the scheme (4.9.47), as a rule, pays off.

Continued consideration of methods of construction of finite-difference analogs and methods for solving the resulting algebraic equations systems is difficult because of the very wide variety of suitable methods. For one-dimensional parabolic equations (9.4.32), presented schemes give results that differ little from other existing schemes. For the equation (9.4.32), we recommend the scheme (9.4.46), because it provides a second-order accuracy in time and space.

The product of solutions $\overline{s}\left(x_1, x_2, t\right)$ of the problem (4.9.30) – (4.9.32) and $s_3\left(x_3, t\right)$ of the problem (9.4.32) – (9.4.34), defined by (9.4.35) and (4.9.47), respectively, provides a solution to the split of the problem (9.4.1) – (9.4.3) of impurity scattering from the instantaneous source in the "near field". This construction of the proposed computational algorithm can solve the problem of modeling of transport and diffusion of impurities in the atmosphere from a point source on almost acceptable the grids, on which hydrodynamic model of the atmosphere is numerically implemented.

If the problem of the salvo emissions and can be considered as instantaneous point source, the application of this concept to problems of nuclear or powerful chemical explosion, obviously, can lead to significant errors. Indeed, as a result of a powerful nuclear or chemical explosion in the atmosphere stands out a significant amount of energy that forms a spherical region of the explosion products. This region in the form of a spherical shock wave propagates with the radial velocity of generally higher than speed of sound. Therefore, in a very short time, the geometric dimensions of the spherical area reached such a magnitude that, in principle, cannot be identified with the notion of a point source. And therefore, in this case, the problem of matching grids when sharing the numerical solution of problems of modeling of atmospheric processes and the propagation of a powerful nuclear or chemical explosion products does not exist.

SOLVING THE PROBLEM OF AN IMPURITY SCATTERING IN THE "NEAR ZONE" IN THE CASE OF A STATIONARY POINT SOURCE

Actual surveillance of continuously operating inert emission sources of almost weightless impurities (gases and particle diameter less than 20 microns) show [Oke T. R. (1978). Boundary layer climates. London. Methuen & Co LTD. 358 p], that at a distance from the source 100 - 10,000 m in the direction of the wind at a fixed time instance, the plan view of the torch (the distribution of the impurity concentration) typically has a bell-shape curve with a sharp maximum in the torch axis. In the following times, as a result of the time averaged concentration, smoother curves are formed with the filtered short periodical fluctuations. The growing band width of perturbations does distribution curve a flat, with maximum value decreasing. A similar bell-shape of the impurity concentration distribution curve is traced in a vertical plane. However, as the width of the range for the perturbation of these planes are usually different, the turbulence intensity, and, therefore, the torch width in the vertical and horizontal transverse directions are not the same.

We use the formula (5.3.24c) to determine the effective displacement of a permanent source coordinates, that was obtained on the basis of extensive experimental materials Schetz (1980) on research of depending active part of the jet axis blown at an angle of 90° in respect to the razing flow. Effective coordinates of the beginning of the passive portion of the torch $x_1^{eff} = x_1^0 + \Delta x_1$, $x_2^{eff} = x_2^0 + \Delta x_2$, $x_3^{eff} = x_3^0 + \Delta x_3$ are determined by the formula (5.3.24c) from the condition $d\left(\Delta x_3\right) / dr \to 0$. Therefore, in modeling of processes of formation of fields of concentration from a stationary point source emissions, one needs to set the coordinates and the physical parameters of the primary source by the passport data of industrial enterprises, and then calculate by (5.3.24s) the effective coordinates, that are taken as the initial data for passive torch area.

Since the total amount of impurities per unit time passing through the vertical section of the torch, that is perpendicular to the average wind direction, should be equal to the intensity of the source, and this cross section area is gradually increased, it is necessary that the average concentration of impurity decreased with increasing distance from the source. This also means that the area under the bell-shape concentration curve should remain constant at any distance from the source. And if the width is increased, the maximum value of the impurity concentration decreases. At any distance from the source, concentration on the torch axis is greater than on its borders.

Therefore, we can assume that random perturbations due to atmospheric turbulence, scatter an impurity so that its concentration is distributed over the binormal law relative to the central axis of the torch (i.e., the normal distribution law in the horizontal and vertical planes). The mathematical description of these curves allows numerically simulate the dispersion of the torch perturbation, i.e., calculate the concentration from permanently emitting sources for inert, almost weightless impurities by formula

$$s\left(x_1, x_2, x_3\right) = \frac{M}{2\pi\phi_2\phi_3 v_1} \exp\left\{-\frac{1}{2}\left[\left(x_2 - x_2^{eff}\right)^2 / \phi_2^2 + \left(x_3 - x_3^{eff}\right)^2 / \phi_3^2\right]\right\}, \tag{9.4.48}$$

where M is the source strength (kg/s); ϕ_2 and ϕ_3 are mean square deviations of the impurity (m) by the coordinate axes x_2 and x_3 respectively; v_1 is an average wind speed (m/s) in the direction of the coordinate axis x_1; $x_j^{eff}\left(j = 1,2,3\right)$ are effective coordinates of the source (m) defined by the formula (5.3.24c), which allows to find the height of the torch rise. Therefore, the value of s is measured in units of mass density (kg/m^3).

The expression (9.4.48) takes into account all the factors affecting the formation of the impurity concentration from the point source. So, obviously, the impurity concentration should be proportional to the intensity M of the source and inversely proportional to the wind speed v_1 because of the stretching torch along its direction ($s \approx M / v_1$). It is also known that changes in the concentration of the flame have a random, Gaussian character. This is shown in equation (9.4.49) by introducing appropriate mean square deviation related to the transverse dimensions of the torch. They grow under the influence of turbulence and, therefore, are functions of the distance from the source and atmospheric stratification. The concentration is inversely proportional to ϕ_2 and ϕ_3, since large values of these characteristics meet the advanced diffusion. Finally, as it was earlier found, the concentration on the underlying surface at a fixed distance from the source decreases with an increase in its effective height x_3^{eff}. In the relation (9.4.48), it is also assumed that when the impurity reaches the underlying surface, it is not absorbed and reflected back into the atmosphere. This is due to the presence of the last term in (9.4.48).

The experimental values ϕ_2 and ϕ_3 can be derived from the field of instrumental observations, and their theoretical assessment is based on the results of the statistical theory of turbulence. The following dependency has enough wide application

$$\phi_j^2 = 0.5\nu_j x_1^{(2-n)},$$

where the parameter *n* is determined by the conditions of stability of the atmosphere that is associated with the vertical profile of the wind speed by the relation

$$v_1 = v_g \left(\frac{x_3}{h_b} \right)^{\left(\frac{n}{2-n} \right)}.$$

Here, v_g is the geostrophic wind speed taken at the upper edge of the boundary layer of the atmosphere h_b.

The above model is often used to analyze the radiation environment near the planned nuclear power plants and for the evaluation of air pollution from industrial and chemical enterprises when selecting a place for its construction (AUND-86, 1987).

The equation (4.9.49) is simplified by writing it for the concentration at the level of the underlying surface

(x_3=0):

$$s\left(x_1, x_2, 0\right) = \frac{M}{\pi\phi_2\phi_3} \exp\left\{ -\frac{1}{2} \left[\left(x_2 - x_2^{eff}\right)^2 / \phi_2^2 + \left(x_3^{eff}\right)^2 / \phi_3^2 \right] \right\}, \tag{9.4.49a}$$

and if, in addition, the source is not raised, and there is no torch rise ($x_3^{eff} = 0$), then

$$s\left(x_1, x_2, 0\right) = \frac{M}{\pi\phi_2\phi_3} \exp\left\{ -\frac{1}{2} \left[\left(x_2 - x_2^{eff}\right)^2 / \phi_2^2 \right] \right\}. \tag{9.4.49b}$$

This model may be modified when there is a need to take into account the special conditions such as:

- Conformity of the rate of spreading impurities to the power law distribution $v_1\left(x_3\right) = Cx_3^n$.
- The presence of elevated inversion, limiting the spread of impurities up.
- The effect of terrain heterogeneity of the underlying surface.

The simple model described by (9.4.48) cannot be used for more detailed study of the migration process and turbulent diffusion of impurities from a permanent source of emissions, because it is based on a large number of simplifying assumptions, which in the real atmosphere is never completely satisfied. If one approaches more strictly or interested in the distribution of concentrations in the atmosphere at the actual weather conditions, it is necessary to turn to more complex convection-diffusion model of the scattering gas and aerosol emissions, which can be solved in the general case only with numerical methods.

In the theoretical study of the propagation of passive impurities ejected from permanent sources into the moving turbulent atmosphere, it is essential to coordinate the advection-diffusion convection process with the process of the considered scalar substance, i.e., with a method for determining the average and the time scales of considered phenomena.

If there was no turbulent diffusion, the impurity under investigation would be tolerated by a current line coincided with the direction of the wind. In fact, as is well known from the field observations, there is always some cross-blurring impurities along the main path. From the point where it was thrown impurity produced, some diverging cone arises with variable impurity concentration over the cross section. The nature of this erosion is closely related to small-scale turbulent fluctuations of wind speed. Thus, the advection-convective transport determines the average speed of the wind, and the diffusion process defines the small-scale turbulence of surrounding environment in which the passive impurity is distributed. At a fully developed turbulence in the atmosphere, there is a wide spectrum of turbulent scales and frequencies. Relatively small vortices determine instantaneous values of velocity and concentration fluctuations. When averaging operation is performed in the diffusion equation (5.3.3), in fact, only their impact is taken into account. Therefore, the determinant parameters of the problem v_j (j=1,2,3), appearing in the turbulent diffusion equation (5.3.6), belong to the interval of time τ, which is practically amounts to almost 2–3 minutes. Consequently, the concentration calculated by this equation should give the average values for the interval τ. When $t \le \tau$, the transport processes and diffusion of impurities in the atmosphere from stationary industrial sources can be regarded as constant, i.e., $\partial s / \partial t = 0$.

Further, if the axis x_1 is oriented in the direction of the wind, then the following assumptions will hold at the distance from the source $L \le \tau v_1$: $v_2=0$, $u = v_1 \approx const$, $w = v_3 \approx const$, $\mu = \nu_1 / \xi \approx \nu_2 / \xi \approx \nu_3 / \xi \approx const$ (homogeneous and isotropic turbulence) and $\mu \partial^2 s / \partial x_1^2 \approx 0$. After the above simplifications, the initial equation (5.3.6), the initial and boundary conditions take the form:

$$u\frac{\partial s}{\partial x} + w\frac{\partial s}{\partial z} = \mu\left(\frac{\partial^2 s}{\partial y^2} + \frac{\partial^2 s}{\partial z^2}\right) - \varrho s, \tag{9.4.50}$$

$$0 \le x \le \tau u, -\infty \le z \le +\infty, -\infty \le y \le +\infty,$$

$$us = M\delta\left(y - y_0\right)\delta\left(z - z_0\right) when\ x = 0,$$

$$s \to 0\ when\ y \to \pm\infty, z \to \pm\infty,$$

where $x=x_1$, $y=x_2$, $z=x_3$; M is the release of substances from the source per unit of time. It further assumes that the origin $\left(x_0, y_0, z_0\right)$ addressing the problem (9.4.50) coincides with the effective coordinates of the point source, $x_0 = x_1^{eff} = 0, y_0 = x_2^{eff}, z_0 = x_3^{eff}$.

We represent the solution of the problem (9.4.50) as a product of two factors

$$s(x,y,z) = s_{\tau}(x,y,z)e^{-\varrho x}, \tag{9.4.51}$$

where s_{τ} is an averaged distribution of the impurity concentration over the interval $t \le \tau$ due to convection and diffusion. If we substitute this expression for the impurity concentration in the formula (9.4.50), after performing appropriate transformations, we obtain:

$$u\frac{\partial s_{\tau}}{\partial x} + w\frac{\partial s_{\tau}}{\partial z} = \mu\left(\frac{\partial^2 s_{\tau}}{\partial y^2} + \frac{\partial^2 s_{\tau}}{\partial z^2}\right), \tag{9.4.52}$$

$$0 \le x \le \tau u, -\infty \le z \le +\infty, -\infty \le y \le +\infty,$$

$$us_{\tau} = M\delta(y - y_0)\delta(z - z_0)\, when\ x = 0,$$

$$s_{\tau} \to 0\ when\ y \to \pm\infty, z \to \pm\infty,$$

The new task (9.4.52) almost exactly coincides with (9.4.50), but it lacks the term $-\varrho s$. Thus, instead of solving the problem (9.4.50), we solve the transformed problem (9.4.52), and then multiply the resulting solution by $e^{-\varrho x}$.

Replace the independent variables (x,y,z) by the new variables (η,ζ,τ) by the formulas

$$\eta = y, \zeta = z - wx / u, \tau = x.$$

Note that the symbols τ and x, and η and y denote the same magnitude. These designations are introduced in order to avoid confusion in the following material. We use the formulas of transition to the new system of coordinates (η,ζ,τ) for the partial derivatives:

$$\frac{\partial s_{\tau}}{\partial x} = \frac{\partial s_{\tau}}{\partial \tau}\frac{\partial \tau}{\partial x} + \frac{\partial s_{\tau}}{\partial \zeta}\frac{\partial \zeta}{\partial x} + \frac{\partial s_{\tau}}{\partial \eta}\frac{\partial \eta}{\partial x} = \frac{\partial s_{\tau}}{\partial \tau} - \frac{w}{u}\frac{\partial s_{\tau}}{\partial \zeta},$$

$$\frac{\partial s_{\tau}}{\partial y} = \frac{\partial s_{\tau}}{\partial \tau}\frac{\partial \tau}{\partial y} + \frac{\partial s_{\tau}}{\partial \zeta}\frac{\partial \zeta}{\partial y} + \frac{\partial s_{\tau}}{\partial \eta}\frac{\partial \eta}{\partial y} = \frac{\partial s_{\tau}}{\partial \eta},$$

$$\frac{\partial^2 s_{\tau}}{\partial y^2} = \frac{\partial}{\partial y}\left(\frac{\partial s_{\tau}}{\partial \eta}\right) = \frac{\partial^2 s_{\tau}}{\partial \eta^2},$$

$$\frac{\partial s_\tau}{\partial z} = \frac{\partial s_\tau}{\partial \tau}\frac{\partial \tau}{\partial z} + \frac{\partial s_\tau}{\partial \zeta}\frac{\partial \zeta}{\partial z} + \frac{\partial s_\tau}{\partial \eta}\frac{\partial \eta}{\partial z} = \frac{\partial s_\tau}{\partial \zeta},$$

$$\frac{\partial^2 s_\tau}{\partial z^2} = \frac{\partial}{\partial z}\left(\frac{\partial s_\tau}{\partial \zeta}\right) = \frac{\partial^2 s_\tau}{\partial \zeta^2}.$$

Substituting the expressions for $\partial s_\tau / \partial x$, $\partial s_\tau / \partial y$, $\partial^2 s_\tau / \partial y^2$, $\partial s_\tau / \partial z$, $\partial^2 s_\tau / \partial z^2$ in equation (9.4.52), we obtain

$$\frac{\partial s_\tau}{\partial \tau} = \frac{\mu}{u}\left(\frac{\partial^2 s_\tau}{\partial \eta^2} + \frac{\partial^2 s_\tau}{\partial \zeta^2}\right).$$

Thus, the new challenge for s_τ in terms of the new variables (η, ζ, τ), which is called the initial-boundary value problem or the task with the initial conditions for the heat equation, has the form

$$\frac{\partial s_\tau}{\partial \tau} = \frac{\mu}{u}\left(\frac{\partial^2 s_\tau}{\partial \eta^2} + \frac{\partial^2 s_\tau}{\partial \zeta^2}\right). \tag{9.4.53}$$

$$us_\tau = M\delta(\eta - \eta_0)\delta(\zeta - \zeta_0)\, when\ \tau = 0,$$

$$s_\tau(\eta, \zeta, \tau) \to 0 when\ \eta, \zeta \to \pm\infty.$$

Note that the spatial variables ζ and η vary from -∞ to +∞, so the solution of (9.4.53) can be obtained by applying the Fourier transform. We make it to obtain the solution for an easy suspended impurity in the form of:

$$s_\tau(\eta, \zeta, \tau) = \frac{M}{4\pi\mu\tau}\exp\left[-u\frac{(\eta - \eta_0)^2 + (\zeta - \zeta_0)^2}{4\mu\tau}\right]. \tag{9.4.54a}$$

The solution (9.4.54a) describes the concentration distribution s_τ only at short distances from the source, while keeping the symmetry of the impurity of the torch relative to the source level, i.e. yet there is no influence of the underlying surface. This fact is not taken into account in a number of works, and therefore is one of the major causes of discrepancies between the calculated concentration and the observed data, especially at some distance from the source. The terms of interaction of impurities stream with the underlying surface can be taken into account if to reformulate the boundary conditions (9.4.53) for the semi-infinite line $0 \le \zeta < +\infty$, excluding the cases of study atmospheric diffusion over rough terrain and over a heterogeneous underlying surface. Using the principle of odd extension (9.4.27), it is easy to ensure that the solution of the first boundary value problem on the semi-infinite line $0 \le \zeta < +\infty$ ($s_\tau = 0$ when $\zeta = 0$) is given by

$$s_\tau\left(\eta,\zeta,\tau\right) = \frac{M}{4\pi\mu\tau}\exp\left[-u\frac{\left(\eta-\eta_0\right)^2}{4\mu\tau}\right]\left\{\exp\left[-u\frac{\left(\zeta-\zeta_0\right)^2}{4\mu\tau}\right] - \exp\left[-u\frac{\left(\zeta+\zeta_0\right)^2}{4\mu\tau}\right]\right\}, \tag{9.4.54b}$$

and with a homogeneous boundary condition of the second kind ($\partial s_\tau / \partial\zeta = 0$ when $\zeta = 0$) is given by the formula

$$s_\tau\left(\eta,\zeta,\tau\right) = \frac{M}{4\pi\mu\tau}\exp\left[-u\frac{\left(\eta-\eta_0\right)^2}{4\mu\tau}\right]\left\{\exp\left[-u\frac{\left(\zeta-\zeta_0\right)^2}{4\mu\tau}\right] + \exp\left[-u\frac{\left(\zeta+\zeta_0\right)^2}{4\mu\tau}\right]\right\}, \tag{9.4.54c}$$

The boundary condition on the underlying surface $s_\tau = 0$ identifies the cases in which an impurity is distributed over the water surface. Since water absorbs impurity, its concentration is just above its surface is zero. The soil surface usually interacts with an impurity in more complicated way. Once on the ground surface, the impurity is not accumulated because some part of it is again returned to the atmosphere by air currents. Therefore, it is assumed, to simplify the task, that the average flow of the impurities at the surface is small, that is $\partial s_\tau / \partial\zeta = 0$.

It remains to make the final step – to write the solution of our problem in the original coordinate system (x,y,z). Substituting $\eta = y, \zeta = z - wx / u, \tau = x$ in (9.4.54a) – (9.4.54c) and taking into account (9.4.51), we finally obtain

$$s\left(x,y,z\right) = \frac{M}{4\pi\mu x}\exp\left\{-\frac{\varrho}{u}x - \left[-u\frac{\left(y-y_0\right)^2 + \left(z-z_0-\frac{w}{u}x\right)^2}{4\mu x}\right]\right\}, \tag{9.4.55a}$$

$$s\left(x,y,z\right) = \frac{M}{4\pi\mu x}\exp\left[-\frac{\varrho}{u}x - u\frac{\left(y-y_0\right)^2}{4\mu x}\right]\left\{\exp\left[-u\frac{\left(z-z_0-\frac{w}{u}x\right)^2}{4\mu x}\right] - \exp\left[-u\frac{\left(z+z_0-\frac{w}{u}x\right)^2}{4\mu x}\right]\right\} \tag{9.4.55b}$$

$$s\left(x,y,z\right)=\frac{M}{4\pi\mu x}\exp\left[-\frac{\varrho}{u}x-u\frac{\left(y-y_0\right)^2}{4\mu x}\right]\left\{\exp\left[-u\frac{\left(z-z_0-\frac{w}{u}x\right)^2}{4\mu x}\right]+\exp\left[-u\frac{\left(z+z_0-\frac{w}{u}x\right)^2}{4\mu x}\right]\right\}, \tag{9.4.55c}$$

It is easily seen that for solutions (9.4.55a), (9.4.55c), the following expression is valid

$$\iiint_{\mathfrak{H}} sdydzdx = M ,$$

and for (9.4.55b), holds the expression

$$\iiint_{\mathfrak{H}} sdydzdx = 0 .$$

The integration is performed over the entire space $\mathfrak{H}$ of the impurity distribution. These integral equalities are a natural consequence of the fact that the impurity either is completely absorbed by the underlying surface (the condition s=0), or completely retained in the atmosphere (condition $\partial s / \partial z = 0$).

The special assessment conducted in (Prusov, Doroshenko, 2006) have shown that an increase in the duration of the sampling up to $t_s \approx 20 \div 30$ minute results in significantly lower the experimental data than the number obtained on the basis of the analytical methods the solution of problem (9.4.50) – (4.9.52) for $x=ut_s$ at averaged actual meteorological values v_j (j=1,2,3). This divergence is the more significant, the greater the distance from the source. I.e., there is the effect of "false" growth of transverse dimensions of the torch. This is because the average wind direction over time, continuously varies, together with the position of the torch axis of a distributed impurity. This result can be interpreted as "mesoscale effect of averaging of concentration". Its assessment should be related to taking into account the characteristics of fluctuations in wind direction caused, in particular, by the action of atmospheric vortices of all sizes. These are set according to the analysis of continuous recording material for a long time. Changing the concentration versus time averaging, to which it refers, it was seen in a number of studies mainly based on qualitative arguments. In order to eliminate differences that occur between the calculated and measured data, there are often introduced empirical coefficients. They reduce the value of the calculated concentrations by several times. So, for the concentrations s_τ and s_t, averaged over time intervals of duration τ and t_s, the empirical relation $s_t = s_\tau\left(\tau / t\right)^m$ at $m \approx 0.2 \div 0.4$ are sometimes used. The undeniable disadvantage of this relation is that it does not take into account the specific weather conditions. And the only using it in the empirical relationships to analyze averaging naturally narrows the use of this relation. It seems interesting theoretical study of a dependency of fluctuations dispersion of wind direction on the inner τ and outer t_s τ intervals of averaging based on generalizations of observational data (Berlyand, M.E. (1975)).

According to meteorological observations in the spectrum of fluctuations of meteorological variables in the real atmosphere, there is a failure in the area of perturbations with the period of $t_s \approx 60$ minutes. Therefore, meteorological values obtained by instrumental observations in the course of the four-dimentional objective analysis are subjected to averaging. The results of this averaging are equivalent to the data given on the smoothed synoptic charts. This allows to extend the results obtained on the basis of analytical problem solving techniques (09.04.51) – (04.09.52), over large intervals ($t_s \approx 60$ min), provided that the calculated average concentration over a time interval $t_s > \tau$ will be subjected to further clarification, taking into account the existence of a real large-scale unsteadiness of wind fluctuations in magnitude and direction. These fluctuations, such as the direction in the horizontal plane, cause substantial deviations of the torch axis from its central position and further contribute to scattering of impurity.

There are two alternative approaches to the calculation of the concentration for the time interval $t_s > \tau$. The first of these, less strict, is associated with the "freezing" of the mean values of turbulent diffusion coefficients v_j $\left(j = 1,2,3\right)$ in the studied time interval t_s. The second way is more rigorous, it takes large-scale fluctuations into account by statistical averaging of concentrations for a period t_s by the ensemble of implementation for the period τ.

For the averaged values of the impurity concentration at an arbitrary point $\mathrm{B}\left(x,y,z\right)$ on the time interval t_s, we assume that in this period of time the wind fluctuations by the direction occur at random, and the probability is described by a Gaussian distribution function

$$f\left(\alpha\right) = \frac{1}{\phi_\alpha \sqrt{2\pi}} \exp\left[-\frac{\left(\Delta\alpha\right)^2}{2}\right]. \tag{9.4.56}$$

Here, the deviation angle $\Delta\alpha = \alpha - \bar{\alpha}$, ($\alpha$ is the mean wind direction in the interval τ, τ is the same interval on t_s); ϕ_α is the standard deviation of the wind direction that is determined on the basis of processing the input data set according to the formula for wind

$$\phi_\alpha^2 = \frac{1}{N-1}\sum_{i=1}^{N}\left(\alpha_i - \bar{\alpha}\right)^2, \tag{9.4.57}$$

where N is a sample size.

Processing of observations (Berlyand, 1975) has identified the presence of dependence ϕ_α^2 on the outside of the averaging interval t_s at the internal interval $\tau = 2.5$ min for cases of unstable and stable atmospheric conditions. From these data it follows that with a change in averaging period from 20 to 40 minutes, the value ϕ_α increases by 20% for unstable stratification and by 35% at a stable one. Increasing the averaging period of from 40 minutes to 60 leads to an increase ϕ_α respectively by 15% and 25% depending on the stability of the atmosphere.

In accordance with the known fundamentals of the probability theory, the averaged concentration value for the period t_s is determined by the expression

$$s_t = \int_{\bar{\alpha}-\pi/2}^{\bar{\alpha}+\pi/2} s_\tau(\alpha) f(\alpha) d\alpha, \tag{9.4.58}$$

where $s_\tau(\alpha)$ is a solution of the diffusion problem for the direction α.

Substituting (9.4.56) to (9.4.58) and integrating, we obtain

$$s_t = \frac{s_\tau(x - x_0, y - y_0, z - z_0)}{2\pi(x - x_0)(\bar{\phi}^2 - \phi_\alpha^2)} \exp\left(-\frac{\alpha_B^2}{2\phi_\alpha^2}\right). \tag{9.4.59}$$

Here, α_B is the angle between the mean wind direction $\bar{\alpha}$ and the direction into the computed point .

Matching experimental and theoretical results allows to extend the proposed algorithm over longer time intervals t_s (up to 2 hours, and sometimes more) and different weather conditions.

In the analytic solution of the problem of atmospheric diffusion for a time interval $t_s \geq 60$ min, one must take into account that the thickness of the air layer in which an impurity is scattered from high sources ($z_0 = 200 \div 300$) can be significant. In this layer, the change of wind speed and components of the coefficient of turbulent exchange are often complex. The study of atmospheric diffusion in rough terrain and over an heterogeneous underlying surface also requires consideration of complex changes of meteorological variables in the vertical direction. In such cases, the solution of the problem is possible only with the assistance of modern numerical methods.

In the study of transport processes and turbulent diffusion of a passive impurity in the area $x \leq ut_s$, where the typical transverse torch dimensions do not exceed a few hundred meters, we shall continue to neglect the variability of meteorological parameters (speed, temperature, coefficient of turbulent diffusion, and others.) in the horizontal plane and to consider them, in the first approximation, independent on x and y. For this area, one can take the assumption of isotropy of horizontal turbulence, i.e., to assume that $v_1 = v_2 = v$. This approximation allows to search the solution of (9.4.50) to be represented as:

$$s(x, y, z) = s_1(x, y) \cdot s_2(x, \sigma), \tag{9.4.60}$$

where the functions $s_1(x, y), s_2(x, \sigma)$ must satisfy the equations

$$u\frac{\partial s_1}{\partial x} = \frac{\nu}{\xi}\frac{\partial^2 s_1}{\partial y^2} - \varrho s_1, \tag{9.4.61}$$

$$u\frac{\partial s_2}{\partial x} + (\bar{w} + w_g)\frac{\partial s_2}{\partial z} = \left(\frac{H}{H - F}\right)^2 \frac{\partial}{\partial z}\left(\frac{\nu_3}{\xi}\frac{\partial s_2}{\partial z}\right) \tag{9.4.62}$$

and initial and boundary conditions:

- For the equation (9.4.61):

$$s_1 = M\delta\left(y - y_0\right) when\ x = 0\,,$$

$$s_1\left(x, y\right) \to 0\ when\ y \to \pm\infty\,. \tag{9.4.63}$$

- For the equation (4.9.63):

$$s_2\left(x, z\right) = 1\ when\ z \to z_0, x = 0\,,$$

$$s_2\left(x, z\right) \to 0\ when\ z \to H\,,$$

$$\frac{\nu_3}{\xi}\frac{\partial s_2}{\partial z} = \beta_0 s_2\ when\ z = 0\,. \tag{9.4.64}$$

In (9.4.60) – (9.4.64), it is put into consideration the vertical coordinate (see Section 5.1.)

$$\sigma = H\left[x_3 - F\left(x, y\right)\right] / \left[H - F\left(x, y\right)\right]$$

and the reduced vertical speed

$$\overline{w} = \frac{H}{H - F\left(x_1, x_2\right)}\left[w - \left(1 - \frac{x_3}{H}\right)u\frac{\partial F\left(x_1, x_2\right)}{\partial x_1}\right],$$

associated with terrain *F*(*x*,*y*). Here, the functions $\overline{w} = \overline{w}\left(\sigma\right)$, $\nu_3 = \nu\left(\sigma\right)$ depend on the vertical coordinate σ; w_g is own deposition rate of heavy impurities (gravitational velocity of the particles); *H* is the height of the area of solving the problem; β_0 is a constant characterizing the interaction of impurities with the underlying surface.

The solution of the problem (9.4.61), (9.4.63) can be obtained analytically similar to the problem (9.4.50). Introducing the notation $\mu = \nu / \xi$ and leaving in the expression (9.4.55a) the terms depending only on *x* and *y*, we obtain the relation describing an averaged distribution of impurity concentration over the interval $t \leq \tau$:

$$s_\tau\left(x, y\right) = \frac{M}{4\pi\mu x}\exp\left\{-\frac{\varrho}{u}x - \left[u\frac{\left(y - y_0\right)^2}{4\mu x}\right]\right\}. \tag{9.4.65}$$

To obtain the averaged values of the impurity concentration at an arbitrary point $B(x,y)$ in the time interval t_s, we use dependence (9.4.59). As a result, we have

$$s_1 = \frac{s_\tau \left(x, y - y_0\right)}{2\pi \left(\overline{\phi}^2 - \phi_\alpha^2\right)} \exp\left(-\frac{\alpha_B^2}{2\phi_\alpha^2}\right). \tag{9.4.66}$$

We now turn to the solution of the problem (9.4.62), (9.4.64). It can only be obtained by numerical method, since the vertical component of the wind speed w and the coefficient of turbulent diffusion ν_3 are functions of the vertical coordinate in the layer of the atmosphere.

Build an uneven grid (9.2.30) with steps $\zeta_k = \sigma_k - \sigma_{k-1}$ (k=1,2,…,K) along $0 \le \sigma \le H$. Approximate spatial derivatives in (9.4.62) with three-point difference relations

$$\left(\overline{w} + w_g\right)\frac{\partial s_2}{\partial \sigma} \approx \left(\overline{w} + w_g\right)\frac{\zeta_k^2\left[\left(s_2\right)_{k+1} - \left(s_2\right)_k\right] + \zeta_{k+1}^2\left[\left(s_2\right)_k - \left(s_2\right)_{k-1}\right]}{\zeta_k\zeta_{k+1}\left(\zeta_k + \zeta_{k+1}\right)},$$

$$\frac{\partial}{\partial \sigma}\left(\mu_3 \frac{\partial s_2}{\partial \sigma}\right) \approx \frac{\zeta_k\left[\left(\mu_3\right)_{k+1} + \left(\mu_3\right)_k\right]\left[\left(s_2\right)_{k+1} - \left(s_2\right)_k\right]}{\zeta_k\zeta_{k+1}\left(\zeta_k + \zeta_{k+1}\right)} - \frac{\zeta_{k+1}\left[\left(\mu_3\right)_{k+1} + \left(\mu_3\right)_k\right]\left[\left(s_2\right)_k - \left(s_2\right)_{k-1}\right]}{\zeta_k\zeta_{k+1}\left(\zeta_k + \zeta_{k+1}\right)},$$

where the notation $\mu_3 = \nu_3 / \xi$ is used and replace the differential equation with partial derivatives (9.4.62) in all internal nodes of the selected vertical grid by the following difference equations:

$$\begin{aligned}
\frac{\left(s_2\right)_k^{n+1} - \left(s_2\right)_k^n}{\ell} = &-\frac{1}{2}\left(\overline{w}_k^{n+1} + w_g\right)\left\{\frac{\zeta_k^2\left[\left(s_2\right)_{k+1}^{n+1} - \left(s_2\right)_k^{n+1}\right]}{\zeta_k\zeta_{k+1}\left(\zeta_k + \zeta_{k+1}\right)} + \frac{\zeta_{k+1}^2\left[\left(s_2\right)_k^{n+1} - \left(s_2\right)_{k-1}^{n+1}\right]}{\zeta_k\zeta_{k+1}\left(\zeta_k + \zeta_{k+1}\right)}\right\} \\
&+\left(\overline{w}_k^n + w_g\right)\left\{\frac{\zeta_k^2\left[\left(s_3\right)_{k+1}^n - \left(s_3\right)_k^n\right]}{\zeta_k\zeta_{k+1}\left(\zeta_k + \zeta_{k+1}\right)} + \frac{\zeta_{k+1}^2\left[\left(s_3\right)_k^n - \left(s_3\right)_{k-1}^n\right]}{\zeta_k\zeta_{k+1}\left(\zeta_k + \zeta_{k+1}\right)}\right\}, \\
&+\frac{1}{2}\left(\frac{H}{H-F}\right)^2\left\{\begin{array}{l}\dfrac{\zeta_k\left[\left(\mu_3\right)_{k+1}^{n+1} + \left(\mu_3\right)_k^{n+1}\right]\left[\left(s_2\right)_{k+1}^{n+1} - \left(s_2\right)_k^{n+1}\right]}{\zeta_k\zeta_{k+1}\left(\zeta_k + \zeta_{k+1}\right)} + \dfrac{\zeta_k\left[\left(\mu_3\right)_{k+1}^n + \left(\mu_3\right)_k^n\right]\left[\left(s_2\right)_{k+1}^n - \left(s_2\right)_k^n\right]}{\zeta_k\zeta_{k+1}\left(\zeta_k + \zeta_{k+1}\right)} \\ -\dfrac{\zeta_{k+1}\left[\left(\mu_3\right)_k^{n+1} + \left(\mu_3\right)_{k-1}^{n+1}\right]\left[\left(s_2\right)_k^{n+1} - \left(s_2\right)_{k-1}^{n+1}\right]}{\zeta_k\zeta_{k+1}\left(\zeta_k + \zeta_{k+1}\right)} - \dfrac{\zeta_{k+1}\left[\left(\mu_3\right)_k^n + \left(\mu_3\right)_{k-1}^n\right]\left[\left(s_2\right)_k^n - \left(s_2\right)_{k-1}^n\right]}{\zeta_k\zeta_{k+1}\left(\zeta_k + \zeta_{k+1}\right)}\end{array}\right\}
\end{aligned} \tag{9.4.67}$$

$2 \le k \le K-1$.

The scheme (9.4.67) is a system of algebraic equations with a tridiagonal matrix. Its solution at each time step is realized by a sweep (see the section 8.1.3.).

The product (9.4.60) of the solutions that meet the problems (9.4.61), (9.4.63) and (9.4.62), (9.4.64), will meet the original problem (9.4.50). It sets the averaged concentration of the passive impurity $s(x_1, x_2, x_3, t)$ at the point of the torch coordinates $x \le ut_s$, y and σ from a stationary point source of the power M with effective coordinates $x_0 = x_1^{eff}, y_0 = x_2^{eff}, z_0 = x_3^{eff}$. The torch part with horizontal scale $L \ge ut_s$ (e.g., $L \ge 3600$ m at $u \approx 10$ m/s and $t_s = 60$ min), which is modeled with the assumptions above to obtain a theoretical solution of the problem (9.4.50) in the closed form, will be called below as "near zone".

Solving the Problem of the Impurity Scattering in the "Far Zone"

Behind the "near zone" ($x \le ut_s$), as the distance grows from both instant and a stationary point source, some subsequent changes of the cloud scattering structure (torch) occur by the influence of two factors:

- Transfer as a whole over rugged terrain and uneven the underlying surface in a direction and at a speed matching the direction and the speed of the air flow that is variable in time and space.
- Anisotropic diffusion of atmospheric turbulence integral consequence of which is the formation of smoke clubs of finite size, the gradual expansion of the cloud (torch) and changes in its geometry.

Therefore, it can be assumed that behind the "near zone" should be "far zone" in which the radius of the cross-cloud size (torch) is several tens of meters. This conditional division of the problem area allows to apply to a "far zone" the numerical methods that are free of restrictions adopted in the "near zone".

Spreading some aerosol substance emitted by a point source in the atmosphere in the "far zone" is due to the motion of an elementary volume of air at a speed $\mathbf{V} = (v_1, v_2, v_3)$, and diffusion is due to atmospheric turbulence with an exchange ratio $v = (v_1, v_2, v_3)$. We formulate a mathematical problem of migration of the aerosol substance with the concentration $s(x_1, x_2, x_3, t)$ in the "far zone" as follows:

$$\frac{\partial s}{\partial t} + v_j \frac{\partial s}{\partial x_j} + \varrho s = Q_s + \frac{\partial}{\partial x_j}\left(\frac{\nu_j}{\xi}\frac{\partial s}{\partial x_j}\right), j = 1,2,3, t > t_s, \qquad (9.4.68)$$

$$s(x_1, x_2, x_3, t)\Big|_{t=t_s} = s_{nz}(x_1, x_2, x_3),$$

$$s(x_1, x_2, x_3, t) \to s_{bg} \ when \ x_1 \to \pm\infty, x_2 \to \pm\infty,$$

$$\alpha\frac{\partial s}{\partial x_3}+\beta s=\gamma \text{ when } x_3=0.$$

In the equation (4.9.69), there is provided a summation over repeated indices. The projection of the speed v_3 can consist of two components $v_3 = v_3 + w_g$, where w_g is the absolute value of the vertical velocity of the particles under the influence of gravitational force. The member Q_s characterizes the outflow ($Q_s < 0$) or inflow ($Q_s > 0$) of the aerosol substance s through the boundary surface of the solving problem area; ϱs is an internal source (sink) due to impurities transformations as a result of leaching, chemical reactions, etc.

The initial conditions (9.4.68) at $t=t_s$ are given based on the fact that impurity concentration on the windward side of the "far zone" from the point source power the M designed for the "near zone" is maintained throughout the computation time. Outside of the "near zone", the background values of the impurity concentration are taken as the boundary conditions or the atmosphere is considered to be "clean" and free from impurities s_{bg}=0.

Consider some of the most common difference schemes for the numerical solution of the diffusion equation (9.4.68). Assume the notation:

$$\vec{\delta}_i^n s=\frac{s_{i+1,j,k}^n-s_{i,j,k}^n}{\Delta x_{1i}},\overleftarrow{\delta}_i^n s=\frac{s_{i,j,k}^n-s_{i-1,j,k}^n}{\Delta x_{1i-1}},$$

$$\vec{\delta}_j^n s=\frac{s_{i,j+1,k}^n-s_{i,j,k}^n}{\Delta x_{2j}},\overleftarrow{\delta}_j^n s=\frac{s_{i,j,k}^n-s_{i,j-1,k}^n}{\Delta x_{2j-1}},$$

$$\vec{\delta}_k^n s=\frac{s_{i,j,k+1}^n-s_{i,j,k}^n}{\Delta x_{3k}},\overleftarrow{\delta}_k^n s=\frac{s_{i,j,k}^n-s_{i,j,k-1}^n}{\Delta x_{3k-1}},$$

$$\vec{\mu}_i^n=\frac{\nu_{i+1,j,k}^n+\nu_{i,j,k}^n}{\xi},\overleftarrow{\mu}_i^n=\frac{\nu_{i,j,k}^n+\nu_{i-1,j,k}^n}{\xi},$$

$$\vec{\mu}_j^n=\frac{\nu_{i,j+1,k}^n+\nu_{i,j,k}^n}{\xi},\overleftarrow{\mu}_j^n=\frac{\nu_{i,j,k}^n+V_{i,j-1,k}^n}{\xi},$$

$$\vec{\mu}_k^n=\frac{\nu_{i,j,k+1}^n+\nu_{i,j,k}^n}{\xi},\overleftarrow{\mu}_k^n=\frac{\nu_{i,j,k}^n+\nu_{i,j,k-1}^n}{\xi}.$$

Simple Explicit Method

Using the operator record of derivatives approximation formulas in the equation (9.4.68), we write the explicit one-step scheme

$$\frac{s_{i,j,k}^{n+1}-s_{i,j,k}^{n}}{\ell}=Q_{i,j,k}-\varrho s_{i,j,k}^{n}+\sum_{m}\frac{1}{\Delta x_{m}+\Delta x_{m-1}}\begin{bmatrix}\vec{\mu}_{m}^{n}\vec{\delta}_{m}^{n}-\overleftarrow{\mu}_{m}^{n}\overleftarrow{\delta}_{m}^{n}\\-v_{m}\left(\Delta x_{m}\vec{\delta}_{m}^{n}+\Delta x_{m-1}\overleftarrow{\delta}_{m}^{n}\right)\end{bmatrix}s, m=\left\{i,j,k\right\}, \tag{9.4.69}$$

having an order of approximation $O\left[\ell,\left(\Delta x_{1}\right)^{2},\left(\Delta x_{2}\right)^{2}\left(\Delta x_{3}\right)^{2}\right]$. The algorithm implementation of the scheme (9.4.69) is quite simple. The solution is obtained by consistent progress (march) from the layer where the initial data are set.

For conditions of the stability of the difference scheme (9.4.69), we use Fourier-analysis (by Neumann) in the space:

$$s\left(x_{1},x_{2},x_{3},t\right)=\overline{s}\left(t\right)e^{i\left(l_{1}x_{1}+l_{2}x_{2}+l_{3}x_{3}\right)}.$$

Substitute this function in the form of homogeneous difference equation (9.4.69). Given that

$$i\sin\left(l_{m}\Delta x_{m}\right)=\frac{e^{il_{m}\Delta x_{m}}-e^{-il_{m}\Delta x_{m}}}{2},$$

$$\cos\left(l_{m}\Delta x_{m}\right)=\frac{e^{il_{m}\Delta x_{m}}+e^{-il_{m}\Delta x_{m}}}{2}, m=\left\{i,j,k\right\},$$

we get

$$\overline{s}^{n+1}=\overline{s}^{n}\left\{1+\sum_{m}\left[2\frac{\nu}{\xi}\frac{\ell}{\left(\Delta x_{m}\right)^{2}}\left\langle\cos\left(l_{m}\Delta x_{m}\right)-1\right\rangle-iv_{m}\frac{\ell}{\Delta x_{m}}\sin\left(l_{m}\Delta x_{m}\right)\right]\right\}, m=\left\{i,j,k\right\}.$$

The resulting expression binds Fourier amplitude modes in successive layers of time. The member in the braces is called the transition coefficient. The equation for transition coefficient

$$\lambda=1+2r\left(\cos\theta-1\right)-i\varsigma\sin\theta,$$

where we have introduced the notation $h=\min\Delta x_{m}, r=\ell\nu/\left(\xi h^{2}\right), \varsigma=\ell v_{m}/h, \theta=lh, m=\left\{i,j,k\right\}$, describes an ellipse with center in point (1-2*r*) of the real axis. The major and minor half-axis of the ellipse are equal to $2r$ and ς respectively. Moreover, the ellipse touches the unit circle at the intersection of that circle with the positive real axis. A necessary condition for the stability of $\left|\lambda\right| \leq 1$ is equivalent to the requirement that an ellipse is completely contained inside the unit circle. This condition imposes the following restrictions on the length of the major and minor half-axes of the ellipse: $\varsigma \leq 1$, $2r \leq 1$. However, even under these conditions the difference scheme (9.4.69) may be unstable. Full stability condition can be found by examining in the usual manner an expression for the transition rate module $\left|\lambda\right| \leq 1$. Having made such an analysis, we get

$$\varsigma^2 \leq 2r, r \leq 1/2. \tag{9.4.70}$$

As shown by numerical experiments, the solution (9.4.69) with $r=1/2$ is characterized by strong dissipation at high values of the parameter θ. Only when $r \leq 1/6$, there is a match of the transition coefficient λ with its exact value.

A useful parameter used in the investigation of the stability of the numerical solution of the diffusion equation is the Reynolds grid number, determined from the relation $\mathrm{Re}_{\Delta x_m} = \xi v_m \Delta x_m / \nu$. This dimensionless parameter, which reflects the ratio of convection to diffusion, plays an important role in determining the nature of the solution of equation (9.4.68). Reynolds grid number can be expressed with $r_m = (\nu / \xi)\left[\ell / (\Delta x_m)^2\right]$ and $\varsigma_m = \ell v_m / \Delta x_m$ as follows:

$$\mathrm{Re}_{\Delta x_m} = \frac{\xi v_m \Delta x_m}{\nu} = \frac{\ell v_m}{\Delta x_m} \frac{\xi (\Delta x_m)^2}{\ell \nu} = \frac{\varsigma_m}{r_m}.$$

The stability conditions (9.4.70) impose the following restrictions on the value of Reynolds grid number $\mathrm{Re}_{\Delta x} = \min\left(\mathrm{Re}_{\Delta x_m}\right)$:

$$2\varsigma \leq \mathrm{Re}_{\Delta x} \leq 2/\varsigma. \tag{9.4.71}$$

Condition (9.4.71) is a very severe restriction on the ratio of steps in time and space. Consequently, the use of explicit method in practice is not advisable. Furthermore, it follows from (9.4.69) that the solution on the layer t^{n+1} is not dependent on boundary conditions. At the same time, the solution of the diffusion equation (9.4.68) on the layer t^{n+1} should depend on the boundary conditions, since the characteristics of parabolic equations have the form $t=const$. Thus, a simple explicit scheme (with a finite ℓ) simulates the physical characteristics of partial differential equations of parabolic type incorrectly. Therefore, to solve the diffusion equation (9.4.68), it is better to use implicit methods, since they take into account all known information on the characteristic $t=const$ and under it.

Simple Implicit Method

The corresponding difference scheme is written as

$$\frac{s^{n+1}_{i,j,k}-s^{n}_{i,j,k}}{\ell}=Q_{i,j,k}-\varrho s^{n+1}_{i,j,k}+\sum_{m}\frac{1}{\Delta x_m+\Delta x_{m-1}}\left[\begin{matrix}\vec{\mu}^{n+1}_m\vec{\delta}^{n+1}_m-\overleftarrow{\mu}^{n+1}_m\overleftarrow{\delta}^{n+1}_m\\-v_m\left(\Delta x_m\vec{\delta}^{n+1}_m+\Delta x_{m-1}\overleftarrow{\delta}^{n+1}_m\right)\end{matrix}\right]s,m=\{i,j,k\},\tag{9.4.72}$$

The scheme under consideration is the first order of accuracy with approximation error $O\left[\ell,(\Delta x_1)^2,(\Delta x_2)^2,(\Delta x_3)^2\right]$ and is absolutely stable, since the transition coefficient in this case is equal to $\lambda=\left[1+2r\left(\cos\theta-1\right)-i\varsigma\sin\theta\right]^{-1}$.

Unfortunately, as it follows from (9.4.72), the system of linear algebraic equations which is formed on the (*n*+1). -th time step is 7-diagonal matrix, since the difference equation includes seven unknowns $s^{n+1}_{i+1,j,k}$, $s^{n+1}_{i-1,j,k}$, $s^{n+1}_{i,j+1,k}$, $s^{n+1}_{i,j-1,k}$, $s^{n+1}_{i,j,k+1}$, $s^{n+1}_{i,j,k-1}$, $s^{n+1}_{i,j,k}$. Typically, such a system of equations is solved by iterative methods which require much more computer time than to solve a system of equations with a tridiagonal matrix. This problem has led to the creation of cost-effective methods of implicit variable directions. They are closely related and sometimes coincide with the splitting methods, or, as they are sometimes called, the fractional step method. The main idea of these methods is to split the three-dimensional finite-difference operator into one-dimensional operators.

To solve the problem (9.4.68), one can use a number of different existing methods of splitting. Choosing the best method is quite difficult, because the difference in the quality of solutions obtained using methods with similar characteristics (order of approximation, stability, convergence, efficiency, etc..) is often negligible. Therefore, there is not attempt here to describe all the known splitting methods to solve the diffusion equation. They can be found in the literature on the numerical solution of equations of mathematical physics. From the set of the existing splitting numerical methods, we focus attention on unconditionally stable implicit scheme of alternating directions having a second order accuracy.

Three-Step Implicit Method

We generalize the well-known scheme of Crank-Nicholson in the case of three-dimensional diffusion equation. As a result, we obtain the following three-step scheme:

- The first step:

$$\frac{s^{*}_{i,j,k}-s^{n}_{i,j,k}}{\ell}=Q_{i,j,k}-\frac{\varrho}{2}\left(s^{*}_{i,j,k}+s^{n}_{i,j,k}\right)+\frac{0.5}{\Delta x_i+\Delta x_{i-1}}\left[\begin{matrix}\vec{\mu}^{n}_i\left(\vec{\delta}^{*}_i+\vec{\delta}^{n}_i\right)-\overleftarrow{\mu}^{n}_i\left(\overleftarrow{\delta}^{*}_i+\overleftarrow{\delta}^{n}_i\right)\\-v_i\Delta x_i\left(\vec{\delta}^{*}_i+\vec{\delta}^{n}_i\right)+\Delta x_{i-1}\left(\overleftarrow{\delta}^{*}_i+\overleftarrow{\delta}^{n}_i\right)\end{matrix}\right]s$$
$$+\frac{1}{\Delta x_j+\Delta x_{j-1}}\left[\begin{matrix}\vec{\mu}^{n}_j\vec{\delta}^{n}_j-\overleftarrow{\mu}^{n}_j\overleftarrow{\delta}^{n}_j\\-v_j\left(\Delta x_j\vec{\delta}^{n}_j+\Delta x_{j-1}\overleftarrow{\delta}^{n}_j\right)\end{matrix}\right]s+\frac{1}{\Delta x_k+\Delta x_{k-1}}\left[\begin{matrix}\vec{\mu}^{n}_k\vec{\delta}^{n}_k-\overleftarrow{\mu}^{n}_k\overleftarrow{\delta}^{n}_k\\-v_k\left(\Delta x_k\vec{\delta}^{n}_k+\Delta x_{k-1}\overleftarrow{\delta}^{n}_k\right)\end{matrix}\right]s\tag{9.4.73}$$

- The second step:

$$\frac{s_{i,j,k}^{**} - s_{i,j,k}^{n}}{\ell} = Q_{i,j,k} - \frac{\varrho}{2}\left(s_{i,j,k}^{**} + s_{i,j,k}^{n}\right) + \frac{0.5}{\Delta x_i + \Delta x_{i-1}}\begin{bmatrix} \vec{\mu}_i^n\left(\vec{\delta}_i^* + \vec{\delta}_i^n\right) - \overleftarrow{\mu}_i^n\left(\overleftarrow{\delta}_i^* + \overleftarrow{\delta}_i^n\right) \\ -v_i\Delta x_i\left(\vec{\delta}_i^* + \vec{\delta}_i^n\right) + \Delta x_{i-1}\left(\overleftarrow{\delta}_i^* + \overleftarrow{\delta}_i^n\right) \end{bmatrix} s$$
$$+ \frac{0.5}{\Delta x_j + \Delta x_{j-1}}\left[\vec{\mu}_j^n\left(\vec{\delta}_j^{**} + \vec{\delta}_j^n\right) - \overleftarrow{\mu}_j^n\left(\overleftarrow{\delta}_j^{**} + \overleftarrow{\delta}_j^n\right) - v_j\Delta x_j\left(\vec{\delta}_j^{**} + \vec{\delta}_j^n\right) + \Delta x_{j-1}\left(\overleftarrow{\delta}_j^{**} + \overleftarrow{\delta}_j^n\right)\right] s \quad .$$
$$+ \frac{1}{\Delta x_k + \Delta x_{k-1}}\left[\vec{\mu}_k^n\vec{\delta}_k^n - \overleftarrow{\mu}_k^n\overleftarrow{\delta}_k^n - v_k\left(\Delta x_k\vec{\delta}_k^n + \Delta x_{k-1}\overleftarrow{\delta}_k^n\right)\right] s$$

- The third step:

$$\frac{s_{i,j,k}^{n+1} - s_{i,j,k}^{n}}{\ell} = Q_{i,j,k} - \frac{\varrho}{2}\left(s_{i,j,k}^{n+1} + s_{i,j,k}^{n}\right) + \frac{0.5}{\Delta x_i + \Delta x_{i-1}}\begin{bmatrix} \vec{\mu}_i^n\left(\vec{\delta}_i^* + \vec{\delta}_i^n\right) - \overleftarrow{\mu}_i^n\left(\overleftarrow{\delta}_i^* + \overleftarrow{\delta}_i^n\right) \\ -v_i\Delta x_i\left(\vec{\delta}_i^* + \vec{\delta}_i^n\right) + \Delta x_{i-1}\left(\overleftarrow{\delta}_i^* + \overleftarrow{\delta}_i^n\right) \end{bmatrix} s$$
$$+ \frac{0.5}{\Delta x_j + \Delta x_{j-1}}\left[\vec{\mu}_j^n\left(\vec{\delta}_j^{**} + \vec{\delta}_j^n\right) - \overleftarrow{\mu}_j^n\left(\overleftarrow{\delta}_j^{**} + \overleftarrow{\delta}_j^n\right) - v_j\Delta x_j\left(\vec{\delta}_j^{**} + \vec{\delta}_j^n\right) + \Delta x_{j-1}\left(\overleftarrow{\delta}_j^{**} + \overleftarrow{\delta}_j^n\right)\right] s \quad .$$
$$+ \frac{0.5}{\Delta x_k + \Delta x_{k-1}}\left[\vec{\mu}_k^n\left(\vec{\delta}_k^{n+1} + \vec{\delta}_k^n\right) - \overleftarrow{\mu}_k^n\left(\overleftarrow{\delta}_k^{n+1} + \overleftarrow{\delta}_k^n\right) - v_k\Delta x_k\left(\vec{\delta}_k^{n+1} + \vec{\delta}_k^n\right) + \Delta x_{k-1}\left(\overleftarrow{\delta}_k^{n+1} + \overleftarrow{\delta}_k^n\right)\right] s$$

Here, the superscripts "*" and "**" represent intermediate values.

As a result of splitting, the problem is reduced to solving three systems of linear algebraic equations with tridiagonal matrices at the (n+1)-th time level. For each step, the system of linear algebraic equations with a tridiagonal matrix is solved by the sweep, the algorithm of which is given in Sec. 8.1.3. The effectiveness of the difference scheme (9.4.73) of three-dimensional diffusion equations is determined by its physical conditionality, unconditional stability, and the second order of accuracy and efficiency.

Thus, the concept of conditional separation of the region of impurity scattering from a point raised source of emissions in the "near zone" and "far zone" allows calculating the concentration of the field throughout the three-dimensional space surrounding the source of impurity, and therefore getting a general solution of the problem of modeling mesoscale processes in the atmosphere and the formation of pollution fields from instant or fixed point source of emissions under real weather conditions.

Examples of impurity scattering simulation tasks from the elevated and ground instantaneous point sources are shown in Figures 6 and 7.

Figure 6 shows the hourly results of modeling the formation and movement of clouds in the atmosphere for 14 hours after salvo emission (Dovgiy, Prusov, Kopeika, 2000). Initial data for weather forecasting tasks was compiled using objective analysis within 12 hours of 17.01.1996 (the background), the data of objective analysis of at 00 o'clock of 18.01.1996 (the current time), and forecast data for 12 o'clock of 18.01.1996. The solution of the problem of propagation of impurities in the atmosphere (9.4.73) in the "near zone" was carried out with the aid of (9.4.21) and (9.4.28), and "far zone" with the help of the difference scheme (9.4.73).

Figure 6 shows the hourly process of forming a track on the underlying surface for eight hours after the hypothetical salvo emission (Prusov, Kondra, Kopiiko, Mikitarenko et al., 1997). Integral estimate of risk under wind action upon structures of the encasement at Chernobyl atomic power plant. - Genova: Italy, 2nd EACWE. P. 1833-1839.]. The initial data for weather forecasting tasks have been compiled

Figure 6. The track on the earth's surface from the salvo emission: ☼ is an ejection site; a), b), ..., f) forecasts at 3, 4, ..., 8 hours after the salvo emission

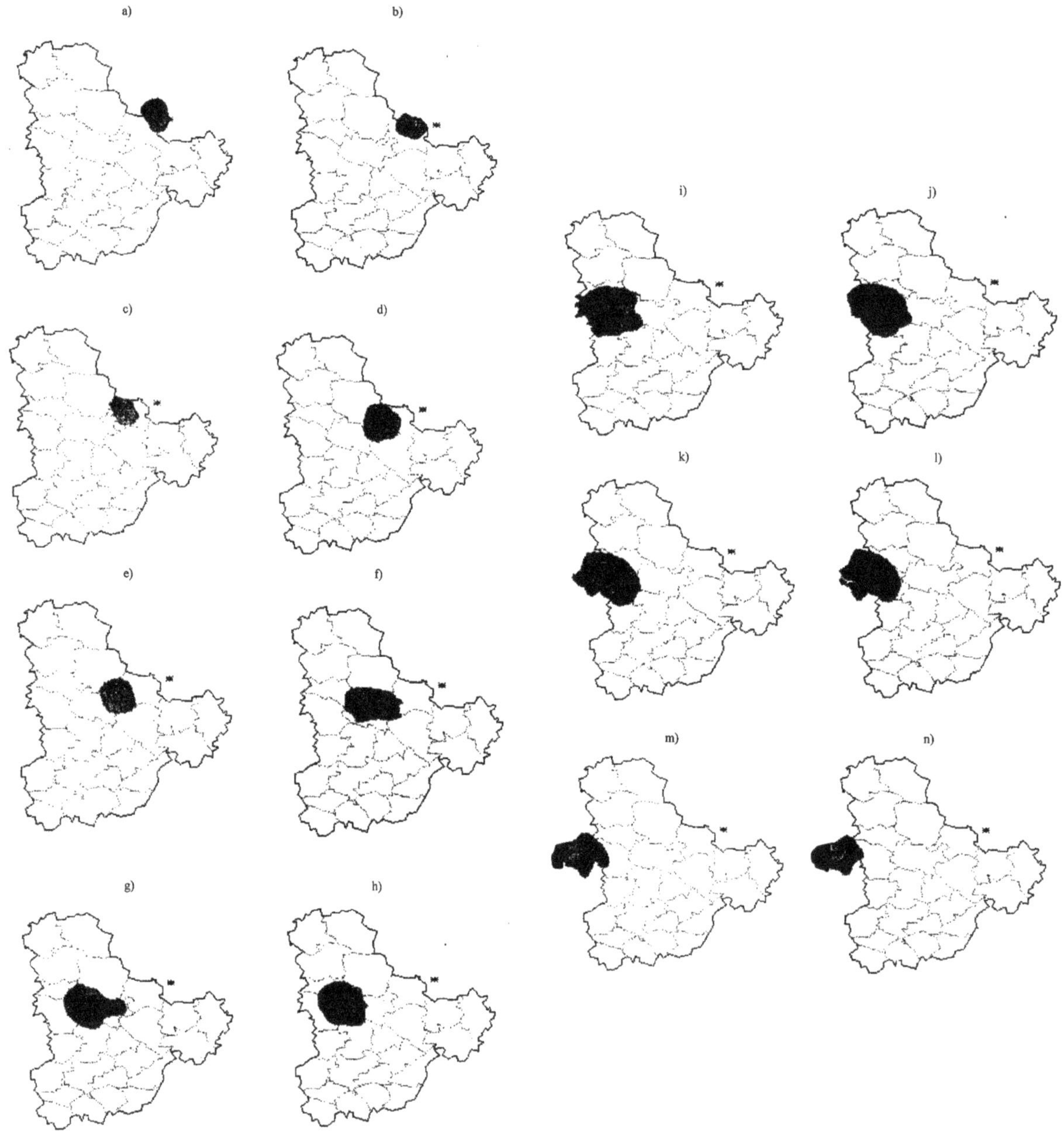

using objective analysis within 12 hours of 05.02.1997 (the background), the data of objective analysis on 00 o'clock of 06.02.1997 (the current time), and forecast data on 12 o'clock of 06.02.1997. The solution of the problem of propagation of impurities in the atmosphere (9.4.73) in the "near zone" has been carried out with the aid of (9.4.21) and the difference scheme (9.4.46), and "far zone" with the help of the difference scheme (9.4.73).

Solving the Problem of an Impurity Scattering from a Group of Sources

Exploiting methods and algorithms described above for calculating the concentrations of impurities diffuse from the group area for emission sources has considerable difficulties. They consist in the need to consider the field of concentrations of arbitrarily arranged emission sources of different geometrical and physical parameters for different speeds and wind directions. A formal approach to solving this problem is to calculate the field of meteorological variables and the impurity concentration in the nodes of a sufficiently dense grid of points, which correspond to concentrations from each individual source. Modern computer resources do not allow for the spatial discretization of the overall ecological and meteorological problems with such a resolution, that scattering of impurities in the vicinity of single emission sources can be described by a mathematical model of atmospheric processes and do not fall into the class of the processes "subgrid" scale. Using spatially heterogeneous grids condensed near unevenly distributed sources algorithmically is hard to implement.

Solving this problem is proved to be very effective as described in section 9.1 of this book with algorithms and methods of "one-way interaction". They enable calculation of the concentration of the impurities from the group scattered across the territory of emission sources under real weather conditions in the complex model (9.1.24) of the atmosphere state (Dovgiy, Prusov, Kopeika, 2000; Prusov, Doroshenko, 2006).

As the meteorological observations show, there is a failure in the field perturbation with the period t_s=1 h in the spectrum of fluctuations of meteorological variables in the real atmosphere. Therefore, the fields of meteorological variables (pressure, temperature, wind speed and direction, humidity), obtained by instrumental observations in the course of a four-dimensional objective analysis undergo averaging per hour and a priori assumed to be unchanged during this hour. The results of this averaging are equivalent to fixed data, which are usually listed on the smoothed synoptic charts.

Consequently, during t_s=1 hour, the actual and prognostic meteorological quantities relating to specific time instants $t = t^n \left(n = 0, 1, 2, \ldots\right)$, presented at the macro-scale grid and used as initial and boundary conditions in the regional model the atmosphere, can be considered constant. If we add the assumption of a uniform velocity field **V** and isotropic turbulence ν in the vicinity of each emission source, then one is able relatively easy, with the help of analytical solutions (9.4.21), (9.4.28) atmospheric diffusion problem (9.4.1) – (9.4.4), to calculate the concentration fields in the nodes mesoscale grid $\bar{\varpi}_h$ of the domain of the regional atmospheric model $\bar{\wp}\left(r\right)$ at time $t=t^n$ (n=0,1,2,...). Obviously, the calculation results based on the analytical methods of solution (9.4.21), (9.4.28) of the problem (9.4.1) – (9.4.4), will be in accordance with certain factual data for relatively small distances from the source.

Thus, the concept of conditional division of the solution area for the problem of scattering impurities from point source emissions by "near zone" and "far zone" allows to specify, on the basis of analytical methods (9.4.21), (9.4.28) for the problem (9.4.1) - (9.4.4), the concentration fields from a group of unevenly distributed sources on the mesoscale grid $\bar{\varpi}_h$ for each time $t=t^n$ (n=0,1,2,...). Next, these calculated concentration fields on an equal footing with the meteorological data (pressure, temperature, humidity, wind speed, and direction), set up on the same time instances $t=t^n$ (n=0,1,2,...) in the nodes of macro-scale grid, being interpolated into mesoscale spatial grid $\bar{\varpi}_h$ of the regional area $\bar{\wp}\left(r\right)$, can participate in solving complex ecological and meteorological problem (9.1.62), (9.1.62) using the interpolation method with multiple nodes (9.1.20).

Figure 7. The track on the earth's surface from a group of stationary sources of emissions (stand for coordinates of the emission source): a) the concentration distribution forecast on 19 o'clock of 04.03.2005, b) the concentration distribution forecast on 11 o'clock of 05.03.2005*

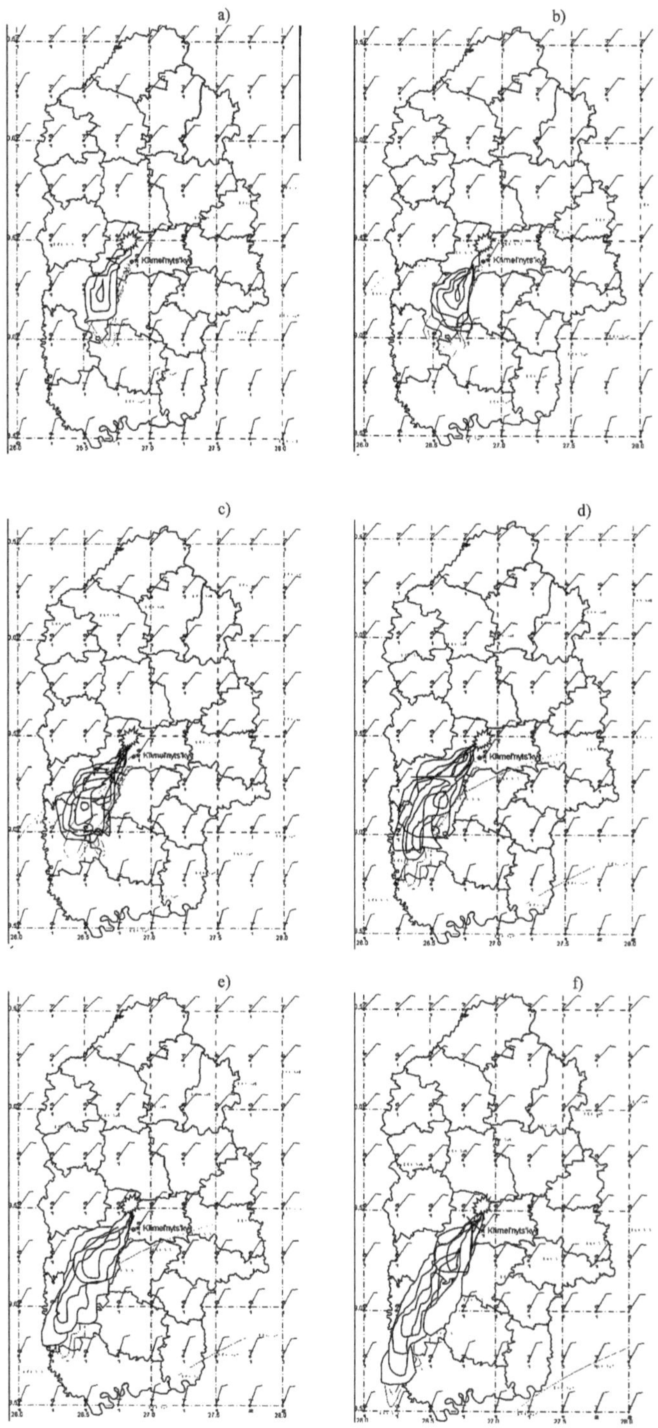

Figure 7 shows the results of numerical experiments on the effect of wind speed, ceteris paribus on the distribution of impurities in the air from a group of permanently acting elevated emission sources. Initial data for weather forecasting tasks have been compiled from objective analysis on the 12 o'clock of 03.03.2005 (the background), the data of objective analysis on the time of 00 o'clock of 04.03.2005 (the current time), and the prognostic information on 12 o'clock of 4.03.2005, forecast data on the 00 o'clock of 05.03.2005, and forecast data on 12 o'clock of 05.03.2005. The solution of the problem of spreading impurities in the atmosphere (9.4.73) in the "near zone" for the formation of concentration fields from each source of emissions on the terms above has been carried out with the aid of (9.4.21) and the difference scheme (9.4.46). The forecast has been implemented on the basis of the solution of complex ecological and meteorological problem (9.1.62), (9.1.62) using the interpolation method with multiple nodes (9.1.20).

As follows from the drawings, the diffusion and transfer of impurities in the atmosphere at a horizontal distance from the source are determined mainly by the magnitude of wind speed. The higher wind speed is, the larger volume of air passes from the source of emission per time unit and the smaller becomes the impurity concentration. It depends on the wind speed what a distance the impurity is transferred at, and a torch shape caused by the impurity concentration in it and when it is pulled along the transfer flow. The wind speed exceeding 15 m/s, and the roughness of the underlying surface cause the spots on the underlying surface.

As shown by numerical experiments, mathematical models and methods described above allow solving the problems of state control and monitoring for the atmosphere, experienced from the impact

Figure 8.

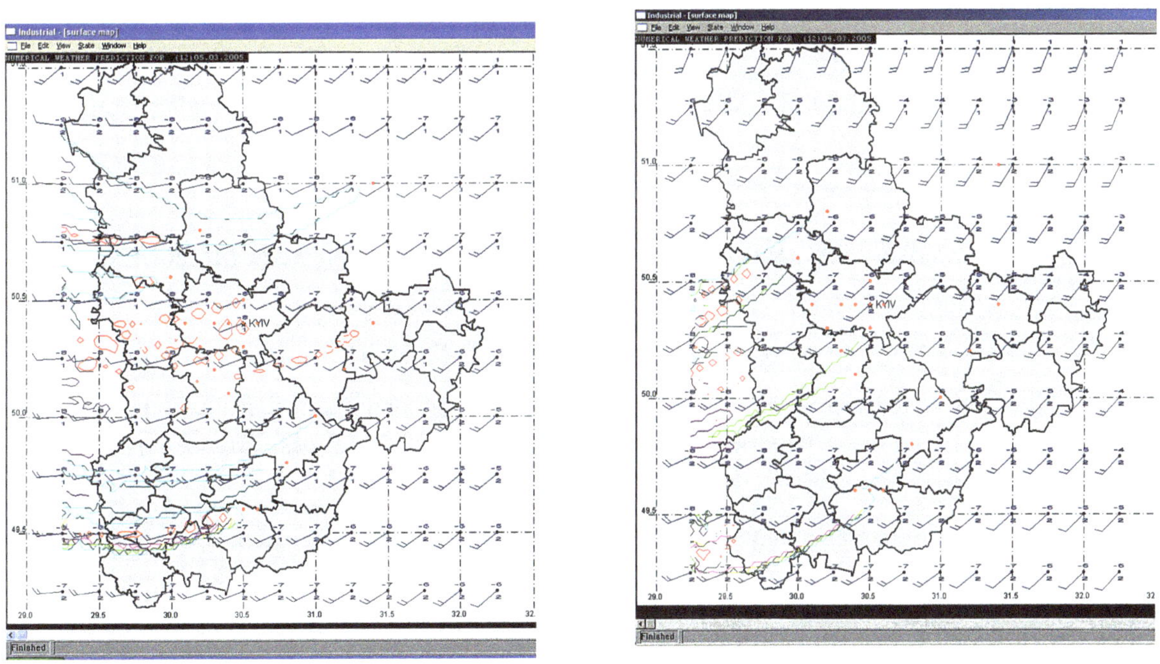

a) b)

of man-made industrial enterprises provided that they are under normal operation and in the case of accidents. These models and methods can be used in the control of transboundary transport of pollutants in the atmosphere.

SUMMARY

In this chapter, we give a systematic exposition of construction, research, and application of numerical methods developed by the authors for effective regional solutions of boundary value problems with a prehistory on the basis of the "one-way interaction" approach. Within this approach, we construct our own method for filling the data, given on the macro-scale grid nodes, in the mesoscale network based on spline interpolation and precise (the fourth order of accuracy) numerical method for the approximation of the first and second order differential equations. Also, we propose a numerical method for solving the non-stationary problem with a prehistory based on interpolation with multiple knots, and absolutely stable implicit difference scheme, that is as effective as explicit scheme, for solving the convective diffusion problem. The results of numerical experiments are exposed on modeling and forecasting of weather conditions in the regions of all sizes and with varying lead times.

At the end of the chapter, we consider an application of our method for solving evolutionary problems of transfer and diffusion of dust-gas mixture injected into the atmosphere by industrial enterprises under the conditions of the project mode of operation and during accidents. There are presented computer scenarios of technogenic pollution of the atmosphere in real meteorological conditions, industrial emissions in normal operation, as well as in fires and accidents.

REFERENCES

Asselin, R. (1972). Integration of a semi-implicit model with time-dependent boundary conditions. *Atmosphere (Toronto)*, *10*, 44–55.

Bakhvalov, N. S. (1975). *Numerical methods*. Moscow: Nauka. (in Russian)

Bakhvalov, N. S., Zhidkov, N. P., & Kobelkov, G. M. (1987). *Numerical methods*. Moscow: Nauka. (in Russian)

Banach, S. (1932). *Theorie des operations lineaires*. Hafner Publishing Co.

Crank, J., Nicolson, P., & Hartree, D. R. (1947). A practical method for numerical evaluation of solutions of partial differential equations of the heat-conduction type. *Mathematical Proceedings of the Cambridge Philosophical Society*, *43*(1), 50–67. doi:10.1017/S0305004100023197

Davies, H. C. (1976). A lateral boundary condition for multi-level prediction models. *Quarterly Journal of the Royal Meteorological Society*, *102*(432), 405–418.

Dovgiy, S. A., Prusov, V. A., & Kopeika, O. V. (2000). *Use of geoinformation technologies in environmental systems environment and natural resources research*. Kiev: Naukova Dumka. (in Ukrainian)

Gustafsson, B. (1975). The convergence rate for difference approximations to mixed initial boundary value problems. *Mathematics of Computation*, *29*(130), 396–406. doi:10.1090/S0025-5718-1975-0386296-7

Hill, G. E. (1968). Grid telescoping in numerical weather prediction. *Journal of Applied Meteorology*, *7*(1), 29–38. doi:10.1175/1520-0450(1968)007<0029:GTINWP>2.0.CO;2

Issues in Atmospheric and Oceanic Modeling. (1985). Orlando, FL: Academic Press.

Kreiss, H. Q. (1970). Initial boundary value problems for hyperbolic systems. *Communications on Pure and Applied Mathematics*, *23*(3), 277–298. doi:10.1002/cpa.3160230304

Kreiss, H. Q. (1980). Problems with different time scales for partial differential equations. *Communications on Pure and Applied Mathematics*, *33*(3), 399–440. doi:10.1002/cpa.3160330310

Miyakoda, K., & Rosati, A. (1977). One way nested grid models: The interface condition and the numerical accuracy. *Monthly Weather Review*, *105*(9), 1092–1107. doi:10.1175/1520-0493(1977)105<1092:OWNGMT>2.0.CO;2

Nicholas, S. (1977). *Approximation of functions of several variables and embedding theorems*. Moscow: Nauka. (in Russian)

Oke, T. R. (1978). *Boundary layer climates*. New York: John Wiley & Sons. doi:10.4324/9780203407219

Oliger, J., & Sundstrom, A. (1978). Theoretical and practical aspects of some initial-boundary value problems in fluid dynamics. *SIAM Journal on Applied Mathematics*, *35*(3), 419–446. doi:10.1137/0135035

Perkey, D. J., & Kreitzberg, C. W. (1976). A time-dependent lateral boundary scheme for limited-area primitive equation model. *Monthly Weather Review*, *105*(6), 744–755. doi:10.1175/1520-0493(1976)104<0744:ATDLBS>2.0.CO;2

Prusov, V., & Doroshenko, A. (2002). Studying variations of pollution levels in a given region of Ukraine during a long time-period. *Systems Analysis Modelling and Simulation*, *39*, 297–311.

Prusov, V., & Doroshenko, A. (2003). Modeling and Forecasting Atmospheric Pollution over Region. *Annales Univ. Sci.*, *46*, 55–73.

Prusov, V. A. (2001). On the issue of forecasting of natural and man-made phenomena with different earliness [in Ukrainian]. *Ecology and Natural Resources*, *1*, 79–96.

Prusov, V. A. (2002). Modeling and forecasting the region atmospheric processes. In *Problems of the technical meteorology* (pp. 67–72). Lviv, Ukraine: Prostir-M.

Prusov, V. A., & Doroshenko, A. Yu. (2005). Methods of efficient modeling and forecasting regional atmospheric processes. In *Advances in Air Pollution Modeling for Environmental Security* (pp. 143–152). Leiden: Springer-Netherlands.

Prusov, V. A., & Doroshenko, A. Yu. (2006). *Modeling Natural and Technogenic Atmospheric Processes*. Kyiv: Naukova Dumka. (in Ukrainian)

Prusov, V. A., & Doroshenko, A. Yu. (2006). *Modeling Natural and Technogenic Atmospheric Processes*. Kyiv: Naukova Dumka. (in Ukrainian)

Prusov, V. A., Grymov, R. Y., & Kopiiko, O. V. (1996). The prediction of environmental pollution in the case of an accident at atomic power plants or in the field of chemical industry. In The problems of the Chernobyl exclusion zone-1996. Kyiv: Administartion of Chernobyl Exclusion Zone. (in Ukrainian)

Prusov, V. A., Kondra, M. P., Kopiiko, O. V., Mikitarenko, M. A., & Perelmuter, A. V. (1997). Integral estimate of risk under wind action upon structures of the encasement at Chernobyl atomic power plant (pp. 1833-1839). In *Proc. 2-nd European-African Conference on Wind Engineering*. Genova, Italy: IAWE.

Prusov, V. A., Kopiiko, O. V., & Grymov, R. Y. (1998). The hydrodynamic mesoscale model of impurity propagation of instant and stationary point sources. In The problems of the Chernobyl exclusion zone-1998. Kyiv: Administartion of Chernobyl Exclusion Zone. (in Ukrainian)

Reinsch, R. (1967). Smoothing by spline functions. *Numer. Mat.*, *10*(3), 77–183. doi:10.1007/BF02162161

Richtmyer, R. D., & Morton, K. W. (1967). *Difference methods for initial-value problems*. New York: Interscience Publishers.

Shapiro, M. A., & OBrien, J. J. (1970). Boundary conditions for fine mesh limited area forecasts. *Journal of Applied Meteorology*, *9*(3), 345–349. doi:10.1175/1520-0450(1970)009<0345:BCFFML>2.0.CO;2

Sundstrom, A. (1986). Boundary conditions for limited-aria integration of the viscous forecast equations. *Beitrage zur Physik der Atmosphare*, *50*, 218–227.

Trenogin, V. A. (1980). *Functional analysis*. Moscow: Nauka.

Williamson, D. L., & Browning, G. L. (1974). Formulation of the lateral boundary conditions for the NCAR limited-area model. *Journal of Applied Meteorology*, *13*(1), 8–16. doi:10.1175/1520-0450(1974)013<0008:FOTLBC>2.0.CO;2

Conclusion

CURRENT PROBLEMS OF METEOROLOGICAL FORECASTING

Over the last 25-30 years an undeniable progress has been made in the development and operational use of modern numerical hydrodynamic methods of weather forecasting. It has become possible due to the development of more sophisticated and realistic models, as well as a substantial increase in the volume of initial meteorological data.

Today, improving the quality of forecasts takes place in two directions: increasing the accuracy of short-term forecast and increasing its earliness. The short-term forecast is mainly determined by the initial state of the atmosphere. The medium-term forecasts need both a detailed knowledge of the primary fields of meteorological quantities and the ability to describe the influence of external factors, leading to a new state of equilibrium of the predicted system. Long-term forecasts are less dependent on the initial state of the atmosphere and incorporate elements of climate modeling. Increased earliness of forecasts suggests the need to expand the range of integration up to cover the entire surface of the globe, highlighting the factor of time, calculation performance, and computing power. In the development of long-term weather prediction models, one must take into account that an increase in the earliness time of the forecast suggests an increase in the role of stochastic processes in the atmosphere. Slight uncertainty in the initial time due to the internal dynamics of the process may lead to erroneous results in an area of considerable size. In this regard for long-term weather forecasts, the traditional equations of a general circulation model of the atmosphere must be modified to include the so-called stochastic-dynamic model and to develop new, more advanced methods of numerical integration.

Up to 1990s, the operational weather services of the developed countries have focused on models of forecasting of synoptic processes in the hemisphere, despite some evidence that the accuracy of such forecasts remained almost constant. One reason for the increased interest in large-scale global models probably was of a political nature, as developed countries felt the responsibility to create a national model. Thus, the most advanced computational resources required for the implementation of the operational model, have been concentrated in specially established national meteorological centers.

In recent years, many national, regional, and global meteorological services have established the regional models and software packages of numerical forecasting for both individual weather events and the weather in general. These models are used to describe local processes in relatively small areas and will give a greater effect in the future than the traditional weather forecast. However, in the development and operationalization of regional models, there appear very difficult technological problems that have become challenges for the science. The main factors determining the behavior of many of mesoscale phenomena in the atmosphere are three-dimensional turbulence and heterogeneity of the underlying surface. Theoretical research of the turbulence and analysis of the measured spectra come to conclusion that the inevitable errors and uncertainties in the initial fields of small-scale motions are passed to motions of large scale and expose regional disturbances before the synoptic forecasts, thereby making the first less predictable. The heterogeneity of the underlying surface, including topography, albedo, heat and the moisture content in the soil, generate such mesoscale phenomena as mountain waves, sea breeze effects, convection and determine their future behavior. Reducing disturbances modulated by these factors in the projected mesoscale fields can be achieved by means of increasing the density of the computational grid, and this, in its turn, increases the computational complexity greatly. It is easy to show that the doubling of the number of nodes in a three-dimensional computational grid leads to a 16-fold increase in computation time and required volumes of computer memory.

Development of new and improvement of existing methods of mesoscale numerical weather forecasts depends primarily on the following factors:

- The accuracy of the description of physical processes responsible for changes in time of macroscale and mesoscale characteristics of the atmosphere under the influence of the relief.
- The adequacy of the actual physical processes and simulated sub-grid scale processes.
- Precision of setting initial and boundary conditions.
- The speed of numerical integration of the original equations.

Prediction of the directions of movement of well-developed vortex structures (e.g., cyclones) and their intensity in mountain areas is a challenge. In a quiet environment, cyclone moves slowly under the influence of advection of planetary vorticity, and, of course, non-uniform thermal conditions on the underlying surface. And when a cyclone comes to the area of the ridge of high steep mountains and begins to interact with the other eddy formations, it is often very difficult to determine its trajectory and velocity of the movement. Studies of such situations with the help of numerical models is of great importance in the light of current practical problems and forecasting hazardous meteorological phenomena (long intense precipitation, squalls, tornadoes, and as their results – floods, mudslides, avalanches). The key issue here is the very high accuracy of the numerical solution of the model equations and reliable description with the help of fields of meteorological variables, primarily related to the non-adiabatic processes parameterization.

The importance of calculations of radiative fluxes for short-term and medium-term weather forecasts is obvious. Heat influx plays an important role in the energy supply of the primary air mass flow and wave formations, arising due to the inconsistency of direction and different types of interactions. An essential feature of the areas of the ascending (descending) streams is to have a balance between radiative heating (cooling) and adiabatic cooling (heating) when lifting (lowering) of the air mass. The estimate of the value order of the determining factor is that the intensity of radiation cooling is from 1 to 30°C/day. If one does not take this process into account, the error in the prediction of temperature and wind speed component will be significant. Uncertainties in the forecast of vertical profiles of temperature, presence and height of the inversion (delay) layers of the atmosphere cause, of course, large errors in the environmental forecasts. Existing versions of multilevel global finite difference and spectral models use a very complex parameterization of radiative processes. It includes calculations of the long-wave and short-wave radiation in the absence and presence of clouds given changes of solar zenith angle and the heat balance of the underlying surface, with the definition of the albedo of the surface as a function of soil moisture content and relative humidity. More modest successes are in the parameterization of radiative processes for regional atmospheric models. Here the problem of improving numerical weather prediction and extension of relevant scientific and experimental work are highly relevant.

One of the main features of the mid-latitude weather maps is the presence of atmospheric fronts that are breaking surfaces separating air masses of different densities, i.e., zones of sharper gradients of temperature, pressure, humidity and wind speed components. Atmospheric fronts have a wave nature. The horizontal scale of the front waves is more by an order of magnitude than mesoscale features of circulation in the plane that is perpendicular to the main stream. These features of the transverse circulation in a vertical plane are connected with the rise of humid air above the ground-up front with a sliding passage within the frontal zone, entrained into the secondary circulation at average levels. The maxima of the horizontal temperature gradient and vertical velocities are near the Earth's surface. The synchronous three-dimensional structure of the formed mesoscale fields of temperature and wind speed causes the processes that generate, maintain, and eventually destroy the larger temperature gradients at the front surfaces. Thus, the modeling of these subtle and complex mesoscale physical processes of a frontogenesis becomes impossible due to the loss of precision in the numerical integration of the equations of a complete model of atmospheric circulation with a horizontal grid spacing of more than 50 km. Modeling such essential for frontogenesis processes, as the impact of humidity on the near edge environment, an interaction of convective systems with frontal circulation and influence of the planetary boundary layer is very difficult. The issue of the extent to which they can be formalized and included in the general model with a simple parameterization, at least to achieve greater realism to this model, still remains unresolved, indicating the still insufficient understanding of the processes determining the dynamics of fronts.

Calculating the intensity of precipitation tends to have the probabilistic character as the evolution of cloud systems is closely connected to the wet convection, and parameterization of this in numerical models uses inevitably a variety of hypotheses or statistical description of the physics of clouds and their dynamics. To improve the parameterization of the physical processes occurring in the atmosphere under the influence of convective processes, one needs considering dynamics of clouds in the calculation scheme, including friction in cumulus clouds and microphysics (the formation of ice crystals, melting processes, the impact of upstream and downstream). The moist convection parameterization also cannot avoid the issue of rejection of the hydrostatic approximation used in almost all existing operational numerical models.

Thus, the main causes of errors of the operational short- and medium-term weather forecasts are the errors of perturbation of the synoptic scale directly in the initial data of meteorological variables fields and insufficiently correct parameterization of non-adiabatic processes and processes of a sub-grid scale. So, the initial improvement in the quality field with sufficient resolution and enhancing the realism of existing models of the hydrodynamic atmosphere circulation would significantly reduce a forecast error.

AREAS OF APPLICATIONS OF THE NUMERICAL FORECASTING TECHNIQUES

Methods of dynamic weather forecasts and individual weather events are very effective and accurate enough to provide their application in practice of meteorological investigation despite the above problems and drawbacks of the existing mathematical models of atmospheric circulation and numerical methods of their implementation. This justify their long-term use in operational forecasting work of many meteorological centers.

Another general area of mathematical modeling is to use it in the experiment, i.e., to obtain a set of predictive scenarios promoting validity, objectivity, and effectiveness of developed decisions. Without exaggeration, one can say that the lab of this direction of meteorology (scenarios forecast process or phenomenon given varying definitive parameters) is a computer lab. While research in other sciences is usually implemented by laboratory experiments, many meteorological and environmental problems, which are still considered to be intractable, may be resolvable based on numerical simulations. Of course, this is not a new trend but just a classic method of theoretical science which obtains new opportunities.

Improving the earliness and accuracy of weather forecasts provides exceptional opportunities and prospects of mathematical modeling and its practical application in the prevention and reduce of the negative consequences of natural and technogenic emergency situations. Mathematical modeling at the present stage of scientific achievements and technical equipment level (numerical experiment, a computer scenario, and hydrodynamic forecast) is one of the most important and effective information management systems, expert analysis, a study of organizational solutions and improve the scientific level of management. Development and implementation of continuous forecasting systems are now becoming more and more relevant with the introduction of automated risk assessment systems, decision-making, and management.

Numerical weather modeling and forecasting is also important to use in modern weather dependent areas like satellite technologies, wind engineering and renewable energy development, forecasting the output power of solar systems, and other green energy technologies.

CONCLUDING REMARK

In this book, modern problems of dynamic meteorology, atmospheric diffusion, and air pollution are set on a uniform basis. A wide range of issues of a theoretical study of the spread of air pollutants from stationary and instant sources of emissions is considered. The meteorological aspects of air pollution and the closely related problems of physics of the atmospheric boundary layer meteorology and other topics are taken into account. Solving such problems lays a common basis for the study of air movement and the impurity transfer processes in the atmosphere since the laws of thermodynamics and vortex mixing mainly determine them. The most clearly this commonality is manifested in the solution of the transport equation and the turbulent diffusion at the appropriate initial and boundary conditions. In future studies, it will be necessary anyway to know the distribution of meteorological variables and turbulent characteristics, determining the mechanism of exchange in the atmosphere.

Appendix

BASIC NOTATION ADOPTED IN THE BOOK

ρ = The density of air,
ρ_d = The density of air dry,
ρ_s = The density of steam,
$\mathcal{R} = \ln(\rho)$ = The special function,
R = The gas constant of dry air,
R_s = The gas constant of steam,
A_p = The specific heat at constant pressure,
A_v = The specific heat at constant volume,
p = The atmospheric pressure,
$p_{00} = 1000$ HPa = The standard pressure,
p_B = The pressure at the underlying surface,
Φ = The geopotential,
g = The gravitational acceleration,
$\gamma_a = g / A_p$ = The vertical adiabatic gradient,
p_d = The pressure of air dry,
e = The pressure of steam,
T = The absolute temperature $(^\circ K)$,
L = The specific heat of vaporization or condensation,
$\theta = T\left(p_{00} / p\right)^{R/c_p}$ = The potential temperature,
F_R = The radiation energy flow density,
q = The specific humidity of wet air,
q_S = The saturation specific humidity,
Q_H = The rate of change of the specific humidity in the atmosphere where the humidity reaches value q_S,
$T_v \equiv T(1 + 0.6078q)$ = The virtual temperature,
δ = The concentration of the condensed (cloudy) a moisture,
s = The concentration of aerosol impurity in the atmosphere,

f_{cr} = The critical relative humidity,
h_c = The height of the condensation level,
θ_{lbc} = The temperature at the cloud base,
θ_{ubc} = The temperature at the cloud top,
I = The intensity of precipitation,
S_k, S_u = The processes of sub-grid scale associated with the rate of formation the cloud and precipitation,
S_s = The rate of change of the impurity concentration in the atmosphere where the chemical reactions take place,
J_s = The rate of leaching contaminants by precipitation,
t = The time,
$\mathbf{X} \equiv (x_1, x_2, x_3) = (x, y, z)$ = The rectangular Cartesian coordinate system,
$\mathbf{X} \equiv (x_1, x_2, x_3) = (\lambda, \varphi, z)$ = The spherical Cartesian coordinate system,
$\mathbf{V} = v_1\mathbf{i} + v_2\mathbf{j} + v_3\mathbf{k}$ = The wind velocity,
$\mathbf{i}, \mathbf{j}, \mathbf{k}$ = The unit vectors along the axes x_1, x_2, x_3 respectively,
$\varpi = dp / dt$ = The vertical velocity in *p*-coordinate system,
$\dot{\sigma} = d\sigma / dt$ = The analog of vertical speed in the coordinate σ-system,
$\sigma = p / p_B$ = The vertical coordinate in σ-system of coordinates,
x_1^{eff}, x_2^{eff}, x_3^{eff} = The effective coordinates of impurities emission source,
$\mathbf{\Omega} = \mathbf{\Omega}_1\mathbf{i} + \mathbf{\Omega}_2\mathbf{j} + \mathbf{\Omega}_3\mathbf{k}$ = The angular speed of the rotational coordinate system,
$\boldsymbol{\nu} = \nu_1\mathbf{i} + \nu_2\mathbf{j} + \nu_3\mathbf{k}$ = The coefficient of eddy viscosity,
$d = \nu / \xi$ = The coefficient of eddy diffusion,
ξ = The Prandtl number for the heat, or the Schmidt number for the mass,
Ri = The Richardson number,
$\tau_n = \tau_x \cos(n, x) + \tau_y \cos(n, y) + \tau_z \cos(n, z)$ = The stresses vector,

$$\left.\begin{aligned} \tau_x &= \tau_{xx}\mathbf{i} + \tau_{xy}\mathbf{j} + \tau_{xz}\mathbf{k}, \\ \tau_y &= \tau_{yx}\mathbf{i} + \tau_{yy}\mathbf{j} + \tau_{yz}\mathbf{k}, \\ \tau_z &= \tau_{zx}\mathbf{i} + \tau_{zy}\mathbf{j} + \tau_{zz}\mathbf{k}, \end{aligned}\right. = \text{The generalized vector function,}$$

$\boldsymbol{\Re} = \Re_1\mathbf{i} + \Re_2\mathbf{j} + \Re_3\mathbf{k}$
$\Im$ = The generalized scalar function,
k = The kinetic energy density of turbulent pulsating movement,
ε = The dissipation rate of turbulent kinetic energy,
ℓ = The characteristic length scale (mixing path) in turbulent flow,
ϕ_U, ϕ_θ, ϕ_q = The universal function for wind speed, temperature, and humidity, respectively,
ϱ_s = The shortwave constituents of radiation,
$\varrho_{s\downarrow}, \varrho_{s\uparrow}$ = The flows of downward and upward shortwave constituents of radiation,
ϱ_l = The longwave constituents of radiation,
$\varrho_{l\downarrow}, \varrho_{l\uparrow}$ = The flows of downward and upward longwave constituents of radiation

d / dt = The absolute (substantial) acceleration,

$\nabla \cdot$ = The Hamilton differential operator, it is used with a point $(\cdot)$ to refer to the divergence of a vector function,

∇ = The Hamilton differential operator, it is used without a point $(\cdot)$ to refer to the gradient of a scalar function,

$\wp$ = The macro-scale domain of solving the problem,

S = The side surface the domain $\wp$,

ϖ_h = The large-scale grid in $\wp$

$\sigma = \sigma_j \ (j = 1, 2, \ldots, N)$ = The node of grid,

$h_j = \sigma_j - \sigma_{j-1}$ = The grid steps in ϖ_h,

$\bar{\wp}$ = The regional domain of solving the problem,

$F(x, y)$ = The equations underlying the bottom surface,

$H(x, y)$ = An altitude adopted as the upper boundary of the computational domain,

$\bar{\varpi}_h$ = The small-scale grid in $\bar{\wp}$,

$\eta = \eta_j \ (j = 1, 2, \ldots, J)$ = The node of grid,

$\bar{h}_j = \eta_j - \eta_{j-1}$ = The grid steps in $\bar{\varpi}_h$.

About the Authors

Vitaliy A. Prusov has been graduated from the National Technical University of Ukraine "Igor Sikorsky Kyiv Polytechnic Institute". He is the Doctor of Sciences in Physics and Mathematics. His research interests are associated with the experimental and theoretical problems of mechanics of liquid and gas, numerical methods for solving equations of mathematical physics, mathematical modeling of the circulation of the atmosphere and the anthropogenic impact, and numerical methods of weather forecasting. Since 1983, V.A. Prusov has been the leader of more than 10 research projects of the National Academy of Sciences of Ukraine related to the development of numerical weather forecasting systems for different regions and periods. Since 2003 he is a professor of the Department of Meteorology and Climatology of the Kyiv National University by Taras Shevchenko. He has more than 70 scientific publications, including 5 monographs and 4 textbooks. He is affiliated with Kyiv National University by Taras Shevchenko and Ukrainian Hydrometeorological Institute of the National Academy of Sciences (Kyiv, Ukraine).

Anatoliy Yu. Doroshenko has been graduated from Kyiv National University by Taras Shevchenko (Kyiv, Ukraine). He is the Doctor of Sciences in Physics and Mathematics. His research interests include high performance computing algorithms, programming numerical methods, algebraic algorithmic models, and formal methods and adaptive programming techniques for automated software design. Since 1997, A.Yu. Doroshenko has been the leader of several research projects of the National Academy of Sciences of Ukraine related to the development of numerical weather forecasting systems for different regions and periods. Since 2005, he is a Professor of the Department of Informatics and Computer Engineering of the National Technical University of Ukraine "Igor Sikorsky Kyiv Polytechnic Institute". He is the author of 4 monographs and about 200 research papers. He is a member of the ACM. He is affiliated with National Technical University of Ukraine "Igor Sikorsky Kyiv Polytechnic Institute" and Institute of Software Systems of the National Academy of Sciences of Ukraine.

Index

I

M

N

O

P

S

T

www.ingramcontent.com/pod-product-compliance
Lightning Source LLC
LaVergne TN
LVHW081416021026
849161LV00011BA/2734